Invasiv Alien Species in Seychelles

Why and how to eliminate them? Identification and management of priority species

Gérard Rocamora & Elvina Henriette

Island Biodiversity and Conservation centre
University of Seychelles

Inventaires & biodiversité series
Biotope – Muséum national d'Histoire naturelle
2015

Invasive plants such as the Merremia creeper may cover large extensions, particularly in areas disturbed by human activities (here, a roadside landscape near Port Launay, Mahé).

Gérard Rocamora

Table of contents

■ *Box*
■ *Case study*

This book is to be quoted as follows: Rocamora G. & Henriette E. 2015. *Invasive Alien Species in Seychelles: Why and how to eliminate them? Identification and management of priority species.* Island Biodiversity & Conservation centre, University of Seychelles. Biotope, Mèze; Muséum national d'Histoire naturelle, Paris (Inventaires & biodiversité series), 384 p.

EDITORIAL COMMITTEE

Didier Dogley (Minister of Environment, Energy and Climate Change), Andrew Grieserjohns (Project Coordinating Unit, UNDP-GEF Biosecurity Project), Jean-Philippe Siblet (MNHN), Jean-Yves Kernel (Biotope éditions) & Gérard Rocamora (IBC-University of Seychelles).

TEXT AUTHORS

Part 1. Why and how to eliminate invasive alien species. Impacts and management options: Gérard Rocamora

Part 2. Identification and management of priority species:
Mammals and Birds: Gérard Rocamora
Reptiles, Invertebrates, Plants and Fungal diseases: Elvina Henriette

Bibliography: Elvina Henriette
Editing of case studies: Gérard Rocamora & Elvina Henriette

CASE STUDIES

Wilna Accouche, Nancy Bunbury, Jessica Moumou, Nick Page, Janske van de Crommenacker,Wilna Accouche and Frauke Fleisher-Dogley (SIF); Environment Department (ED); Greg Canning (Frégate Island); Will Dogley (SAA); Chris Feare (Wildwings), Rene Gaigher (University of Stellenbosch), John Nevill (Consultant), Pat Matyot, Pierre-André Adam, Melinda Curran & Gérard Rocamora (ICS).

MAPS

Elyn Albert (Environment Information & Data section; Climate Affairs, Adaptation & Information Department; Ministry of Environment & Energy)

TEXT REVIEW

*Members of ISSG or other specialist groups of the IUCN-SSC.

All manuscript: Jeanne A. Mortimer* (IBC-UniSey) & Antoine-Marie Moustache (SAA/PCU).

Part 1 (general text): Jean-Philippe Siblet (MNHN, Paris), Souad Boudjelas* (PII & ISSG-SSC of IUCN), Jane Beachy (University of Hawaii), Aston Berry (IBC-University of Seychelles) and Dan Simberloff (University of Tennessee).

Part 2 (species accounts): Mammals: James Russell* (University of Auckland; New Zealand); Birds: Chris Feare * (University of Leeds, UK) & Adrian Skerrett (ICS, Seychelles); Reptiles: Jeanne A. Mortimer* (IBC-UniSey) & Ronley Fanchette (ED); Ants: Rene Gaigher (Stellenbosch University, South Africa); Snails: Justin Gerlach* (NPTS); Tiger mosquito: Vincent Robert (IRD, France); Plants & Fungal diseases: Christopher Kaiser-Bunbury* (University of Darmstadt, Germany), Katy Beaver (PCA), Bruno Senterre (IBC-UniSey) & Jane Beachy (US Army/University of Hawaii).

ACKNOWLEDGEMENTS

Individuals: Wilna Accouche, Elyn Albert, Pierre-André Adam, Riaz Aumeeruddy, Willy André, Laurent Bagny, Sam Balderson, Delphine Paugam-Baudouin, Jane Beachy, Katy Beaver, Colin Bell, Laurence Bénichou, Annie Bermon, Bernard Bijoux, Nicolas Borowiec, Robbie Bresson, David Brown, Nancy Bunbury, Sarah Caceres, Licia Calabrese, Greg Canning, Jason Carpin, James Chang-Tave, Tanya Leibrick, Lindsay Chong-Seng, Gwénaëlle Chavassieu, Gideon Climo, Nik Cole, Philip de Comarmond, Antonio (Mazarin) Constance, Perley Constance, Javier Cotin, Franck Courchamp, Norbert et Valérie Couvreur, Melinda Curran, Hélène Delatte, Gilles-David Derand, Derrick Derjacques, Paul Desnousse, Joachim Didon, Will Dogley, Ahab Downer, Aurélie Duhec, Andre Dufrenne, Ronley Fanchette, Chris Feare, Frauke Fleisher-Dogley, Karl Fleischmann, Antoine Franck, Jude Gédéon, Piero Genovesi, René Gaigher, Justin Gerlach, Luc Gigord, Irene & Roy Mac Grath, Xavier Heinen, Steve Hill, Dennis Hardy, Kelly Hoareau, Nick Holmes, Jean-Noël Jasmin, Simon Julienne, Gemma Jessie, Terry Jules (†), Tony Jupiter (†), Daniela Jupiter, Christopher Kaiser-Bunbury, Christoph Küffer, Jan Kömdeur, André Labiche, Victorin Laboudallon, Erwan Lagadec, Olivier Langrand, Lucia Latorre, Christophe Lavergne, Paul Lavigne,Thomas Le Bourgeois, Gildas Le Minter, Denis Matatiken, Pat Matyot, Ray Maria, Lucy Martin, Dane Marx, Sue Maturin, Jimmy Mélanie, Jean-Yves Meyer, Ralph Meyer, Cédric Morel, Jeanne A. Mortimer, James Mougal, Mervyn Mounac, Antoine-Marie Moustache, Don Merton (†), Aurélien Nahaboo, John Nevill, Pep Nogués, Roland Nolin, Shyama Pagad, Rolph Payet (Minister of Environment and Energy 2012-2014), Monique et Jean-Luc Pereira, Justin Prosper, Serge Quilici (†), Ronny Renaud, Tatiana Raposo de Rezende, Kervyn Rayeroux, David Richardson, Serge Robert, Todd Robinson, David Rowat, James Russell, Soufou Said, Gilles Saout, Araceli Samaniego, Glenny Savy, Richard Schumann, Bruno Senterre, Jean-Philippe Siblet, Nick Skerrett, Joachim Soler, Bill Simmons, Helena Sims, Adrian Skerrett, Joachim Soler, Eric Sophola, Yohann Soubeyran, Johnnie Souieleh, Jacques de Spéville, Marc Stickler, Angela Street, Elke Talma, Vikash Tatayah, Julien Triolo, Markus Ultsch-Unrath, Linda Vanherck, Terence Velle, Michel Vielle, Rainer von Brandis, Janske van de Crommenacker, Bill Waldman, Rowana Watson, Greg Wepener, Nicolas Zuel.

Organisations: Miguel Torres, Animal Control Products (NZ), Bird Island, Centre de Recherche et de Veille sur les maladies émergentes de l'Océan Indien (La Réunion); Centre de coopération internationale en recherche agronomique pour le développement - La Réunion & Montpellier (Pl@ntInvasion); Comité français de l'UICN; Conservation des Espèces et Populations Animales, Conservatoire Botanique National des Mascarins (La Réunion), Cousine Island, D'Arros Research Centre, Denis Island, Délégation à la recherche de la Polynésie Française, Durell Wildlife Conservation Trust, ETF University of Zurich (Switzerland), Frégate Island, Sisters Ltd. / Chateau de Feuilles, Green Islands Foundation, Grupo de Ecología y Conservación de Islas (Mexico); Invasive Species Specialist Group of the Species Survival Commision of IUCN, Institut national de la Recherche agronomique (PACA, France), International Union for the Conservation of Nature, Island Conservation (USA), Island Conservation Society (Mahé-Aride-Alphonse-Desroches-Farquhar-Silhouette), Islands Development Company, Institut de Recherche pour le Développement (La Réunion), Ministry of Environment and Energy - Environment Department, & Climate Affairs, Adaptation & Information Department, Management of President of United Arab Emirates Affairs, Marine Conservation Society of Seychelles; Muséum national d'histoire naturelle, Paris (Service du patrimoine naturel et Service des publications scientifiques); North Island, Nature Protection Trust of Seychelles, Ministry of Health - Public Health Department, Mauritian Wildlife Foundation, Nature Seychelles (Mahé- Praslin-Cousin), Office national des forêts (La Réunion), Pangaea boat, Plant Conservation Action Group, Réserve Naturelle de Mbouzi (Mayotte), Seychelles Agricultural Agency and Veterinary Section (Ministry of Investment, Natural Resources and Industry), Seychelles National Parks Authority, Seychelles Islands Foundation (Vallée de Mai & Aldabra), Stop Insectes (La Réunion); Terrestrial Restoration Action Society of Seychelles, Comité français de l'Union internationale pour la conservation de la nature - Groupe Outre mer, University of Auckland (New Zealand), University of East Anglia (UK), University of Gröningen (Holland), University of Darmstadt (Germany), Stellenbosch University (South-Africa), US Army Garrison, Environmental Division (Hawai'i-USA).

And all other colleagues who, or organisations that, have helped one way or another with the production of this book and that we might have forgotten!

The authors are also grateful to their families, Aurora Solé, Magali & Miriam Rocamora i Solé, Mercè Escoda and Serge Rocamora; François Payet and Marie-Nella Henriette for their constant support despite all they had to endure during the four years it took us to write and publish this book!

The preparation and publishing of this book was funded by the Government of Seychelles, UNDP and Global Environment Facility, under the GOS-UNDP-GEF Mainstreaming Biosecurity project.

Special thanks go to the Fondation Total (France), the Sainte Anne Beachcomber Resort & Spa (Seychelles), the Fundació Miguel Torres (Catalonia-Spain) and Pestoff (New Zealand) for providing additional sponsorship, and to Air Seychelles for kindly bringing the books from Paris to Seychelles.

IMPORTANT NOTES

Island names: Island names used in this book are those normally in use by the inhabitants of Seychelles, as they appear in the list of islands of the Constitution of Seychelles. For practical reasons, the word 'Ile' (Island in French) that appears in front of certain island names has sometimes been omitted, and 'North Island' is sometimes used for 'Ile du Nord'.

Species names: Common species names are normally provided in English. However, when these do not exist or are not commonly used in Seychelles, we have used Creole or local French names (e.g. 'Kalis-di-pap' or 'Calice du Pape'). Scientific names appear in italics in the text for non-Seychellois readers, as common names often differ between countries. These are not given for invasive species treated in case-studies, or in species accounts in the second part of the book, nor repeated for native species when already given in the chapter text or appearing elsewhere in tables.

Foot referencing: Superscript numbers in the text correspond to references written at the bottom page of each subchapter, box or case study. Full references are provided in the bibliography at the end of the book. To facilitate reading, the following code has been used in the text:
- 1, 4 means reference 1 plus reference 4,
- 1-4 means all references between reference 1 and reference 4 (= 1, 2, 3, 4).

Foreword

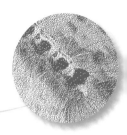

Prof. Daniel
Simberloff
Director of the
Institute for
Biological Invasions,
University of
Tennessee, USA.
Board member of
Island Conservation
(Santa Cruz,
California, USA)

Invasive alien species are now widely recognised, along with a warming climate, as one of the great global changes transforming the face of the earth at an unprecedented rate. Most scientists did not realise the scope of this change until the 1980s, but even before then, citizens of various islands recognised the great damage wrought by some invaders. The first comprehensive book on the global spread of invasive species, Charles S. Elton's *The ecology of invasions by animals and plants* (1958), pointed to the particular devastation visited on islands such as Hawaiian Islands, New Zealand, and Tristan da Cunha by myriad invaders, including the Common myna, Giant African snail, Black rat, Big-headed ant, and European rabbit, all present in Seychelles.

Islands are particularly prone to damage by invasive non-native species for the same reason that islands are the repositories of a hugely disproportionate fraction of the earth's biodiversity. Not counting Australia, islands comprise only 3% of the earth's land surface area but a far greater percentage of most groups of organisms – for instance, over 15% of plant and bird species are restricted to islands (Whittaker and Fernández-Palacios 2007). This is because their isolation permits rapid evolution of island populations, with little or no genetic influx from elsewhere, eventually leading to new species. Species found only on particular islands – "island endemics" – characterize isolated islands, and this is especially true of isolated archipelagos consisting of many islands, like Seychelles. Seychelles, isolated for at least 60 million years, abounds in striking endemic plants and animals, including five bat species and a whole family of frogs. Alfred Russel Wallace, co-discoverer of the theory of evolution, highlighted the animals of Seychelles in his classic book *Island Life* (1880).

However, the same isolation that fosters the proliferation of endemic species means that island species are not subjected to interactions with the great variety of types of species that inhabit continents. Island species are superbly adapted to the environments they evolved in, but those environments did not include the many species introduced deliberately or inadvertently by humans over the last 500 years, and particularly over the last century. Thus, for example, many islands, such as those of Seychelles, lack land mammals except for bats. Birds therefore do not evolve defensive habits against predators like cats and rats, as they do on continents. Most plants do not evolve chemical defenses or physical ones against grazing and browsing mammals, because these had not been present (although young palms have spines possibly as a protection against Giant tortoises). Island isolation also means that island species have not had to evolve resistance to many of the pathogens that abound on continents. When humans and the animals and plants they bring arrive with the diseases to which they themselves are resistant, island species encountering them for the first time are often devastated.

Invasive alien species and habitat destruction are the two main causes of the hecatomb that afflicts biodiversity on so many islands. Island plant species are 2.5 times as likely as continental plant species to be classified as endangered, and island bird species are 2.7 times as likely to be classified as endangered as their continental counterparts. Roughly 75% of documented extinctions during the last five centuries are of island species, and a plethora of recent discoveries of fossil and sub-fossil remains on islands suggests that even that number is probably an underestimate of the true value. Islands tend to have far more introduced species relative to natives than do continental regions, partly because of trade and travel, and sometimes because new residents deliberately introduce plants and animals that they miss from home. Seychelles inner islands have about twice as many introduced plant species as native ones. Among introduced animals, Seychelles has at least nine vertebrates as well as five invertebrates on the list of 100 of the world's worst invaders published by the IUCN. In addition, eight non-native plants are on that list.

Against this background of rich but fragile biodiversity and increasing threat from bio-logical invasions, several islands have been in the vanguard of an international effort to stem the tide of new invasions. New Zealand with its biosecurity laws and rapid response mechanisms and Australia with its weed risk assessment procedures are the most prominent. In Europe the island nation of Great Britain is far ahead of the continental nations in aggressive measures to counter invasions, while in the United States the island state of Hawaii is both the most heavily afflicted state and joins Florida as states with the most vigorous defensive measures. However, the nation of Seychelles, comprising over 100 islands and with fewer than 100,000 people, has become an inspiration to the world in its ambitious program to protect the native species and ecosystems that make it part of one of the world's recognised hotspots of biodiversity. With a minimum of 46 eradications of at least a dozen vertebrate species, including such globally high-profile invaders as the black rat, brown rat, myna, house cat, and house mouse, Seychelles is among the world leaders. On many islands of the Seychelles, invasive plant populations have been brought under sufficient control – and in some cases eradicated – so that they no longer threaten native species, even though the situation remains challenging on the larger inhabited islands.

Furthermore, and to its great credit, Seychelles has undertaken these eradication and control measures in the framework of restoring native island ecosystems to a status similar to what they would be if the invasions had not happened in the first place. In other regions, successful eradication or control of a particular invasive species sometimes simply paves the way for another invasion by the same or different species – the "treadmill effect" – because no restoration plan was implemented. In Seychelles, restoration has often involved reintroducing species (or near relatives) to islands from which they had been eliminated. The translocation of giant tortoises from Aldabra to several other islands is well known to conservation scientists and the public worldwide. Should the comprehensive Biosecurity Act in preparation become part of the national legislation, Seychelles would join New Zealand as the only nation to take a truly holistic approach to dealing with invasions.

The publication of *Invasive Alien Species in Seychelles* is a milestone in this remarkable struggle to protect native species and ecosystems from biological invasions. For no other nation, not even leaders such as the United States, New Zealand, Australia, and South Africa, is there such a comprehensive single volume describing all aspects of the issue – history of the invasions by both animals and plants, ecological and socioeconomic impacts, sociological and geographic context, and management efforts and their outcomes. It will be a required reference for invasion scientists. Further, this book will serve the lay public well, with a primer embedding the invasion story in Seychelles in the larger saga of the global change being wrought by invasions, and with descriptions and illustrations of the key invaders of the archipelago. For managers in the Seychelles and beyond, it will be a very useful handbook of methods to attack particular invasive species – what worked and what did not. Gérard Rocamora and Elvina Henriette are to be congratulated for conceiving this project and bringing it to fruition.

Knoxville, Tennessee, USA

June 12, 2015

Bibliography

Elton, C.S. 1958. *The ecology of invasions by animals and plants*. London: Methuen.

Wallace, A.R. 1880. *Island life*. London: Macmillan.

Whittaker, R.J., and J.M. Fernández-Palacios. 2007. *Island biogeography* (2nd ed.). Oxford: Oxford University Press.

Preface

Didier Dogley
Minister for
Environment,
Energy and
Climate Change
of the Republic of
Seychelles

Introduced weeds, pests and destructive microbes, formally known today as Invasive Alien Species (IAS), have long been recognised as major threats to agriculture, forestry, fisheries and biodiversity. Invasive species remain the greatest threat to biodiversity in vulnerable, geographically and evolutionarily isolated ecosystems such as small islands. The Convention on Biological Diversity defines them as "An alien species whose introduction and/or spread threatens biological diversity", which is understood as a foreign species that, once established into a new ecosystem, impacts negatively by hindering the growth and development of native or existing organisms and changes their natural environment to enhance its own dominance. There are many such alien invasive animals and plants in Seychelles today and their impact on both the natural world and the economy is significant.

Historically, scientific reports produced by the Percy Sladen expedition, David Stoddart, Francis Friedmann and others help to shed light on the changes that have occurred since they undertook their inventories of the fauna and flora of our islands. The colonial Agricultural Reports of Seychelles, which covered both agriculture and forestry activities, had a section dedicated to the control and eradication of specifically targeted pests and weeds on the main granitic islands. The reports highlighted the importance that the colonial agricultural administration placed on the control of weeds and pests in an effort to minimise and reverse the losses they were recording at that time in these sectors.

Rapid development associated with an increase in both tourism arrival numbers and changes in consumption patterns by the local population has made it necessary for Seychelles to increase the volume and diversity of goods it imports. Today, we import not only from our traditional trade partners and neighbouring countries but also from new emerging production centres in Latin America, Africa and Asia, thereby increasing the range and diversity of IAS our islands are exposed to.

Furthermore, in recent years, flights to destinations within and outside the region have become more affordable to the local population, hence the observed increase in frequency and number of Seychellois travelling to other countries. This coupled with a strong tradition in horticulture and an emerging trade in pets especially aquarium fish has made it crucial to put appropriate policy, legislative and administrative measures in place. With it the authorities have recorded a higher level of legal and illegal importation of foreign plants and animals. All this has prompted the government to develop, with the help of the UNDP and the GEF, a comprehensive project addressing invasive species that would mainstream Biosecurity into the main production sectors and develop the first Biosecurity Strategy for Seychelles. Other major achievements of this project include the drafting in 2010 of a comprehensive Biosecurity Act, the production of an operational manual and booklet for customs, the installation in 2012 of an X-ray machine at the international airport capable of detecting organic matter, the establishment in 2015 of a national Biosecurity Committee involving representatives from six ministries, parastatal agencies, civil society and media, and finally the publication of this remarkable book 'Invasive Alien Species in Seychelles'.

Collaborative action between the Department of Environment, international and local NGOs, private islands owners and managers, island foundations and biodiversity experts initiated in the 1980s has resulted in a significant number of invasive species control and eradication programmes, with sometimes mixed but overall positive results. Seychelles has become for example a world leader in the eradication of rats and feral cats on islands within the tropics.

This book reveals that today Seychelles is a world leader after Australia for its total number of alien vertebrate eradications in tropical islands, and ranks fifth when all islands are considered. The main reason for investing in the elimination and control of invasive species is to create new and additional suitable habitats for native and endemic species of animals and plants, particularly those globally threatened. Over the last few decades, more than a dozen species threatened with extinction have greatly benefited from IAS management and other direct measures in Seychelles, and have seen their conservation status greatly improved. Today, more than 20 of our islands have benefited from ecosystem rehabilitation and active conservation programmes that involve the management of invasive species. Motivated managers, scientists and field teams work together on these islands to progressively restore habitats and to advance the high level of technology and science that is indispensable to succeed in these operations.

These achievements have been possible thanks to exemplary partnerships between the above mentioned stakeholders, and the help of international donors such as the FFEM, the GEF, the EU and others that have acknowledged our efforts and the progress done by our small nation in the protection of its environment. Seychelles has now over 50% of its land surface protected by law, an unmatched proportion which reinforces its leading position in the field of nature conservation.

Dr Gérard Rocamora and Dr Elvina Henriette have worked for many years for the Seychelles Department of Environment and are today closely involved with the University of Seychelles; they have also worked with many private islands and NGOs and accumulated an extensive experience in invasive species management over the years. They have patiently compiled and synthesised in this manual as much information as possible from the experience available in Seychelles in the field of invasive alien species, with the collaboration of the many acknowledged contributors. This includes a summary of the information and knowledge acquired during the course of the Mainstreaming Biosecurity project, undertaken since 2007 by the Government of Seychelles, UNDP and GEF. The very existence of this book should be seen as an additional and very valuable output of the above referred spirit of partnership and collaboration that the government wishes to further promote.

This book intends to present both general and practical information to environmental practitioners, managers and members of the public on why and how to fight invasive alien species, with detailed management recommendations for 44 species identified as the most problematic in our country. I am convinced that this will be also of high interest beyond the borders of Seychelles to many island conservationists, administrations, experts and scientists from elsewhere in the Indian Ocean and around the world who have to deal with similar IAS problems, and often the very same species.

Victoria, Mahé, Seychelles
June 15, 2015

A tribute to Don Merton (1939-2011)

Gérard Rocamora

Dr Don Merton was a leading conservationist from New Zealand who made an outstanding contribution to the field of island conservation and restoration, in particular the recovery of critically endangered species and the eradication of rats and other invasive alien mammals. His pioneering work was also fundamental to the key conservation advances made in Seychelles in recent years.

Don started his career in the late 1950s, and worked until he retired in 2005 with the Department of Conservation of New Zealand, and also as a private consultant. He quickly recognised the devastating impact that rats and other introduced animals had on island ecosystems, and worked to eradicate these on many off-shore islands in his country. He became famous for rescuing the Chatham Island Black Robin, the world population of which had gone down to just four males and one female, and for the innovative management and translocation techniques that he developed to save several other endangered species from extinction, including the Kakapo, a flightless emblematic parrot of New Zealand, and the South Island Saddleback. His talent and creativity were quickly recognised internationally, and his involvement was sought in Mauritius and Seychelles to eradicate rats and cats from islands and to advise on the recovery of endangered birds.

Don first came to Seychelles in 1996 with his wife Margaret to conduct, on Bird Island, the first rat eradication ever attempted in Seychelles. This ground operation was a success, resulting not only in the total removal of rats but also of rabbits. He returned in 1998 to assess the potential for eradication of four other islands, and again in 2000-2001 to lead the first aerial rat eradication attempts using a helicopter, which resulted in the successful eradication of rats and mice on Frégate, and cats on Curieuse and Denis (rats were unfortunately found again later on these two islands).

This was a great opportunity for some of us in Seychelles to work with him and to learn from such a motivated and experienced person. Firmly believing how crucial the elimination of rats, other introduced mammals and invasive alien species in general was for the recovery of endangered endemic species and island ecosystems, I became inspired by Don's example. I decided to try and follow his work on the elimination of rats and other alien invasive species from more islands, in order to provide suitable habitats for all those endangered native species that survive only on a handful of predator-free islands.

Although Dr Don Merton was never to return to Seychelles, we kept regular contact and he continued to encourage me and provide valuable advice on the rat eradication projects that followed in Seychelles. In 2003, I decided to conduct my first rat eradication on the small island of Anonyme, using a permanent grid of bait stations to first eradicate the rats and then prevent their recolonisation from Mahé, only 500m away. Don was very excited and fully supportive of this challenging approach, which proved to be successful.

As an Honorary member of the Island Conservation Society since its creation in 2001, Don continued to provide regular advice between 2005 and 2009 while serving as a member of the scientific committee of the ICS FFEM project 'Rehabilitation of Island Ecosystems'. During this project, we eradicated rats from four more islands and two islets and increased by nearly 50% the rat free area of the granitic archipelago. Having also benefited from the experience of Don's former assistant Gideon Climo during the eradication of rats on Ile du Nord (North Island) in 2005; we were able by 2007 to form an entirely local team by partnering with Helicopter Seychelles – and later ZilAir – to successfully eradicate rats from Conception and three islands of the remote Cosmoledo atoll. We then succeeded the eradication of rats from Grande Soeur and Petite Soeur in 2010, much to Don's satisfaction. Despite the distance and the years, he remained deeply interested with Seychelles, and he was most delighted to know we were progressing well with the work that he first began on the tiny Ruapuke/Maria Island (1ha) in 1960, the first island in the world where rats are known to have been eradicated.

'Richard Henry' (RH) held by Don Merton, November 2010. Don played a key role in the recovery of this New Zealand flightless parrot species. Believed to be c.100 years old, RH was captured in 1975 by Don and was the last known survivor of this endemic species from mainland New Zealand. RH spent 35 years on islands free of alien predators, and died in December 2010.
Margaret Merton

Unfortunately, Don Merton passed away on April 10th 2011, at the age of 72. He received a unanimous tribute by many conservation organisations around the world. With his impressive tally of more than 50 island eradications of invasive animals that he was involved with, Don has been an inspiration to many and a book dedicated to his life achievements has been published. He received many national and international awards for his massive contribution to advance island conservation and restoration techniques, including the UNEP Global 500. In 2011, the New Zealand government created the Don Merton Conservation Pioneer Award.

Don was a model of what I believe an ideal conservationist should be: a hands-on practitioner with solid scientific knowledge, physical endurance in the field and mental tenacity, making things happen and ready to share his knowledge with others. Modest and unassuming, he was a very kind and easy person to work with. Don was also a sensitive person with regards to treatment to animals. I will never forget the day when I found him next to a young wild cat we had caught in a cage-trap on Thérèse Island, making gentle soft sounds to calm and comfort the animal. Don had eliminated so many cats in his life, I was a bit surprised. "What do you want to do with this one?" I asked; "Nothing... we have to kill it" he replied. That day, I could feel he too did not like killing any animal unless absolutely necessary. Elimination of invasive alien species should be well understood and done for a purpose, not just for the sake of doing so.

Like many others, I lost a much respected and supportive friend. But Don's pioneering work is now firmly embedded in island restoration and conservation management around the world; he has left behind a strong legacy that will continue benefiting us in major ways. The achievements of Don and others in Seychelles and around the world have proven vital for many species and habitats. Nevertheless, lack of vigilance and poor biosecurity could undo so much of this success. We hope that this book will help people to realise how crucial the fight against invasive alien species, combined with other biosecurity measures, are to preserve the biodiversity of Seychelles and other islands around the planet. May it encourage all of us to keep pushing in the right direction, and to make sure that these legacies and achievements are preserved and enhanced.

Anse Royale, Mahé, Seychelles
May 8, 2015

Important definitions

The following definitions pertain to situations involving invasive alien species. But in many cases the terminology also applies to more general situations but with slightly broader definitions.

Abatement measures: A series of strict procedures, pertaining to aspects of shipping and aerial transport (such as storage, loading, unloading, pest-proof rooms) aimed at reducing to nearly zero the risk of introducing or reintroducing invasive species. Abatement measures are preventive measures to minimise (abate) the risks of invasion or reinvasion.

Biodiversity: Same as *biological diversity*, i.e. the degree of variation of life forms inhabiting terrestrial, marine and other ecosystems and the ecological complexes they are part of. This includes diversity within and between species and ecosystems.

Biosecurity: Efforts to prevent the spread of invasive species to islands and sites that are currently free of them, and to reduce risks to the economy, environment and human health through measures that involve prevention, surveillance, exclusion, incursion response, mitigation, adaptation, control and eradication.

Commensalism: An association between two species in which one benefits and the other derives neither benefit nor harm (e.g. rodents are considered commensals of humans), in contrast to species establishing parasitic or mutualistic relationships.

Containment: Keeping an invasive species within a defined area.

Contingency: Response to incursion of an invasive species, usually involving a plan to ensure that all needed actions, requirements and resources are in place to eradicate it or bring it under control.

Control: Reducing the density and/or distribution of an invasive species below a pre-defined acceptable level.

Endemic: A native species, *race* (subspecies) or other plant or animal category that is naturally restricted and unique to a particular geographic region or island.

Eradication: The removal of the entire population of an invasive species from an island.

Establishment: The situation in which members of a species are engaging in successful reproduction in a new habitat/on a new island, sufficient to ensure the continued survival of the population.

Feral: Living in the wild, untamed. Applied to animals or plants that have gone from a domesticated condition back into the wild (locally called 'maron' in Creole and local French).

Habitat: The place or type of site where an organism or population is able to survive.

Herbicide: A chemical substance toxic to plants that is used to kill or reduce unwanted vegetation. There are 'contact herbicides' that kill the plant cells they are directly in contact with, 'residual herbicides' that are sprayed on the ground where they remain active, and 'systemic herbicides' that penetrate and travel to all organs inside the plant.

Indigenous: Refers to a native taxon that occurs naturally – but not exclusively – at the geographical area under consideration.

Introduction / Introduced species: Plants, animals and other organisms taken beyond their natural range by people, deliberately or unintentionally; this includes any part, gametes, seeds, eggs, or propagule that might survive and subsequently reproduce. This movement can be within a country, between islands, between countries or continents.

Invasion: The establishment and successful reproduction of an introduced species, and its spread in areas distant from its sites of introduction; this normally has a detrimental effect on native species when it affects natural habitats.

Mitigation: Reducing the likelihood that a risk occurs and/or reducing the impact of the risk if it has already occurred.

Monitoring: Repeated measurement of an indicator in order to assess how it is changing with time (e.g. abundance and distribution of a species, success of management projects).

Native species: Plants, animals and other life forms that occur naturally on an island or in a specified area, having evolved there or arrived there without human intervention.

Naturalised: Applies to non-native (introduced) organisms which survive in the wild and reproduce successfully to self-sustain their populations. Naturalised species become invasive when they spread into areas distant from its sites of introduction.

Non-target species: Species not meant to be affected by control or eradication operations.

Niche: A term that defines the ecological function of a species within a particular ecosystem complex, and the position it occupies in relation to interactions with other species and the different natural habitats it uses.

Pathway: The means (e.g. aircraft, vessel), purpose or activity (e.g. agriculture, forestry, horticulture), or a commodity (e.g. timber) by which an alien species may be transported to a new location, either intentionally or unintentionally.

Pest: Normally used in an agricultural sense to refer to an invasive plant (weed) or animal.

Quarantine: Official confinement of living organisms in complete isolation for further inspection, testing and/or treatment as part of biosecurity measures.

Rehabilitation / Restoration: Both terms refer to the active process of assisting the recovery of an ecosystem that has been degraded or destroyed. 'Restoration' is normally defined as a process that aims to re-establish conditions identical to those that prevailed in the past, whereas 'rehabilitation' only aims to re-create a productive, functional ecosystem whose species composition and characteristics may differ significantly from the original.

Reintroduction: An attempt to establish a species in an area that was once part of its historical range, but from where it has disappeared.

Reinvasion: The re-establishment of an invasive species that had been eradicated.

Secondary poisoning: The situation in which a predator or a scavenger has ingested one or more animals that contain levels of active poison in their bodies as a result of eating bait; may lead to death when the toxicity in the body of the consumer reaches a certain critical level.

Stakeholder: Any person or organisation that contributes to or is affected, either positively or negatively, by an operation such as a control or an eradication project.

Surveillance: A routine monitoring process to detect the arrival of new incursions of invasive species, as part of biosecurity.

Surfactant: A compound added to an herbicide mix that, by reducing water surface tension, improves the spread of the herbicide spray on a plant surface.

Species: A biological unit of classification that corresponds to a group of organisms that share a high degree of physical and genetic similarity, capable of interbreeding and producing fertile offspring, and that show persistent differences across generations. Presence of specific locally adapted traits may further subdivide into subspecies.

LIST OF ACRONYMS

CBD: Convention on Biological Diversity

CEPA: Conservation des Espèces et Populations Animales (France)

DoE: Department (or Division) of Environment (now Environment Department)

DGPS: Differential Global Positioning System

DRC1339: 3-chloro-4 methyl benzenamine hydrochloride

ED: Environment Department of MEE (formerly DoE)

ETF: Environment Trust Fund (Seychelles)

ETH: Swiss Federal Institute of Technology, Zurich

FAO: Food and Agriculture Organisation of the United Nations

FFEM: Fonds Français pour l'Environnement Mondial (= French GEF)

GEF: Global Environment Facility

GIS: Geographic Information System

GIF: Green Islands Foundation

GISD: Global Invasive Species Database

GoS: Government of Seychelles

GPS: Global Positioning System

GVI: Global Vision International

IAS: Invasive Alien Species

IBA: Important Bird Area

IBC: Island Biodiversity and Conservation centre (UniSey).

ICBP: International Council for Bird Preservation - now BirdLife International

IC: Island Conservation – Conservación de Islas (USA-Canada-Mexico)

ICS: Island Conservation Society – Fondation pour la Conservation des Îles (Seychelles)

IDC: Islands Development Company

ISSG: Invasive Species Specialist Group

IUCN: International Union for the Conservation of Nature

LD50: The dose required to kill 50% of a pest population

MINRI: Ministry of Investment, Natural Resources and Industry

MNHN: Muséum National d'Histoire Naturelle (Paris)

MEE: Ministry of Environment and Energy

MoH: Ministry of Health

MWF: Mauritian Wildlife Foundation

NBGF: National Botanical Gardens Foundation

NBSAP: National Biodiversity Strategy and Action Plan

NGO: Non-Governmental Organisation

NPTS: Nature Protection Trust of Seychelles

NZ: New Zealand

PII: Pacific Invasives Initiative

PPM (ppm): Parts Per Million

PCA: Plant Conservation Action group

PVC: PolyVinyl Chloride

SAA: Seychelles Agricultural Agency

SIDS: Small Island Developing States

SIF: Seychelles Islands Foundation

SNPA: Seychelles National Parks Authority

SR: Seychelles Rupee

SBRC: Seychelles Birds Record Committee

S4S: Sustainability for Seychelles

TRASS: Terrestrial Restoration Action Society of Seychelles

UNDP: United Nations Development Programme

UNESCO: United Nations Education, Scientific and Cultural Organisation

UNISEY: University of Seychelles

WWF: World Wildlife Fund / World Wide Fund for Nature

What is an
Invasive Alien Species?

Invasive Alien Species (IAS) are plants, animals or other organisms taken beyond their natural range of occurrence by people, either deliberately or unintentionally, the spread of which adversely affects the environment or human interests. The **International Union for the Conservation of Nature** (IUCN) also defines them as:

'Non-native species whose introduction, establishment and propagation threaten the ecosystems, habitats or indigenous species with negative environmental, economic or sanitary consequences' (general)

'Animals, plants or other organisms introduced by humans into places out of their natural range of distribution, where they become established and disperse, generating a negative impact on the local ecosystems and species' (environment)

Alien species are also known as **non-native**, **non-indigenous**, introduced, foreign or **exotic** species. They have been introduced beyond their natural past or present distribution by people or human-related activities. Some of these species have become – or have the potential to become – **invasive species** established in Seychelles. This means that they can survive, reproduce and disperse into natural ecosystems, such as forests, rivers or wetlands, but also into man-made habitats such as farmland, residential areas and cities, and create serious adverse impacts. Introduced plants that are able to survive and to reproduce successfully are said to be 'naturalised'; and it is only when they spread into areas distant from introduction sites that they are considered to have become invasive. Spread in natural habitats normally generates a negative effect on native species and ecosystems, but this may not be the case in already altered man-made habitats such as agricultural land. Several definitions exist for words such as 'naturalised' or 'invasion' (see Richardson *et al.* 2000, and our definitions p. 16).

Most of what we consider invasive species are **plant or animal forms** (mainly species, but also subspecies / races) **that can spread into the environment and create negative impacts to local species** of plants and animals. Examples include 'Western Indian lantana' ('Vieille fille', *Lantana camara*[a]), Cocoplum ('Prune de France'), Albizia or Cinnamon ('Cannelier') trees, once introduced in Seychelles for ornamental, soil retention, wood and spice production respectively, but which have since spread into much of Seychelles natural and man-modified habitats at the expense of native plants. **Exotic mammals** such as rats and cats, and alien birds such as Barn owls or Common mynas, were introduced by people, either unintentionally or deliberately (as pets or to control other invasives). Upon their release, however, they competed with and preyed upon native animals and plants, and have had an extremely negative impact on our island ecosystems.

Invasive species can also include pathogens that cause **fungal diseases**, such as the Takamaka wilt, responsible for killing many trees and severely impacting coastal woodlands and forests. Invasive alien species may also be detrimental to human activities **such as agriculture, fisheries, industry, tourism and other sectors of the**

a Scientific names provided only for species not treated in species accounts or case studies; see "Important notes", p. 9

economy (see "Socio-economic impacts", p. 33). A case in point is that of rats, which negatively impact not only the environment and native species, but also cause extensive and costly damage to agricultural crops, telephone and electric infrastructure, the hotel industry and human health. The Spiralling white-fly, a small insect, affects both native plants in natural habitats and cultivated plants in orchards and gardens. The Melon fruit fly (see case study p.114) is one of many alien invasives that specifically affect agricultural production. The recent invasion (February 2015) by hairy caterpillars provoking skin rash and 'spreading havoc' across the country shows how our small islands are increasingly vulnerable to the arrival of new invasive species (see case study p. 42).These few examples illustrate how important the impact of invasive species can be on our everyday lives and on the economy of the country.

It is also possible that some native species may proliferate and become invasive (see also box p. 183), usually in response to environmental changes (weather, agricultural development, habitat clearing, colonisation by alien species, etc.). In agriculture, the terms 'pests' or 'weeds' often refer to invasive alien species but can also designate native organisms that are detrimental to crops.

References / Further reading

- Richardson *et al.* 2000; SCBD 2002; Mooney *et al.* 2005; Soubeyran 2010; Simberloff & Rejmánek 2011, IUCN 2012; Simberloff 2013a.
- Global Invasive Species Programme: http://www.diversitas-international.org/activities/past-projects/global-invasive-species-programme-gisp
- IUCN/SSC Invasive Species Specialist Group: http://www.issg.org/
- CABI – Invasive Species Compendium / Glossary http://www.cabi.org/isc/glossary

Papaya Mealy bug (*Paracoccus marginatus*) symptoms.
Randy Stravens / SAA

About this book

1. Mainstreaming Prevention and Control Measures for Invasive Alien Species into Trade, Transport and Travel across the Production Landscape.

This book was produced as a component of an Integrated Ecosystem Management Programme called '*Mainstreaming Biosecurity*'[1], developed jointly by the Government of Seychelles, the United Nations Development Programme (UNDP), and the Global Environment Facility (GEF). It is meant to be accessible to both professionals and non-specialists, and to provide guidance in the identification of important Invasive Alien Species (IAS), and in the implementation of practical management measures, including prevention, control and eradication.

Although the book primarily targets a Seychelles audience, many of the same invasive species that affect Seychelles also impact other island territories and nations in the wider Indian Ocean region and around the world. It follows that virtually any-one trying to address the negative impacts of invasive species can benefit from the Seychelles experience.

The first part of this book is a general section that summarises information on the overall global problems caused by invasive alien species, and why oceanic islands like Seychelles are so vulnerable to them. After a brief historical background, we provide examples illustrating major impacts of invasive species on ecosystems, public health, conservation medicine, and the various sectors of the economy (tourism, agriculture, infrastructure, etc.). We also place emphasis on the key role of invasive species management in island conservation and restoration programmes, for which Seychelles has gained wide international recognition, and review the progress made during the last decades. Management options for controlling or eradicating invasive species and general recommendations to set up such programmes are presented. The importance of preventing the introduction and establishment of new invasive species as well as reinvasion by those already eradicated, is highlighted. The concept of integrated management is presented, together with the '*Seychelles Biosecurity Strategy*'. This comprises actions as diverse as public information, school education, customs and health procedures at ports of entry, preventive measures and contingency plans that allow early detection and rapid action, and effective control and eradica-tion programmes. We also provide some indicative costs of control and eradication programmes. A practical list of actions that you as a citizen can undertake to fight invasive species and prevent new infestations is given. A list of invasive species identi-fied as the most problematic for Seychelles concludes this section.

The second part of the book is a manual that provides for each of the selected prior-ity species an account illustrated with pictures. Each text includes a description and a summary of information pertaining to the species, including its origin, biology and ecology, its distribution and abundance in Seychelles, and the damage and threats it causes to the environment and different sectors of the economy. For each species there is a section summarising all control and eradication experiences in Seychelles - and in some cases also abroad. This is followed by management recommendations of best practices (insofar as they have been established), and also warnings about practices to avoid, in order to control, eradicate, limit the spread and prevent rein-festation of each particular species. Photographs facilitate the identification of each invasive species and, wherever possible, illustrate primary methods of control. A list of references for further reading, including technical publications, toolkits and websites available from the internet are provided for each species in footnote references and in a general bibliography presented at the end of the book.

The species chosen for illustration are those known to impact the environment and in many cases other sectors of the economy; they have been selected from priority lists previously established by stakeholders through surveys and workshops. For species and diseases that affect agriculture, a booklet is already available (Dogley W., 2004). These are not treated in our book, except for some that are known to also affect the environment. Invasive species currently being eradicated from their last stronghold in Seychelles (Feral goat, Red-whiskered bulbul, Ring-necked parakeet) have been treated through case studies rather than species accounts.

The Spiralling white fly is responsible for huge agricultural losses in many tropical countries, and is also known to affect native plant species in Seychelles.
Elvina Henriette

Introduction

The tendency of humans, intentional or not, to carry plants and animals with them across oceans and continents has resulted in the considerable spread of a vast number of species beyond their natural range of distribution. Transportation of living creatures for agricultural purposes has accompanied the expansion of numerous civilisations (e.g. Polynesian, Indonesian, and Middle-Eastern peoples in the Pacific and Indian Oceans; and Europeans across the entire globe during colonial times). Such transport has played a major role in the economic development of human societies around the world. Today, almost anywhere on the planet human survival depends largely on food production from non-indigenous plants and animals. In a minority of cases, however, the translocated species, which include not only plants and animals, but also fungi and bacteria responsible for diseases, have expanded rapidly and negatively impacted native ecosystems, as well as human societies (when they impact crops, livestock or people's lives).

This problem has been particularly acute on oceanic islands around the world, where invasive alien species have caused terrible ecological damage and have been a major cause of habitat degradation and species extinctions. Over the past few hundred years, the rate of species extinction is estimated to have increased by about 1,000 percent, and invasive species have played an essential role in this process. With the considerable development of trade, transport and tourism, the number of pathways for their introduction has increased, and invasive species have now become an accelerating problem responsible for the homogenisation and banalisation of biodiversity worldwide. When combined with climate change, they may now pose an even more devastating threat to the biodiversity of islands – and to the planet's biodiversity in general.

Countering the impacts of invasive species has become an urgent necessity, particularly in small island states. In the past, most attempts to mitigate, control or eradicate invasive species in Seychelles were directed at those invasive species that negatively impacted economically important activities such as agriculture (crops, coconut production, farming) or threatened public health (through mosquito or rat-transmitted diseases).

However, during the last 15 to 20 years, efforts have also been directed towards invasive species that threaten native biodiversity, particularly those affecting species found only in Seychelles and threatened with global extinction. During this period, island rehabilitation activities have been undertaken on a minimum of 23 small and middle sized islands, and have involved the eradication of introduced predatory animals, control of invasive plants and replanting of native trees. As a result of this process, individuals of rare endemic species that had been restricted to only a handful of predator-free islands, or those whose natural habitat was vanishing could be transferred to the rehabilitated islands, thereby creating alternative populations that prevented their extinction. The conservation successes achieved by restoring small islands are all based either on the effective eradication or exclusion of introduced predators and competitors, and on the control and elimination of invasive plants. Seychelles has gained considerable experience and is now considered a leading country in this field in the tropical world.

In October 2003, a regional workshop on invasive alien species and terrestrial eco-system rehabilitation was organised in Seychelles by the Indian Ocean Commission. This provided a first opportunity to share experiences, identify priorities and define joint actions for the western Indian Ocean islands. Since 2007, a number of other important workshops and activities have been conducted in an effort to mainstream biosecurity measures for invasive species into the main sectors of the economy; and key documents such as a national biosecurity strategy have been produced. In June 2009, a national Island Restoration Workshop took place as part of the ICS-FFEM project. In January 2012, another workshop organised by IUCN France on Mayotte gathered representatives from western Indian Ocean countries and French overseas territories; and this provided an opportunity to measure regional progress made in terms of IAS management and to identify the great challenges still ahead of us.

In recent years, various controversies[1] have arisen in regard to invasive alien species and biological invasions in general. These include claims that most biological invasions are inconsequential, that invasive species increase local biodiversity, that fighting them is a form of xenophobia, and that there is little we can do to prevent or control invasions. In addition, proponents of animal rights object to some of the management practices. These questions have fuelled an intense debate amongst scientists[2]. Some concerns can be easily countered and others can be addressed in a practical and pragmatic way.

In Seychelles, our day-to-day experience with the negative impacts of invasive species, and the positive results that have been produced through their management, demonstrate that the problem is real and very serious, but also that something can be done about it. That so many plants, vertebrates and invertebrates, are still on the verge of global extinction primarily because of invasive species, strongly advocates that this problem be treated as a global priority.

This book intends to build upon the substantial experience developed in Seychelles with invasive species. By summarising and sharing the current available knowledge (albeit sometimes very limited), it aims to generate further interest in the fight against these species, and to stimulate a new momentum for the conservation and restoration of islands in Seychelles and beyond, for the lasting preservation of their unique biodiversity.

Further reading
Mauremootoo 2003; Comité français de l'UICN 2012; Island Conservation 2012.
1 (See Davis *et al.* 2011; Simberloff 2011; Lambertini *et al.* 2011; Valéry *et al.* 2013; Blondel *et al.* 2013; Simberloff & Vitule 2014); 2 (Simberloff 2013b, Tassin 2014).

Part 1
Why and how to eliminate
Invasive Alien Species?
Impacts and management
options

As part of its environmental programme, the Sainte Anne Beachcomber resort conducts invasive species control operations that contribute to the recovery of native wildlife on the island.
A. Issock / Ste Anne Resort

Why fight Invasive Alien Species?

Chapter *1*

Invasive species have a considerable impact
on island ecosystems, public health and the
national economy. The bay of Beau Vallon,
on Mahé, is one of the most touristic and
populated areas in Seychelles. Landscape
vegetation is largely dominated by exotic invasive
species, such as Cinnamon (foreground).
Gérard Larose

Invasive alien species:
A global problem

First of all, it is essential to properly understand the reasons why we should fight invasive species.

Elimination of invasive alien species should not be done just because they are invasive, alien, or both. There are very practical reasons to fight them, such as ensuring the survival of endangered species, protecting our health, and preserving our economic interests. This chapter provides an insight to the global problem of IAS; it explains why Seychelles and other island states are particularly affected, and describes their impact on native animals, plants and ecosystems, and in our everyday lives.

The problem of invasive alien species stems from the ability of humans to transport living creatures with them, to distant places that they would normally have little or no chance to reach using natural means of dispersal by wind, ocean currents, the flow of rivers or natural transportation by wild animals or plant propagules (in the case of parasites). These alien species have been carried either on purpose, to meet the needs of human societies (for food, timber, farming, etc.), or unintentionally as undetected stowaways aboard increasingly sophisticated means of transport including boats, cars, trains and planes[1-3]. The development of human transportation and exchanges over the centuries, and the dramatic acceleration of economic trade and all kinds of movement by people around the world during the last decades have resulted in the problem of invasive species being one of the major causes of global environmental change[4].

Invasive alien species have now become a global problem, affecting to varying degrees, every single country in the world, although islands are by far the most vulnerable (see next section). **Invasive species are now considered to be the second biggest threat to the biodiversity** of the planet (after habitat destruction)[5, 6], and their consequences represent a huge cost to human societies worldwide[7, 18]. The impacts of invasive species can be extremely severe and often irreversible. The environmental damage is huge; they displace and eliminate native species, and alter the functioning of ecosystems, a problem that is particularly acute in protected areas[8]. **Invasive species have been responsible for the extinction or the decline of many species worldwide**[9], particularly on oceanic islands such as New Zealand, Hawaii, La Réunion, Mauritius, overseas French territories, etc.[10]. A majority of all 717 global animal extinctions (55 to 67% according to IUCN and IC) recorded to date have probably been caused by invasive predatory mammals such as rats and cats, or introduced birds carrying deadly diseases (e.g. avian malaria that caused extinctions of endemic birds in Hawaii), and these continue to threaten many more species[4a, 4b]. Over 90% of the 134 recorded bird extinctions known since 1500 have occurred on islands and often as a result of impacts from IAS, such as introduced mammals which prey on them or destroy their habitat[11, 12]. An aggravating factor is that human activities (i.e., forest logging, cultivation, roads, constructions, pollution, fire, etc.) have increasingly altered natural ecosystems, making them more vulnerable to invasion by a great variety of alien species and facilitating their establishment and spread. In 2014, an IUCN survey of the existing 228 natural World Heritage Sites showed that 104 (46%) were affected by invasive alien species[21].

Invasive species also have a huge impact on the economy and public health of communities around the globe. The spread across borders of **agricultural alien pests**, including weeds, insects, fungi or bacteria is responsible for billions of US dollars in loss of production[13], estimated at about 5% of annual global GDP[7]. In Maldives, rats reduce yields by at least 40% in coconut plantations[22, 23]. In Europe, during the second half of the 19th century, the introduction of Grape phylloxera *Phylloxera vastatrix*, an insect native to North America that affects commercial grapewines, is one of the worst biological invasions on record for its dramatic economic consequences. This epidemic destroyed almost all the European vineyards, which had to be entirely replanted with European vinegrapes grafted on American rootstock[19]. This triggered massive migrations of ruined families from southern Europe to other parts of the world, as it happened for example in rural areas of Catalonia that have remained depopulated since then[20].

Grape phylloxera symptoms on infested leafs.
Miguel Torres wineries

Introduced animals such as rats, which transmit diseases like leptospirosis and hepatitis, or invasive mosquitoes responsible for spreading diseases like dengue, chickungunya and malaria significantly **impact human lives** and national economies[7, 14]. Some invasive species are also responsible for significant **structural damages**, such as that caused by invasive termites (affecting buildings and wood infrastructure) or rats (electricity and telephone wires). Curative treatments, prevention, mitigation, control or eradication of invasives (including those related to the human or animal health sectors) generate additional **high costs in terms of biosecurity and management** (although these are normally cost-effective compared to doing nothing)[7, 15]. **Loss of ecosystem services** is an additional cost often not taken into account although it can be very high[16]. By affecting ecosystems and landscapes, invasive species impact entire communities or societies which depend on the ecosystems for their living, or that value them as part of their cultural heritage. This can often induce significant socio-economic changes[14].

The translocation and global spread of many plant or animal species for cultivation, timber, farming or other needs, have provided immense benefits and facilitated the development of many traditional or modern human societies[14]. However, these benefits are often associated with the negative burden of invasive species, which have now become one of the greatest challenges of modern human societies and the ever increasing process of '*globalisation*'. Never before in human history have numbers of invasive species, quantity of invaders and resulting biological invasions, and their ecological and economic consequences been so important[6]. In addition, **climatic change** is likely to magnify the effect by eliminating ecological/climatic barriers and increasing the potential range of certain invasive species, and by facilitating their spread in conjunction with an increased frequency in natural disasters (floods, cyclones, storms, fires, etc.)[17].

References / Further reading

1 Mc Neely *et al.*, 2001; 2 Mc Neely, 2001; 3 Millennium Ecosystem Assessment, 2005; 4a Clavero & García-Berthou 2005; 4b IUCN (undated); 5 Mooney *et al.*, 2005; 6 Simberloff, 2013a; 7 Pimentel *et al.*, 2001; 8 Foxcroft *et al.*, 2013; 9 Vié *et al.* 2009; 10 Soubeyran 2008; 11 Island Conservation, 2012; 12 Lambertini *et al.*, 2011; 13 Oerke, 2006; 14 Reaser *et al.* 2007; 15 Wittenberg & Cock, 2001; 16 Vilà *et al.*, 2009; 17 Tassin, 2010; 18 Emerton & Howard, 2008; 19 Legros & Argelès, 1986; 20 Escoda, 1996; 21 IUCN data; RocamoraMagali , 2015; 22 Fiedler, 1984; 23 Dolbeer *et al.*, 1988.

The impacts of invasive species in Seychelles

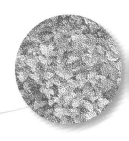

The many invasive species already present in Seychelles are having a tremendous impact on both the natural environment and important sectors of the economy.

LOSS OF NATIVE BIODIVERSITY AND ECOSYSTEM DISRUPTION

Introduced invasive species have seriously impacted Seychelles ecosystems. The ecological trauma that occurred on the archipelago following human colonisation, which started only in 1770, is in great part due to the simultaneous arrival of humans and invasive species[1, 2]. The fact that much of the original forests of Mahé had already been cleared by 1819 must have greatly facilitated the spread of new plant colonisers.

Some statistics on the proportion of exotic species in the flora of Seychelles and its main island groups exist but require reviewing and updating. In 2013, a minimum of 913 alien plant species had been registered, representing 54% of all the plant species (about 1,700) recorded in the whole of Seychelles[3]. This number keeps increasing. In the inner islands alone, there are 790 exotic vascular plants representing 65% of the total. A previous study[4] indicated that 72% (c. 265) of the 370 woody plants recorded from the granitic islands were exotic. The percentage of land area dominated by alien plants is probably of the same order of magnitude and clearly represents the majority of the land area of the country.

In the coralline outer islands, however, the proportion of alien species in the vascular flora (particularly woody plants) is significantly less, especially on the raised limestone southern atolls of the Aldabra group where much of the land surface is still covered by native vegetation[3]. In contrast, nowadays much of the lowland and mid-altitude forests of the granitic islands are covered with invasives such as Cinnamon, Albizia, White Cedar (Calice du Pape), Devil Tree (Bois jaune) or Cocoplum (Prune de France) (see species accounts page 308 and following)[4, 5]. These species were originally introduced for spice production, timber or erosion control but then became invasive. Some of these rapidly growing plant species form dense thickets; others develop strong below ground interactions that affect other plants; and some invasives (like Cinnamon) even emit chemicals that prevent regeneration of native plants[6].

The arrivals of so many new plant competitors, the combined negative effects of introduced insects that cause increased defoliation or compete with natural pollinators[7, 32], and invasive seed-eating rodents, have considerably disturbed the biology of the plant communities. Apart from outcompeting native vegetation, alien invasive plants indirectly affect native animals too (herbivores, frugivores or insectivores). Some native insects, for example, may have drastically declined because of the extreme rarefaction of their host plants. Invasive plants or animals may also constitute a food source for a predator that will increase in numbers and provoke the decline of native prey species. These trophic cascading effects can generate tremendous indirect impacts on many species and on the functioning of the whole ecosystem[8]. This phenomenon must have played a key role in the drastic reduction in range and numbers of many endemic plant or animal species, some of which now survive only in small habitat refuges. In 1996, of the estimated 250 woody native flowering plant species present in the granitic Seychelles, up to 21% (54) were listed as threatened under the criteria of IUCN[9]. Several species such as the Jelly fish tree (Bois méduse

Medusagyne oppositifolia) are still on the verge of extinction; the endemic orchid *Oeceoclades sechellarum* and the 'Gerivit andemik' *Vernonia sechellensis* are considered globally extinct, and some other species may have actually become extinct even before they could be discovered or described.

The impact of introduced invasive mammals such as rats and cats, which are known to be responsible for many extinctions and population declines of native species of vertebrates on islands worldwide[10-12, 22] has also been tremendous in Seychelles. A number of endemic birds (e.g. Seychelles magpie-robin *Copsychus sechellarum*, Seychelles warbler *Acrocephalus sechellensis*, Seychelles fody *Foudia sechellarum*), once reported to be abundant on the larger granitic islands by early explorers, later became endangered and restricted to only one or a few small rat and cat-free islands. The same situation probably applies to several endemic invertebrates (Whip spider *Phrynichus scaber* – also found on Silhouette, Giant tenebrionid beetle *Polposipus herculeanus*) and reptiles (Wright's skink) with a similarly restricted distribution. On Aldabra, the flightless Aldabra rail *Dryolimnas* (*cuvieri*) *aldabranus* is only found on cat-free islands, and in the Mahé group, the Seychelles white-eye *Zosterops modestus* survives in viable numbers only on Conception, the only island without Black rats. Trials with artificial nests containing quail, wax and plasticine eggs suggested that the main predators responsible for the high failure rate observed with Seychelles white-eye nests on Mahé, and Aldabra Drongo *Dicrurus aldabranus* nests on Aldabra, were rats[13, 14]. The three recorded extinctions of endemic birds in Seychelles (Chesnut-flanked white-eye *Zosterops semiflavus*, Seychelles parakeet *Psittacula wardi*, Aldabra warbler *Nesillas aldabranus*) were probably also caused by rat and cat predation, along with other factors such as habitat change and persecution. It is likely that some native endemic birds and other animals too became extinct before even they were ever described. The fact that all six Important Bird Areas (IBAs) for seabirds in the inner islands are small islands that were never (or only for brief periods) invaded by rats and cats, is also indicative of the very high negative impact these mammals (which prey on eggs, chicks and adults), can have on seabird colonies[30].

The African barn owl was introduced in the 1950s to control rodents, but it spread dramatically to all islands and made the situation worse by preying on seabirds. The Indian (or Common) myna may have been introduced as a cage bird or to control large insects; but it colonised most islands in Seychelles and turned into a fierce competitor for food and cavity nest sites, and also into a nest predator of many native birds and reptiles, including the endemic chameleons and geckos. Generally speaking, introduced birds may carry diseases such as avian malaria or the 'psittacine beak and feather disease' (PBFD), which could decimate our endemic Seychelles black parrot *Coracopsis barklyi* (see case study p. 276). The presence of rats, cats and introduced bird predators, is one of the main threats and limiting factors to the recovery of the majority of the eight globally threatened birds in Seychelles, including the Seychelles black parrot. The introduction of the Tenrec (Tang) had a very negative impact on large invertebrates (such as the Giant millipede, now restricted to islands without Tenrecs), and probably also on endemic amphibians (small frogs and caecilians). On Aldabra, feral goats have severely impacted the native vegetation and may have been a primary cause of the decline of the endemic Giant tortoises between the 1970s and the 1990s, as goats destroyed shade trees on which the tortoises are dependent, and restricted their access to food[16, 17] (although probably to a minor extent as Giant tortoises are grazers whereas goats are primarily browsers).

It is conceivable that these invasive plants and animals considerably altered the functioning of the ecosystems of Seychelles, by changing the forest structure, water budget, nutrient availability, food chains, and pollination and seed dispersal processes; but few studies have been conducted to quantify these impacts. Some invasive plants seem to also increase the risks for erosion or fire. Certain invasive species can interact between them and develop mutualisms that boost their development and exacerbate dramatically their impact on the ecosystem. For example, on certain islands the existing symbiosis between invasive ants, aphids or mealybugs (see case study p. 289) has generated an explosion of both species and severe negative effects on both plants and animals; this kind of synergy between invasive species has been qualified as 'invasional meltdown'[8]. On the other hand, some invasive species may also have positive impacts such as protecting water catchments, or colonising and protecting the soil of eroded lands (as is the case for Cocoplum). In some situations, invasive species will substitute for native species in providing food resources to native animals, or in acting as pollinators or seed dispersers for native plants which original mutualists have become too rare to play their role efficiently. This has been documented for Cinnamon or *Lantana camara* ('Vieille fille'), the berries of which are consumed by the Seychelles white-eye. Trees such as the White cedar ('Calice du Pape') or Cinnamon also host relatively high densities of leaf invertebrates that are consumed by this bird, although insects tend to be more abundant in native trees[13]. Introduced Honey bees *Apis mellifera*, although they compete with native pollinators, are now amongst the most frequent pollinators of many native plants in Seychelles[7].

The early introduction and spread of Cinnamon in Seychelles probably limited the expansion of more harmful invasives such as Chinese guava or Jambrosa, and nowadays its thick dense root mat on the topsoil also serves to protect against the establishment of woody invasives or creepers that would be more dangerous for the ecosystem[5]. Landscapes heavily dominated by invasives such as Cinnamon in Seychelles are now called 'novel ecosystems'. A novel ecosystem is defined as an ecosystem that is so far removed from its original state that it has now 'flipped' into a very different state, with different organisms and interactions, and perhaps even physical and chemical characteristics[18].

The extent and degree of dominance by (and probably a few other invasive trees) has been such for so long that it is probably no longer feasible to eliminate these forests. In fact, the elimination of large extents of mature, functional forest, could have very negative consequences for certain species of native fauna that have somewhat adapted to them, and also for the functioning of the ecosystem in general (in terms of water and nutrient retention, etc.)[5, 19]. Apart from the fact that it would take many years for a new forest to mature, there is a risk that other invasive species, the impacts of which are far worse, might well replace them. In simple words, a forest that is dominated by invasives is far from ideal but it is probably better than to have no forest at all. Novel ecosystems may still provide important ecosystem services and habitats for a number of native species. For example, Cinnamon provides an abundant source of berries for the endangered Seychelles white-eye; and several large invasive trees such as Sandragon *Pterocarpus indicus* or Albizia *Falcataria moluccana* seem to play an important role in the habitat and ecology of the endangered Seychelles scops-owl *Otus insularis*. These 'novel ecosystems' are dominated by alien plants but they also allow the survival of pockets of native plants. Restoring entirely these ecosystems is not realistically feasible on a large scale. However, it is possible to help small stands

Except on mountain tops, Mahé hill landscapes are heavily dominated by invasive trees such as Cinnamon or Albizia.
Gérard Rocamora

of native species (such as those found around inselbergs) to become larger, and to create new ones[5]. This strategy can be effective to restore ecosystem functionalities such as pollinating networks for rare native species in protected areas[19, 20].

SOCIO-ECONOMIC IMPACTS

One of the human activities most impacted by invasive species is **agriculture**. Alien invasive weeds are mainly creepers, herbaceous plants and some shrubs such as *Lantana camara* which lead to losses in crop and agricultural land[21]. Little data is available on estimated economic costs of invasive species in Seychelles[23]. For example, rats alone can inflict losses (both before and after harvest) of 15 to 30% for fruit, vegetable and root crop production in Seychelles, estimated at up to US$1.3 million per year, and possibly more[24, 25]. In the case of the Melon fruit fly (see case study p. 114), annual losses in fruits and vegetables were estimated at more than US$2.2 million in 1999, and could potentially affect up to 60% of the sensitive crops (cucumbers, pumpkins, melons, squash, gourds, etc.)[25, 26].

Invasive species are also detrimental to **public health**, as many infectious diseases for humans (or animals) are themselves alien invasives, or carried by invasive animals. The Tiger mosquito *Stegomya albopicta* is an important vector for the transmission of many pathogens and viruses, including chikungunya, yellow fever encephalitis and dengue fever[31]. The 2005-2008 chikungunya outbreak cost the Seychelles an estimated total of US$1.9 million in lost revenue in terms of the gross domestic product, and medical and disease control costs[27, 28]. Rodents are vectors of diseases such as leptospirosis, salmonellosis, hepatitis, plagues, murine typhus or rat-bite fever. Every year, Seychelles records 30 to 70 persons infected with leptospirosis (85% of whom are men) (see Fig. 1), including over half a dozen fatal cases; this is one of the highest rates of infection in the world (1/1,000 inhabitants) and it causes more fatalities than AIDS[31]! The Ministry of Health estimated in 2008 that the total cost of medical treatment for rat-transmitted diseases was nearly US$1.5 million[25].

In addition, rodents cause on-going and important **structural damage** by gnawing cables (electrical but also telephone, television optic fibre, internet cables, vehicle wires, etc.). This results in significant disruption to power supply and increasing fire

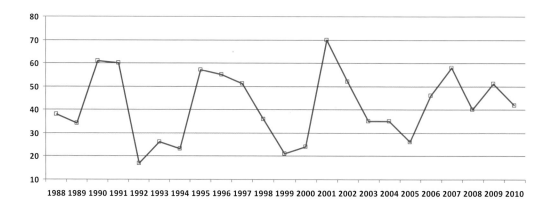

Figure 1: Number of confirmed cases of leptospirosis per year (source: Epidemiology & Statistics Unit - Division of Health Surveillance and Response - Department of Public Health). Cyclical outbreaks are not explained but could be linked with climatic conditions and/or rodent abundance.

risk. Annual cost of rat damage to infrastructure was estimated for telecommunication alone at between US$0.1[25] and US$0.4 (SR2M)[29] in 1994.

Tourism is also affected by invasive species like rodents, especially on small islands where outbreaks may have devastating impacts on the image of a hotel. This fact has played a significant role in the decision to eradicate rodents from private islands that host hotel resorts or are involved with other forms of tourism, along with the obvious environmental and sanitary gains derived from eradication. **Costs**, such as expenditure for preventive measures, which amounted to US$65,000 per year for quarantine and border control in Seychelles, need also to be taken into account[25]. During the mid-2000s, the total annual cost of pesticides ranged from US$1 to US$4 million. **Economic impacts of environmental damages** also add to the indirect costs incurred from invasive species. For example, there would be an expected loss in revenue from visitors to islands or nature reserves that are currently predator-free or creeper free, were these sites to be reinvaded[23-25]. The overall economic direct impact of rodents alone on the economy of Seychelles was estimated at US$2 million in 1994[29], and extrapolated to possibly US$3 million in 2007. In 2009, a study suggested that the overall economic damage caused by six major IAS (rodents, feral cats, melon fruit fly, invasive woody plants, creepers and the Takamaka wilt disease) to agriculture, human health, infrastructure, plus biodiversity and conservation sectors may have amounted to US$23.8 million plus an extra US$7 million spent on efforts to limit damage (US$21 for invasive mammals: cats, rats, goats and pigs and US$0.25 to limit damage), i.e. a total of US$21 to 31 million per year[24, 25]. Subsequent cost-analysis showed that prevention is by far the most cost-effective strategy, compared to eradication (when feasible), but that the latter is a better option than control.

References / Further reading

1 Lionnet, 1984; 2 Mauremootoo, 2003; 3 Senterre et al., 2013; 4 Kueffer & Vos, 2004; 5 Kueffer et al., 2013; 6 Kueffer et al., 2007, 7 Kaiser-Bunbury et al., 2010; 8 Simberloff, 2013a; 9 Carlström, 1996; 10 Towns et al., 2006; 11 Medina et al., 2011; 12 Harper & Bunbury, 2015; 13 Rocamora & François, 2000; 14 Rocamora & Yeatman-Berthelot, 2009; 15 Rocamora & Skerrett, 2001; 16 Coblentz & Van Vuren, 1987; 17 Rainbolt & Coblentz, 1999; 18 Hobbs et al., 2013; 19 Kueffer & Kaiser, 2014; 20 Foxcroft et al., 2013; 21 GoS/UNDP/GEF (undated); 22 Courchamp et al., 2003; 23 Murray & Henri, 2005; 24 Mwebaze et al., 2009; 25 Mwebaze et al., 2010; 26 Ikin & Dogley, 2009; 27 Ministry of Health, unpublished data; 28 Henriette & Julienne, 2009; 29 Ministry of Health, 1994; 30 Mulder et al., 2011; 31 Bovet et al., 2013; 32 Kaiser et al., 2011.

HOW DO ALIEN SPECIES ARRIVE AND BECOME INVASIVE?

The arrival of invasive alien species to Seychelles is exclusively linked with human activity. Even before the French established their first official permanent human settlement on Ste Anne in 1770, early travellers (including passing merchant vessels, explorers and pirates) had visited the islands and established camps, and had purposefully set free a number of domestic animals such as goats, cattle, pigs or guinea fowl, to serve as potential future sources of food[2]. Those early visitors were also responsible for the first unintentional introductions of commensal animals such as Black rats, which thrived on board ships, and which were already reported on Mahé in 1773[1]. Cultivated plants, timber trees, and even weeds would have been amongst the first invaders. The question of whether Coconut palms and Casuarina trees (locally called 'Filao' or 'Sed'), the natural distributions of which are centred in the Pacific, came naturally or were brought in by early travellers such as Indonesians or Arabs is still being debated.

From that time onward, human activities have dramatically increased opportunities for alien species to travel to new places and become invasive. These introductions have continued through an endless variety of 'pathways', mainly linked to international Trade, Travel, Transport or Tourism (the four Ts), now culminating with the process of globalisation of the economy[3-6]. Nowadays there are many possible ways for alien species to arrive in Seychelles, and below are the main pathways that are (or may be) responsible for their introduction (including pathogens and parasites carried by non-native species)[5-7].

Intentional introductions (legally authorised or through smuggling):

• Food products (plants, animals, fish products)

• Agriculture (crops and livestock)

• Release of pets or other domestic animals (including aquatic species)

• Horticulture and gardening (ornamental plants)

• Hunting and fishing (game species and live fish)

• Landscaping (use of non-native plants in hotels, etc.)

• Mariculture and aquaculture (fish, molluscs, and crustaceans introduced for production)

Unintentional introductions by means of:

• Trade and movement of goods (alien species translocated in/on containers, untreated timber and wood packaging, food products, thatching for roofs, etc.)

• Movement of people (travel and tourism through air and sea transport)

• Movement of equipment / vehicles between islands

• Shipping and boating (ballast water, sediment, hull fouling, anchors)

• Aviation (in cargo, and on or in the aircraft itself)

• Postal services (including living material purchased via the internet)

• Aquaria (discharge of organisms with waste water)

The invasion process

The three main phases of the invasion process are: Introduction, Establishment, and Spread[4]. Establishment

Transport, Travel, Trade and Tourism (the 4 'T's) are the main pathways for invasive species.
G. Rocamora

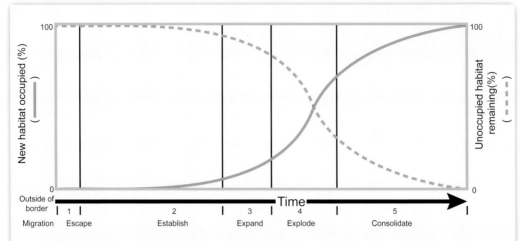

Figure 2: The invasion process (from Emerton & Howard, 2008; after Williams, 2003). Phases are expressed as a proportion of the habitat occupied by an invading species.

means that the introduced species survives in the new environment and reproduces successfully until it establishes a self-sustaining population.

The 'Spread' phase can be further divided between Naturalisation and Invasiveness, although different definitions have been given[8]. In general, naturalisation applies to non-native (introduced) organisms which survive in the wild and reproduce successfully to self-sustain their populations. Naturalised species become invasive when they spread into areas distant from its sites of introduction and become detrimental to native species and ecosystems. The diagram below summarises the invasion process. Here the 'Spread' is divided between 'Expansion', 'Explosion' and a longer phase of 'Consolidation'.

Invasiveness

The process of how and why any particular exotic species becomes invasive is not well known; and various factors or hypothesis can be proposed to explain this phenomenon[9]. Invasiveness may be linked to the relatively high levels of competitiveness that characterise alien species in comparison to island natives, such as the absence of natural enemies from their countries of origin including parasites and pathogens (called 'enemy release' effect); the likelihood that vacant ecological *niches* exist in the simplified island ecosystems ('empty niche' hypothesis), the ability to prevent other plants to grow (rhizochemical dominance hypothesis), an extremely abundant production of seeds (propagule pressure hypothesis), amongst others[14]. In some cases, it may be linked to changes in the environment, such as the sudden availability of habitats suitable for a particular species (see p. 183 the case of Merremia), but also to changes in the ecology of the species, which becomes suddenly adapted to colonise new habitats[9]. There is strong evidence that mutations can also intervene in the process of producing an invasive genotype. Crossings from very different source populations leading to new recombined genotypes, or the arrival of a new strain with a greater ecological plasticity may both intervene in the acquisition of invasiveness[10]. Various degrees of potential invasiness can be defined. Rapid growth rates, large reproductive capacity, strong dispersal capabilities, and broad tolerance for varied environmental conditions (moisture, temperature, acidity, etc.) will indicate predisposition, as well as invasiveness in other countries with ecological similarities. Introduced species could be affected, at least in theory, by 'inbreeding depression' resulting from the limited genetic variability carried by their initially very small numbers of invaders. However, invasive species demonstrate that this problem does not affect systematically all small populations in the same way[9, 10]. It can be compensated by a number of favourable or predisposing factors, the conjunction of which may trigger at some point the process of invasiness. The invasion risk of species that are already in the country (or in a particular island) can be assessed in the field based on monitoring and observation. Alien plants can be classified to be (at a given point in time) more or less likely to become rapidly invasive depending on what stage (phase) of the invasion process they are found, by recording whether they are reproducing successfully, spreading away or not from mother plants,

Supply boats and planes are the main pathways for invasive species.
G. Rocamora

their dominance, their impact on other species and the ecosystem in general, etc.[8]. This exercise has already been conducted for creepers in Seychelles[11].

Not all alien species become invasive

Only some alien species become invasive after becoming established, although they may retain the potential to do so for a long time, and undergo a lag phase before suddenly spreading[12, 14]. For example, of an estimated minimum of 265 species of exotic shrubs and trees present in the granitic Seychelles[13], no more than 17 (6%), to 31 (12%) if we consider those that show signs of invasiveness and/or are known to be invasive in other countries (see tables 12 and 13 p. 181-182), have been reported as invasive so far. But even if these represent a proportionally small number, each of these few invasive exotic species can cause very significant damage; hence all exotic species should be viewed as potential biological 'bombs'. The Global Invasive Species Database provides a list of inventoried invasive species; these include 32 plants that are amongst the 100 world's worst invasive species[15].

References / Further reading
1 Lionnet 1984; 2 Fauvel 1909; 3 Macdonald et al., 2003; 4 Emerton & Howard, 2008; 5 Dogley W., 2009; 6 Ikin and Dogley W., 2009; 7 DWCT / MWF, 2010; 8 Richardson et al., 2000; 9 Simberloff, 2013a; 10 Simberloff, 2013b; 11 Senterre, 2009. 12 Kueffer et al., 2010a; 13 Kueffer & Vos, 2004; 14 Holzmueller & Jose, 2009; 15 GISD, 2014; 16 Torchin & Mitchell, 2004.

Why are Seychelles and islands in general so vulnerable?

VULNERABILITY OF ISLANDS TO BIOLOGICAL INVASIONS

Oceanic islands are particularly vulnerable to invasion by exotic species. Because of their isolation during millions of years, they have developed assemblages of unique species, called endemics, and ecosystems which have evolved in complete isolation from continental floras and faunas. The large extension of surrounding marine environment and the restricted land area and habitats available have considerably limited the number of living organisms able to reach and colonise these islands. This explains the great fragility of islands and the ecological trauma they all suffered after their colonisation by humans, in particular because of the arrival of alien species from the outside world[1-3].

Compared to mainland species, terrestrial island creatures generally have reduced dispersal abilities (for example flightless birds like the Aldabra rail), and their populations are often limited in size, making them particularly vulnerable to extinction[4]. This is why most of the 804 extinctions of plants and animals recorded to date on earth since 1500 have occurred on islands (68 to 95% depending on the taxonomic group), and why an important proportion of threatened species (e.g. 58% of threatened birds) are on islands (IUCN data). Coming on ships as unwanted passengers, or deliberately introduced for human subsistence, mainland species are adapted to live in complex ecosystems. When these aliens reach the islands, and find opportunities to get into new ecological *'niches'*, they can be very successful as they tend to be better competitors than their local counterparts, and so may become invasive.

Island plants and animals, which have evolved within simple island ecosystems that comprise relatively small numbers of species, are not adapted to cope with the impact of efficient predators such rats and cats, herbivores such as rabbits and goats, depredative insects, competitive weeds or diseases brought in from continental areas. This is why oceanic islands are very prone to biological invasions. Invasive species have been the first cause of population declines and species extinctions on islands[2-6]. Moreover, the limited human and financial resources available on small islands exacerbate the problem. The fact that the economies of Small Island Developing States (SIDS) rely more on importations than do those of continental countries (43% of GDP compared to 27%, respectively; IMF data) further increases the risk of new invasive species introductions. The fragile economies of island states makes them much more vulnerable to socio-economic impacts from invasive species compared to mainland countries[7].

VULNERABILITY OF THE BIODIVERSITY AND ECONOMY OF SEYCHELLES

As is the case for other remote oceanic islands, the terrestrial biodiversity of Seychelles is very susceptible to invasive alien species. Seychelles belongs to the hotspot for global biodiversity comprising Madagascar and neighbouring islands, and hosts a high level of terrestrial endemism, particularly on its granitic archipelago which has been isolated from land masses for over 60 million years[8]. About 20% of the native plants species of Seychelles (a minimum of 136 over 707 native species) are endemic[9]. Endemism reaches 26% for all vascular plants (and 29% for the flowering plants in the granitic archipelago). As many as 86 plants from Seychelles are listed as threatened according to the IUCN criteria[15]. Endemic animals include 14 species (and 16 subspecies) of birds, five species of bats, at least 16 species of reptiles, 12 species of amphibians (including an endemic family of frogs), two species of freshwater fishes, plus several thousand invertebrates (the majority of which are insects). The majority of the endemic species of vertebrates are rare and globally threatened, particularly in the case of birds and amphibians. In total, Seychelles hosts about 120 terrestrial Globally Threatened Species, including 17 land vertebrates[6].

Several species of endemic birds, reptiles and invertebrates are restricted or have their strongholds on a handful of predator-free islands. The number of such strongholds has increased considerably during recent decades thanks to island restoration programmes which have combined the eradication or control of invasive predators, the rehabilitation of habitats through replanting of native trees, and successful island translocations of threatened animals (see following chapter). Some of these predator free islands are also home to large seabird colonies and have been designated as Important Bird Areas (IBA) by BirdLife International[10]. These small island refuges are home to a significant part of the biodiversity of Seychelles and represent an important source of income as major tourism attractions[11]; but they are also extremely vulnerable to reinvasion by rats and other dangerous invasive species[12] such as Crazy ants, Barn owls, creepers, Takamaka wilt, etc.

In particular, there are currently no legal restrictions preventing boats from landing on rat-free islands, because the Constitution of Seychelles grants free public access to all beaches, with the possible exceptions of protected areas (nature reserves such as Aride and Cousin Islands; Aldabra atoll, and marine national parks that include beaches). This makes these islands and their precious biodiversity extremely vulnerable to rat and other IAS invasion. It also puts them under constant risk of losing the benefits acquired through years of collective efforts to restore their ecosystems and the populations of flagship threatened species such as the Seychelles magpie-robin, the Seychelles white-eye, the Wright skink, the Giant tenebrionid beetle and many others. Regulations under the Animal and Plant Biosecurity Act 2014 should address this problem and condition access to the shores of islands of high biodiversity value to the implementation of biosecurity protocols (for vessels, goods and people; already in place for certain protected areas).

Additional factors that make Seychelles vulnerable to invasive species include the impossibility of controlling its immense maritime borders and boat movements between islands (about 155 main islands scattered within an immense Exclusive Economic Zone of 1,374,000 km²). It addition, its tropical / equatorial climate provides favourable conditions for a myriad of organisms originating from other tropical and sub-tropical regions.

Indicators of international exchange such as tonnage of fresh agricultural products (animal and plants) imported, quantity of air cargo, number of containers landed, timber imports, etc. are all increasing, as is the number of yachts, cargo ships, international flights and tourist arrivals[13]. These provide the main pathways for invasive species into the country[14]. In recent years Seychelles has everyday become more threatened by invasive species. This is problematic for a country that depends primarily on its natural resources and biodiversity for its income (fisheries, tourism)[11, 12]. Seychelles economy is very vulnerable to environmental, agricultural or public health damage provoked by invasive alien species.

References / Further reading

1 Simberloff & Rejmánek, 2011; 2 Island Conservation 2012; 3 Simberloff 2013a; 4 MacArthur & Wilson 1967; 5 Lambertini *et al.*, 2011; 6 IUCN 2014; 7 Reaser *et al.* 2007; 8 CEPF 2014; 9 updated from Senterre *et al.* 2013; 10 Rocamora & Skerrett 2001; 11 Murray & Henri, 2005; 12 Mwebaze *et al.*, 2010; 13 Ikin & Dogley W. 2009; 14 Dogley W. 2009; 15 Bruno Senterre, unpublished; 16 DWCT / MWF 2010.

The globally threatened Seychelles black parrot *Coracopsis barklyi* (IUCN Category Vulnerable) only breeds on Praslin Island. Many Seychelles endemic species are threatened because of a small population and a distribution restricted to a few islands.
Irene & Roy Mc Grath

Two Sheath-tailed bats *Coleura seychellensis* (adult and young). This Critically Endangered species endemic to Seychelles is probably the most threatened species of bat (and mammal) on the planet, with a global population of not more than 60 individuals known in 2015 (only from Mahé and Silhouette).

Vincent Robert / IRD

D'Arros Island, Amirantes group. Small islands are very vulnerable to biological invasions, as well as to climatic change (extremely flat coralline island may be less than 2m).

Gérard Rocamora

The hairy caterpillar (?*Euproctis* sp.) invasion in Seychelles – novel threats from a highly resistant pest

by **Pat Matyot** (Island Biodiversity & Conservation center, University of Seychelles)

On 12th February 2015 the Ministry of Health in Seychelles issued a communiqué entitled *Caterpillar Skin Rash*: *"The Disease Surveillance & Response Unit in the Ministry of Health has been receiving several reports since Friday 6th February 2015 of clients from Ramo Estate reporting to the Anse Aux Pins health centre with itchy red rash which was linked to contact with a certain hairy caterpillar. An investigation by Public Health Officers was conducted to assess the situation and identify the possible causative agents. It was noted that there was infestation of a specific plant, locally known as "Tantan"* [the Castor-oil plant, *Ricinus communis*], *with a specific caterpillar covered in hairs, yellowish in colour with black spots... Several other districts since then have also reported sightings of such caterpillars. A public health officer who came into contact with the caterpillar also developed rash..."*[11].

Sensational titles in the press on the invasion of hairy caterpillars.

Rapid spread

Within a week people who had developed skin irritation after having been in the vicinity of the caterpillar were seeking medical assistance at health centres all over Mahé, with the majority of cases being recorded in the northern and central parts of the island. However, the Seychelles Agricultural Agency reported that the presence of the caterpillar had been confirmed on Praslin Island as well, and that it had been observed on other host plants, including Indian almond or "Bodanmyen" (*Terminalia catappa*), Pomegranate or "Grenad" (*Punica granatum*) and Golden apple or "Frisiter" (*Spondias cytherea*). The Division of Risk and Disaster Management (DRDM) stepped in to coordinate efforts to tackle the spread of the pest[2], and produced an action plan involving a variety of agencies, from the Ministry of Health, the Department of Environment and the Seychelles Agricultural Agency to the Ministry of Education. Indeed, nine schools had reported caterpillar outbreaks, especially on Indian almond trees on school grounds, with Mont Fleuri Secondary School being described as "the most infested". By the end of March 2015, it was known that the hairy caterpillar had reached the islands of Ste Anne, Cerf, Moyenne and La Digue as well, and was feeding on at least 12 different plant species, including Mango (*Mangifera indica*). The DRDM had received 205 requests for assistance to deal with caterpillar infestations. Sixteen schools had been fumigated[3].

Taxonomic uncertainty

The hairy caterpillar, about 2½ cm long when fully grown, yellow with black tubercles, as well as the pure white moth that it develops into, was identified as a species of *Euproctis* or some closely related genus such as *Orvasca* or *Sphrageidus* (family Erebidae of the order Lepidoptera), but the precise identity of the species remains a mystery. This taxonomic impediment is due to the fact that there has been no recent review of the large genus *Euproctis* and its close relatives (more than 650 species): many museum specimens await

Left: Adult moth in resting position.
Pat Matyot

Right: Egg plaque on undersurface of papaya leaf.
Pat Matyot

description, and the faunas of many countries, including those around the Indian Ocean, need to be further investigated[4]. However, the species causing "panic" in Seychelles is clearly an introduced one (it is not the presumably endemic *E. pectinata*, not found again since it was first collected in the mist forest of Silhouette Island in 1908), and it had never been observed anywhere in Seychelles before January 2015 (the first known sighting on Pomegranate at La Rosière, near Victoria, on 31st January, before it was realised that its hairs had urticating properties). In the Maldives, *Euproctis fraterna* regularly reaches outbreak levels[3]. Lepidoptera specialists from Australia are reported to be currently working on the taxonomy of the species at the request of the SAA[5].

Venom

The skin irritation or dermatitis caused by the caterpillar is similar to that resulting from contact with tiny "dart hairs" or spicules (not the clearly visible long hairs) from the caterpillars of other *Euproctis* species in other parts of the world, such as *E. chrysorrhoea* (the brown-tail moth) in Europe, *E. fraterna* in the Maldives, *E. edwardsii* in Australia, *E. subflava* (=*Artaxa subflava*) in Japan, etc. These species are known to have barbed urticating hairs that are, in effect, hollow microneedles filled with venomous fluid containing proteases, phospholipase A (PLA) and other enzymes that are irritants to vertebrates[6,7]. De Long[8] calculated that a single caterpillar of *E. similis* (China) had up to 2 million such hairs. On penetrating the skin, the spicules liberate their venomous contents. Contrary to the situation elsewhere, there have been no known cases of irritation of the eyes (conjunctivitis) or respiratory distress caused by the caterpillars in Seychelles.

In the case of those *Euproctis* species that have been studied, direct contact with living caterpillars is not necessary: urticating hairs may be aerosolised ("floating" in the air) and transported airborne, even indoors, by wind; they may even settle on clothing that has been put out to dry. Also, dead caterpillars and empty larval skins can still shed toxic spicules, and these can retain their injurious properties for several months or even years. Furthermore, with other *Euproctis* species, larval hairs are spun into the cocoon that surrounds the pupa, and some of these are picked up by the anal tuft of the adult moth as it emerges from the cocoon – giving the moth, too, the ability to liberate urticating spicules, including when ovipositing so as to provide the eggs with a protective covering[9]. All this is true of the mysterious hairy caterpillar that has appeared in Seychelles. Fortunately, the skin irritation caused by the envenoming hairs from egg plaques, caterpillars, pupal cocoons and adult moths – mostly on the arms and torso – usually does not last more than three to four days[20]. However, there have been cases of susceptible persons having to receive medical treatment.

An adaptable coloniser

The mode of entry of the unidentified ?*Euproctis* into Seychelles is not known. Eggs and/or larvae (caterpillars) may have arrived on introduced plant material, but the hitchhiking of egg masses or pupae in cocoons stuck to cargo (e.g. in crevices in wood) or even containers is also possible. Caterpillars have been observed dangling from strands of silk and becoming airborne ("ballooning"), and this would explain at least in part the rapid spread of the species throughout the granitic islands (Table 1). The distance travelled by ballooning caterpillars depends on wind velocity, height of release point, size of caterpillar, size of silk strand, and presence of barriers[10]. The adult moth is not prone to long-distance active flight, but may be transported by wind.

Island	First recorded sighting
Mahé	January 2015
Ste Anne	February 2015
Ile au Cerf	February 2015
Moyenne	February 2015
Praslin	February 2015
La Digue	March 2015
Silhouette	May 2015
Ile Aride	May 2015
Cousin	May 2015
Cousine	May 2015

Table 1: Islands of Seychelles where the stinging caterpillar (?*Euproctis* sp.) has been sighted.

The caterpillars are extremely polyphagous, with at least 16 reported food plants (Table 2), the Castor oil plant *Ricinus communis* and Indian almond *Terminalia catappa* (in that order) being clearly their preferred hosts, presumably because of differences in nutritional quality and/or the presence/absence of secondary metabolites acting as chemical defences in the leaves. On some plants (Pomegranate, Guava and Mango), they feed not only on leaves but also on fruits, while on others (e.g. *Ixora coccinea*) they consume the flowers. No data has been collected on survival rates of caterpillars on different plants, varying in chemistry, nutritional content and architecture.

Table 2: Known food plants of the hairy caterpillar (?*Euproctis* sp.) in Seychelles. Sources: 1 = personal observation; 2 = Seychelles Nation of 30th March 2015; 3 = various informants, personal communications). The caterpillar has also been observed on an unidentified tiny plant with yellow flowers" (2) and on the flowers of an ornamental (exotic) palm (3).

English name	Creole name	Latin name
Indian almond (1, 2)	Bodanmyen	*Terminalia catappa*
Rangoon creeper (1)	Santonin	*Quisqualis indica*
Castor-oil plant (1, 2)	Tantan	*Ricinus communis*
Mango (1, 2)	Mang	*Mangifera indica*
Golden apple (1, 2)	Frisiter	*Spondias cytherea*
Cashew (1)	Kazou	*Anacardium occidentale*
Pomegranate (1, 2)	Grenad	*Punica granatum*
Java apple (2)	Zanmalak	*Syzygium samarangense*
Guava (1, 2)	Gro gouyav	*Psidium guajava*
Leucaena (2)	Kasi	*Leucaena leucocephala*
Cat's claw(3)	Kanpes	*Pithecellobium unguis-cati*
Passion flower(2)	Bonbonplim	*Passiflora foetida*
Drumstick tree (2)	Bred mouroung	*Moringa oleifera*
Amaranth (1)	Bred paryater	*Amaranthus dubius*
Pink oleander (1)	Lorye roz	*Nerium oleander*
Mulberry (3)	Mirye	*Morus alba*
Ixora (1)	Bison	*Ixora coccinea*
Fish poison tree (3)	Bonnen kare	*Baringtonia asiatica*
Banana (1, 2)	Bannann	*Musa ×paradisiaca*

One particularly surprising observation is that on the underside of papaya (*Carica papaya*) leaves the caterpillars apparently feed on the invasive papaya mealybug (*Paracoccus marginatus*)[11], although it is possible that they actually ingest only honeydew or wax secreted by the mealybugs. When plants of Amaranth or "Bred paryater" (*Amaranthus dubius*) infested with the caterpillars were cut and put out to dry in the sun, a few days later it was observed

Pupa removed from cocoon.
Pat Matyot

Hairy caterpillars on pomegranate fruit.
Pat Matyot

that the caterpillars had bored inside the cut stems, feeding on the internal tissue and using the hollowed stems to shelter from the sun. As for the adult moths, like other members of the family Erebidae, they have greatly reduced mouthparts and do not feed, living on food reserves built up during the caterpillar stage. They are therefore able, for the entire duration of their lives, to concentrate on mating and oviposition. One adult female moth that emerged from a pupa in captivity lived for seven days in captivity, laying presumably infertile eggs before it died[20].

Threats

So far, the stinging caterpillar and the pupa and moth that it develops into have been primarily a menace to public health because of the itchy "urticarial papules" caused by their spicules or "dart hairs". This discomfort has caused varying degrees of destabilisation to households, schools and places of work, with vulnerable affected persons in some cases having to receive symptomatic and supportive treatment with external and intramuscular antipruritics and anti-inflammatory agents. No statistics have been released yet by the Ministry of Health, but the sanitary threat is a regular topic of discussion in the social media and elsewhere. The fact that the adult moths are attracted to artificial lighting in homes and public places increases the number of people who are at risk of exposure to the urticating hairs. Because of this and because of the large number of Indian almond trees in coastal areas, tourism establishments are potentially under threat.

Castor-oil plants (bushes) growing in waste places and parts of Indian almond and other trees are sometimes almost completely defoliated by the caterpillars. These feed communally, with up to at least 150 per infested Indian almond leaf[20]. However, there have been no instances of whole-tree defoliation as have been reported in the Maldives (caused by *E. fraterna*) and in Central Asia (caused by *E. kargalika*). Many establishments and individuals have had infested trees drastically trimmed or even felled to eliminate sources of caterpillar hairs from the environment. In some cases these were prize specimens and their removal has meant the loss of valuable fruit trees or an aesthetic loss. For now, the caterpillar is not considered to be a serious agricultural pest, nor is there any record yet of it feeding on any of the threatened endemic plants of

Seychelles. But conservationists must remain alert to the possibility that it may adapt to more native plants (*Barringtonia asiatica* is considered to be native, and *Terminalia catappa* is treated as a likely native).

It is too early to ascertain what effect ?*Euproctis* sp. will have on animal biodiversity in the islands. The urticating hairs and warning coloration (aposematism) of the larvae of *Euproctis* species and their relatives are considered to be an effective repellent defence against vertebrate predators – the gregariousness of the caterpillars presumably increasing the signalling effect of warning coloration, black and yellow in ?*Euproctis* sp.[12]. In Europe it is known that herbivorous animals can be severely affected by eating vegetation contaminated with the hairs of similar species[13]. *Euproctis* caterpillars are among the lepidopteran larvae, presumably eaten in forage or fodder, that have been implicated as the cause of abortions and stillbirths in horses in North America and Australia[14]. Phisalix[15] mentions reports of "serious cases of enteritis" among chickens and ducks that had fed on hairy caterpillars, though not necessarily *Euproctis* species. However, in Europe, some passerine birds such as the Great tit (*Parus major*), are able to consume the soft tissues of the caterpillars without ingesting the hairs[16]. Cuckoos (family Cuculidae), on the other hand, are able to ingest the larvae whole: the hairs end up piercing the thick gizzard lining, which subsequently peels off in patches and is regurgitated together with the hairs[17]. In Seychelles, there have been unconfirmed reports that the introduced Indian mynah (*Acridotheres tristis*) has been observed feeding on the hairy caterpillars[18].

Natural factors that could regulate ?*Euproctis* populations

Research is needed to elucidate the developmental cycle, the population and infestation dynamics (establishment, growth and survival) of ?*Euproctis* sp. in Seychelles. Foodplant distribution, meteorological conditions, etc. need to be taken into consideration. The pest is reported to have a life cycle of 28 days[19]. With one specimen in laboratory conditions, the pupal stage lasted eight days[20]. Predators, parasites and pathogens, when present, would control the populations of this species. The Potter wasp *Delta alluaudi* and the Yellow paper wasp *Polistes olivaceus* normally catch small caterpillars but have not been observed hunting the larvae of ?*Euproctis* sp. Grobler[21], in his account of the biology and ecology of *Euproctis terminalis* in South Africa, reported that parasitic "insects are of minor importance, however, and the main control agent is a strain of the fungus *Entomophthora (Empusa) grylli*, though this is effective only when host populations become very high and have caused heavy defoliation". The activity of control agents would be affected by weather conditions. Hostetter and Bell[22] stress: "Undoubtedly, wind and rain are important factors in the dispersal and translocation of baculoviruses [one of the main groups of viruses responsible for disease in insects] within the environment and perform a very necessary function in their maintenance within ecosystems." In this respect, it is interesting to note that in many parts of the world outbreaks of *Euproctis* and related species have tended to follow in the wake of droughts, i.e. presumably unfavourable conditions for pathogens and other control agents[23]. In Seychelles, the early part of the outbreak (February 2015) coincided with unusually dry weather. It will be interesting to see if wet weather has any impact on the population dynamics of the species. Heavy rainfall could dislodge caterpillars from host plants as well as favour the development and transmission of natural control agents like fungi. On the other hand, the leaves of the Indian almond lose their chlorophyll and become red with anthocyanins (implying important chemical changes) before being shed during the dry season of the south-east trade winds, something that may also affect the species[24].

Control of ?*Euproctis* sp.

The National Pest Management Technical Committee (NPMTC), operating in collaboration with the DRDM, published a leaflet with "control tips", and the same information has been disseminated over the broadcast media[25]. The advice to the Seychellois public includes the following:

- The need for protective clothing when inspecting plants to see if hairy caterpillars are present.
- To eliminate small-scale infestations in gardens, etc. a mixture of Javel solution (one part Javel, three parts water) or strong soap solution and cooking oil to be applied to infested plants.

- Leaves and branches where caterpillars are present may be pruned and buried (not thrown into public dustbins).
- Bio-pesticides (Nilinsect, i.e. *Bacillus thuringiensis*; Neemik or Neembaan, main active ingredient Azadirachtin, all available from SAA Requisites Store) to be used preferentially, including for tall trees and for repeated treatments within inhabited areas.
- Licensed pest control operators to be hired for large-scale infestations, especially when fogging with insecticides (e.g. pyrethroids such as Decis) is required to stop the infestation.
- Spraying not to be done in the middle of the day when the sun is strong as the sprayed pesticide may burn the plants (phytotoxicity), as well as break down too rapidly (photodegradation).

An additional recommendation, not included in the latest brochure but mentioned in a press release of 12th March 2015, is to conduct light trapping of adult moths using an energy saver bulb above a container of soap solution, in which the moths drown. Research has shown that shorter wavelength lighting (which appears whiter and emits ultraviolet light) is more attractive to moths than longer wavelength (more yellow) lighting[26]. Non-target species would also be attracted to light-traps.

As far as prophylactic spraying in public areas is concerned, the strategy of the NPMTC has been to fog sites with large-scale infestation using Decis (deltamethrin), a broad-spectrum pyrethroid insecticide, with a view to rapidly bring down ?*Euproctis* sp. populations so that the microbial insecticide *Bacillus thuringiensis* can then be used to further suppress the lowered population[26]. In April-May 2015, fogging operations were conducted in the most densely inhabited areas of Mahé (Anse Royale, Au Cap, Anse aux Pins, Hermitage, Forêt Noire, Corgate Estate, Belvédère, Rochon, Plaisance, Mont Fleuri, Mont Signal, Belvédère, St Louis, La Gogue, Maldive, Glacis, La Retraite, Village Pascal, and between La Misère and Cap Ternay).

Decis (deltamethrin, a neurotoxin) is a broad-spectrum pesticide that is effective against larvae at all stages and kills on contact. However, concerns regarding the fogging operations are that (a) there are inevitable overspray and drift (airborne movement of pesticide outside the target area); (b) non-target species are killed – not only tree-living, leaf-feeding and flower-visiting invertebrates (with potential repercussions on populations of birds and the critically endangered Sheath-tailed bat that feed on insects) but also aquatic fauna; and (c) there are risks to human health. On 19th February 2015, students from Mont Fleuri Secondary School suffered from nausea and headaches after their school was fumigated late the previous day; it was then decided to fumigate affected schools on Fridays to allow the chemicals used to disperse over the weekend[27]. The National Pest Management Technical Committee (NPMTC) has identified a number of areas where chemical treatment should not be carried out. These include localities within nature reserves such as the Morne Seychellois National Park – hence along the main road from Bel Air and Sans Souci to Port Glaud. In such areas *Bacillus thuringiensis* (*Bt*), also used to treat vegetation in the 28 schools affected by hairy caterpillar infestations, should be applied[28].

Bt is toxic to cells in the midgut epithelium of insect larvae. For its application to be successful, (a) it must be of the right strain – in the United States, the browntail moth *Euproctis chrysorrhoea* is most susceptible to the CrylAc toxin[29]; (b) the caterpillars must ingest contaminated foliage; (c) it must target early instar larvae – *Bt* is less effective on larger larvae[30]; (d) it must not be exposed to rain during the first 24 hours, and there must be repeat applications even if it does not rain; and (e) during sunny weather it must be applied in the late afternoon to reduce exposure to UV light. Both rain and sunlight are known to reduce the effectivness of *Bt* application - sunlight inactivates spores and crystals and they can be washed off by rain[31]. It is not clear yet if treatment with *Bt* has been very effective (assuming it was correctly applied and regularly repeated) – but in early June 2015 hairy caterpillars were again causing dermatitis among pupils at at least two schools on Mahé[32]. At the same time, "collateral damage" is possible: the results of studies in the United States indicate that *B. thuringiensis* sprays are toxic to some non-target Lepidoptera for at least 30 days after the spray[33].

Pheromone trapping is contemplated as an alternative control method[26]. As a first step, it would be interesting to see the effectiveness of traps baited with live females. In the genus *Euproctis*, sex pheromones have

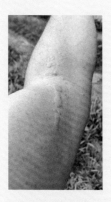

Skin rash caused by the hairy caterpillar.
Paul Desnousse

been identified for *E. similis xanthocampa* in China, *E. chrysorrhoea* in the United States, *E. similis* (subspecies not specified), *E. pseudoconspersa*, *E. taiwana* and *E. pulverea* in Japan – but they do not appear to contain any common compound that would be characteristic of the genus *Euproctis*[34]. However, compounds similar to those present in the pheromone of *E. pseudoconspersa* have been found in that of *Artaxa subflava*, previously considered to be *Euproctis subflava*[35]. Synthetic pheromone has been used to trap males of *E. pseudoconspersa* in China: larval and egg densities in a treated field were reduced by 27.87-50.85% and 38.89-51.11%, respectively, compared with an untreated field[36].

The option of resorting to biological control against the hairy caterpillar and its moth has not yet been openly discussed, possibly because of the risk of adverse effects on non-target species. In the United States, from 1906 onwards, the tachinid fly *Compsilura concinnata*, a polyphagous parasitoid of Lepidoptera, was introduced from Europe to fight the Gypsy moth (*Lymantria dispar*) and the Browntail moth (*Euproctis chrysorrhoea*). However, there is evidence that *C. concinnata* is a persistent and substantial source of mortality for native Lepidoptera, and that it could be seriously harming some species[37]. A narrow host range nucleopolyhedrovirus (NPV) was found in the southern U.K. for the control of *E. chrysorrhoea,* and claimed to be monospecific, and unlikely to present a risk to any non-target species[38]. Seychelles does not currently have sufficient research capacity to develop on its own a biological control method, and even with support from other countries this process would take a number of years[39]. Before any predator or parasite could be introduced, it would need to be certified totally host specific and not capable of turning against native Lepidoptera or other native organisms of Seychelles.

Conclusion

The hairy caterpillar ?*Euproctis* sp. is a very recent pest in Seychelles, with no record of it having been sighted before January 2015. Within a matter of weeks, however, it became a nuisance and public health threat to the extent that it very quickly earned a Creole name, "Senir plim" (meaning hairy caterpillar) while many other invasive species still do not have a Creole name, years after their introduction. A properly constituted study is needed to elucidate its taxonomic status, biology, ecology and population dynamics. This information is necessary to better understand if its eradication is feasible and, if not, how its population can be lowered to an acceptable level. As long as it is not successfully managed, there is the real danger that it will spread to yet more islands, and the risk of even more important outbreaks of such an adaptable and resistant species cannot be ruled out.

References / Further reading

1 Ministry of Health, 2015; 2 Uranie, 2015; 3 Lablache & Mériton-Jean, 2015; 4 Alberto Zilli, Natural History Museum, London, & Alexander Schintlmeister, Dresden, *pers. comm.*; 5 Bob Petrousse, *pers. comm.* to G. Rocamora; 6 De Jong *et al.,* 1976; 7 Diaz, 2005; 8 De Long 1981; 9 Eltringham, 1913; 10 Zalucki *et al.,* 2002; 11 James Payet, Anse à la Mouche, *pers. comm.* & Pat Matyot, *pers. obs.*; 12 Gamberale & Tullberg, 1998; 13 Sellier *et al.,* 1975; 14 Mullen, 2009; 15 Phisalix,1922; 16 González-Cano, 1981; 17 Meise & Schifter, 1972; 18 Ronley Fanchette, Environment Department, *pers. comm.*; 19 National Pest Management Technical Committee, 2015; 20 Pat Matyot, *pers. obs.*; 21 Grobler, 1957; 22 Hostetter & Bell,1985; 23 see for example Csóka, 1996; 24.G. Rocamora, *pers. comm.*; 25 NPMTC, 2015; 26 Somers-Yeates et al., 2013; 26 Randy Stravens, *pers. comm.*; 27 Anonymous, 2015a; 28 Anonymous, 2015b; 29 Dubois *et al.,* 2001; 30 Parks & Townsend, 2011; 31 Behle *et al.,*1997; 32 Anonymous, 2015b; 33 Johnson *et al.,* 1995; 34 Yasuda *et al.,* 1995; 35 Wakamura *et al.,* 2007; 36 Yonmo *et al.,* 2005; 37 Louda *et al.,* 2003; 38 Cory *et al.,* 2000; 39 Bernard Raynaud, CIRAD-La Réunion, *pers. comm.*

HYBRIDISATION: A SOURCE OF GENETIC ENHANCEMENT FOR INVADERS AND A MAJOR THREAT FOR NATIVE BIODIVERSITY

Invasive Alien Species can sometimes hybridise with closely related native forms. This becomes a source of genetic pollution for the native species, and it may eventually provoke their 'genetic extinction' as a species[15]. This is particularly the case when hybrids can backcross into the original native populations. After a certain time, the latter will become progressively genetically 'diluted' within a much larger population of alien individuals and hybrids.

One of the most famous examples is the case of the globally endangered White-headed duck *Oxyura leucocephala* (native to southern Europe) that became threatened with hybridisation by the introduced Ruddy duck *Oxyura jamaicensis* (native to the Americas) during the 1990s[1]. The latter had escaped from captivity in the 1970s and had become very abundant in England, and some individuals had started to migrate to Southern Europe and to interbreed with the congeneric White-headed duck. Ruddy ducks and hybrid males were found to be aggressive and dominant over both males and females of the native duck.A shooting campaign was set up in the 1990s in France and Spain to eliminate the invasive duck and any existing hybrids, and also in UK to control the introduced breeding population[2]. In Spain, hybrids were shown to be fertile and produced viable offspring in back-crosses with both parental species[3]. Genetic techniques played a crucial role to determine the non-natural origin of the birds, the fact that they belong to two different species, and the non-introgression of Ruddy duck genes into birds identified morphologically as White-headed ducks[4].

In Seychelles, the introduced grey-headed subspecies of Madagascar turtle dove *Nesoenas picturata picturata*, originally from Madagascar, has interbred with the Seychelles red-headed form *N. (picturata) rostrata* endemic to the granitic Seychelles. Although they are not completely identical to the Madagascar subspecies, the hybrids *picturata × rostrata* primarily resemble the Madagascar form and normally have grey heads, whilst red heads reappear extremely rarely in the population. Over the years, grey-headed hybrids have back-crossed with any red-heads that were left and have progressively invaded most of the inner islands and the Amirantes[5]. The red-headed Seychelles form, currently identified as an endemic subspecies but which could perhaps be attributed full species status pending taxonomic revision[16], has been genetically diluted to the point that pure *rostrata* genotypes probably no longer exist. As a result, red-headed turtle doves, which still occurred in good numbers on a few islands (Cousin, Cousine, Aride, Bird) until the 1980s and mid 1990s, have now almost all disappeared[6, 7]. Although recovering this form may no longer be possible, we are proposing to 'reconstruct' a population of red-headed phenotypes by crossing the recessive red-headed individuals and selecting offspring with the characteristics of the original Seychelles turtle dove *rostrata* form. This would need to be done at a very remote place with no turtle doves and out of flying reach of hybrids, such as Alphonse atoll[19], in the extreme South of the Amirantes.

Other examples of hybridisation in Seychelles between native and invasive alien species involve the Madagascar fody *Foudia madagascariensis*, which we consider an introduced species[8]. The first documented hybridisation, with the Seychelles fody *Foudia sechellarum*, was reported on Aride in 1993[9] with the presence of at least two hybrids. This resulted from the unassisted movement to Aride (which prior to 2002 had no Seychelles fodies) of one female Seychelles fody from Cousin. It paired with a male Madagascar fody from Aride and produced hybrid offspring. The next case occurred on D'Arros, in 2003, where individuals showing intermediate physical characteristics between these same two species were discovered and later confirmed to be hybrids[10, 12]. Hybrid males from D'Arros often look like Seychelles fodies but during the breeding season some show unique colouration patterns (including bright red, orange or large patches of yellow on the head). These patterns normally never appear on this species plumage and are reminiscent of the Madagascar fody. Hybrid females or males in dull interbreeding plumage are much more difficult to distinguish from pure Seychelles fodies. The origin of this hybridisation at D'Arros may be linked to the very small number (five) of Seychelles fodies transferred from Cousin to D'Arros in 1965, and the unbalanced sex-ratio (three females and two males[11]) that left one female with no possibility to find a mate of her own species. Today, a significant percentage of the hundreds of birds on D'Arros (there were c. 850 in 2005[13]) appear to have genes from both species[12], and removal of all

Hydrids between Madagascar (*Nesoenas picturata picturata*) and Seychelles (*N. p. rostrata*) subspecies of Madagascar turtle-doves. Left: Grey-headed phenotype.
Gérard Rocamora

Right: A rare reddish-headed phenotype.
Roy Mac Grath

the individuals that are not genetically pure Seychelles fodies is probably impossible. These 'D'Arros fodies' may now have to be considered a hybrid population genetically distinct from the pure Seychelles fody populations present on the inner islands. Although hybridisation may sometimes represent a natural mechanism for the evolution of new species (as it has been for Galapagos finches[17]), this is a case of human-induced genetic evolution resulting from hybridisation of a native species with an introduced one. Because of the remoteness and isolation of D'Arros, the five other island populations of Seychelles fodies do not come into contact with this hybrid population. Interestingly, Madagascar fodies are present in each of the five islands currently occupied by the Seychelles fodies but no cases of hybridisation have been reported when both species have well established populations (including when some may be qualified as rare, as for Madagascar fodies on Cousin[13], or Aride). There were also no cases of hybridisation reported after 64 and 47 Seychelles fodies where translocated to Aride and Denis respectively. In these situations, each bird appears not to be tempted to breed with individuals from the other species, as it can find mating partners, or join non-breeding bird groups, within its own species population.

After the discovery in 2012 on Grande Terre, Aldabra, of breeding Madagascar fodies that probably came from neighbouring Assomption, a significant number of Aldabra fodies *Foudia (eminentissima) aldabrana* showing morphological characteristics reminiscent of the Madagascar fody were found and identified as hybrids[14]. An effort was made to eliminate both hybrids that could be detected from their physical characteristics and also pure Madagascar fodies (over 200 in total). Unfortunately, some birds that appeared externally to be pure Aldabra fodies were found to actually retain genes from the Madagascar fody[14]. It is hoped that the Madagascar fody can be declared eradicated on Aldabra by the end of 2016 (see case study p. 273).

In Mauritius, some cases of hybridisation between the introduced Madagascar fody and the endangered endemic Mauritius fody *Foudia rubra* have also been recorded, and research with captive birds is being conducted to determine the viability, fertility and potential for these hybrids to backcross into the population of the Mauritius fody[18]. To prevent future cases of hybridisation between fodies, for example during future island translocations, great care needs to be taken to avoid situations whereby very small numbers of one species are left in the presence of another, much more numerous population of a different species of fody. This also highlights the importance of being vigilant and conducting surveillance on islands free of Madagascar fody, so that early detection and rapid elimination of invaders can take place before hybridisation with native fodies happens and becomes widespread.

Cases of hybridisation between native and invasive forms (species or subspecies) have been recorded all around the world for many other animal and plant groups (minks, wildcats, freshwater fishes, insects, etc.)[15]. In most cases if not all, these turn out to be detrimental to the native forms and many cases of extinction have been recorded.

Hybridisation between introduced and native populations of the same species, or between two or more different introduced populations may result in genetic enhancement and the creation of new, even more invasive forms. It has been demonstrated that mixing populations from different sources and the resulting increase in genetic variation can boost the invasiveness of particular species[15]. Genetic analysis has demonstrated or at least strongly suggested multiple introductions of diverse origins for many invasive species around the world (e.g. Cheatgrass Bromus tectorum in the Canary Islands; Malaysian trumpet

Upper left: The introduced Madagascar fody can hybridise with endemic fodies.
Monique Pereira

Upper right: Seychelles fody (male, Cousin).
Gérard Rocamora

Lower row: Two hybrids between Madagascar fody and Seychelles fody (D'Arros Island).
Gérard Rocamora

snail *Melanoides tuberculata* in La Martinique, the Zebra mussel *Dreissena polymorpha* in North America, etc.). Similarly, Reed canary grass *Phalaris arundinacea*, introduced from various parts of Europe into North America where it is also native, to revegetate shorelines and as a forage crop, has become a major wetland pest in North America where it is now genetically more diverse than it is in its European native range[15, 20]. Two entirely new species of salsify *Tragopogon* sp., an invasive meadow weed, have appeared in the United States following hybridisation between different species of the same genus that had been introduced separately[20].

In Seychelles, the extraordinary degree of invasiveness demonstrated during the last 10 to 15 years by *Merremia peltata* may well be due also to multiple introductions, or to hybridisation between native populations and recently introduced ones. This creeper, found in the Pacific and also considered native on other islands of the western Indian Ocean, is considered native in Seychelles by some authors[21, 22]. Old specimens can be found deep inside mature secondary and undisturbed forests where it colonises open areas that appear after large trees collapse, but without showing evidence of invasiveness. Historically, extensive areas of non-natural landscapes have existed in Seychelles as a result of forest clearance, extensive cinnamon exploitation and agricultural activities, and these areas were not colonised by this

The invasiveness of native *Merremia peltata* could be due to hybridisation with introduced populations.
Gérard Rocamora

creeper. Only during the last two decades has Merremia begun to heavily invade disturbed areas mainly along roadsides, forest edges, abandoned agricultural lands and human habitations. It is therefore possible that exotic populations of Merremia introduced to Seychelles may have interbred with those already present, triggering an invasive process. These recently introduced populations, or the ones resulting from hybridisation with local populations, may have then adapted to the considerable expansion of open landscapes created by the alteration of natural habitats by humans (see species account p. 337). Confirmation of this hypothesis will only be possible through molecular genetic analysis.

References/further reading

1 GISD, 2014; 2 Rocamora & Yeatman-Berthelot, 2009; 3 Muñoz-Fuentes *et al.*, 2007; 4 Muñoz-Fuentes *et al.*, 2013; 5 Diamond & Feare, 1984; 6 Rocamora & Skerrett, 2001; 7 Feare & Gill, 1995; 8 Cheke & Rocamora, 2014; 9 Lucking, 1997; 10 Rocamora, 2003a, 2003b; 11 Dawson, 2003; 12 Richardson & Rocamora, 2004; 13 Vega, 2013; 14 J. de Crommenacker / SIF, unpublished data; 15 Simberloff, 2013; 16 N. Collar / BirdLife International, *pers. comm.*; 17 Grant & Grant, 2002; 18 Lucy Garrett and Nicolas Zuel / MWF, *pers. comm.*; 19 Rocamora in Betts, 2009; 20 Simberloff & Rejmánek, 2011; 21 Senterre, 2009; 22 Gerlach, 2006.

MARINE INVASIVE SPECIES IN SEYCHELLES

This book focuses mainly on invasive terrestrial species, but aquatic species may also pose serious threats to biodiversity. A few species of plants and animals living in freshwater are treated in the second part of this book, and a few more are listed in Chapter 7 as species that require particular attention. These include two species of fish: Tilapia *Oreochromis mossambicus* introduced for aquaculture; and *Poecilia immaculata / reticulata*, one of the most popular freshwater aquarium fish species[1].

This green exotic algae belonging to genus *Caulerpa*, known to be invasive in other places, has been found at several locations in Seychelles and could cause important environmental damage.
Pierre-André Adam

Some marine invasive species have also been detected in Seychelles. In 2005, as part of a project initiated by IUCN in partnership with Marine Parks Authority (today SNPA) and the Department of Environment, a survey was conducted around Victoria harbour and Ste Anne Marine Park, and three marine invasive species were identified[2-4]. These include: *Ericthonius braziliensis*, a tube dwelling amphipod originating from the Atlantic and the Mediterranean that lives in marine sediments and amongst algal filaments and most likely transported on the hull of vessels; Stenothoe valida, a small amphipod resembling a shrimp most likely transported in ship's ballast water; and *Mycale* cf. *cecilia*, a sponge from the Pacific that is very common in shallow waters and believed to have been transported to Seychelles as a fouling organism on the hulls of ships.

This demonstrated that the threat from marine invasive species, although not easily visible to the general public, is a real problem requiring preventive measures. Some recommendations were made to enable detection and early action against marine invasive species. As a result, ships are now restricted from discharging their ballast water within 100 km of Port Victoria in order to allow currents to transport any alien organisms away from the granitic islands. In addition, a campaign was conducted in 2006 to create public awareness about marine invasive species. This involved exhibitions, radio and television programmes, school competitions, a supply of educational materials and a workshop to identify possible solutions[5].

Marine plants can also become invasive. In 2002 a green algae belonging to the genus *Caulerpa*, probably originating from the Caribbean, was found at Astove atoll at depths of 15-65 m. This plant is characterised by a rhizome-like base that spreads over the substrate. Although the species identification has not yet been confirmed at Astove, it is believed to be the same species, *C. bikinensis*, commonly used in aquaria, that was released by mistake into the Mediterranean where it has since caused enormous damage[6].

Certain native marine species are also known to become invasive and problematic in association with certain anthropogenic factors. At very high densities, the Crown of thorn starfish *Acanthaster planci*, a very large spiny starfish, can destroy coral reefs by feeding on coral[7]; and the Black Spined Urchin *Diadema* sp. that normally feeds on reef algae can damage the reef substrate and prevent reef regeneration by feeding on coral recruits[8]. Both species can be controlled by physical removal, and chemical methods are also available. A small native snail *Drupella* sp. (particularly *D. cornus*) feeds on live corals and can also cause widespread damage to hard corals during population explosions[7].

More information including management recommendations on marine invasive species can be found on the website www.biofoulingsolutions.com.au.

References / Further reading

1 Nevill, 2009; 2 IUCN / GoS, 2006; 3 Abdulla *et al.*, 2007; 4 IUCN, 2009; 5 Soubeyran, 2010; 6 Wendling *et al.*, 2003; 7 Beaver & Mougal, 2009; 8 Wendling *et al.*, 2004.

Restoring islands to save species: The Seychelles experience

Chapter 2

Satellite islands of the Praslin group
(Petite Soeur, Grande Soeur, and
Marianne in the background).
Raymond Sahuquet / STB

This chapter describes the achievements that have been made possible in Seychelles in terms of biodiversity conservation, by combining invasive species management, propagation of native vegetation, and recovery of threatened species on small islands.

Over the last few decades, the restoration or rehabilitation of small islands has been promoted and used as an effective conservation tool in Seychelles to create sanctuaries for native biodiversity[1-5]. This has been achieved by eradicating or controlling invasive alien predators and competitors from islands, and by recreating native habitats through the elimination of alien invasive trees and shrubs and the replanting of native vegetation. Into these rehabilitated islands, globally threatened species of endemic birds and other native wildlife have been translocated (reintroduced or introduced with a conservation purpose), leading to their subsequent recovery[29-32]. These successful conservation actions have been inspired by similar operations also taking place on other oceanic islands such as New Zealand[6].

In the following sections, the terms 'restoration' and 'rehabilitation' will be used interchangeably in their broad sense, with both terms referring to the active process of assisting the recovery of an ecosystem that has been degraded or destroyed. Some authors, however, use more restrictive definitions (see ours p. 16). 'Restoration' is normally defined as a process that aims to re-establish conditions identical to those that prevailed in the past, whereas 'rehabilitation' only aims to re-create a productive, functional ecosystem whose species composition and characteristics may differ significantly from the original[27]. 'Rehabilitated' ecosystems (*sensu stricto*) will not be identical to those that existed before the arrival of humans and therefore will not be completely free of alien species, but the process will allow the re-establishment of functional ecosystems, even though conservation management may always be required, particularly on small islands[28]. Dominated by native species, restored or rehabilitated ecosystems will host habitats that are suitable for a number of native animals including rare vanishing species with nowhere else to go.

Bought by the Cadbury family to become a Nature Reserve in 1973, Aride Island has benefitted from long-term restoration activities.
Colin Bell

THE BIOLOGICAL AND CONSERVATION VALUE OF SMALL ISLANDS[1]

The diversity of living creatures, also called biodiversity, is not equally distributed on Earth. During the last 15 years, international conservation organisations have demonstrated that a high proportion of global bio-diversity is concentrated in a limited number of hotspots[2]. Seychelles belongs to one of them (along with Madagascar and neighbouring islands) and tropical oceanic islands in general are important to biodiversity[3]. Within their limited land masses, islands have produced assemblages of unique ('endemic') species, and ecosystems which have evolved in complete isolation away from continental faunas. This explains the great fragility and vulnerability of islands to the ecological trauma that followed their colonisation by humans.

Islands are home to about 20% of the known global diversity of plants and animals but account for less than 5% of the area of the planet[4]. Because they usually host few or no natural predators, small oceanic islands often support large concentrations of seabirds[6], turtles or marine mammals. But most oceanic islands around the world have been subjected to ecological trauma caused by: human predation upon fearless and defenceless animals; forest clearing and wetland reclamation that destroy natural habitats; and introduction of exotic plants, animals, and diseases that decimate native faunas through predation or competition. As a result, over 80% of all recorded species extinctions have taken place on islands[4]. Moreover, islands host a large proportion, about 45%, of the species (birds, reptiles, amphibians, plants, etc.) that are considered to be threatened by extinction[5].

There are over 100,000 islands on the planet[5]. Large islands are important biodiversity hotspots that need to be preserved; but small islands can also play a crucial role in the conservation of nature at a global level and can help to partly reverse the negative impact of humans. Because of their limited size, small islands are relatively easy to restore or rehabilitate[7, 8] by the process of eradicating or control-ling invasive species of animals and plants, replanting native trees, and (re)introducing native animals. This process creates natural sanctuaries where threatened species and concentrations of marine ani-mals can thrive once again. This is why most of the small islands in Seychelles, the Indian Ocean and around the world are particularly precious and represent a potential biological treasure; and it is why they should all be wisely managed and preserved as, or to become, biodiversity refuges for the future.

References / Further reading

1 modified from Rocamora 2010a; 2 Mittermeier *et al.* 2004; 3 Conservation International (Madagascar) 2014; 4 Island Conservation 2012; 5 IUCN 2014; 6 Mulder *et al.* 2011; 7 see definitions p. 16; 8 SERI, 2004.

Boudeuse (Amirantes) is a very small island free of invasive predators and home to a large colony of Masked boobies.
Gérard Rocamora/Pangaea boat

Small predator-free islands: Biodiversity refuges protecting species from extinction

Seychelles exemplifies the immense value of small islands to protect native biodiversity from the dramatic impact of invasive predatory species, particularly rats and cats. In the inner islands, large seabird colonies are now only found on a handful of small granitic islands of more than 10 ha that remained naturally free of rats: Aride, Cousin, Cousine, Ile aux Récifs, and Frégate (occupied by rats only between 1995 and 2000), plus the coralline island of Bird where rats were present during approximately 30 years (1967 to 1996) but where a Sooty tern *Onychoprion fuscatus* colony survived[7, 8]. These six islands are the only ones verifying criteria of international importance for seabirds in the inner islands, and have been already declared as Important Bird Areas (IBAs)[9], or should be when the inventory is updated (c. 47,000 pairs of Sooty terns and c. 3,300 pairs of Bridled terns *O. anaethetus* are now known to breed on Ile aux Récifs[10, 11]). The same observation can be made in the outer islands where, of the eight existing IBAs for seabirds, all except Aldabra (which hosts significant seabird populations on mangroves and rat-free islets) and part of Cosmoledo (Grand Ile and its huge colony of 1.2 million pairs of Sooty tern) correspond to islands that have naturally remained free of rats (see figures 3-4). St Joseph & St François atolls, two potential IBAs verifying criteria of international importance for seabirds[12-14], are also rat-free. Large seabird colonies have vanished from islands that became infested by rats soon after the arrival of humans, such as Ile du Nord (North Island) in the inner islands, Coëtivy or Assomption in the outer islands, described as hosting large numbers of seabirds by early European navigators[15, 16].

In the inner islands, some species of landbirds, reptiles and invertebrates (see Table 3 below) can only be found on rat-free islands where they thrive in the absence of introduced predators. Some of these endemic species came very close to extinction and continue to be globally threatened today despite having been transferred to other predator-free islands. There were no more than 26 to 29 Seychelles warblers left on Cousin in 1959[17] and as few as 12 to 21 Seychelles magpie-robins on Frégate between 1960 and 1990[18]. Today, the Wright skink (also present on rat-free islets St Pierre, L'Ilot Frégate and Ile aux Fous[21]) is restricted to rat-free refuges, and the Giant tenebrionid beetle only occurs on Frégate Island[22]. However, the Chestnut-flanked white-eye *Zosterops semiflava* was not as lucky and became extinct during the first part of the 20th century after its last known refuge on Marianne Island was invaded by rats and extensively planted with coconuts (as happened to so many other small islands)[20].

Other species that did not become restricted only to rat-free islands have a distribution that also suggests a strong impact of introduced predators on their populations, in addition to other likely negative factors. No more than an estimated 50 Seychelles white-eyes survived on the large rat-infested island of Mahé by 1997 (in areas where both Black rats and Brown rats occurred). However, the species was able to maintain a healthy population on Conception, where only the non-arboreal Brown rat occurred, having probably become extinct on all the other small islands around Mahé, all occupied

Table 3: Natural distribution of endemic species of birds, reptiles and invertebrates of global conservation concern restricted to a few small Seychelles islands having remained free of rats.

Species	IUCN category	Aride	Cousin	Cousine	Mamelles	Ile aux Récifs	Frégate
Seychelles magpie-robin *Copsychus sechellarum*	EN						X
Seychelles warbler *Acrocephalus sechellensis*	VU		X				
Seychelles fody *Foudia sechellarum*	NT (formerly VU)		X	X			X
Wright's skink *Trachylepis wrightii*	VU	X	X	X	X	X	X
Giant tenebrionid beetle *Polposipus herculeanus*	VU						X

by high densities of Black rats[19, 23]. The Indian Ocean whip spider *Phrynichus scaber*, also survives on rat-infested Silhouette, but is otherwise restricted to a few rat and predator free islands where its densities appear to be higher. The Giant millipede, while abundant on all rat-free islands and islets (including Mamelles, Ile Sèche, L'Ilot Frégate, Ile aux Vaches marines[24]), survives at low densities on some rat-infested islands (e.g. Silhouette, La Digue, Marianne; also Conception and Grande Soeur before rats were eradicated[24-26]), but never on islands where Tenrecs are present (Mahé, Thérèse, Anonyme, Praslin). Native plants are well represented on some of these small rat-free islands. Interestingly, Wright's gardenia *Rothmannia annae*, a globally threatened Seychelles endemic, only grows wild on Aride but the reasons for such restricted distribution are unknown.

The fact that these small islands managed to remain rat-free for so long may be explained by a combination of factors including their restricted access to people; the

Species saved from extinction on small rat-free islands.
Only 26 to 29 Seychelles warblers survived on Cousin in 1959, and only 12 to 21 Seychelles magpie-robins were left on Frégate between 1960 and 1990. The Seychelles fody was restricted to Cousin, Cousine and Frégate. The Wright's skink is restricted to a handful of rat-free refuges, and Frégate Island is the only place in the world where the Giant tenebrionid beetle exists.

Clockwise from upper left: Seychelles magpie-robin (Aride); Seychelles warbler (Aride); Wright's skink (Aride); Giant tenebrionid beetle (Frégate Island); Seychelles fody (male; Aride Island).
Javier Cotín; Hadoram Shirihai[a]; Javier Cotín; Tanya Leibrick; Hadoram Shirihai[a]

a. Photo contributed by *Photographic Handbook of Birds of the World*, Jornvall & Shirihai, A & C Black, London ©

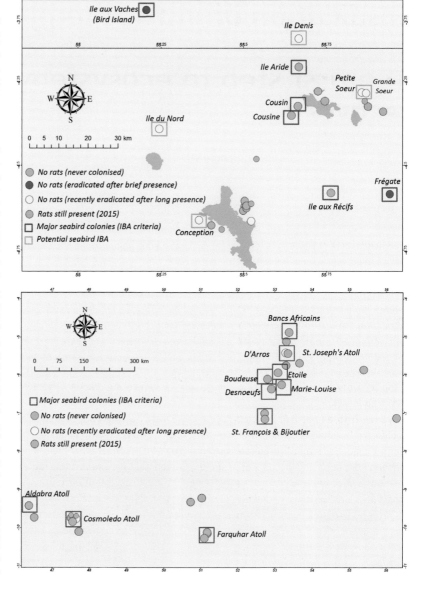

Figure 3: Correspondence between seabird colonies of international importance (verifying Important Bird Areas criteria[9, 10, 11]) and islands (over 1 ha) that remained naturally rat-free in the inner islands of Seychelles. Smaller circles correspond to rat free islets with no large seabird colonies. Other islands where rats were eradicated are being recolonised by seabirds and have the potential to become new IBAs.

Figure 4: Correspondence between seabird colonies of international importance (verifying Important Bird Areas criteria[9, 13, 14]) and islands that remained naturally rat-free in the outer islands of Seychelles. Except for Aldabra, all major seabird colonies (sites inventoried, verifying IBA criteria, or to be confirmed as IBAs) are on rat free islands. Islands where rats have been eradicated are being recolonised by seabirds and have the potential to become new IBAs.

often difficult landing conditions; the care taken by their owners or inhabitants not to introduce rats; and also pure luck. However, cats were brought (probably as pets) to Aride, Cousine and Frégate by early inhabitants and later became feral. This greatly damaged breeding seabird colonies and endemic landbirds such as the Seychelles magpie-robin, which later disappeared from Aride, and almost vanished from its last refuge on Frégate which would have resulted in global extinction.

References / Further reading

1a Merton et al., 2002; 1b Shah, 2001, 2006; 1c Nevill, 2001a; 2 Rocamora, 2010a; 3 Samways et al. 2010b; 4 Nevill, 2011; 5 Asconit & ICS, 2010; 6 Towns et al., 1990; 7 Merton et al. 2002; 8 Feare 1979;9 Rocamora & Skerrett, 2001; 10 Bristol, 2003; 11 Feare et al., 2007; 12 Bristol & Millett, 2002; 13Adam et al., 2009; 14 Kappes et al., 2013; 15 Fauvel, 1909; 16 Cheke, 2010; 17 Crook, 1960; 18 Kömdeur, 1996; 19 Rocamora, 1997b; 20 Cheke 2013; 21 Bowler, 2006; 22 IUCN 2014; 23 Rocamora & François 2000; 24 G. Rocamora, pers. obs.; 25 Hill, 2002; 26 Galman, 2011; 27 SERI, 2004; 28 Kueffer et al., 2013; 29 Kömdeur, 1994; 30 Richardson et al. 2006; 31 Rocamora & Henriette, 2008; 32Shah, 2008; 33 Gerlach, 2007.

Eradication of introduced predators and competitors: The first step to ecosystem recovery

Predation and competition from invasive alien animals have a huge impact on island ecosystems and represent a key limiting factor to their recovery[27-32]. As a result, conservation and restoration activities on small islands often start with the eradication of introduced alien predators such as rats and cats, which have major detrimental effects on ecosystems. Avian predators such as Barn owls and Common mynas also need to be controlled or eradicated, along with invasive alien herbivores such as rabbits or goats which can do considerable damage to the vegetation. Although a rehabilitation process may also be started by removing exotic invasive plants and re-planting native vegetation, experience from the Seychelles and elsewhere shows that eradicating invasive animals first allows the ecosystem to recover more quickly[33, 34]. On large inhabited islands such as Mahé or Praslin where the eradication of introduced predators cannot be envisaged, the presence of these animals prevents the re-establishment or the recovery of native species, including in areas where substantial vegetation restoration has been conducted.

By June 2015, a total of **67 attempts to eradicate invasive animals** have occurred on 23 different Seychelles islands, and five of these operations are still ongoing. This does not include attempts to eradicate invasive animals for purposes other than nature conservation, such as agricultural pests (e.g. Melon fruit fly, see p. 114). Table 5 details the species, islands, and years as well as the outcome of each of these operations. Most of these island eradications targeted mammals (69%) and birds (30%) (Fig. 6). They include 44 mammal, 20 bird and one insect eradication attempts on 21 islands of at least 10 ha, plus two mammal eradications on two islets smaller than one ha. On five occasions, species that were not the main target (Feral cats and Barn owls) were indirectly eradicated in response to the removal of rats, and on three occasions island populations (Feral goats and chickens) died out following control operations. Of the remaining 54 (direct) eradication attempts completed, a total of 39 attempts (72%) succeeded (this includes two repeated rat eradications on Anonyme), and 15 attempts (28%) had a failed outcome[a]. **In all, by October 2015, at least 47 alien vertebrate populations had been removed from islands in Seychelles as a result of eradication programmes**. In addition, of another 5 operations still in progress by October 2005, three are already in their final phase of intensive monitoring to detect the presence of any survivors (Ringed-necked parakeet on Mahé, and Madagascar fody on both Assomption and Aldabra) and should be confirmed successful in the coming months. By 2016, it is very likely that the number of vertebrate populations eradicated on islands in Seychelles will have reached 50.

As of October 2015, **a minimum of 12 species of land vertebrates** have been successfully eradicated from islands in Seychelles, and this total should reach 13 once the Madagascar fody eradications on Assomption and Aldabra are confirmed (Fig. 5).

a. In our statistics, 'success' and 'failure' refer to the final outcome of the eradication attempts, not to the immediate result of the eradication protocols employed. This includes the success of the biosecurity protocols applied (or not) to prevent reinvasion. In 'failed' outcomes, one or more of the following occurred: survivors were present; rapid reinvasion occurred within one year (rodents); or operations were interrupted or lapsed for too long (mynas, goats).

These species are: Black rat (11 successful island eradications, including two on islets of less than one ha), Feral cat (seven, plus three indirectly in response to rat eradication), Feral goat (four, plus two that died out after some control), Common myna (four), Brown rat (three), Feral chicken (two, plus one progressively eliminated by trapping for consumption), House mouse (two), Feral rabbit (one), Barn owl (one, plus two indirectly in response to rat eradications), House crow (one), House sparrow (one) and Red-whiskered bulbul (one). Most of these operations occurred on small to medium size islands (less than 300 ha) as part of ongoing island rehabilitation programmes. The exceptions are those involving the House crow and House sparrow which took place on Mahé, the goat eradications on Aldabra, and the Red-whiskered bulbul eradication on Assomption. Success rates, that take into account the final outcome of these operations[a], are calculated only over the 'direct' eradications, and vary considerably depending on target species: 33% for House mouse, 56% Common myna and 57% for Feral goat, 75% for Black rat and Brown rat, and 100% for the other species. The eradication of Black rats from the two small islets (less than 1 ha) of Ile aux Rats in 2006 (near Anonyme) and Petit Polyte in 2007 (near Grand Polyte, Cosmoledo) is not included in these last calculations.

In addition, two species of semi-feral domestic mammals have also been removed from some islands (pigs and cows from Cousine and Ile du Nord; and possibly pigs from Frégate). In that context, 'semi-feral' means that part of the animals were not completely wild, hence easy to catch, and may have still depended on supplementary food from humans. Each case involved very small numbers[1-3] and in some cases it is unclear if the populations were reproducing in the wild. For example, only half-a-dozen pigs and even smaller numbers of cattle were shot on Ile du Nord. Until more details can be made available, we cannot consider these operations as genuine eradications. Feral cats died out from Picard after the 1970s, but apparently no control or eradication programme was involved. None of these are considered to be proper eradications in our current calculations. However, we did take into account the reported eradication of goats on Aride by shooting before 1920[3], and the removal of feral goats from the islands of Polymnie (475 ha) and Ile Esprit (41 ha), on Aldabra, sometime in the 1970s, probably in response to some localised control (see case study p. 245). The extinction of feral chickens from Desnoeufs in 2007 following repeated captures over the years by local staff for consumption[4] was also considered as eradication. If the removal of semi-feral pigs and cows from certain islands is considered to be eradication, then the total number of vertebrate species eradicated from islands in Seychelles would become 14 (and 15 when the Madagascar Fody eradication on Aldabra and Assomption are confirmed), and the total number of vertebrate populations eradicated from islands would go beyond 50.

Numbers of attempts and success rates have varied over time (Table 4 and Figs 7-8). Only 11 eradication attempts (targeting goats, cats, mynas and crows) were conducted prior to 1995, and 73% of those were successful. During the period 1995-2004, there were 28 attempts to eradicate mammals and birds from islands and these had a success rate of 64%. During the next decade, 2005-2014, the number of eradication attempts dropped to 20, but the 14 completed enjoyed an overall success rate of 86% (12 out of 14 finished operations; nine targeting rats, one goats, and four birds two of which failed). Better success is probably due to the greater experience accumulated by both local practitioners and island managers, which enabled better project selection, field implementation, and post-eradication measures to prevent reinvasions.

Six operations are still ongoing: five involve bird eradications; and one targets an invertebrate, the Big-headed ant.

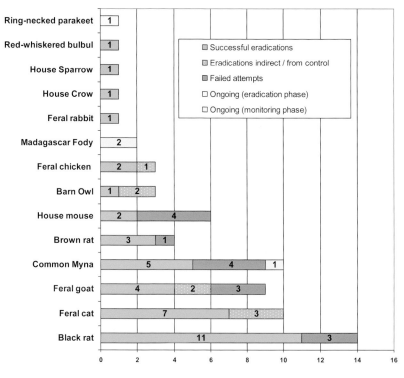

Figure 5: Number of eradication attempts (n = 67), and success outcomes for the 14 species of invasive vertebrates targeted in Seychelles. Of 54 total direct attempts completed, 39 (72.2%) were successful and 15 (27.8%) failed; 5 additional populations were eradicated indirectly (in response to the removal of another species), and 3 more died out after some control. By October 2015, 5 operations are still ongoing. Species are listed in increasing order of successful attempts. Success or failure refers to the final outcome of the operations. If excluding the rat eradications on two islets (< 1ha), success rate is 70.6% and failure 29.4% (n=52).

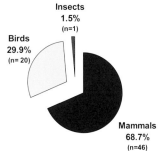

Figure 6: Attempts to eradicate invasive animals in Seychelles per taxonomic group (until October 2015; n = 67; agricultural attempts not included).

Table 4: Temporal distribution of attempts to eradicate invasive animals in Seychelles and success outcomes by the end of 2014 (n = 67).

	Pre-1995	1995-2004	2005-2014	Total attempts
Successful eradications	8	18	12	38
Indirect eradications	2		6	8
Failed outcomes	3	10	2	15
Ongoing attempts			6	6
Total attempts	13	28	26	67

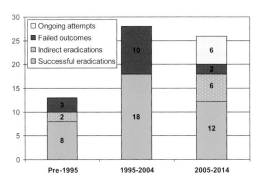

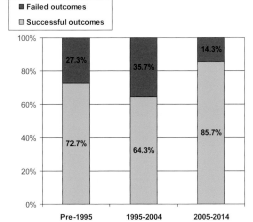

Figure 7: Temporal distribution of attempts to eradicate invasive animals in Seychelles (n = 67; 60 direct eradications + 7 obtained indirectly or following control). Success or failure refers to the final outcome of the operations (see text).

Figure 8: Temporal variation of success rates for the eradication of invasive alien animals in Seychelles (n = 53 direct attempts completed; 11+28+14).

MAMMAL ERADICATIONS

b. Charles Veitch (NZ), David Todd (RSNC) and Victorin Laboudallon (Forest & Conservation Division; Min. of Ag.)

The two first reported eradications in Seychelles targeted goats and cats on Aride, before 1920 and in the 1930s respectively[5]. Cats were next eradicated from Frégate in 1981-82 with the assistance of a New Zealand expert[b] who trained Victorin Laboudallon, a Seychellois government employee[6, 7]. The latter was then employed by Cousine to eradicate cats in 1983-85[8]. On Aldabra, in 1993, SIF initiated a campaign to eliminate feral goats, and by 1995 they had been eradicated from Picard and Malabar (but not Grand Terre)[9]. In 1996, Bird Island hired New Zealand expert, Dr Don Merton (see p. 14) to undertake the first eradication of rats (and rabbits) in Seychelles, which was conducted as a ground operation. This was followed in 2000 by a joint project co-funded by the Dutch Trust Fund, the Department of Environment and two private islands to eradicate Black rats, mice and cats from Curieuse (a National Park) and Denis islands, and Brown rats and mice from Frégate (where a ground based attempt conducted by BirdLife International had failed in 1996)[10, 11]. The technical expertise – which included the aerial spreading of rodenticide bait by helicopter – was

c. Don Merton, Gideon Climo, Bill Simmons and helicopter pilot (NZ), and Victorin Laboudallon (DoE)

provided by a New Zealand team and the formerly trained DoE member of staff[c]. Aerial logistics were provided by Helicopter Seychelles and support for wildlife management by BirdLife Seychelles (today Nature Seychelles). The final outcome of rodent eradications on Curieuse and Denis Islands was not successful due to rapid reinvasion or the survival of some individuals[10]. Denis attempted again in 2002 – and this time was successful in a ground operation. In 2003 D'Arros Island (in a ground operation) and North Island (Ile du Nord; in an aerial operation) attempted to eradicate both

d. Gideon Climo and on Ile du Nord helicopter pilot (NZ) plus Cpt Rick Dooley (Helicopter Seychelles)

rodents and cats; but mice survived on D'Arros and Black rats were found again on Ile du Nord within a year. These eradications were also conducted with technical expertise from New Zealand[d], a Helicopter Seychelles pilot was trained, and Nature Seychelles provided support for wildlife management on D'Arros and Ile du Nord[7, 12].

e. Gérard Rocamora; with André Labiche (ICS) and Perley Constance (DoE) in 2006

In 2003, the Island Conservation Society[e] conducted on Anonyme Island (10ha) the first successful eradication of rats in Seychelles without external expertise, by combining a permanent grid of bait stations and trapping. In 2006, rats re-established briefly on the Anonyme plateau after the island was sold, but were rapidly eradicated again within a few months, and were also eradicated from nearby Ile aux rats[13, 14]. In 2005, as part of a project co-funded by FFEM (see case study p. 96), ICS and the North Island team[15] successfully eradicated Black rats from Ile du Nord (201 ha) through aerial spreading of rodenticide pellets by helicopter and using technical expertise from both Seychelles and New Zealand[f]. In 2007, ICS was able to eradicate Brown rats from Conception and Black rats from three islands of the remote Cosmoledo atoll using the same method[14]; this was done entirely by a local team[g] that included Helicopter Seychelles and IDC for logistical support[16]. In 2010, the owners of Grande Soeur and Petite Soeur used the same local expertise[h] and equipment to successfully eradicate rats on their two islands, and the rare feral cats present were indirectly eliminated in the process[17]. The same had occurred on Cosmoledo where

f. Gérard Rocamora, André Labiche (ICS), Gideon Climo & helicopter pilot (NZ), & Cpt Rick Dooley (HS)

g. Gérard Rocamora, Pierre-André Adam, André Labiche, Roland Nolin, André Dufrenne (ICS) & Cpt Rick Dooley (HS)

h. Gérard Rocamora, André Labiche, Roland Nolin (ICS) & Cpt Rick Dooley (ZilAir)

Rat eradication with aerial rodenticide drop on Grande Soeur and Petite Soeur.
G. Rocamora

Table 5: Details of attempts to eradicate invasive alien animals from islands in Seychelles in 2015.

Total number of attempts = 67 (62 completed + 5 ongoing). Successful attempts (in **bold**) = 47 (39 completed; + 5 indirect eradications induced by rat removal and 3 from control, in brackets). Success or failure refers to the final outcome of the operations. 'Failed' attempts may involve animals that either survived or immediately reinvaded after a technically successful eradication phase. Only the active removal of individuals that had established a substantial wild breeding population is considered a genuine eradication attempt. Removal of small numbers of wild or semi-feral domestic animals (cattle, pigs, chickens), or populations of the same that had died out naturally are not included here. * = ongoing eradications that are in their final, monitoring phase, in the process of being confirmed (Oct. 2015). Islands are listed per chronological order of their first eradication.

Islands / Species (by chronological order of eradications)	Area (ha)	Feral cat	House crow	Common myna	Feral goat	Feral chicken	Feral rabbit	Barn owl	Black rat	Brown rat	House mouse	House sparrow	Red-whiskered bulbul	Madagascar Fody	Ring-necked parakeet	Black-headed ant	Total attempts	Success (Outcome)
Aride	73	1930s		**1993-94** +occ. inv.				1996 +occ. inv								2014-ongoing	4 (+ 1 ongoing)	4
Mahé	15252		1977-94 occ. inv									2003-04 +occ. inv			2012-ongoing		2 (+ 1 ongoing)	2
Frégate	219	**1981-82**		**2010-11** (98-02 failed) +occ. inv.						**2000** 1996 failed							6	4
Cousine	26	**1983-85**		**2001-02** +occ. inv		1996					**2000**					(control 2008)	3	3
Bird Island (Ile aux Vaches)	101						1996-97		1996-97		1996-97 failed						3	2
Cousin	29			**2000-02** +occ. inv													1	1
Picard (Aldabra)	940	(died out since 1970s)			**1993-95**												1	1
Malabar (Aldabra)	2680				**1993-95** 1987-88 failed												2	1
Polymnie (Aldabra)	475				(died out by **1976** after control)												1	(1 from control)
Ile Esprit (Aldabra)	51				(died out by **1976** after control)												1	(1 from control)
Curieuse	289	2000							2000 failed		2000 failed						3	1
Denis	143	2000		2000-01 failed **2010-15**					**2002** 2000 failed		**2002** 2000 failed						7	4
D'Arros	140	2003								**2003** 2003 failed	2003 failed						3	2

Table 5 (cont.): Details of attempts to eradicate invasive alien animals from islands in Seychelles until 2015.

Total number of attempts = 67 (62 completed + 5 ongoing). Successful attempts (in **bold**) = 47 (39 completed + 5 ongoing). Successful attempts (in **bold**) = 47 (39 completed + 3 from control, in brackets). Success or failure refers to the final outcome of the operations. 'Failed' attempts may involve animals that either survived or immediately reinvaded after a technically successful eradication phase. Only the active removal of individuals that had established a substantial wild breeding population is considered a genuine eradication attempt. Removal of small numbers of wild or semi-feral domestic animals (cattle, pigs, chickens), or populations of the same that had died out naturally are not included here. * = ongoing eradications that are in their final, monitoring phase, in the process of being confirmed (Oct. 2015). Islands are listed per chronological order of their first eradication.

Islands / Species (by chronological order of eradications)	Area (ha)	Feral cat	House crow	Common myna	Feral goat	Feral chicken	Feral rabbit	Barn owl	Black rat	Brown rat	House mouse	House sparrow	Red-whiskered bulbul	Madagascar Fody	Ring-necked parakeet	Black-headed ant	Total attempts	Success (Outcome)
Anonyme	10								2003 & 2006								2	2
North Island (Île du Nord)	201	2003		2006-09 failed 2012-ongoing		2003		(2005) occ. inv	2005 2003 failed								6 (+ 1 ongoing)	3 (+ 1 indirect)
Île aux rats	<1								2006								1	1
Desnoeufs	35					(died out by 2007)											1	(1 from control for consumption)
Conception	69									2007							1	1
Grande Ile, Grand Polyte, Petit Polyte (Cosmoledo atoll)	143	(very likely, 2008)							2007								2	1 (+ 1 indirect)
	21	(2007)							2007								2	1 (+ 1 indirect)
	<1								2007								1	1
Grande Sœur	85	(2010)		2011 failed				(2010) occ. inv	2010								4	1 (+ 2 indirect)
Petite Sœur	35								2010								1	1
Grande Terre (Aldabra)	11610				2007-2012 93-97 failed 87-88 failed									2012-15 ongoing *			3 (+ 1 ongoing)	1 (+ 1 being confirmed)
Assomption	1171												2012-2014	2012-15 ongoing *			1 (+ 1 ongoing)	1 (+ 1 being confirmed)
TOTAL ATTEMPTS PER SPECIES		10	1	9 (+1 ongoing)	9	3	1	3	14	4	6	1	1	(2 ongoing)	(1 ongoing)	(1 ongoing)	67 = 62 + 5 ongoing	47 = 39 + 8 indirect / from control
Successful outcomes		7 (+ 3 indirect)	1	5	4 (+ 2 control)	2 (+ 1 control)	1	1 (+ 2 indirect)	11	3	2	1	1				47 = 39 + 8 indirect / from control	–

SEYCHELLES: A LEADING COUNTRY FOR THE ERADICATION OF INVASIVE VERTEBRATES FROM ISLANDS

a. "Island territory" is here defined as an island or an archipelago that is either an independent state (e.g. Seychelles) or an overseas territory belonging to an independent country (e.g. Puerto Rico, Ile de La Réunion, Guam, etc.).

b. This result differs from the 46 vertebrate successful eradications provided in our previous calculations due to small differences in definitions and data quality assessments. Two eradications with poor quality data (goats on Polymnie and Ile Esprit), one with restricted range (House sparrow), and one followed by reinvasion (Black rats on Anonyme in 2003) are excluded.

c. Data are from the Database of Island and Invasive Species Eradications (DIISE 2014), up to December 2014. It includes only whole island eradications (i.e. excluding incursion events and restricted range eradications) with successful outcome (i.e. not reinvaded) where data quality is verified or satisfactory. Tropical islands are those located between 23.4378° North and -23.4378° South of latitude. Note: totals for Australia are under-represented because they do not include additional events reported from (d).

d. Gregory et al. 2014 - http://www.feral.org.au/wp-content/uploads/2014/02/AusEradications_2014.pdf

Seychelles ranks fifth globally amongst island territories[a] for the number of eradications, and third amongst independent countries for tropical islands.

The Database of Island Invasive Species Eradications (DIISE) is a publicly available resource that summarises information on individual eradication projects and provides opportunities to analyse trends in past eradication projects around the world. Up to 2014, there have been over 1,300 attempts to eradicate vertebrates from islands worldwide. The database for Seychelles was updated with the results of our present inventory (Table 5) and compared with data from other territories facilitated by the international NGO Island Conservation. In this database, totals are provided separately for overseas island territories that belong to larger political states (such as La Réunion and French Polynesia for France, British Indian Ocean Territory for UK, Puerto-Rico and Guam for the US, etc.). Hence, comparisons can be made by pooling or not these overseas island territories with their respective countries (i.e. independent states).

The DIISE indicates that, in 2014, with a minimum of 42[b] populations of alien vertebrates eradicated from 23 Seychelles islands, Seychelles ranks fifth amongst world island territories in terms of the total number of alien invasive vertebrate populations eradicated from islands[c] (Fig. 9). New Zealand is by far the highest contributor to this eradication effort, followed by Australia (total currently underestimated[d]), the USA and Mexico. When overseas territories are pooled according to their respective countries, the totals for USA, UK and France become much higher (86, 69 and 62 respectively), dropping Seychelles to the rank of seventh among independent nations.

Nevertheless, even when adding overseas territories to these nations, Seychelles ranks third in the world for the eradication of invasive vertebrate populations from tropical islands, after Australia and the USA (Fig. 10). When we first conducted this exercise in 2012, Seychelles was ranking second, just after Australia and before the USA, where the number of eradications in overseas territories has been recently increasing. When overseas island territories are considered separately to the countries they belong to, Seychelles still ranks second after Australia for the number of vertebrate eradications on tropical islands. The graphs below illustrate these results.

The Seychelles has also seen 14 different species of invasive alien vertebrates eradicated from islands (including feral pigs and cattle). Other countries with more than 10 species eradicated include Australia (27[d]), USA (22 sp.), New Zealand(21), France(13), Mexico(12), UK(11) and Ecuador(10).

These vertebrate eradications, and accompanying habitat restoration measures, have enabled about 20 island reintroductions or conservation introductions of eight rare and threatened species in Seychelles. As a result, four have already been downlisted (to a less threatened status) in the IUCN Red List. Other native wildlife such as seabirds, reptiles, invertebrates or plants have also benefited (see "Ecosystem recovery after the eradication of rats and cats", p. 169). The achievements of Seychelles in terms of invasive species eradications and prevention of extinctions are exemplary for many other island-rich nations – small and large.

The Database of Island Invasive Species Eradications, DIISE, (http://diise.island-conservation.org) is developed by the international NGO Island Conservation, the Coastal Conservation Action Laboratory at the University of California, the IUCN SSC Invasive Species Specialist Group, the University of Auckland and Maanaki Whenua-Landcare Research, New Zealand.

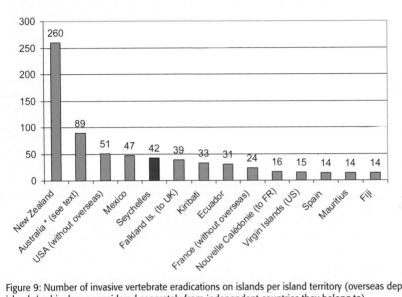

Figure 9: Number of invasive vertebrate eradications on islands per island territory (overseas dependent islands/archipelagos considered separately from independent countries they belong to).
* totals for Australia are under-represented because they do not include additional events reported from Gregory *et al.* 2014.
Data: DIISE/Island Conservation-IUCN database 2014; present work for Seychelles

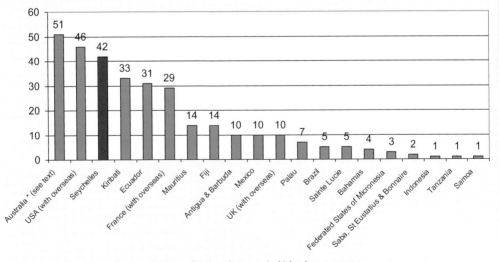

Figure 10: Number of invasive vertebrate eradications from tropical islands per country (independent countries including all their overseas islands/archipelagos dependencies).
* totals for Australia are under-represented because they do not include additional events reported from Gregory *et al.* 2014.
Data: DIISE/Island Conservation-IUCN database 2014; present work for Seychelles

cats were no longer recorded at Grand Polyte after the rat eradication, and only in much reduced densities at Grand Ile[4, 14] where they had been observed consuming Brodifacoum cereal pellets[17], and have not been recorded since then[17, 18]. The last mammal eradication was achieved on Aldabra by SIF, with the last goats removed from Grande Terre in 2012[19].

BIRD ERADICATIONS

The first bird eradication in Seychelles targeted the Indian house crow on Mahé between 1977 and 1984[7, 20, 21]. On Aride, a small breeding population of Common mynas was exterminated in 1993-94, and a small breeding population of Barn owls in 1996[21, 22]. Other small colonising populations of mynas were also targeted successfully on Cousin and Cousine[1, 23]. In 2003-04, a small colonising population of House sparrows was eliminated by DoE on Mahé[24]. In 2011, Frégate Island conducted the world's first successful island eradication of an established myna population of more than 700 birds[25] (see case study p. 259). Barn owls, which were permanently established on both Ile du Nord and Grande Soeur, were (indirectly) removed in 2005 and 2010 respectively following secondary poisoning and subsequent lack of food after the rat eradications[14]. Despite the successes, all these islands continue to be visited by such occasional invaders, some of which like House crow or House sparrow are closely monitored by the Seychelles Bird Records Committee[26]. This is thanks to the information regularly provided to the public by the Ministry or Environment and the local media, which helps to detect invaders early and eliminate them before they start breeding.

There have been no confirmed island eradications of invasive reptiles in Seychelles, the Red-eared slider was not reported since 2006 on Mahé but was observed again in 2012. Eradication of the Crested-tree lizard may ultimately be successful on Ste Anne, where occasional captures of the few remaining individuals are taking place.

As of October 2015, five eradication attempts were still ongoing in Seychelles: two begun in 2012 to eliminate the Madagascar fody on both Assomption and Aldabra (Grande Terre) where they hybridise with the Aldabra fody (all two at final monitoring stage); one operation on Ile du Nord to eliminate mynas, one to eradicate the Ring-necked parakeet from Mahé, and one targeting the Big-headed ant on Aride. More details on all these operations can be found in the species accounts and case studies in the second part of this book.

References / Further reading

1 Samways et al., 2010; 2 Bruce Simpson/North Island, pers. comm.); 3 Victorin Laboudallon, pers.comm.; 4 Roland Nolin, pers. comm.; 5 Warman & Todd, 1984; 6 Watson et al., 1992; 7 Beaver & Mougal, 2009; 8 Laboudallon V. 1987; 9 Rainbolt & Coblentz, 1999; 10 Merton et al., 2002; 11 Thorsen et al., 2000; 12 Climo, 2004a, 2004b; 13 Rocamora et al., 2005; 14 Rocamora & Jean-Louis, 2009; 15 Climo & Rocamora, 2006; 16 Rocamora, 2010a; 17 G. Rocamora, pers. obs.; 18 Martin & Pinchart, 2015; 19 Bunbury et al. 2013; 20 Ryall, 1986; 21Skerrett et al., 2001; 22 Malcom Nicoll in Betts, 1997; 23 Dunlop et al., 2005; 24 Fanchette, 2003; 25 Canning, 2011; 26 Skerrett et al., 2007; 27 Towns et al., 2006; 28 Rocamora, 2007c; 29 Towns et al. 2011; 30 Russell et al., 2011; 31 Medina et al., 2011; 32 Harper & Bunbury 2015; 33 Towns et al., 1990; 34 Chapuis et al., 1995.

Removal of exotic invasive plants and trees to replant native vegetation

Rehabilitating the ecosystem on an island is a long-term aim. It starts by initiating a process whereby the alien species are progressively removed, and natural habitats (coastal forests, marshes, rocky inselbergs, etc.) are recreated by replanting native plants and trees in order to make them once again suitable for native animal species. Once recovered or reintroduced, native insects, reptiles, landbirds or bats will re-establish pollination and seed dispersal networks with the native plants[1], whereas other animals such as seabirds, millipedes, worms and other detritivorous species will contribute to soil enrichment[2, 3].

Control of invasive plants has been conducted on many islands of Seychelles as one of the first steps in efforts to restore the ecosystem and protect native biodiversity. In 1996, some projects were initiated by the Forestry Section of the Department of Environment on 5 ha of Mahé. These consisted in the removal of exotic plants such as Cinnamon, Cocoplum or Rose Apple and their replacement by native species in mid and high altitude forest (above 400 m of altitude) of the Morne Seychellois National Park, where a total of 2.25 ha were replanted with 1,280 endemic saplings per hectare. Between 1999 and 2001, 26 ha of forest gaps heavily invaded by creepers were cleared, and 4.8 ha were later replanted in 2002-03 with endemic palms at approximately 1,000 saplings/ha in two different sites: one close to sea level (Cap Ternay) and one in mid-altitude (L'Exile)[4]. Small areas were restored at Mare aux Cochons (MSNP)[25]. Experimental inselberg restoration is also taking place on Mahé since 2011[23].

However, most of invasive plant control activities in Seychelles have taken place in the small and medium sized granitic islands. This is part of ongoing programmes aimed at restoring lowland coastal forests dominated by exotic species and abandoned coconut plantations. The goal is to progressively increase the area dominated by native species.

These operations do not normally aim for a full restoration of native habitats, which in many cases would be difficult to achieve as no-one knows their exact composition before they disappeared. Instead, an intermediate and more compromised 'rehabilitation' (*sensu stricto*[5]) is targeted. This involves the inclusion of some widespread naturalised invasive exotics (such as Cinnamon in some instances; see p. 31) that allow the regeneration and development of certain native species whilst preventing the establishment of more invasive ones[6, 7]. Such a system may be tolerated in particular areas or during a certain period of time, or used as a long-term permanent solution. This kind of rehabilitation may represent a first step in the restoration process and has the advantage of maintaining the ecosystem function while the proportion of native species can be progressively increased. For example, if all exotic trees are removed at once from an existing forest where they are dominant, the forest may completely disappear. This could pose a problem for the survival of particular species of animals and plants. Even if an exotic mature forest is not ideal, for example in terms of richness in invertebrates and berries for certain landbirds, it is preferable to no forest at all for the 8-10 years that a new native young forest will take to grow.

In Seychelles, most of the control and clearing of exotic plants has been done by mechanical means, using machetes and chainsaws for woody plants, or pulling by hand for creepers. In the case of great expanses of invasive bushes or small trees that need to be removed, heavy machinery is sometimes used (as on Frégate or Ile du Nord). Chemical treatments have been rarely used in Seychelles to eliminate invasive species, although some trials have been and are still being conducted on several islands (see recommended methods for control of broadleaf exotic trees and creepers, p. 326 and 340 respectively).

On some islands such as Frégate or Ile du Nord, important nurseries dedicated exclusively to propagating Seychelles endemic and native plants and trees have been created. These have benefited from collaboration with the nurseries of the Department of Environment (Botanical Gardens and Barbarons Biodiversity Centre, now managed by the NBGF) and Grande Anse, Sans Souci and Fond Boffay-Praslin (now managed by the SNPA). Private islands nurseries have been successful in multiplying most of the 85 endemic plants of Seychelles. They produce tens of thousands of saplings that have been replanted over the years. Precise figures for the total areas that have been rehabilitated are not available for all islands, but we estimate that a total of at least 220 ha has been actively cleared of alien invasives, replanted with native trees and maintained in at least 20 islands (Table 6bis, p. 81). This figure may be as high as 300 ha if we include areas partially restored (i.e. rehabilitated), in which coconut trees and some other invasives (but not all) are removed to help the long-term spontaneous regrowth of native species. This total reaches 465 ha when including native woodland recovery. Such habitat rehabilitation was initiated on Aride and Cousin in the 1970s[11, 12], and have since the mid-1990s been implemented on Frégate, Ile du Nord, Félicité, Denis and Cousine, and to a minor extent on D'Arros, Conception, Moyenne, Desroches and Ste Anne. Only relatively small areas have been rehabilitated on the larger islands of Mahé (including coastal fringe vegetation; e.g. at Grande Anse), Praslin, Silhouette and La Digue compared to what has been done on the small and medium-sized islands. Nevertheless, on Praslin, an intensive EU funded programme was started in 2014 to eliminate Cinnamon and other invasive broadleaf trees from the Vallée de Mai Nature Reserve (45 ha)[8]. At Aldabra, some rehabilitation activities have also taken place since the 1970s on Picard, Polymnie and Ile Michel to control, and since 2013 to eradicate Sisal[9], an operation expected to be completed during 2015[10].

Elimination of coconut trees inland of the beach crest has been done at most of the rehabilitated islands that were extensively planted with coconut trees and exploited for copra in the past. They are being replaced by native and endemic species, through spontaneous regrowth or replanting. Coconut trees and other invasives were first removed from Cousin and Aride during the 1970s and 1980s and the native vegetation was left to regrow spontaneously while seedlings from exotic species were removed[11, 12]. On the 7 ha of lowland plateau of Aride, active replanting of native trees was undertaken until the early 1990s[13]. In the mid-1990s, Frégate engaged in serious restoration efforts that continued until the 2010s. On Frégate, the total area rehabilitated now probably exceeds 60 ha (about half of which is fully restored). A minimum of 75,000 saplings have been produced and planted around the island to replace the exotics that were removed[14].

On Cousine (25 ha), the entire island has been similarly rehabilitated. During 1995-2005, when much of the plateau (less than 10 ha) was reforested, 25 alien species were totally eradicated and over 3,300 saplings planted[15]. On Denis Island, intensive restoration efforts were undertaken in 2002 to transform 35 ha of former coconut

Removal of exotic species is an essential step to rehabilitate island ecosystems.
SIF/Vallée de Mai

plantation with remnants of indigenous forest into a native-dominated broadleaf forest; and 3,000 young trees were planted[16].

On Ile du Nord (North Island), a nursery was established in 2003. It has since produced over 50,000 saplings which have been planted on 45 to 50 ha (half of which is fully restored) after exotic plants and coconut trees had been removed, mostly during the ICS FFEM project (see case study p. 96, and pictures p. 361)[17]. On Conception Island, between 2001 and 2009, the equivalent of 6 ha of old coconut plantation was rehabilitated, with 1800 native saplings planted. On Anonyme, in 2005, about 0.5 ha was cleared of exotic plants and 400 saplings planted[18]. In Moyenne National Park small areas have also been replanted with natives[19].

Vegetation Management Plans have been prepared and are periodically revised for some of these islands where habitats are being restored (e.g. North Island)[20]. It is particularly important to choose the type of trees to be replanted well in advance. This involves a comprehensive assessment of island potential in terms of the areas to be restored and animals to be reintroduced. The latter will require specific plant species to be multiplied and planted (e.g. berry producing trees for the Seychelles White-eye). A detailed Vegetation Management Plan needs to be prepared accordingly, and should include an assessment of which invasive plant species can be eradicated. It is essential to provide maintenance to plantations for at least two years[14, 17].

Very little vegetation restoration has taken place in the outer islands. On D'Arros Island, promising protocols have been established, and since 2009 some 11 ha of former coconut plantations have progressively been replaced by plantations of young native trees[21, 22]. On Alphonse (c. 1 ha) and Desroches (c. 12 ha), since 2006 and 2009 respectively, small areas have been cleared of exotics and replanted; and vegetation rehabilitation plans have been adopted for each island[24].

References / Further reading

1 Kaiser-Bunbury & Mougal, 2014; 2 Mulder *et al.* 2011; 3 Mulder *et al.* 2009; 4 Kueffer & Vos, 2004; 5 SERI, 2004; 6 Kueffer *et al.* 2013; 7 Kueffer *et al.* 2010b; 8 SIF, 2015; 9 van Dinther *et al.* 2015; 10 N. Bunbury/SIF, *pers. comm.*; 11 Warman & Todd, 1984; 12 Kömdeur & Pel, 2005; 13 P. & H. Carty/RSNC, *pers. comm.*; 14 Steve Hill/FIP, *pers. comm.*; 15 Samways *et al.* 2010b; 16 Nevill 2011; 17 PCA/North Island, 2009; 18 Rocamora & Jean-Louis, 2009; 19 K. Beaver, *pers. comm.*; 20 Beaver *et al.*, 2007; 21 von Brandis, 2012; 22 von Brandis, *pers. comm.*; 23 Kaiser *et al.*, 2015; 24 ICS, unpublished; 25 Valentin *et al.* 2008, in Simara *et al.* 2008.

Species translocations and recolonisations to rehabilitated islands

*The recovery of native fauna and flora on rehabilitated islands where introduced preda-
tors and competitors have been controlled or eradicated has already been observed on
many islands around the world[1-3]. This is also occurring in Seychelles, where numerous cases of unassisted colonisations by birds have been recorded. In addition, for rare
and threatened species that require active conservation management, translocations
of individuals to (re)create new populations have been conducted.*

RECOVERY AND (RE)COLONISATION

After the eradication of introduced predators and competitors, and the rehabilitation of a sufficient amount of habitat, some species that had become cryptic start to
reappear. This happened for example on Conception, where the presence of living
endemic Giant millipedes *Sechelleptus sechellarum* and snails *Stylodonta unidentata*
was discovered in 2009, two years after the eradication of Brown rats[4]. Other native
species such as reptiles (skinks and geckos) and landbirds (blue pigeons, moorhens,
etc.) usually increase rapidly in numbers (see "Ecosystem recovery after the eradication of rats and cats", p. 169).

Animals with a strong capacity for dispersal, such as seabirds or certain common
landbirds, may recolonise spontaneously rehabilitated islands within a few years. This
happened for example on Ile du Nord (North Island) where White-tailed tropicbirds
were first found breeding on the ground in 2008, less than three years after the rat
eradication[5]. Likewise, on D'Arros Lesser noddies also started breeding in 2008, five
years after the last rats and cats were removed[6]. On Petite Soeur, a small number
of breeding Wedge-tailed shearwaters had established themselves only a year after
Black rats were eradicated. On Ste Anne, Desroches and Alphonse, small colonies of
the same species became rapidly established following rat and cat control. However,
the process of recolonisation by seabirds can sometimes be slow.

Seabirds already present on these rehabilitated islands such as Fairy terns, White-
tailed tropicbirds, Wedge-tailed shearwaters, Lesser noddies and Brown noddies
have been increasing in numbers (see fig. 24bis, p. 175). Before the eradications,
some of these species had been breeding occasionally or in very small numbers in
the presence of nest predators.

The process of (re)colonisation by native species on islands freed of introduced
mammals can be assisted. In particular, the return of seabirds may be speeded by the
use of decoys and call recordings to attract passing adults looking for suitable nesting
sites[7]. This has only been used in Seychelles for the Sooty tern on Denis Island, but
with little success so far[8]. On Cousine, now freed of cats and no longer exploited for
agriculture and seabirds but as an ecotourism resort, favourable habitat management
was undertaken in 1998 and by 2003, Sooty terns had returned to breed after more
than three decades of interruption[9, 10].

TRANSLOCATIONS OF RARE AND THREATENED SPECIES

To accelerate or to make possible a process of colonisation that may otherwise either take decades or never happen in natural conditions, the translocation of a small group of individuals may be employed to create or re-create new populations of a particular species. This includes reintroductions to islands where the species was formerly present, and also 'conservation introductions' or assisted colonisations in islands outside its known historical range. This conservation management technique is particularly well suited for rare and threatened animals and for other species that have limited possibilities of dispersal, and it has been widely employed around the world[11]. However, these operations are costly and require a high level of preparation and preliminary studies listed in the new IUCN *Guidelines for Reintroductions and Other Conservation Translocations* (IUCN 2013)[12]. Each operation requires, for example, an assessment of the suitability of natural habitats found on the destination island, and an assessment of the size of the source populations and of the genetic diversity of the animals to be transferred. Health screening also needs to be conducted in both the source populations and at the destination island. Detailed logistical plans need to be prepared for the transfers, along with the assurance that the transferred population can be adequately monitored after its release (see for example the project proposals for the transfer of Seychelles white-eyes to Frégate or to North Island[13-16]).

Rare endemic landbirds and other native animals, especially those that cannot fly such as land reptiles or certain invertebrates, need to be assisted if new populations are to become established. Seabird translocations have been employed in the Pacific and are currently being trialled in Mauritius, whereby chicks ready to fledge are transferred and released onto a new island where they can be expected to return to breed and form a new colony after a period of several years, once they have reached adulthood[7, 17].

Table 6 lists all the translocations to predator-free islands of rare and threatened species that have been conducted in Seychelles. Translocations of Aldabra giant tortoises (IUCN Red List category 'Vulnerable') are not considered here as these animals have been moved freely and introduced successfully to many islands regardless of the presence or absence of introduced predators[18]. By 2014, no fewer than **19 translocations of eight rare and threatened species had taken place on nine different rehabilitated islands**. This includes six species of endemic landbirds (listed in the table and in the next paragraph), one species of reptile (Seychelles Black-mud terrapin) and one species of invertebrate (Seychelles leaf insect). Of these 19 translocations (nine reintroductions and 10 conservation introductions), 17 were successful. In addition, the Seychelles sunbird, a common non-threatened landbird, was transferred and established successfully on Bird Island in 2006[19]. This brings the total to 20 animal translocations (of which 18 were successful) of nine species to ten different islands that have benefited from restoration activities. However, the translocations of Seychelles leaf-insects to Conception and of Seychelles white-eyes to Cousine failed[4, 20]. The total number of successful island translocations to predator-free islands enabled by the rat and cat eradications is 16 (seven reintroductions and nine conservation introductions, see table 6 and p. 173). The transfer of Seychelles fodies to D'Arros happened at a time when the island was still infested by Brown rats and cats, whereas Seychelles magpie-robins were translocated to Cousin that was never invaded by mammals. Therefore these have not been counted as translocations to rehabilitated islands where introduced

Table 6: Translocations of rare and threatened species (other than Aldabra giant tortoises) and some common species to rehabilitated islands in Seychelles. X = naturally present; **Bold**: reintroduction; **Bold underlined**: conservation introduction; *Italic*: failed attempt; *: initial transfers to Aride between 1978 and 1995 were unsuccessful; ©: cats eradicated; ®: rats eradicated; species are listed per chronological order of translocation; Brackets: IUCN status of rare species requiring revision.

	IUCN Threat Status	Cousin -	Aride ©	Cousine ©	Frégate ©®	Denis ©®	Ile du Nord ©®	Conception ®	D'Arros ®©	Picard (Aldabra) ©	Bird ®	Total
Seychelles fody *Foudia sechellarum*	NT	X	2002	X		2004			1968			3
Seychelles warbler *Acrocephalus sechellensis*	VU	X	1988	1990	2011	2004						4
Seychelles magpie-robin *Copsychus sechellarum*	EN	1994-95	2002*	1995-96	X	2008						4
Seychelles white-eye *Zosterops modestus*	EN			2007	2001		2007	X				3
Aldabra rail *Dryolimnas (cuvieri) aldabranus*	(LC)								1999			1
Sey. black paradise flycatcher *Terpsiphone corvina*	CR					2008						1
Seychelles black-mud terrapin *Pelusios subniger parietalis*	(LC)		2012			2008						2
Seychelles leaf insect *Phyllum bioculatum*	(LC)							2010				1
Number of translocations of rare and threatened species		1	3	3 (1 failed)	2	4	2	1 (failed)	1	1	0	19 (2 failed)
Seychelles sunbird *Cynniris dussumieri*	LC										2006	1
Number of translocations per island (all native species)		1	3	2	2	4	2	0	1	1	1	20 (2 failed)

predators had been eradicated. The island with the highest number of successful translocations is Denis (four), followed by Aride (three), then Cousine, Frégate and Ile du Nord (two), all other islands having benefited from only one species translocation.

The first documented bird conservation introduction attempt in Seychelles was the release on D'Arros of five Seychelles fodies from Cousin in 1965 by the Bristol University Seychelles Expedition[21]. Although a population of about 850 has now developed as a result[22], the very small numbers introduced and the unbalanced sex-ratio (three females and two males) probably explains the hybridisation now apparent between the Seychelles fodies and the Madagascar fodies that had previously been introduced to D'Arros[23].

In 1988, after the number of Seychelles warblers had increased and reached carrying capacity on Cousin, 29 of them were transferred to Aride, where the population has now grown to over 2,000 birds, and in 1990 a similar number was transferred to Cousine where numbers have stabilised at c.180 birds[24]. More recently, additional transfers involving 58 birds in each case to maximise the transfer of genetic variation, were conducted from Cousin to Denis Island in 2004[25], and to Frégate in December 2011. This very successful programme[a] has increased the warbler's population from 29 birds on a single island in 1959 to c.3,000 and rising on four islands in 2014 (Fig. 11)[26].

In 1995, as part of the Seychelles Magpie-robin Recovery Programme[b], small numbers of these birds (6) were transferred from their only remaining population on Frégate to both Cousin and Cousine Islands. After some unsuccessful trials, the species also became established on Aride in 2002, and then on Denis in 2008. As a consequence, the numbers of endemic magpie-robins increased from just 15-20 birds

a. Led initially by ICBP/ BirdLife International, and later Nature Seychelles, in partnership with the University of Gröningen and the University of East Anglia.

b. Initiated by ICBP/ BirdLife International and the RSPB, now led by Nature Seychelles and conducted in partnership with all islands and managing organisations involved.

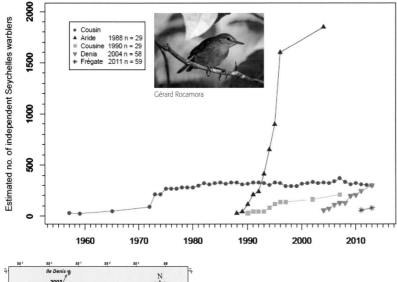

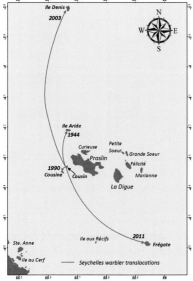

Figure 11: Back from the brink of extinction: Population growth of the Seychelles warbler obtained through habitat restoration and translocations in four islands freed of predatory mammals. Population census estimates for the five existing populations. Embedded legend gives year and number of founders translocated from the original population on Cousin where only 29 individuals survived in 1959. Current population estimates per island are Cousin 320, Aride 1850, Cousine 210, Denis 300 and Frégate 80 (Wright et al 2014), with a global population size estimate of 2760 birds. Figure courtesy of David Richardson.

remaining on Frégate in 1964 (and no more than 21 birds in 1990) to over 268 birds in five islands in 2014[27, 28].

The Seychelles fody has also been introduced from Cousin to Aride (64 birds) and from Frégate to Denis (47 birds) in 2002 and 2008 respectively, and now has viable populations on six islands instead of three originally[22]. The Seychelles black paradise flycatcher, originally restricted to La Digue was transferred to Denis in 2008 (23 birds); and by 2014 it had established a small population of over 40 individuals[29].

In 2001-2003, as part of the Seychelles White-eye Recovery Programme[c], 37 Seychelles white-eyes were transferred from their unique and recently discovered stronghold on Conception Island to Frégate Island, where a new healthy population of more than 150 birds had developed by 2010[30]. This operation was repeated in 2007[f] with a transfer of 25 white-eyes to Ile du Nord (North Island), where they are currently

c. Initiated by the Department of Environment and later coordinated by the Island Conservation Society.

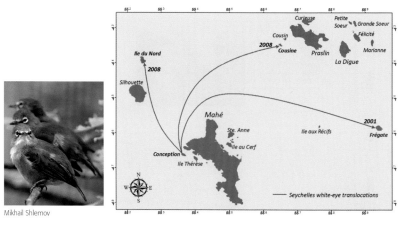

Mikhail Shlemov

Figure 12: Increases in the Seychelles white-eye population obtained through habitat restoration and translocations in islands freed of predatory mammals, plus control or eradication of rats in source populations of Mahé and Conception respectively. Embedded legend gives year and number of founders translocated from Conception. Population estimates per island appear in the graph, global population has doubled from an initial 320 to c. 650 birds. Data: G. Rocamora, E. Henriette, A. Labiche/ICS, DoE & North Island, unpublished. Frégate estimate extrapolated. 23 birds transferred to Cousine in 2007 (3 of which from Mahé) failed to establish and the species died out in 2014.

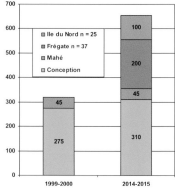

thriving (c. 100 birds in 2014)[31-33]. Another 23 birds were transferred to Cousine in 2007 where they did not succeed in establishing a viable population and the last white-eyes died out in 2014[20]. Island transfers have enabled the global Seychelles white-eye population to approximately double from 320 birds on two islands in 1999 to c. 650 birds on four islands in 2014 (see Fig. 12).

In the outer islands, the endemic flightless Aldabra rail, whose taxonomy is being reviewed, may soon be designated a full species[34]. In the past, the presence of feral cats had restricted it to only three cat-free islands on the Aldabra atoll (Polymnie, Malabar and Ile aux Cèdres). In 1999, 18 rails were transferred to Picard Island (Aldabra[d] resulting in the establishment of a breeding population[35] that by 2011 comprised over 1,100 pairs inhabiting the entire island[36].

Some rehabilitated islands have also benefited in recent years from transfers of animals other than birds. The local subspecies of Black mud terrapin (*Pelusios niger parietalis*, described as endemic to Seychelles but the status of which is now questioned) was transferred from Mahé (15 ind.) and Ile au Cerf (5 ind.) to Ile du Nord in 2008, and from Frégate to Aride (15 ind.) in 2012[f, 37, 38]. Between 2002 and 2010, Black mud terrapins (29 ind. in total) were also reintroduced to Silhouette and the relictual population of Yellow-bellied terrapin on that island (*Pelusios castanoides intergularis*, also described as a Seychelles endemic subspecies) was supplemented[e, 39]. The rare Seychelles leaf insect *Phyllum bioculatum* was transferred from Mahé to Conception in 2010 (30 ind.); this was the first documented invertebrate translocation in the western Indian Ocean, which unfortunately failed[f, 4].

d. Conducted by the Seychelles Islands Foundation in partnership with the Cape Town University Fitzpatrick Institute of Ornithology.

e. Developed on Silhouette by the Nature Protection Trust of Seychelles with IDC support between 1997 and 2010.

f. As part of the Island Conservation Society-led FFEM project.

Giant tortoises (*Aldabrachelys gigantea*) are probably the first animals in Seychelles to have been transferred between islands for reintroduction. After the Seychelles giant tortoises naturally present on most of the granitic islands had been overexploited and driven to extinction[40], a minimum of eight granitic islands where tortoises are still present today have been repopulated since 1850 with wild animals taken from Aldabra or captive stock[g]. These include Frégate, Curieuse and Cousin where tortoises were reintroduced before 1950, Moyenne (probably in the 1970s), and Ile du Nord, Cousine, Grande Soeur and Silhouette that were last repopulated during the period 1993-2012[18]. In 2011, some 172 juveniles and six adults from a breeding programme[e] aiming at recreating two morphotypes described from the granitic islands were released on three islands with already existing stock (Ile du Nord, Cousine and Frégate)[41]. Aldabra giant tortoises have also been introduced or reintroduced[h] to at least 10 coralline islands during the 20th century (i.e. Bird, Denis, D'Arros, Desroches, Rémire, Alphonse, Farquhar, Providence, Assomption, Cosmoledo) during the last 25 to 50 years, although some of these populations are small and of uncertain long-term viability[18]. These reintroductions or conservation introductions of mega-herbivores that have roamed and dominated the terrestrial ecosystems of Seychelles for millions of years should be regarded as an essential step to complete the island restoration processes. This in view of the likely very important (but still poorly known) role that these animals fulfil in the ecosystem in terms of seed dispersal, germination and fertilising of native plants, and interacting with soil invertebrate communities through their dung. These mega-herbivores can be used as analogs to replace extinct tortoises and as a tool to restore island ecosystems, as it is done actively in Mauritius[42-45]. In Seychelles, a minimum of 18 successful translocations of Aldabra giant tortoises are known to have taken place, including four to rat and predator free islands (Cousin, Cousine, Frégate

Translocation to islands where invasive predators were removed and native habitats re-established have allowed several globally threatened species to recover (Seychelles white-eye transfer to Ile du Nord).
Gérard Rocamora/ICS

g. Conducted by private island owners, NGOs and Department of Environment.

h. Conducted mainly by IDC and private island owners under the authority of Department of Environment.

Birds are transferred to another island in the sound proof, ventilated 'Helibird' box (design: G. Rocamora).
Mike Meyers/North Island

and Bird). The reintroduction of giant tortoises to Aride in 1933-34 is not accounted for as the animals were later removed and brought to Cousin in 1951[46]. Giant tortoises must have benefited from predator eradication and/or habitat restoration activities on Cousine, Frégate, Ile du Nord, Grande Soeur, Bird, Cousin, Denis, D'Arros and Picard; although signs of population increase have only been noted in the first four islands[18]. Taking into account the reintroductions of Giant tortoises, **the total number of species translocations between islands having taken place in Seychelles appears to be of at least 38. These transfers have benefitted to at least 24 islands.**

Population translocations to islands that have benefited from predator eradications and ongoing restoration processes have proved to be a very powerful tool to improve the conservation status of endemic species threatened with global extinction in Seychelles. Seven successful translocations of five globally threatened and one non-threatened endemic birds, plus one translocation of a rare non-globally threatened reptile were conducted to three islands from which both rats and cats were removed (Denis, Ile du Nord and Frégate Islands). Another five populations of globally threatened birds were successfully translocated to two islands where only cats had been removed (Aride, Cousine). In addition, the endemic Aldabra rail, the last flightless landbird of the Indian Ocean, (currently not considered globally threatened), was reintroduced to Picard Island after the island became cat free and the species is now present on four islands. Hence no less than **12 new island populations of five globally threatened bird species** have been created on five islands where rats or cats have been eradicated, plus another four populations (two birds and two chelonians) of non-globally threatened species on an additional two islands (Bird and Picard).

As a result of all the translocations undertaken and the essential accompanying work of habitat restoration, **four globally threatened birds have been downlisted on the IUCN Red List:** the Seychelles warbler from Critically Endangered to Vulnerable; the Seychelles magpie-robin and Seychelles white-eye from Critically Endangered to Endangered; and the Seychelles fody from Vulnerable to Near-Threatened. A fifth threatened species, the Seychelles black paradise flycatcher, which was transferred to Denis Island, is still considered Critically Endangered. A sixth globally threatened species, the Aldabra giant tortoise (Vulnerable), has benefited from translocations that reinforced its existing populations on three islands (Ile du Nord, Cousine and Frégate), and from population increase on a fourth island (Grande Soeur) where restoration activities had been undertaken.

References / Further reading

1 Mulder et al., 2011; 2 Veitch et al., 2011; 3 Russell et Holmes, 2015; 4 Galman, 2011; 5 Rocamora & Jean-Louis, 2009; 6 R. von Brandis, pers. comm.; 7 Jones & Kress, 2012; 8 Feare et al., 2015; 9 Feare et al., 2007; 10 Samways et al. 2010a, 2010b; 11 Seddon et al., 2007; 12 IUCN, 2013; 13 Rocamora, 2001; 14 Rocamora et al., 2001; 15 Rocamora et al., 2006; 16 Mac Gregor, 2007; 17 Kappes & Jones, 2014; 18 Gerlach et al., 2013; 19 Savy, 2006; 20 Julie Gane / Cousine Island, pers. comm.; 21 Dawson, 2003; 22 Vega, 2013; 23 G. Rocamora, J.M. Pons & D. Richardson, unpublished data; 24 Kömdeur, 1994; 25 Richardson et al., 2006; 26 Wright et al., 2014; 27 Lucking, 2013; 28 Burt et al., in prep.; 29 A. de Gröene & A. Labiche / GIF, pers. comm.; 30 Henriette, 2011; 31 Rocamora & Henriette-Payet, 2008; 32 Rocamora, 2010b; 33 Henriette & Dine, 2014; 34 Janske van de Crommenacker / SIF, pers. comm.; 35 Wanless et al., 2002; 36 Sur et al. 2013; 37 Rocamora et al., 2009; 38 G. Rocamora / ICS, unpublished data;39 Gerlach & Gerlach, 2011b; 40 Fauvel, 1909; 41 Gerlach & Gerlach, 2011a; 42 Hansen et al., 2010; 43 Griffiths et al., 2010; 44 Kaiser-Bunbury et al., 2010; 45 Griffith, 2014; 46 Warman & Todd, 1984.

Replanting of native species on Praslin with local community.
Elvina Henriette / TRASS

(a)	Planting after exotic sp. removal	Woodland recovery (Pisonia dominated)
Frégate	60	
Ile du Nord	45-50	
Félicité	40	
Denis	37	4
Curieuse	18-20	
Aride	7	62
Cousin	?	27
Cousine	10	16
Bird	< 1	35
Conception	2 (coconut removal)	> 1
TOTAL	c. 225	c. 145

(b)	Planting after exotic sp. removal	Woodland /Shrubland rehabilitation
Praslin	25 (inc. 10 bare land)	20
Mahé	15-20	
Desroches	12	
D'Arros	11	
Aldabra	(Sisal removal)	<5
Alphonse	1	
St Anne	1	
Silhouette	< 1	
Moyenne	0,5	
Anonyme	0,5	
TOTAL	c. 75	c. 25

Table 6 bis: Islands where vegetation management has been undertaken and approximate areas rehabilitated (c. 300 ha /465 ha inc. native woodland recovery). **(a)** small to medium-sized inner islands **(b)** Mahé, Praslin, and other islands.
Source: published information, site managers, ED

Creating more island refuges for native biodiversity

Since the 1970s, ecosystem restoration has progressively taken place at more than 20 small and medium sized islands of Seychelles (see Fig. 13). The result has been the multiplication of island refuges for native biodiversity and particularly for rare emblematic species highly threatened with extinction.

The first steps towards the rehabilitation and legal protection of islands of high bio-diversity value in Seychelles were taken by several foreign NGOs, the International Council for Bird Preservation (today BirdLife International), the Royal Society for Nature Conservation presided over by Christopher Cadbury, and the WWF in collaboration with the Seychelles' Government. In 1968, funds were raised internationally to purchase Cousin Island from its private owners in order to save the Seychelles warbler from extinction[1]. In 1973, Christopher Cadbury, who had also provided almost two thirds of the funds to purchase Cousin, purchased Aride Island and donated it to RSNC to protect its outstanding seabird populations[2]; and both islands were then declared as Special Reserves by the government in 1975.

The successful restoration activities that ensued in these two nature reserves provided a clear example for other private and government islands to follow. During the 1980s, while Cousin and Aride were being restored, active management of invasive species and rehabilitation activities were promoted by ICBP with the support of the Division for Forest and Conservation of the Department of Agriculture, and extended to two private islands and rat-free biodiversity hotspots: Frégate[3] and Cousine[4]. Another private island, Bird (Ile aux Vaches) decided to engage in some rehabilitation activities in the 1990s. During the following decade, with the help of local environmental NGOs such as Nature Seychelles, Island Conservation Society or Plant Conservation Action Group,

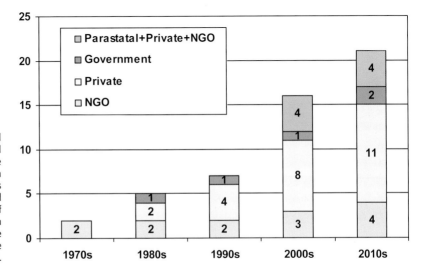

Figure 13: Total number of small islands that have benefited from invasive species management and other forms of ecological restoration in Seychelles since the 1970s, and type of management.

more private islands such as Denis, North Island (Ile du Nord), D'Arros, Anonyme and Conception followed[5-8]. ICS initiated small scale rehabilitation activities on the outer islands of Desroches and Alphonse in collaboration with IDC and hotel partners, and rat and cat eradications were also started at Cosmoledo atoll. Since 2010, Grande Soeur and Petite Soeur have joined the club of rat and cat free islands, although no habitat restoration has taken place. After Aldabra atoll became a Special Reserve and a UNESCO World Heritage Site in 1982, the Seychelles Islands Foundation, a public trust, initiated invasive species management activities on several islands of the atoll, and since 2011 it has been implementing several new and successful programmes on Aldabra, and also on neighbouring Assomption in collaboration with ICS and IDC, as well as at the other UNESCO World Heritage Site, the Vallée de Mai (see case-studies).

With about 30 small and middle-sized islands free of rats and cats (see Table 7, p. 87), Seychelles is probably one of the countries that has proportionally more territory free of invasive predatory mammals. Rats and cats have been completely removed from a total of 11 islands larger than 10 ha (eight in the inner islands and three in the outer islands). Between 1996 and 2011, the number of islands (of 10 ha or more) free of rats has increased from only four (Aride, Cousin, Cousine, Ile aux Récifs) to 12 in the inner islands, and from nine to 12 in the outer islands. During that period, the total rat free area of Seychelles has increased by a factor of more than three (1.3% to **3.9%** of the country's total land area; see Fig. 14). For the inner islands, the percentage of the land surface without rats and cats has been multiplied by over six times (0.7% to **4.4%** of land surface), and for the outer islands the rat free area has almost doubled (1.9% to **3.3%** of land surface).

However, there is great scope for more eradication operations that would benefit both wildlife and humans (see Table 8, p. 87). If we consider all the islands where it would be possible to eradicate rats given current available techniques, and also consider at which of those sites it would be realistically feasible to prevent reinvasion, there are another 20 more islands where such eradications could take place. Five more granitic islands (Curieuse, Félicité, Marianne, Thérèse and Ronde de Praslin) could be made free of predatory mammals in future along with the small islands from the Ste Anne group (Longue, Moyenne and Ronde). However, the latter would require a permanent grid of bait stations (as it exists on Anonyme; see Table 17, p. 198). This would prevent the establishment of rats that can arrive by swimming ashore from nearby infested Ile au Cerf Island (only some 0.5 km away).

The outer islands have a higher potential with at least 10 and possibly 12 more islands where rats could be eradicated. But this would depend on the number, size and types of management envisaged for future developments on Coëtivy and Assomption. It follows that the maximum potential area would be 1,733 ha or 7.3% of land surface free of rats (and cats) in the inner islands, and 5,265 ha or 23.9% of land surface in the outer islands, and **for the whole of Seychelles 6,998 ha or 15.4% of the total area of the country that could be made free of alien predatory mammals**. Clearing rats and cats from these additional 20 islands would be a great achievement for Seychelles and would open huge potentialities in terms of ecosystem rehabilitation and population recovery for many species (landbirds, seabirds, reptiles, amphibians, invertebrates and rare plants).

IAS management to promote biodiversity has been undertaken in Seychelles to various degrees at about 20 small or medium sized islands (c. 10 to 1,000 ha), and in six of them (Aride, Cousin, Cousine, Frégate, Ile du Nord, Denis) as part of long-term rehabilitation programmes based on intensive management of vegetation and animal species, including conservation introductions of threatened birds and reptiles. With more than 50% of its land surface designated as protected (Nature Reserves or National Parks), Seychelles probably has the highest proportion of its land surface legally protected of any country in the world.

It is therefore surprising that this 50% does not include any of the predator-free private island refuges that are devoted to ecotourism (apart from Aride and Cousin, the two NGO-owned islands that are nature reserves), as these have no legal protected status despite the fact that they host numerous endemic species either threatened or of high conservation concern. In fact, some of the populations of greatest concern (e.g. Seychelles white-eye, Seychelles magpie-robin, Giant tenebrionid beetle, Whip spider) occur at such sites without legal protection. To protect the tremendous investments that have been made by private owners, NGOs and international donors in the rehabilitation of the ecosystems of these islands, legal protected status that limits possibilities of future developments on these islands and allows their owners to control access to their shores by visiting vessels would be highly desirable to reduce the present high risk of reinvasion by rodents. This may be soon possible under the new Protection Areas legislation expected to be passed in 2015. In the absence of a legal protection status, the obligation to abide by certain landing protocols to prevent rat infestation whilst respecting public constitutional rights of access to beaches, represents the minimum that should be put in place. Regulations being prepared under the Animal and Plant Biosecurity Act 2014 should also be dealing with these aspects.

References / Further reading

1 Stoddard, 1984; 2 Sands, 2012; 3 Watson *et al.*, 1992; 4 Samways *et al.* 2010; 5 Merton *et al.*, 2002; 6 Rocamora & Jean-Louis, 2009; 7 Nevill, 2011; 8 von Brandis, 2012.

Saint François atoll, a predator-free island in the Alphonse group.
Colin Bell

New perspectives and challenges in invasive species management and restoration strategies

The creation of small island refuges where invasive species can be adequately managed to enable the (re)introduction of species that cannot survive in the presence of predatory mammals has proven so far extremely efficient in improving the conservation status of several endemic birds of Seychelles – some of which had come very close to global extinction. However, several factors limit the development of this conservation strategy.

AVAILABILITY OF ADDITIONAL ISLANDS SUITABLE FOR RESTORATION AND PRESERVATION

Although Seychelles still has a considerable potential to increase its surface free of alien predatory mammals, the number of islands where such operations can be conducted is limited. In the inner islands, there are currently only five to eight islands left which could be realistically made and kept free of rats and cats. Unfortunately, most of these have ongoing or planned development projects or do not presently fulfil the required strict management conditions to prevent reinvasion in the long term.

CHALLENGES TO ERADICATE RATS FROM LARGER TROPICAL ISLANDS

Of the 650 *Rattus* spp. eradications with sufficient data quality reported in DIISE (http://diise.islandconservation.org/ accessed 14 September 2014), 69 have failed. The failure rate outside the tropics is 8.5%, but in the tropics it is 18.8%. When including eradications where reinvasion occurred (which may be misdiagnosed failure), the failure rates drop to 6.3% and 16.1% respectively[1]. Although it is likely that techniques to eradicate rats from tropical islands will progressively improve in future, the currently available techniques, based on the spreading of bait pellets containing an anticoagulant substance, are better adapted to the temperate climate and are more difficult to implement in the tropical world[2, 3]. This is particularly the case in the equatorial climates of the inner Seychelles where rains (that can seriously affect the attractiveness and palatability of rat pellets) and the natural food supply (which reduces the likelihood that rats will eat the bait) are abundant during a large part of the year.

The techniques initially established in New Zealand have been progressively improved and applied to much larger temperate or subantarctic islands. Hence the records for the largest islands where rat eradication has been possible are: Macquarie Island, of 12,872 ha, off Tasmania (Australia) since 2014 for Black rat; and Campbell Island, of 11,300 ha (New Zealand) since 2001 for Brown rat[4]. Both islands are subantarctic territories. In the tropics, the largest island-wide eradication of Black rats (up to 2014) occurred at Hermite Island in Australia, of only 1,020 ha[a, 5], and the largest island-wide eradication of Brown rats occurred on Frégate Island, of 219 ha. In the western Indian Ocean, the record for Black rats is Ile Platte in Mauritius, of 253 ha, followed by Ile du Nord in Seychelles, of 201 ha.

a. rat eradication was also achieved on the large nearby Barrow Island, of 24,103 ha, but by baiting only an infested area of 245 ha[31], hence we do not consider it as an island-wide eradication.

Limiting factors specific to tropical islands also include the presence of mangroves, where efficient methods to eradicate Black rats from large areas do not exist yet[1, 2]. Any bait spread on mangroves by helicopter at low tide will not remain on the ground for a sufficiently long time before it is destroyed by the next high tide; although small areas of mangrove can be dealt with 'collars' of bait blocks or bait stations placed in trees (as we did in Grand Ile, Cosmoledo, see Table 17, p. 198) or by using 'bolas' made of rodenticide blocks as it has been done in Micronesia[b, 6]. This is currently the main obstacle to an eradication of Black rats from Menai (268 ha), the last remaining infested island on Cosmoledo atoll. It is also the main limiting factor for envisaging a large scale eradication on Aldabra (15,380 ha)[7].

b. Extra baiting with rodenticide blocks tied to mangrove vegetation has also been used in Mexico in the challenging operation of Cayo Centro (539 ha, Chinchorro bank), which has large extensions of mangrove and wetlands[32, 33].

An added difficulty in tropical islands is the presence of dense vegetation that provides an abundant source of food that rats may prefer over the bait[1, 2, 10]. This is especially problematic for the arboreal Black rat. Rat eradications on islands with high densities of coconut trees (e.g. former coconut plantations), and where nuts still provide an abundant source of food to rats may also prove challenging (see for example the failed eradication attempt of Eagle Island in the Chagos in 2008)[8]. Black rats often live in the crown of coconut trees, and can have restricted home ranges. On Denis Island, in 2002, one radio-tracked rat that came down to the ground every night never went beyond a short distance of about 15 m from its coconut tree during more than a week (Gideon Climo, *pers. comm.*). This means that bait spreading in areas dominated by coconut trees must be done very evenly for it to be available to all rats.

Unpredictable rainfall patterns and the year-round high primary productivity of Seychelles ecosystems – at least in the northern archipelagos of the inner islands and the Amirantes – are another added difficulty to conduct rat eradications. Despite the existence in Seychelles of a relative 'dry' season between June and October, where the probability of rainfall is reduced, our tropical climate remains largely unpredictable and episodes of high rainfall may occur during this period. Apart from allowing a high production of food available to rats that translates into their breeding year round[1, 9], rainfall can seriously affect the texture and appetence of bait. Our experience shows that, when rainfall occurs after a rodenticide application (dropped from helicopter or hand spread), certain baits (such as cereal pellets Pestoff Rodent Bait 20R) can resist rains of up to 25 mm, and conserve some effectiveness after they dry up; but the next rodenticide drop then needs to be done earlier than originally planned[10].

At certain islands the extreme abundance of crabs can also present a great difficulty[3, 11-13]. Crabs can eat huge quantities of bait without suffering any toxicity. As a result the bait will not be available to the rodents. The integration in rodent bait of a crab deterrent compound that would not affect palatability to rats would help to decrease current high application rates that are required to take into account crab consumption; this would likely increase the chances of success of eradications on tropical islands[14].

All these factors explain – at least in part – why the chance for a rat eradication to fail is about 2.5 times higher in the tropics than in temperate climates[1, 12].

SUITABLE HABITATS AT RESTORED ISLANDS
NON-EXISTENT OR TOO LIMITED FOR SOME SPECIES

A major limitation of the *small island restoration* model lies in the fact that some rare and threatened species require very specific habitat types that may not necessarily be found in small or medium sized islands. Examples include the critically endangered

	Area (ha)	Feral cat	House mouse
Inner Islands			
Frégate	219		
North Island (Ile du Nord)	201		
Denis	143		
Bird Island (Ile aux Vaches)	101		X
Grande Sœur	85		
Aride	73		X
Conception	69		
Petite Sœur	35		
Cousin*	29		
Cousine*	26		
Ile aux Récifs*	20		
Anonyme	10		
*Mamelles**	*9*		
*Ile aux vaches marines**	*5*		
*Ile aux Cocos**	*2*		
Ile aux rats	*1*		
Outer Islands			
Grande Ile (Cosmoledo)	143	§	
D'Arros	140	X	
St Joseph atoll*	122		
Bancs du Sud (Providence)*	71		
Marie Louise*	53		X
Desnoeufs*	35		X
Ile du Sud-Ouest (Cosmoledo)*	30		
Bancs Africains*	31		
Goëlettes (Farquhar)*	25		
Grand Polyte (Cosmoledo)	21		
St Francois (Alphonse)*	17		
Ile du Nord (Cosmoledo)*	11		
*Ile du Nord-Est (Cosmoledo)**	*9*		
*Banc de sable (Farquhar)**	*7*		
*Pagode (Cosmoledo)**	*6*		
*Goëlettes (Cosmoledo)**	*5*		

Table 7: Main islands free of rats and other predatory mammals in Seychelles (2015).

Islands are listed in order of decreasing size. Small islands of less than 10 ha in italics.

In black = islands where rats have been eradicated.
In blue * = naturally free of rodents and cats
X = islands with cats or mice still present;
§ = cat-free status being confirmed

Table 8. Additional islands in Seychelles where rodents and cats could be eradicated and their reinvasion prevented. X = presence. Moyenne, Longue and Ronde (Ste Anne group) would require a concerted operation.

	Area (ha)	Black rat	House mouse	Feral cat
Inner Islands				
Marianne	100	X		X
Félicité	268	X		X
Curieuse	286	X	X	
Ronde (Praslin)	19	X		
Thérèse	74	X		X
Longue	*17*	X	?	X
Moyenne	*9*	X		
Ronde (Mahé)	*2*	X		
Outer Islands				
Assomption	1171	X		X
Coétivy	931	X		X
Astove	660	X		X
Ile du Sud (Farquhar)	400	X		X
Desroches	394	X		X
Ile du Nord (Farquhar)	300	X		X
Poivre	255	X		X
Alphonse	174	X		X
Providence	157	X		X
Platte	54	X		X
Rémire	27	X		X
Manahas (Farquhar)	10	X		X
Marie-Louise	53		X	
Desnoeuf	35		X	

Seychelles sheath-tailed bat *Coleura seychellensis* (the most threatened vertebrate in Seychelles with a world population size of less than 100 individuals), and the vulnerable Seychelles swiftlet *Aerodramus elaphrus*, both of which breed inside caves and feed on flying insects. Both species currently occur only on the larger granitic islands.

Such a limitation also applies to endemic plants and animals (reptiles, amphibians, invertebrates) found exclusively at altitudes greater than 300 to 400 m on Mahé or Silhouette (where the majority of the terrestrial diversity of Seychelles is concentrated[15]). Most of these species would probably not survive in the low-lying habitats of the small islands, where the climatic conditions differ from the humid and colder bioclimates that characterise the middle and high altitudes. The endangered Seychelles scops-owl *Otus insularis*, currently confined to the middle and high altitude forests of Mahé, is another species that may not adapt to restored small low-lying islands.

Some species may also require large expanses of specific habitats that would not be available on small islands. The Seychelles black parrot *Coracopsis barklyi* (VU), currently restricted to the island of Praslin, could perhaps be transferred to a predator-free medium sized island such as Frégate or Ile du Nord and survive; but it is unlikely to form viable populations there[16]. Even if a substantial amount (10-15ha) of palm dominated forest, a habitat that plays a key role in the bird's ecology[17], was to be created on these islands, this would probably not be sufficient. Therefore the future of this species relies on large islands such as Praslin, and possibly Silhouette where large extensions of palms exist and where a translocation of the Black Parrot is being envisaged. The Seychelles magpie-robin, which no longer exists in the larger granitics and is currently confined to five small and medium-sized islands in numbers varying between 20 and 120 birds, may never be able to form long-term viable populations on any of these large islands where permanent predator control and species management would be required. The same applies to the Seychelles Black paradise flycatcher *Terpsiphone corvina*, whose only remaining viable population is on La Digue. There, each pair requires approximately 1 ha of coastal native forest. The amount of rehabilitated coastal forest on small/medium islands (such as Denis Island where it was transferred to in 2008) may not be sufficient enough for the species to form sufficiently viable alternative populations[18].

INCREASED INTERSPECIFIC INTERACTIONS ON SMALL ISLANDS WITH MULTIPLE REINTRODUCTIONS

As predicted by the island biogeography theory of Mc Arthur and Wilson[19], the number of species that can coexist on an island is limited by its land surface area, and is related to the amount and diversity of habitats available on the island (as part of a dynamic equilibrium that also depends to the distance to mainland sources). The survival of any new species introduced (or reintroduced) to an island will therefore depend not only on the quantity and quality of suitable habitats available for that species at that time, but also on the interactions that will develop between this newly arrived species and the ones already there[20].

Although most of the island transfers that have been conducted in Seychelles to create alternative populations of threatened species have so far proved successful, some examples remind us that there is probably a limit to the number of (re)introduced species that a small island can host. On Conception, the Seychelles leaf insects introduced in 2009 appear to have been preyed upon shortly after the transfer (possibly by Seychelles skinks and/or large unknown invertebrates); no survivors were found during the subsequent months[21].

On Cousine (26 ha), the 23 Seychelles white-eyes translocated there in 2008 did establish a small breeding population, but predation of nests and fledglings by other species, along with high adult mortality, did not allow this small population to grow despite considerable management efforts. Remarkably, some of the highest mortality was caused by another introduced species, the Seychelles magpie-robin[22]. Such problems were not observed after the introduction of white-eyes to two larger islands, North Island (Ile du Nord, 201 ha) and Frégate (219 ha), although a large population of Seychelles magpie-robins is present on the latter. At those sites, the numbers of introduced Seychelles white-eyes have multiplied several times during the same period of time[23, 24].

Problems of interspecific competition or negative interactions between species after conservation introductions seem more likely to occur on small well restored islands. Islands such as Cousine (which has never been invaded by rats) host higher densities of many native species (landbirds, reptiles, invertebrates), either naturally

On Ste Anne Island, a 40 m grid of 150 rodenticide bait stations combined with trapping aims at keeping very low rat densities over 30 ha of natural habitats around the touristic resort , and a small seabird colony.
© Digital Globe/Google Earth. Data SIO, NOAA, U.S. Navy, NGA, Gebco

present or introduced, than would a larger island that is just starting its process of rehabilitation, such as Ile du Nord. There, rats had been present until 2005, so the number of native species that might negatively interact with white-eyes (the first bird to be introduced) was very limited. Hence, although each of the (re)introduced species has its own niche that may not significantly overlap with the niches of other species already present, increasing limitations related to interactions between species should be expected as the number of conservation introductions on a particular island increases. In summary, it **probably becomes more difficult to 'squeeze in' new species as the ecosystems of a small restored island become increasingly saturated**.

THE NEED TO DEVELOP ALTERNATIVE CONSERVATION APPROACHES SUCH AS 'MAINLAND ISLANDS' OR PREDATOR-PROOF FENCES

One of the great advantages of the conservation management strategy widely adopted in Seychelles, which consists in making conservation introductions of threatened species to small rehabilitated islands where invasive alien species have been eradicated or are kept under control, is that it corresponds to an *in situ* approach. With this approach animals remain in the wild and continue to interact with other native plants and animals and the ecosystem in general. This contrasts with more intensive and costly *ex-situ* management approaches, in which animals are taken away from their natural environment to be bred in captivity. *Ex-situ* management always requires the delicate additional step of readapting the captive reared animals into the wild. Nevertheless, this approach has been extensively used with much success by the Mauritian Wildlife Foundation (under the leadership of Prof. Carl Jones) working in partnership with the Mauritian Government and international organisations such as the Durell Wildlife Conservation Trust, and also by the Department of Conservation and its partners in New Zealand[25].

In Seychelles, the availability of many small islands suitable for rehabilitation, the presence of a variety of environmental NGOs, and private island owners willing to develop ecotourism (see next section) has probably favoured the *in situ* approach to restoring small islands. However, in view of the limits of this strategy, and whilst continuing to move forward in the process of restoring other small and medium islands (particularly in the outer islands), more efficient restoration programmes now need to be further developed especially for the large islands. These should involve the

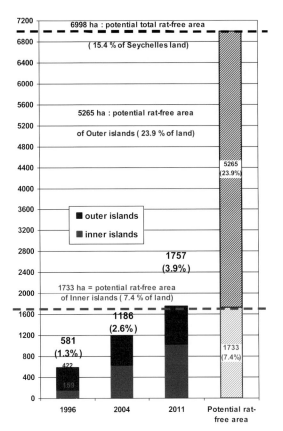

Figure 14: Evolution of the rat free area in the inner islands, the outer islands and the whole of Seychelles (reported in hectares and as % of the land surface). The total land surface that could potentially be freed of predatory alien mammals (rats and cats) is also indicated.

following: the permanent control – rather than the eradication – of invasive alien predatory verte-brates; the use of preda-tor fences in peninsulas or inland sites to create rat and cat free refuges; vegetation management to limit introduced inva-sive plants and favour native species; man-agement programmes for existing rare plants and animals; and where appropriate, conservation introductions of threat-ened species. The general concept would be to create *biological islands* – also called *mainland islands* – where ecological con-ditions are controlled to enable a number of native species to continue to thrive, and /or others to be reintroduced[26]. Such programmes are being conducted in the currently most advanced island countries/ territories in terms of IAS management such as New Zealand, Australia and Hawaii. For the time being, only the main breeding areas of the endangered Seychelles white-eye on Mahé (25 to 50 ha depending on years) have benefited from permanent rat control since 2006. This type of mainland island operation is also taking place over 30 ha on Ste Anne since 2011 (see p. 89). Developing such projects locally (see also Predator-fencing p. 150) could bring many of the rarest emblematic wildlife species of Seychelles, now tucked away in remote small island sanctuaries, back to the main islands where they once lived. This has been sometimes requested by members of the public arguing that these species, such as the Seychelles magpie-robin and other rare animals, are nowadays mainly enjoyed by private islands and foreign visitors. By providing a better access to these species and native wildlife in general, such mainland-island projects would be very beneficial to environmental education programmes for the general public and school children.

Other island territories of the western Indian Ocean (Mauritius, Rodrigues and La Réunion) are developing and conducting rehabilitation programmes on large conser-vation areas that are devoted to establishing such *mainland islands*. Some of these make remarkable use of Aldabra giant tortoises as ecological analogs to replace their own extinct native giant tortoises[27-29]. Similar operations should be conducted on a number of selected priority sites for conservation – in particular on Silhouette and Mahé. The development of innovative IAS management techniques – in particular to control invasive predatory species – will be key to the success of such rehabilitation programmes. Generally speaking, a more integrated *ecosystem approach* should be sought, aiming to rehabilitate entire habitats and communities (including those of invertebrates)[30], rather than focusing only on single flagship threatened species, as it has often been done until now.

References / Further reading

1 Russell & Holmes, 2015; 2 Varnham, 2010; 3 Keitt *et al.*, 2015; 4 DIISE, 2014; 5 Burbidge, 2011; 6 Harper *et al.*, 2015; 7 Wegmann *et al.*, 2008a; 2008b; 8 Daltry *et al.*, 2007; 9 Merton *et al.*, 2002; 10 Climo & Rocamora, 2006; 11 Griffiths *et al.*, 2011; 12 Holmes *et al.*, 2015; 13 Wegmann, 2008; 14 Wegmann *et al.*, 2011; 15 Senterre *et al.*, 2013; 16 Rocamora & Laboudallon, 2013; 17 Reuleaux *et al.*, 2015; 18 Safford & Hawkins, 2013; 19 MacArthur and Wilson,1967; 20 Blondel 1979, 1986; 21 Galman 2011; 22 Jolliffe *et al.*, 2008; 23 Henriette & Rocamora, 2011; 24 Henriette & Dine, 2014; 25 Jones & Merton, 2012; 26 Innes & Saunders, 2011; 27 Griffiths *et al.*, 2009; 28 Hansen *et al.* 2010; 29 Griffiths, 2014; 30 Kaiser-Bunbury *et al.* 2010; 31 Morris 2002; 32 Conservacion de Islas, 2015; 33 Samaniego *et al.*, 2015.

Mangroves are a major challenge and a limiting factor to eradicate rats from large tropical islands (Menai Island, Cosmoledo atoll).
Gérard Rocamora

Developing partnerships and ecotourism to fund ecosystem restoration

Control and eradication of invasive species has a cost (see p. 162), so financing mechanisms need to be established to support the costs of the restoration activities. Partnerships develop when different stakeholders whose interests converge decide to join forces to combine available expertise, manpower, equipment, or capacity to mobilise funds. Such partnerships can efficiently achieve good results and lower costs in the long-term process of ecosystem rehabilitation. The resulting synergies are an important asset with which to meet the technical and ecological challenges of these complex operations[1].

Seychelles provides some of the finest examples of collaborations developed between private island owners, parastatals, government agencies and NGOs in order to achieve successful control or eradication of invasive species and to develop ecosystem reha-bilitation programmes. These have been taking place mainly on small to medium sized islands, some privately and some government owned. The numerous conservation successes reviewed in the previous sections have in most cases been achieved through such collaborations[2, 3] (see also next case study p. 96).

The table below details the funding sources available to the different types of islands where invasive species management and rehabilitation activities are taking place in Seychelles. It shows that the revenue available to a majority of islands, irrespective of their ownership, management or protective status, comes mainly from tourism activi-ties, one of the main pillars of Seychelles economy[4]. This can be in the form of entrance fees from day visitors (Aride, Cousin, Grande Soeur, Curieuse), or revenue from small exclusive luxury resorts or larger hotels. Private islands or hotel operators that have established partnerships with NGOs may use their own private funds or also benefit indirectly (and sometimes directly) from programmes funded by international donors.

On government islands under the jurisdiction of the Island Development Com-pany where the Island Conservation Society (ICS) operates, island Foundations have been formally constituted whose membership includes the various partners (public, parastatal, private, NGO, local community, etc.) involved in the management of the environment of the island[5]. The partners, which usually include hotel and tourism operators, provide funding for conservation activities. They, in turn, benefit from sig-nificant in-kind contributions provided by ICS and other partners implementing the conservation activities on site. This model is intended to be progressively developed on remote islands with high biodiversity value, such as Cosmoledo atoll[6].

But there are numerous islands that have no tourism activities and therefore cannot count on regular revenue to fund the core operating costs of maintaining permanent conservation teams on the ground. This is a strong limitation. On large inhabited islands, ecosystem rehabilitation in key biodiversity areas is mainly supported by government-related bodies; but other sources of funding can also be made available. Private companies can make contributions through the government Environment Trust

Table 9: Funding sources available to the different types of islands in Seychelles where invasive species management and rehabilitation activities are taking place.

Protected area status corresponds to broad categories proposed under the Seychelles Protected Area Policy, in view of the forthcoming new Seychelles Protected Area legislation.

Ownership	Management	Protection status	Funding sources potentially available for rehabilitation activities	Islands concerned (*proposed by Government to be partly protected)
NGO	NGO	Nature Reserve (terrestrial/marine) or National Park (terrestrial/marine)	• Entrance fees and retail sales to tourists • Private & corporate donations (local & foreign) • Endowment fund • Bilateral donors (developed countries) • International donors (e.g. GEF, FFEM) • Corporate Social Responsibility (government tax) • In-kind contributions from volunteers, (University) students, NGO & private experts	• Aride • Cousin • Moyenne
Private	Private	None (some private islands are considering partly protected area status)	• Small island resort • Private funding • International or bilateral donors through NGOs • In-kind contributions from NGOs, University students, volunteers & private experts • Entrance fees	• Bird Island (Ile aux Vaches) • Cousine • Denis • Frégate • Grande Soeur & Petite Soeur - Ile du Nord (North Island)
Private	NGO	National Park (terrestrial/marine) (partly)	• Private funding • Endowment funds?	• D'Arros • St. Joseph
Government	Parastatal or Public trust	Nature Reserve (terrestrial/marine) or National Park (terrestrial/marine)	• Entrance fees (people & boats) • Private & corporate donations (local & foreign) • Bilateral donors (developed countries) • International donors (e.g. GEF, FFEM) • In-kind contributions from University students, volunteers & private experts	• Aldabra • Curieuse • Vallée de Mai
Government	Parastatal, private & NGO	Partly National Park (terrestrial/marine) or Protected land/ marine area	• Conservation levies on visitors • Hotel or small resort (+ diving, boat& fly-fishing) • Private villas • Endowment fund through island foundation • Contributions from IDC • International donors (e.g. GEF, FFEM) through NGO or island foundation • In-kind contributions from NGOs, University students, volunteers & private experts	• Silhouette • Alphonse group* • Desroches* • Assomption* • Cosmoledo* • Astove • Poivre* • Farquhar* • Ile aux Cocos • Ste Anne

Fund, and hotels and tourism operators may also collaborate with NGOs and provide funding to control invasive species in sites of special interest (mangroves, wetlands, etc.) or for the benefit of rare native species of animals or plants.

Since 2013, NGOs have also been able to benefit from Corporate Social Responsibility (CSR), a social income tax of 0.5% contributed by all middle-sized and large companies operating in Seychelles. Half of this tax can be directly donated to accredited trusts or charitable entities. Generally speaking, although long-term revenue to fund invasive species management and conservation activities in general appears to have been secured for many islands, an effort needs to be made to try and reduce the high dependence of existing funding mechanisms on tourism by developing alternative financial sources. The creation of a national endowment fund (Seychelles Conservation Trust Fund) is being envisaged with the support of the GEF, The Nature Conservancy and other international donors, with the aim of providing regular income to Seychelles protected areas. This project is linked to a national debt swab scheme, and one of the objectives is to extend the existing protected areas network to 50% of the national terrestrial territory and up to 30% of the marine Economic Exclusive Zone, in line with recent Government conservation commitments.

Collaborations are not always easy and never straightforward, especially when they involve multiple stakeholders with totally different backgrounds and constraints. These might include island and hotel managers, tourism operators, conservationists and scientists working together on the same island with logistical constraints. Inevitable tensions that may arise during any collaboration should be overcome by remembering how important and positive the collective gain for nature conservation will be. This applies in particular to challenging eradication operations with important human, logistic and financial demands. Such operations can be expected to have positive outcomes that will largely reward and compensate all partners for all the efforts involved. Partnerships need to be built and shaped over time until the best possible mutual benefits are achieved for each partner, just like a symbiosis between living creatures in the wild. Effective partnerships require good communication, compromises, and understanding in times of difficulties[1].

The successful record of Seychelles in regard to control of invasive species and ecological restoration of some of the smaller islands is attributable, at least in part, to the fact that these operations can be, if not always profitable, at least economically viable to private owners or investors willing to establish tourism operations and generate revenue[7]. High densities of rats often encountered on small tropical islands are totally incompatible with tourism; and where eradication and subsequent long-term exclusion are technically feasible they are more cost-effective than long-term control.

The progressive rehabilitation of an island ecosystem through elimination of exotic invasive plants, and the conservation introduction of rare and threatened native animals to recreate wildlife sanctuaries with their original fauna can be marketed as an attraction for visitors[8, 9]. It represents an added value and a powerful *raison d'être* for these private islands. Cousine, Frégate, Ile du Nord and Denis Islands are all good examples of successful programmes for which exclusive tourism and restoration of species and habitats have worked hand in hand[2, 10]. High densities of native wildlife such as giant millipedes, lizards, seabirds or landbirds may sometimes pose new problems for high end resorts on rehabilitated islands. However, these animals have always been in Seychelles; humans are the new-comers and they should adapt to the native fauna, rather than the reverse.

Touristic resorts have
become the main
economic activity in
some outer islands
(Desroches Island).
Henry Brink

Eco-tourists visiting
Aride Island
Nature Reserve.
Gérard Rocamora

References / Further reading

1 Rocamora, 2010a; 2 Government of Seychelles, 2011b; 3 Asconit & ICS, 2010a, 2010b; 4 Payet R.A. 2006;
5 Skerrett, 2010; 6 Rocamora & Payet 2002; 7 Nevill, 2004; 8 Rocamora, 2010b; 9 Samways *et al.*, 2010;
10 Kueffer *et al.*, 2013.

Rehabilitation of Island Ecosystems.
A programme combining invasive
species management, habitat
restoration, threatened species
recovery and capacity building

Contributed by **Gérard Rocamora** and the **Island Conservation Society** (ICS).

This programme, conducted during 2005-2012 with co-funding from the Fonds Français pour l'Environnement Mondial (French GEF), provides an example of a very diverse and comprehensive island restoration programme, which operated on nine different islands in Seychelles. It made a significant contribution to the fight against invasive alien species and to ecosystem rehabilitation by carrying out a dozen eradication or control operations on alien animals, substantial habitat restoration, capacity building, plus several conservation introductions of rare and threatened animals.

Among the most remarkable achievements was the eradication of rats from five islands and two islets. The total area of these sites (c. 450 ha) increased the rat free area of the inner islands at that time by about 50%. At each island, preventive measures to minimise the risks of reinvasions were designed and implemented. Rats were eradicated from Ile du Nord (North Island; 201 ha[a]), Conception (69 ha), Anonyme (10 ha) and Ile aux rats (1 ha) in the inner islands; and from Grand Ile (143 ha), Grand Polyte (21 ha) and Petit Polyte (1 ha) at Cosmoledo atoll, in the outer islands. In addition, rats have been kept under control since 2006 through a grid of bait stations and trapping in the two main breeding areas of the endangered Seychelles white-eye on Mahé (over 25 ha, on two properties owned by the Management of President of UAE Affairs Ltd.).

Measures to prevent reinvasion by rats were established on Ile du Nord and reinforced on Aride. Barn Owls were eradicated from both islands, and a first control programme of Common mynas – initially aiming at eradication but which had to be interrupted – was conducted on Ile du Nord. Feral cats disappeared from Grand Polyte apparently as a result of the 2007 rat eradication. Some feral cats survived on Grand Ile in 2008, but their eradication was incomplete due to problems of access to Cosmoledo and the threat of piracy[b]. In 2014, the rat-free status on these islands was reconfirmed and no cats were found also on Grand Ile despite several days of searches and use of feeding tables with sand-patches (see p. 235).

At Ile du Nord, the vegetation was partially or fully rehabilitated on a total of c. 45 ha of habitat. Most exotic trees and bushes were removed and about 14,000 indigenous saplings were planted, mainly by North Island Ltd personnel (40 ha) under the guidance of the Plant Conservation Action Group. In addition, ICS and DoE worked together to rehabilitate vegetation on Conception Island, to remove Coconut and plant native species; at Anonyme, to control Cocoplum;

Black mud terrapins were reintroduced to Ile du Nord and Aride Island. Gérard Rocamora / ICS

....

a.   at that time the largest high island (which excludes Ile Plate from Mauritius) in the western Indian Ocean/Afrotropical region where Black rats had been eradicated.

b.   on 27.03.08, the Indian Ocean Explorer, a ship that provided logistic support to our team on Cosmoledo was captured at sea by pirates, taken to Somalia and its crew members kept hostage for 88 days.

Three training courses opened to all organisations were conducted during the ICS FFEM project. Gérard Rocamora / ICS

and at Grand Polyte, Cosmoledo, to control Sisal. Vegetation management plans were produced for Ile du Nord and updated for Conception:

New populations of rare and threatened endemic animals were produced by the following island translocation projects: Seychelles white-eyes & Black mud terrapins to Ile du Nord; Seychelles leaf insects to Conception; Black mud terrapins to Aride; and Seychelles white-eyes to Cousine (in collaboration with another project). In preparation for these conservation introductions, several thousands of berry-producing trees were planted for white-eyes, and some marsh areas were rehabilitated for terrapins.

Contributions were obtained from the New Zealand Centre for Conservation Medicine for birds, NPTS and SOPTOM (*Station d'Observation pour la Protection des Tortues et de leurs Milieux*; France) for the terrapins. National Species Action Plans 2009-2013 were also produced for two threatened endemic birds—the Seychelles black parrot *Coracopsis barklyi* and the Seychelles white-eye *Zosterops modestus*.

Training courses that dealt with birds, plants and invertebrates and included practical training in the field were conducted and attended by about 35 staff members from various local organisations. An Island Restoration Workshop was also conducted. Local capacity was enhanced by enabling two university students (one Seychellois and one French) to earn diplomas, including a Masters degree from *Université de la Réunion*; a Doctorate (PhD) degree conservation introductions of the Seychelles white-eye *Zosterops modestus* at the *Muséum national d'histoire naturelle* (MNHN, Paris); and a PhD on the impact of rat eradications on invertebrate communities in rehabilitated islands.

Infrastructural improvements for wardening included a rat-proof room on Aride, a small field station and helipad on Conception, and a small boat and campsite at Cosmoledo. Efforts were also made to develop ecotourism and increase revenue for conservation.

Scientific monitoring programmes using standardised methodologies were set up for birds, reptiles, invertebrates and vegetation on each of the islands. Contributions were received from the ETH Geobotanical Institute of Zürich (Switzerland) and PCA. The first signs of recovery of the ecosystems on the rehabilitated islands were documented. Seabirds have increased in numbers and new species such as White-tailed tropicbird *Phaethon lepturus*, Wedge-tailed shearwater *Puffinus pacificus* and Masked booby *Sula dactylatra* have colonised the islands. Other native species including landbirds, reptiles and large invertebrates have also increased

A translocation attempt of captive reared Seychelles leaf-insects was conducted during the FFEM project. Gérard Rocamora / ICS

in abundance. These predator-free islands can further contribute to the conservation of the unique biodiversity of Seychelles by receiving additional transferred populations of threatened endemic animals that are unable to survive in the presence of invasive introduced predators. Future transfers may include emblematic species of the inner islands, such as the Seychelles magpie-robin *Copsychus sechellarum* or the Whip spider *Phrynicus scaber*. Outer island species may include a native turtle dove now restricted to a single island on Cosmoledo atoll, or the flightless rail that might be reintroduced from Aldabra.

Led by the Island Conservation Society – with North Island, Department of Environment and IDC also implementing activities or providing logistics – this programme involved a remarkably high number of 30 local and foreign partners (private island owners, companies, NGOs, governmental agencies and scientific institutions). The total budget (€1,500.00) included 30% co-funding provided by the French GEF (*Fonds Français pour l'Environnement Mondial*). Some targets could not be met on Cosmoledo atoll, but the project pioneered logistically complex operations such as a rat eradication using helicopter in the southern islands, located c. 1,200 km from Mahé.

This project created a momentum in Seychelles that encouraged island rehabilitation and the development of partnerships. It inspired for example other private islands, such as Grande Soeur and Petite Soeur where rats were subsequently eradicated in September 2010 by the same local team, to start their own restoration programmes. A regional project (*An Island Restoration Partnership for the Western Indian Ocean*) supported by UNDP, CEPA (*Conservation des Espèces et des Populations Animales*, France) and Fundació Miguel Torres (Catalonia-Spain) was proposed by the Seychelles government and accepted by the Indian Ocean Commission in 2012, and is still waiting to be funded.

Electronic brochures in French and English, and more details on the ICS FFEM Rehabilitation of Island Ecosystems project are available at www.islandconservationseychelles.com/ffem.html

References / Further reading
Rocamora & Jean-Louis, 2009; Asconit & ICS, 2010a (English version); Asconit & ICS, 2010b (French version).

Partnership is the key to success, particularly for invasive species management

The ICS FFEM project increased by nearly 50% the rat-free area of the inner islands, and more than tripled this area in Cosmoledo atoll.
Gérard Rocamora/Helicopter Seychelles

More than 15,000 native sapplings were propagated and about 40 ha rehabilitated by the North Island Environment department team.
Gérard Rocamora

The Seychelles white-eye was dowlisted from Critically Endangered to Endangered following the successful translocation from Conception to North Island.
Mike Meyers/North Island

Preventing new invasions and defining priorities for action

Chapter *3*

On North Island, preventive measures to prevent colonisation by invasive species include the unloading of cargo into a pest-proof trailer that is brought into a pest-proof room.
Marc Stickler

This chapter focuses on the importance of preventing new biological invasions, and to define priorities for action for invasions that are already happening.

The impacts of introduced invasive species and of human activities in general on the global environment need to be managed sustainably and rationally so that the eco-systems, upon which life on earth depends, including our own existence, continue to function properly. The negative and potentially devastating effects of invasive species on our economies and health also need to be actively countered.

There are a number of ways to fight against the spread of invasive species in order to minimise, and where possible, reverse their impact upon our natural environment. These include not only direct interventions to limit the distribution, abundance and impacts of invasive alien species, but also prevention, mitigation and educational measures. All these are needed for an effective integrated management strategy (see p. 160).

SOME FUNDAMENTAL PRINCIPLES OF INVASIVE ALIEN SPECIES MANAGEMENT

Inspired by the Global Invasive Species Programme IAS toolkit[1], the following basic principles have been adopted by a number of institutions including the Secretariat of the Pacific Regional Environment Programme. We have adapted them to the situation in Seychelles:

- **Prevention is better than cure:** preventing colonisation is more effective and cheaper than management measures after invasives have become established. Border control and strict abatement measures must be the first line of defence.
- **Early detection & rapid reaction:** When preventive measures fail, locating the invaders before they have a chance to establish and spread is key to their successful eradication.
- **Eradication** is more effective and cheaper in the long run than permanent control of a pest population. So it should always be considered as the first response to a past population wherever feasible.
- **Biological control** should be envisaged for those species that cannot be eradicated by conventional means.
- **Containment** within delimited areas should be implemented against species for which eradication – or biological control – is not deemed feasible or appropriate.
- **Permanent control** of an established pest population by chemical and/or physical methods should be considered where eradication and containment are all deemed not feasible with current or achievable resources.
- **Mitigation of impacts** is the last resort to "*live with*" an invasive species when eradication, containment, and control cannot be undertaken, have failed, or are not efficient enough. Actions typically focus on affected native species or habitat management rather than on the invasive species itself, in order to reduce the severity of its impacts.
- **Principle of precaution:** In cases of uncertainty and insufficient scientific knowledge to accurately assess either the risk of a species becoming invasive, or its present or future impact, one should assume that impacts will occur and action should be taken to prevent the species from spreading or becoming established.
- **Science oriented:** Risk assessment, prioritisation and management of invasive species must be based on sound science to maximise effectiveness and value for money.
- **Prioritisation:** Not all introduced species become invasive, and action should be first oriented towards those currently causing, or having the potential to cause, the most harm.

References / Further reading

1 Wittenberg, R. and Cock, M.J.W. (eds.) 2001. Invasive Alien Species: A Toolkit of Best Prevention and Management Practices. CAB International, Wallingford, Oxon, UK, xii – 228.

Prevention is better than cure: Acting on entry points and pathways

Prevention is the most efficient and cost-effective way of fighting invasive species[1, 2a]. It consists in limiting introduction and spread of invasive species through strict and adequate regulations, sensitisation and the development of best practices. Some countries, such as Australia, have put a lot of effort into prevention, and this includes proactive public education and communication strategies.

Introductions of alien species are increasing with the growing volume of trade, broadening of trading partners, and increased travel and tourism that accompany globalisation[3]. A large number of invasive alien species have not yet reached Seychelles, and preventive measures need to be enhanced in view of the serious threats that they pose to Seychelles (biodiversity, economy, human and animal health, etc.)[4].

Exotic species are often introduced voluntarily for commercial agriculture or horticulture, ornamental purposes or the pet trade. Whilst the expected benefits of these introductions are largely for private interests, if the species becomes invasive the cost will usually be borne by the entire community. In order to limit the number of new alien plants and animals imported, the principle of responsibility of the importer in case the imported species becomes invasive should be adopted. In addition, a risk analysis of potential invasiveness should be done for every intentional introduction proposed, and it should include an analysis of costs and benefits[2b, 3]. Unfortunately, voluntary introductions are not always done legally and smuggling of plants, animals, seeds, fruits, vegetables or animal products is a serious problem worldwide. In 2012, an X-ray machine capable of detecting organic matter was installed at the international airport, and an operational manual and a booklet were produced for customs.

Accidental introductions often concern plants (seeds), terrestrial, freshwater and marine invertebrates or pathogens. A list of the main pathways for invasive species to arrive in Seychelles is given on p. 35; the main entry points are the harbours and airports where most of the trade, transport and tourism between Seychelles and the rest of the world occur.

Article 8(h) of the Convention on Biological Diversity (CBD) states that *"Each Contracting Party shall, as far as possible and as appropriate: ...(h) Prevent the introduction of, control or eradicate those alien species which threaten ecosystems, habitats or species."* Prevention of the introduction of alien species should be the first goal.

Border control and quarantine measures are the first line of defence to limit the risk of introduction of new invasive species[2b]. For it to be effective, appropriate legislation must be in place. Positive and negative lists (i.e. *'Black lists'* and *'White lists'* for plants), indicating species banned for importation and those allowed in certain conditions, need to be defined[5, 6]. Identification tools such as manuals, databases, and protocols for surveillance and risk analysis (see below) are also required. In New Zealand, the Department of Conservation has produced an Island Biosecurity Best Practice Manual[7]. In Alaska, for example, the website stoprats.com, put together by a wide

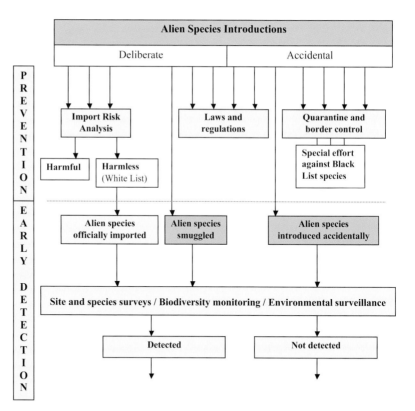

Figure 15: Management options to deal with alien species introductions (after Wittenberg and Cock, 2001). The prevention scheme will result in three different groups of species (authorized, smuggled and accidentally introduced) passing through it and entering the country. Yellow boxes: areas where we can and should act. Pink boxes: areas out of our control.

range of partner organisations, provides detailed protocols and education materials to prevent the spread of rats. **Chemical treatments** (e.g. with methyl bromide or other strong pesticides) are often applied to imported commodities (such as fruits and vegetables) suspected to be contaminated with non-native organisms, or in a systematic preventive way. Their objective is the complete removal of all propagules of all species, with no danger to human health. The systematic fumigation with insecticides of aircraft before take-off is a well known preventive measure to avoid the arrival of new invasive invertebrates.

Preventing inter-island spread of invasive alien species is also of crucial importance for Seychelles. The above principles of prevention also apply at the level of individual islands, particularly those of high conservation value, where managers have invested in a process of ecosystem restoration and dedicated much effort to remove prominent invasives such as rats, cats, mynas or barn owls. Vigilance must remain high to prevent reinvasion of species that have been successfully eradicated or invasion from new invasive species. This requires constant effort and discipline. Priority lists of species deserving special attention and recommended biosecurity protocols for inter-island transportation of vessels, people and goods have being prepared as part of the GoS-UNDP-GEF Biosecurity Project. This includes in particular transportation to protected areas and islands of high biodiversity value [13] (available from www.pcusey.sc).

Biosecurity protocols for protected areas and islands of high biodiversity value should involve three main types of activities (see for example the Aldabra Biosecurity Plan[11]; or biosecurity measures recommended for Aride Island Nature Reserve[13], partly inspired from those recommended for Ile Ronde in Mauritius [14]).

1. Abatement (preventive) measures and 'quarantine' protocols:

WHY DO WE NEED BIOSECURITY PROTOCOLS?

- To prevent introduction of invasive species

'Prevention is better than cure': The risk of introduction can never be totally eliminated; but it can be significantly reduced.

- To prevent survival and establishment of invasive species

'Eliminate invaders quickly'. Once we know invaders have arrived they must be quickly eliminated. This requires improving both detection and monitoring methods, and also eradication techniques.

- Visitors rules (e.g. biosecurity checklist for visitors)
- Inspection and storage prior to transportation: Quarantine room management
- Vessel and loading procedures at island of departure
- Unloading procedures at island of arrival
- Pest-proof room management and procedures at island of arrival
- Special procedures for transportation of construction material, wood and furniture, heavy machinery and equipment
- Importation of plants and animals (for food or farming)
- Translocation of wild species of plants or animals (native and non-native)

2. Surveillance protocols:
- Regular survey of potential entry points (boat landing sites, helipad, airstrip, etc.) and paths used by visitors and island staff in search of new exotic plant species.
- Surveillance kits comprising traps, lures and other detection devices, and manuals to identify invasive species.
- Maintenance and refilling of all bait stations, traps, gnaw-sticks for rats, etc.

3. Emergency procedures in case of an invasive species incursion:
- Incursion response plans describing what to do first, details of resource persons to contact, etc. These plans require clear identification of duties and responsibilities and are normally coordinated by the Biosecurity Officer.
- Incursion kits containing the necessary equipment to provide an initial response to an invasive species report, first to confirm it and then to attempt eradication. The kit can be divided into sections (for example to be used for rodent, bird and invertebrate invasions).

The sooner that biosecurity protocols are implemented in the face of a potential invasion, the higher the chances of intercepting and stopping the invasive species. In other words, it will be much easier to detect and stop invasive plants or animals hidden in goods and commodities while still on Mahé, than during transportation or after the cargo has reached its destination island. This means that it is worth concentrating biosecurity efforts on mainlands or islands that are logistical hubs, like Mahé or Praslin. Biosecurity documents are dynamic and require frequent revision and updating, and recommended procedures often involve checklists and crosschecks. Information and training on invasive species identification, surveillance protocols and emergency procedures need to be provided to island staff and residents on a regular basis.

The strict application of **abatement measures** (i.e. preventive measures to reduce the risk of invasion) such as safe loading, landing and unloading procedures for boats supplying these islands, is key to the preservation of a significant part of Seychelles's biodiversity. Hence all goods and commodities imported need to be subject to the following procedures:

- First, be inspected and screened for any seeds or animals present (including small insects). Chemical treatments may be required to ensure that all cargo is pest-free;
- Be sealed and stored in a pest-proof room on Mahé. Recommended containers to transport goods include plastic boxes and barrels, fish bins, poly-pail type buckets with sealable lids, plastic drums / barrels with screw-on lids[12] ('touque' in French), and heavy PVC water-proof bags. Container with ill-fitting lids, cracks or holes, sealed cardboard boxes and plastic bags are no longer considered appropriate[10];
- Be loaded onto a vessel that is kept permanently rat and pest free. It is essential that piers and jetties for supply boats, and plane or helicopter hangars are kept as pest-free as possible;
- Upon arrival, be unloaded inside a pest-proof room (or a pest-proof trailer that will bring the cargo to the pest-proof room); see Annex for example of procedures).

All rat-free islands, especially those inhabited, need to have permanent bait stations with rodenticide, traps and gnawsticks to prevent reinfestation or to detect it immediately should it occur. When large amounts of cargo need to be brought onto the island (e.g. for a construction), close inspection and fumigation of all cargo inside a sealed container by local pest control companies (e.g. using methyl bromide for 72 hours) is recommended to avoid any unwanted introduction.

In that context, the current absence of legal clauses preventing anyone from landing on the beaches of these islands without restriction or control – as permitted under the constitution of Seychelles – is counterproductive. It puts at risk the outstanding biodiversity of these islands, and the efforts invested over many years by the owners/ managers and their local and foreign partners (NGOs, private experts, Division of Environment, etc.). It is hoped that this problem can be solved under the new **Seychelles Biosecurity Act** in preparation. Continued public debate and outreach campaign about the need to establish and enforce measures that prevent the spread of invasive species is highly desirable.

Risk assessment is a process used to assess the invasiveness of particular species, in order to predict how serious their negative repercussions may be. It may be applied to species that have not yet reached the country, or to species already present and suspected of having the potential to become invasive. The threat associated with an invasive alien species greatly depends on the species itself, the extent of the invasion, and the vulnerability of the ecosystem being invaded. Since it is difficult to predict the behaviour of a species based only on what happened under different circumstances at other islands or countries, there is always a significant uncertainty and subjectivity in such predictions[1,8]. Evaluating the probability that a species may become invasive, the uncertainty of this assessment and the magnitude or cost of the potential damage can also differ significantly between experts depending of their field of expertise (e.g. a conservationist versus an agriculturalist). However, this exercise provides a logical process that facilitates decision making about management actions that may be needed.

Certain models based on the climatic and other physical conditions prevailing in the native – or in the already invaded – range of the species can also be used (e.g. Climex, MaxEnt, MaxLike); but these rarely take into account interaction with other species (predators, competitors or pathogens) or possible adaptation, which limits their interest[8]. Risk assessment should also consider climatic change predictions. The risk assessment exercise has also been extended to assess the various pathways and define the most dangerous for the introduction of new invasive species, and to recommend preventive measures. Its main merit is to consider all possible risks linked to a particular pathway.

Risk assessment should be conducted for all alien plants and animals for which importation into the country is requested, to address potential invasiveness and other relevant risks associated with the proposed introduction. Detailed restrictive procedures have been set up in many island countries and territories such as Australia, New Zealand or Hawaii[2b]. Identification of known and potential invasive species, including information on their pathways of introduction and spread is critical.

The Aichi Biodiversity Target 9 recommends nations to a) identify and prioritise Invasive Alien Species for their management and b) identify and prioritise pathways of introduction in order to implement measures to manage these pathways. In Seychelles, a national database was set up as part of the GoS-UNDP-GEF Biosecurity Project[9] and preliminary risk assessment has been conducted for a number of invasive species and pathways under the GoS-UNDP-GEF Biosecurity project, based on the likelihood of introduction, establishment and spread, and the magnitude of potential damage[3,4].

Risk analysis is a more general concept; it can be considered as a systematic approach to decision making regarding the impact / threat of alien species through hazard identification, risk assessment, risk management, and risk communication. Pest risk management is the evaluation and selection of options to reduce the risk of introduction and spread of a pest[10].

The IUCN SSC Invasive Species Specialist Group (ISSG) is leading the development of the Global Register of Introduced and Invasive Species (GRIIS) within the framework of the Global Invasive Alien Species Information Partnership (GIASIPartnership), a CBD supported initiative. GRIIS provides compiled verified country inventories of known invasive alien and introduced species. ISSG and partners have also been encouraged to work further on the development of a classification/ranking of alien and invasive species based on the nature and magnitude of their impacts[11]. Certain countries like Australia and the United States have produced specific manuals for identification of potential '*black listed*' invasive species. See for example the Australian Weed Risk Assessment (AWRA), and the Hawaii Pacific Weed Risk Assessment[12, 13]. The AWRA is a series of 49 questions on the history, range and biology of candidate species for introduction. Each answer is given a mark between –3 to +5, and the total score for all answers determines whether the species is accepted for importation, rejected, or if more information is requested to evaluate its invasive potential[8].

Public awareness and education about invasive alien species should also be a key component of prevention efforts.

References / Further reading
1 Wittenberg & Cock, 2001; 2a SCBD, 2002; 2b Shine, 2008; 3 Dogley W., 2009; 4 Ikin & Dogley, 2009; 5 Soubeyran, 2010; 6 Kueffer, 2011; 7 DoC, 2003-2010; 8 Simberloff, 2014; 9 Nevill, 2009; 10 IPPC / FAO, 2004, 2005a, 2007; 11 Harper, 2014; 12 Tatayah *et al.*, 2007; 13 Rocamora, 2015; 14 Merton et al. 1989; 15 see Draft decisions from COP 12 <http://www.cbd.int/doc/meetings/cop/cop-12/insession/cop-12-wg2-crp-04-en.pdf; 16 see https://sites.google.com/site/weedriskassessment/home; 17 see http://www.cbd.int/doc/meetings/cop/cop-12/insession/cop-12-wg2-crp-03-en.pdf; http://www.cbd.int/doc/meetings/cop/cop-12/insession/cop-12-wg2-crp-04-en.pdf

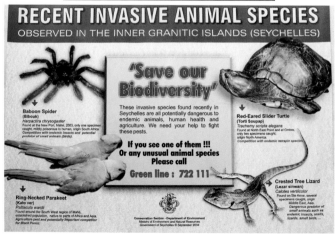

Public sensitisation is key to detect new invasive species early and prevent them to become established

Early detection and rapid reaction

Whenever preventive measures – or 'abatement' measures as often used in the case of a particular island – such as border control have failed in preventing the entry of a certain invasive species into the country, early detection and rapid reaction are essential to prevent its establishment or its spread, and to have a chance to eradicate it at minimal cost[1].

This generally requires an efficient and **operational system of surveillance** across the national territory (for example like ONF has set up in La Réunion[2]), which does not exist yet in Seychelles. In the case of particular islands, there is need for strict abatement measures to prevent the infestation of new invasive species – or the reinvasion of those that have already been eradicated. Such a system is already in place at several private islands of high biodiversity value, including Aride and Cousin Nature Reserves; Cousine; Frégate; Ile du Nord and Denis.

Priority species for early detection should include invasives already present in neighbouring countries or islands that are known to have a high impact on the environment, the economy (agricultural pests, animal diseases), or human health (e.g. mosquitoes, rodents, etc.)[3a, 3b]. Early detection is especially important when these invasives are known to be difficult to control. Such a list of priority species for Seychelles is currently being established as part as of the Biosecurity project and should be available online by mid-2015 (www.pcusey.sc).

Good communication and training, and coordination of efforts are key to the success of early detection (see for example in French overseas territories or in Hawaii[2, 4]). A network of well trained and informed persons is indispensable and needs to be developed in Seychelles. Information and training will need to include personnel responsible for undertaking surveys, but also professionals (conservation officers, rangers, farmers, tour guides, dive instructors, etc.). **Sensitisation and education** activities about the problem of invasive species need also to be directed to young people and the public in general through the media, displays, programmes in school, brochures and publications like this one. We need to develop a general attitude of national responsibility of permanent alert not only from all concerned professionals and stakeholders, but from citizens of Seychelles in general.

Surveys should be organised periodically to look for new invasive species, or may be focused on groups of particular concern (e.g. mosquitoes, coccids, woody plants, weeds, etc.)[2]. This implies the availability of well trained personnel with sufficient knowledge about identification and sampling techniques for the various taxonomic groups, knowing where and how to look for particular priority species. Specific surveys are required regularly on islands or areas of high conservation value (e.g. hosting endangered species), in sensitive habitats (wetlands) or near entry points (airports, harbours).

Rapid response is possible only if the area concerned by the invasion is limited. Shortening the delay between the time when the presence of a new alien is reported, its identity revealed and its invasion risks assessed, and the moment when effective ground action is undertaken to remove it is key to the success of the operation[2, 3].

In cases in which the potential invasiveness of a newly detected alien is already known or can be confirmed rapidly, and if the methods of eliminating it are well known, its eradication should be undertaken immediately.

The establishment of **contingency plans**[5] (i.e. urgent response plans, see also definitions on p. 16) facilitates decisions and communication, and will ensure that everything required in terms of actions and resources will be in place to eradicate or bring the pest species under control[1, 6]. Anticipating the likely arrival – or the reinfestation – of a common widespread invasive (i.e. rat, mouse, Spiralling white-fly, etc.) on islands presently free of these pests is highly recommended. These plans should indicate:
- How to identify the invasive species.
- An estimation of its potential impacts.
- Urgent actions to undertake, and the relevant methods to contain, control or eradicate the species.
- Resource individuals and organisations to be contacted
- Financial and human resources required, and how to mobilise them.

In Seychelles, following a series of workshops focused on Biosecurity, a 'General Emergency Response Plan for Pests, Diseases and Alien Invasive Species' was prepared in 2014 so that a rapid response can be made and coordinated between the different government departments and other stakeholders of private sector and civil society involved (see reference and website link[10]).

The Department of Conservation of New Zealand has produced various such documents for islands with high biodiversity value and these documents are accessible from the internet. Templates for contingency plans for rodents can be found in the toolkit for the eradication of rats and cats of the Pacific Invasives Initiative[7] (see p. 140). Useful guidelines and templates can also be found in a similar toolkit for Invasive Plant Management produced by the PII[8].

In the future, modern technologies should improve and facilitate the early detection of invasive species. Remote sensing using aircrafts or satellite imagery, such as 'LiDAR' (Light Detection and Ranging) and hyperspectral imagery, makes it possible to detect the presence of certain invasives in the middle of native plants, depending on the spatial resolution and spectrum-colour resolutions available, both of which are likely to increase significantly in the near future. Genomic techniques like sequencing of nucleic acids are becoming increasingly used to verify with certainty the identity of a particular species, and this can be used to detect the presence of certain invasives at low densities, for example by analysing DNA in faecal samples or regurgitation pellets; the same may be used to detect a hybridisation process between an alien species and a native one (see p. 49)[9].

References / Further reading
1 Wittenberg & Cock, 2001; 2 Soubeyran, 2010; 3a SCBD, 2002; 3b Shine, 2008; 4 Kueffer, 2011; 5 DoC, 2003-2010; 6 Ikin & Dogley, 2009; 7 PII, 2011a; 8 PII, 2012; 9 Simberloff, 2014; 10 Government of Seychelles, 2014 (available from http://www.pcusey.sc/index.php/pcu-projects/repository/BS/Project-Technical-Reports/Outcome2/TechReports2/General-Emergency-Response-Plan-for-Pests-Diseases-and-Alien-Invasive-Species-2014).

Establishing priorities for action: What to fight against first?

Management actions required for invasive species control are often numerous and tend to be constantly increasing, whereas human and financial resources are limited and need to be optimised. It is therefore crucial to be able to prioritise those situations when intervention is most needed.

DECIDING ON THE MOST IMPORTANT SPECIES TO TARGET

Several important factors should be taken into account[1,2]:

<u>Current or potential extent:</u> Invasive species that are recent colonisers or present over restricted areas should take priority over those that have already become naturalised, become truly invasive and cover large extensions.

<u>Current or potential impact:</u> Target those species first that are known to have the highest impacts.

<u>Value of habitats infested or potentially infested</u>: Measures should be taken first in those habitats having a high ecological, conservation, economic, social or historic value.

<u>Difficulty of control/eradication</u>: Species that can be eradicated or efficiently controlled with the available resources should take priority over those that will be difficult to control due to insufficient technical, financial, human or logistical resources. This means assessing the efficiency of known control methods, and prioritising taxa for which the prognosis of success is highest.

These factors can be used as filters to screen out the worst pest, and it is recommended that the factor 'current or potential extent' be used first[1]. Species can be ranked for each factor (ignoring factors that do not apply to the site considered), and overall rankings for priority of action combining the different factors can be derived. A more detailed, priority-setting system for weeds is presented in the *Handbook for Ranking Exotic Plants for Management and Control*[3], that combines significance of impact (current impact and ability of the species to become a pest), and feasibility of control or management.

In summary, deal first with species that:
- Are localised and have not yet dispersed, hence could become a serious future threat.
- May affect habitats of high value (hence with a high potential impact)
- Are in a situation (i.e. on a small or medium-sized island) such that their eradication or efficient control is deemed feasible.

DECIDE ON WHERE TO ACT FIRST

Action should be favoured where it is likely to be the most efficient and where it is indispensable to reach the desired objectives (conservation, economic or other). Experience

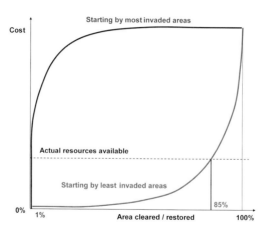

Figure 16: Efficiency and cost effectiveness of removing invasives by clearing the least infested areas first . At the same cost, 85% of the least infested area was cleared compared to 1% of a highly infested area (Brondeau & Triolo, 2007).

shows that it is more efficient and cost effective to concentrate efforts in areas that are less infested and therefore easier to restore, rather than start in areas that are the most infested[4]. Generally speaking, when eradicating invasive plants, it is best to work from the perimeter towards the core of the infestation, as this limits spread[2]. This was exemplified in La Réunion by the *Office national des Forêts*, where the same available resources allowed clearing either 85% of a less infested site compared to only 1% of the most infested areas[5] (See Fig. 16). Other criteria used in La Réunion to select sites to be restored include the conservation value (presence of rare and threatened natural habitats and species), and the accessibility of sites (for sites of similar conservation value)[6]. The cultural value of sites may also be taken into consideration.

Nature reserves or other conservation areas where national biodiversity treasures are at stake must be given priority for invasive species management and restoration work. In Seychelles, restoring small islands by eradicating introduced predators and replacing invasive plants by native vegetation should continue to be seen as a national priority as these islands can then become new refuges for endangered endemic species that have nowhere else to go.

TAKE NATIONAL PRIORITIES INTO ACCOUNT

National priorities also need to be taken into account when planning local actions against invasive species. The Seychelles' Biosecurity Strategy[7] (see p. 160), contains a general framework for action but does not have an Action Plan that prioritises specific actions. The National Biodiversity Strategy and Action Plan (2011-2020)[8] is filling part of this gap.

National priorities for the management of invasive plants and animals need to be defined more precisely in Seychelles. For example, tables listing invasive plants and animals present in Seychelles that indicate whether they are candidates for rapid response, eradication, suppression, biocontrol, etc, would be a useful tool. These should include an assessment of the potential for eradication, containment or control of the different species; and it could guide the establishment of Vegetation Management plans for particular islands or protected areas. Unfortunately, production of such tables is beyond the scope of this book (but see Tables 12-16 on most problematic species in Seychelles, p. 179-184).

In Mauritius, the Invasive Alien Species Strategy and Action Plan establishes three levels of priority (Critical, High and Medium) and lists actions within five hierarchical *Management Elements* and five *Cross-Cutting Elements*[9]. In La Réunion, a collaborative strategy of action against invasive species has been produced as well as a guide for the ecological restoration of native vegetation[10].

THE NEED FOR MORE REGIONAL COOPERATION

Enhanced coordination and international cooperation to counter invasive species are now required. This objective had been declared in past regional workshops for both the western Indian Ocean islands (Mahé, 2003)[11] and the southern African region (Lusaka, 2003)[12]. In 2012, a four year European Union project implemented by IUCN in collaboration with the Indian Ocean Commission started with the aim to enhance systems and strategies to efficiently prevent and manage biological invasions on islands. This included the development of regional cooperation within the western Indian Ocean with the support of key organisations from the Pacific islands. The existing Pacific email network and information exchange platform coordinated by the Pacific Invasives Initiative and the University of Auckland has been extended to the western Indian Ocean and a number of specialised training workshops have been conducted with participants from the various countries. A regional newsletter on IAS has been started and a 'Manual for Prevention and Management of Biological Invasions on Islands' is being prepared. A best practice guide is expected to be produced that will integrate lessons learned by adapting policies and protocols from the Pacific and testing them in pilot sites across our own region.

However, there is still scope to further develop regional cooperation between western Indian Ocean islands to disseminate and build upon local scientific knowledge and technical expertise about invasive species. In particular, technical knowledge that exists in Mauritius, La Réunion and Seychelles now needs to be shared with Madagascar and the Comores to develop capacity. Further networking and joint training, the development of collaborative operations and projects with contributions from different countries, and joint publications to share results and lessons learned should be promoted and facilitated by the invasive alien species component of the biodiversity strategy of the Indian Ocean Commission. There is a need for more concerted action on invasives species at regional level, not only to fight more efficiently parasites and diseases affecting agriculture and human health in the region, but also to prevent the entry and spread within the region of invasives plants, animals and diseases that represent a high danger to the environment of the whole western Indian Ocean biodiversity hotspot (e.g. chytrid fungus[13]). A regional strategy specific to invasive species, and the idea of regional priorities, should be promoted.

References / Further reading
1 Wittenberg & Cock, 2001; 2 Luken & Thieret, 1997; 3 Hiebert & Stubbendieck, 1993; 4 Soubeyran, 2010; 5 Brondeau & Triolo, 2007; 6 Triolo, 2005; 7 Government of Seychelles, 2011a; 8 Prescott *et al.*, 2013; 9 Government of Mauritius, 2010; 10 Anonyme, 2010; 11 McDonald *et al.*, 2003; 12 Mauremootoo, 2003; 13 Labisko *et al.*, 2015.

CASE STUDY The Melon fruit fly, a challenge for Seychelles agriculture

Contributed by **Will Dogley**, Seychelles Agricultural Agency.

A high number of invasive alien species are agricultural pests. The Melon fruit fly (*Bactrocera cucurbitae*) is a good example of a recent biological invasion which has caused substantial economic damages. Native to Asia, it first colonised Mauritius and La Réunion, and was first observed on Mahé in November 1999, near the airport in a surveillance trap. It probably had been introduced by means of an in-flight meal containing fresh fruit or vegetables that had not been properly disposed of. By 2003, it had colonised most of the granitic islands[1].

The Melon fly is considered the most destructive pest of melons and related crops; and has been recorded on more than 125 species of host plants, including cucurbits, tomatoes, and many other vegetables[2]. This rapid invasion was considered a 'national disaster' since all cucurbits – by far the biggest agricultural vegetal production in the country- were severely affected. The production of melon and watermelon (the more sensitive species) became almost impossible, while other somewhat more resistant species such as cucumbers, pumpkins, squash, chayote, zucchini, experienced severe decreases in yields. In late 2000 the economic impact of the fly was estimated at US$2.2 million per year with a loss of up to 60% in the production of the most important cucurbit crops (e.g. pumpkin, cucumber, melon, snake gourd, bitter gourd, calabash gourd)[1]. By then many farmers had decided to abandon these crops altogether because of the high level of damage during production and after harvest.

After a number of trials to control the species in Seychelles[3], the Melon fruit fly was subjected to an eradication attempt that used a combination of bait spraying and male annihilation (MAT) blocks. Control methods that can be used effectively by farmers and gardeners in Seychelles include sanitation (destruction of infested, fallen and overripe fruits, and the proper disposal of crop residues); pheromone traps with Cuelure for monitoring and mass trapping (see trap p. 161); bait sprays (Bait Application Techniques, BAT) with Protein Hydrolysate liquid attractants plus Malathion; use of baited traps or Male Annihilation Technique (MAT) blocks; and wrapping fruits with a protective covering or bags, among other techniques. A new control project in collaboration with the IAEA (International Atomic Energy Agency) envisages the use of the Sterile Insect Technique, in which male insects are mass reared, exposed to radiation and rendered sexually sterile, and then mass released to mate with native females in order to prevent reproduction[4]. Considerable additional research may be required before a method using biological agents can be found, as no method has been reported to be effective so far[3].

Although the eradication programme did not succeed in eradicating the pest before it had established itself and spread to other islands, it nevertheless managed to keep the fruit fly population relatively low, and reduced damage to crops. Moreover the capacity building programmes associated with the eradication programme led to positive socio-economic impacts such as an improvement in the quantity and quality of locally produced crops, better prices for the consumers, a reduction in imports of certain cucurbit crops and higher incomes for farmers. The project also led to increased awareness of modern and improved farming practices, and a renewed motivation to invest in cucurbit crops.

Capacity building included various training sessions for technicians, farmers, gardeners and other key stakeholders, and many valuable manuals, brochures, posters, leaflets, and other publicity materials were produced. An emergency response system has been established to protect the country from other unwanted invasive species, with a network of traps, two incinerators and trained personnel able to fight future invasions by quarantine pests. The pest control methods adopted have been geared to be as environmentally-friendly as possible, and an integrated pest management approach combining preventive measures was put into practice. This involves preventive management practices, appropriate insecticide application including a continuation of the MAT and BAT techniques, and a trapping network to detect newly infested areas[3]. Preventive measures such as wrapping developing fruits in paper bags or polystyrene sleeves, harvesting of early mature green fruits, or burying three feet deep all infested fruits and crops residues, can be effective in controlling the infestation[5]. Integrated pest management is so far the best available option to control the Melon fruit fly. Sustainability and effectiveness of this project will depend on continued government support as well as on partnership between public and private sectors. Lessons learnt could be replicated in other similar situations or adapted to new contexts, particularly in Small Island Developing States (SIDS).

Melon fruit fly identification and biology[2].

Adult: 6 to 8 mm in length. Distinctive wing pattern, long third antennal segment, dorsum of the thorax is reddish yellow with light yellow markings and without black markings, and the head is yellowish with black spots. **Egg:** Pure white, about 2 mm long, elliptical, nearly flat on the ventral surface, more convex on the dorsal, often somewhat curved. **Larva:** White, but may appear coloured by the food within the alimentary canal; has a cylindrical-maggot shape; 7.5 to 12 mm in length (last instar larvae). **Pupa:** 5 to 6 mm in length, dull red or brownish yellow to dull white, according to host.

Life cycle: Development cycle is 12 to 28 days depending on host and weather conditions. The female lays up to 1,000 eggs in young fruit and also in succulent stems or in ripe fruits of certain host plants. Pupation occurs 5 cm deep into the soil, beneath the infested fruit. Adults may live more than a year, feeding upon juices of host plants, nectar, and honeydew from various insects. Produces up to 10 generations per year.

References / Further reading
1 Dogley W., 2007; 2 Dhillon *et al.*, 2005; 3 Beaver & Mougal, 2009; 4 Knight, 2008; 5 Dogley W., 2004.

Adult Melon fruit fly on fruit. Antoine Franck (CIRAD)

How to eliminate invasive alien species

Chapter *4*

Black rats were eradicated in 2007 from Grande
Ile, Grande Polyte and Petit Polyte on Cosmoledo
atoll, and Sisal was controled on Grand Polyte.
Colin Bell

Physical, chemical and biological methods

This chapter presents the main methods used to eliminate invasive species, and describes available techniques to euthanise animals in a way that is considered humane.

There are three main types of methods that can be used to eliminate invasive species: physical/mechanical, chemical or biological. These different techniques can be used one after the other, or simultaneously, during both control and eradication programmes. Details on the methods mentioned in this chapter can be found in the second part of the book (species accounts).

PHYSICAL OR MECHANICAL CONTROL

Physical or mechanical methods consist in physically or mechanically removing invasive plants or animals to reduce their abundance, and when possible, to eradicate them. Physical control generally implies manual action and the use of simple unpowered tools, whereas mechanical means the use of engine-powered tools.

They include the destruction of exotic plants or trees by hand pulling, cutting or the use of machinery, the felling or ring-barking of trees, handpicking of snails, insect trapping (using luminescent or pheromone traps), shooting or trapping animals such as rats or birds, or removal of host plants (as is done for the Takamaka Wilt disease)[4, 12]. A method, called 'soil solarisation' or 'solar heating', used in agriculture for soil disinfestation against fungi, bacteria, nematodes and soil arthropods, consists in mulching the soil and covering it, usually with a transparent polyethylene cover, for duration of a few weeks to up to two years. This elevates soil temperature can be very effective in killing invasive plants too[14], although this will also affect native organisms. Control of vertebrates may also require sophisticated techniques such as the use of dogs, Judas goats (see p. 245) or helicopters[19, 22, 23]. Containment of a newly arrived pest or disease to prevent its establishment also falls into this category.

Physical methods can be used for both control and eradication, although the latter normally applies to restricted areas. Many examples of physical methods are provided in the species accounts for animals and plants in the second part of the book, and in the references provided for further reading[1-6]. The main advantage of physical methods is that they are generally highly selective. Traps, for example, can be extremely effective when well adapted to the morphology and behaviour of the animal to be captured, but it often requires also a good knowledge of the biology of the species and trapping experience from the operator. New trap designs and protocols are constantly proposed and tested to try and improve trapping efficiency.

The common denominator of all these methods is that they are almost always very labour intensive and therefore costly, and they often require experienced and dedicated staff[22]. Recreational activities like hunting, fishing, or other economically viable forms of harvesting (e.g. water hyacinth), may produce an incentive for people to spread the invasive species to new areas not yet colonised[1].

In the case of animals, physical methods such as trapping often present the inconvenience of having to kill the animals caught. The next section presents recommended methods considered humane to euthanise invasive animals captured alive (p. 127).

Left: Black rats trapped in cage traps.
Gérard Rocamora/ICS

Right: Physical removal of plants can be very difficult and time consuming (Grand Polyte, Cosmoledo).
Gérard Rocamora/ICS

Large exotic trees can be ring-barked and left to die but need to be brought down later in areas where they can injur people when falling down (White cedar ring-barked in Vallée de Mai).
Lucia Latorre / SIF

CHEMICAL METHODS

Chemical methods consist in using chemical or toxic products to affect the physiology and vital processes of targeted exotic species in order to eliminate them.

Chemicals may be used to destroy large number of individuals over extensive areas. For example, anticoagulant rodenticides are commonly used to control or eradicate rats, as are other molecules that will knock down numbers of introduced mammals or birds in order to completely eradicate them from particular islands (e.g. cats, rabbits, mynas, etc.)[3, 22]. Herbicides and insecticides developed for controlling pests in agricultural production, and elimination of disease vectors, are also available to

ETHICS, ANIMAL RIGHTS PRINCIPLES AND PUBLIC CONCERNS

Animals trapped in cages or used as decoys must be provided with food, water and shade.
Gérard Rocamora

Animal welfare issues need to be given full consideration as part of any invasive species management programme. Eradication and control programmes of invasive animals normally require the killing of individuals – often in large numbers – through a range of methods, such as trapping, shooting or poisoning. Welfare issues primarily focus on the pain and suffering caused to both target and non-target individual animals during these operations, while ethical concerns require the justification of these programmes. Both points can be discussed and should be weighed against the benefits of minimising the impacts of invasive species which can cause serious environmental and economic damage, including spread of deadly human diseases, and putting at risk the very existence of endangered unique animals or plants.

On the other hand, the individual animals are living creatures, and their suffering must also be taken into consideration. Their elimination must be conducted in full compliance with the existing legislation regarding animal welfare, and particular attention must be paid to reduce their suffering insofar as possible. *Captured animals should always be processed and killed in the most rapid and possible humane way.* This means that all methods aiming at killing animals must first cause an irrevocable loss of consciousness within a set amount of time without pain until death occurs (see p. 127).

No form of cruelty or sadism towards individual captured animals should ever be tolerated. This is important from an ethical perspective, and also because such inappropriate behaviour can undermine public support for the project. Such inappropriate attitudes could create discontent when it is important to keep as wider public support as possible. Some persons may demand that the control or eradication programmes be discontinued if they believe that unnecessary suffering is inflicted on animals, that inhumane methods are being used, or that the losses of non-target species is too high.

Mitigating solutions to minimise possible suffering should always be sought. In the case of eradications, however, these must be carefully considered in case they affect the efficiency of the methods. For example, rubber-protected leg hold traps for cats are a true improvement in terms of animal welfare (classical unprotected traps are now banned in many countries). They reduce animal suffering, but their employment significantly increases the chance that individual cats will escape and subsequently become very difficult to recapture. In the end, this may hamper the success of the eradication, or considerably increase its length and costs.

There may be ethical concerns about the use of strong pesticides and the side-effects these may have on other species, the environment and food chain in general, or on human health. In fact certain potential biological agents can pose a threat to the environment that is greater than the invasive species itself. Another issue is that some stakeholders may view certain invasive species as beneficial (e.g. introduced feral chickens, goats, rabbits or game birds that can be used a source of food or for hunting). The latter point can only be addressed through persuasion, mitigation or compensation. Concerns over the use of pesticides or biological agents need to be addressed with strict respect for legislation, application

Leg-hold (soft catch) trap for cats, with rubber strips on the closing face of the jaws.
Mauritius Wildlife Foundation

of protocols and guidelines in the preparation of these operations. This includes a precautionary attitude involving bibliographic searches on past operations, additional research, tests or trials that may be
required, relevant environmental monitoring before, during and after the operations, and a transparent
communication with regards to the results obtained.

Further reading
Close *et al.*, 1997; Cowan and Warburton, 2011; DWCT / MWF, 2010; Littin & Mellor, 2005; McDonald *et al.*, 2002; PII, 2011; Sharp &
Saunders, 2005; Tassin, 2014.

control invasive species and reduce their level of infestation below an economically
or ecologically acceptable level[7]. Herbicides are applied as a spray on the foliage, on
stumps with a paintbrush, or injected into the plant (see p. 326 and 340). Insecticides
are normally sprayed on infested parts of plants, but can also be injected in certain
cases. Chemicals can be applied to large areas quickly with relatively little labour and
may appear as a relatively cheap option. The main disadvantages of these methods
include impacts to non-target species. These may involve long-term accumulation
of non-lethal levels, risks of persistence in the environment and the possibility that
target species will become resistant[4, 22].

Pesticides need to be well-matched to the targeted pests. Extensive past use of
insecticides such as DDT, other organochlorides and Persistant Organic Pollutants (POPs) in general, had massive detrimental effects on the environment and
on human health because they were non selective and concentrated in the food
chains[1]. For example, for several decades birds of prey (e.g. eagles and falcons)
were decimated around the world because the high concentrations of DDT caused
their eggshells to become too thin and brittle, and this pesticide was banned in the
1980s. But there are still other toxic POPs, including residues of organochlorine
pesticides that continue to be responsible for reproductive failure, deformities, physi-
ological and behavioural dysfunctions in wildlife[7]. Nowadays, such broad-spectrum
and very toxic chemicals are banned and replaced by molecules found naturally in
plants (e.g. pyrethroids) or that are similar to insect hormones. However, there are

Coconut trees formerly planted for agriculture beyond coastal areas may be mechanically removed to recreate natural habitats.
Gérard Rocamora

Cereal baits containing an anticoagulant chemical are most commonly used to control or eradicate rodents (Pestoff Rodent bait 20 R pellets).
Gérard Rocamora

too many pesticides that still have too wide a spectrum and unknown negative potential effects on non-target species and human health. Another problem linked to the widespread use of pesticides is the fact that 520 insect and mite species, nearly 150 plant pathogens and over 270 weed species are now resistant to pesticides. For all these reasons, increasing pesticide use to improve pest management cannot be considered sustainable[8, 9].

Pesticide application requires expertise and skill. Even when applying chemicals considered to be low-risk (for 'unrestricted use'), people need to be trained in the proper control techniques to be used, and especially in regard to safety. Unfortunately, there are many examples of misuse of pesticides leading to impact and mortality of non-target species (e.g. bees and other natural pollinators), environmental pollution, worker exposure and injury, and even death[1].

As do most countries, Seychelles has a list of registered pesticides and authorisation must be sought (currently from the Pesticide Control Board) to import them, and in some instances to use them. To reduce the impact on non-target species, bait stations and modes of application (for example, for rats or cats) need to be adapted to fit as closely as possible the morphology and biology of the target species and to exclude non-target species. Contraceptives mixed in bait are being used to control certain species (e.g. feral pigeons in towns). This technique offers a promising new perspective and is being investigated for other invasive animals, such as rats. Examples of chemical methods are provided in the species accounts for animals and plants in the second part of the book, and in the references provided for further reading[1-6].

Herbicide chemicals can be used with paintbrush on stumps to prevent regrowth of invasive woody plants.
Gérard Rocamora

BIOLOGICAL METHODS

Biological methods consist in using living organisms to control or suppress an invasive pest, making it less abundant and/or less damaging than it would otherwise be. Invasive plants and animals carry with them a minority of parasites and pathogens from their native range (about 25% or less), and the novel parasites that they found in their new locations are generally only a fraction of the number lost[24]. This partly explains why invasive species show increased demographic performances, and provides the possibility of introducing host-specific pathogens from their native range to control them.

In its simplest form (Classical Biological Control), this technique is about introducing natural enemies (predators or parasites) from the countries of origin of the invasive species. An essential aspect of this method is that the biological agents considered for introduction (e.g. pathogens or vectors, predators or parasites) must be host-specific[1]. Otherwise, they may become invasives themselves and negatively impact native species of flora and fauna. This and the fact that there are usually no ways of controlling the newly introduced species once it has been released are the main risks and disadvantages of classical biological control[22].

Examples where generalist predators (e.g. mongoose, toads, birds) or parasites, introduced to fight weeds, invertebrates or vertebrates, have spread and become a serious problem, abound in the literature. In Hawaii, there has been strong controversies regarding insects introduced to fight invasive plants that were later observed on native ones[10]. In Seychelles, the early introduction of cats to virtually all inhabited islands, or the introduction around 1950 of the African barn owl to control rats (see p. 248), are good examples of such negative experiences. Shortly after being introduced, both cats and owls became problematic predators that inflicted considerable damage to native vertebrates (landbird, seabirds, reptiles) and island ecosystems of Seychelles in general.

Such disasters must be avoided at all costs in the future. Nowadays there are modern protocols for biological introductions such as the ones provided the International Plant Protection Convention of FAO[11] ('*Code of Conduct for the Import and Release of Exotic*

Biological Control Agents' and *'Guidelines for the export, shipment, import and release of biological control agents and other beneficial organisms'*. These oblige certification of host-specificity of any agents proposed for introduction. Since such stringent standards have been adopted, there had been by 2005 no reported cases of biological weed control causing significant harm to non-target species or to the environment at large[10].

Biological control has been a key component to many successful integrated pest management programmes around the world[12]. Classical biological control has been used for over 100 years, and by 2009 the introduction of more than 2,600 biological agents has been recorded worldwide, with significant positive impacts for food security and the economy of human populations[13]. In recent decades there has been great scientific progress in the study of the biology and ecology of pathogens and invertebrate predators. This includes the techniques of captive rearing, release and monitoring. Captive rearing is required when the natural enemy, once released, cannot keep its prey at a sufficiently low level, hence humans multiply it and disseminate it when and where they need it, a process called 'augmentative biological control'[14]. There have been very significant successes in biological weed control, particularly in the USA[15]. In Tahiti (French Polynesia), the fungal pathogen *Colletotrichum gloeosporioides forma specialis miconiae* has been used successfully against the invasive alien tree *Miconia calvescens*[26, 27]. Biological control now appears to be the most efficient and cost effective approach to control many invasive species. However, biological control is not suitable for eradication, and should not be expected to completely eliminate the pest.

Such controls can be ecologically safe because of the high specificity of the biological agents used, although thorough investigations and preparations still need to take place to avoid any unforeseen problems. At Aldabra, the captive rearing and release in 1989 of *Rodolia cardinalis* and *Rodolia chermesina*, two ladybirds that had been introduced from La Réunion to Mahé in 1880, appears to have been successful in controlling an invasive coccid *Icerya seychellarum* (a widespread tropical invasive pest species despite its name) that was causing severe damage and threatening very rare plants[16, 17].

Other methods of biological control involve massive releases of **sterilised males** (for example through chemical castration or atomic radiation) so that most females will copulate with them and produce no fertile eggs (a technique planned in Seychelles for the Melon fruit fly, see case study p. 114). **Pheromones**, which are species- or genus-specific substances produced by insects to attract their mates, can also be used in traps or widely disseminated to confuse adults and prevent reproduction. This method, called 'mating disruption' has been widely and successfully used against many species of parasitic moths (Lepidoptera) affecting a great variety of plants and fruits (maize, chestnuts, cucurbits, etc.)[1, 12].

The use of micro-organisms such as fungi, bacteria or viruses that can kill certain pests has become widespread to control pests in agriculture and horticulture, and also represents a promising form of biological control for environmental purposes. *Bacillus thuringiensis* – known as '*Bt*' – can control certain caterpillars, beetles, mosquitoes or flies; and it is used in Seychelles to control mosquito larvae. However, different strains are available and some are more effective than others for the control of particular groups. Strains of Bt are pathogenic to species within the Lepidoptera, Diptera or Coleoptera depending on the type of delta endotoxins that they produce. The CryIA class is specific to Lepidoptera (moths, butterflies & their caterpillars), the CryII class to Lepidoptera and Diptera (flies), the Cry III class to Coleoptera (beetles), and the Cry IV class to Diptera[25]. Entomopathogenic (insect-killing) fungi or nematodes are also becoming increasingly popular in agriculture. In Mauritius and La Réunion a fungus has been successfully used to control an alien beetle parasitising the sugar cane. These 'biopesticides', derived from natural organisms, are environmentally safe, and non-toxic to users and consumers[1]. Some can be very host-specific, only affecting the pest target and closely related organisms.

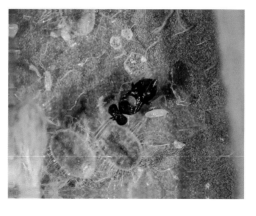

E. guadeloupae, a tiny wasp naturally present in La Réunion, is a parasitoid that can be used for biological control on the Spiralling white-fly.

An interesting form of biological control against alien pathogens of plants is the use of antibiotic-producing bacteria, or hypovirulent strains of the same pathogen, which will contaminate the virulent form and make it harmless or tolerable for the plant. This latter technique has, for example, proven quite successful against the Chestnut blight Cryphonectria (formerly Endothia) parasitica, a pathogenic bark fungus introduced from Asia, on the island of Corsica and southern continental France[12].

Pathogens can also be used against vertebrate pests[18, 19] for example, the myxomatosis virus or the rabbit hemorrhagic diseases used to control European rabbits in Australia. Unfortunately, similar use in Europe to control rabbits has resulted in a cascading ecological disaster: the rarefaction of the rabbit in much of its Mediterranean native range, and of the predators that feed on it, such as the Bonelli's eagle Hieraaetus fasciatus[20]. This demonstrates why extreme caution is required before using pathogens or indeed any type of biological control. All such organisms must be screened for potential risks to other lifeforms (including humans), and a good knowledge of the pathogen-host relationship is required. Other disadvantages and risks associated to pathogen release include poorly efficiency at low density of the target species, and acquisition of immuno-resistance by the latter[22].

In the case of rats, the production of genetically modified animals capable of producing only male offspring, and the use of viruses to disseminate immunocontraception (a process by which the immune system is made to attack reproductive cells and induce sterility) are promising methods currently being investigated[21]. Genetic modifications may include spreading in a wild insect population a gene of susceptibility to a certain pesticide, or sterile triploid individuals[14]. Nature reserves and conservation areas, where pesticides are often prohibited, are particularly suited for the use of biological control. Provided that rigorous precautions are met, Seychelles may in the future provide a stage for advancing and integrating such modern techniques in the process of ecological restoration of islands.

Developing biological controls is expensive, as it often takes years before a suitable agent is found, tested, approved for release, and produced in sufficient quantities for commercial use. In addition, several different types of biocontrols sometimes need to be developed and released in order to bring one target invasive species under control. However, biological control of pests has been very successful around the world, with success rates of more than 80% in many countries, and it represents the most promising control technique for the future[9].

Existing methods and technologies available, whether physical, chemical or biological will continue to improve and will become more effective with time. However, this alone will not reduce the flow of invasive alien species unless it is combined with improved policies to prevent their entry and spread, and more efficient detection measures so that they can be eliminated quickly[14].

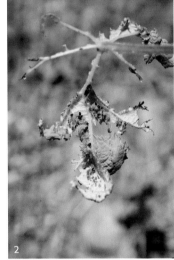

Illustration of the efficiency of the introduced biological control agent *Cibdela janthina* (Tenthredinidae, Hymenoptera) on the invasive *Rubus alceifolius* in La Réunion. 1. Adult on *Rubus.* 2. Larvae feeding on *Rubus* 3 & 4. Five months after their release the stands of *Rubus* appear considerably defoliated and have been reduced very significantly (3: July 2008; 4: Nov. 2008).
Thomas Le Bourgeois / CIRAD - Pl@ntInvasion

References / Further reading

1 Wittenberg & Cock, 2001; 2 Soubeyran, 2010; 3 PII, 2011; 4 PII, 2012; 5 Tye, 2009; 6 McDonald *et al.*, 2003; 7 BirdLife International, 2004; 8 Oerke, 2006; 9 Witt, 2014; 10 Russell *et al.*,2006; 11 IPPC / FAO, 1996, 2005b; 12 CABI, 2014; 13 Shaw *et al.* 2014; 14 Simberloff, 2014; 15 Culliney, 2005; 16 Géry, 1991; 17 Johnson & Threadgold, 1999; 18 Dobson, 1988; 19 Orueta 2007; 20 Rocamora, 1997; 21 Campbell *et al.*, 2015; 22 Courchamp *et al.*, 2003; 23 Orueta & Ramos, 2001; 24 Torchin & Mitchell, 2004; 25 Dubois *et al.*, 1997; 26 Meyer & Fourdrigniez, 2011; 27 Meyer *et al.*, 2012.

Euthanising invasive animals

When it is necessary to kill an animal, 'humane' procedures must always be applied. These procedures must avoid distress, be reliable, and produce rapid loss of consciousness without pain until death occurs. Although no ideal method of euthanasia exists, the procedures used should verify as closely as possible these principles. Killing captured animals in a humane way requires proper training. In particular, proper handling of animals that need to be physically restrained is essential to minimise pain, fear, distress or anxiety, and also to ensure the safety of the operator. In some cases, the prior use of sedative and/or immobilising drugs may therefore be required.

Methods of euthanasia may be chemical or physical. They may cause death through direct or indirect hypoxia (suppression of oxygen intake and delivery into body cells), direct depression of neurons vital for life functions, or physical destruction of brain activity and vital neurons. Insofar as possible, the method should cause minimal emotional impact to the operator and observers. Some physical methods of euthanasia, for example stunning followed by exsanguinations (draining the body of blood), although not aesthetically pleasant may ensure immediate insensitivity to pain and hence be humane. Therefore, "the choice of technique should be made based on the sensibilities of the animal to be killed rather than the sensitivities of the observer or operator, although the latter should not be disregarded".

The following are the main methods considered humane that are recommended to euthanise captured invasive alien animals. More details can be found in the species accounts in regards to the specific use of these methods for different animal species.

INHALATION OF CARBON DIOXIDE (CO_2)

The use of inhalation agents requires a closed chamber or bag to hold the gas until it reaches the concentration needed to provoke loss of consciousness. It is unsuitable for animals that can hold their breath (e.g. diving or burrowing birds or mammals) or that breath at low frequency (amphibians or reptiles). Safety measures must ensure that operators are not exposed to toxic gases. This method is also not suitable for newborn and very young mammals which are physiologically adapted to lack of oxygen. CO_2 is therefore not recommended as the sole method of euthanasia for animals less than 16 weeks old.

- Inhalation of carbon dioxide (CO_2) from gas cylinders within an appropriate chamber or bag is one of the easiest methods of euthanasia. Risks to the operator from carbon dioxide are negligible compared to those of anaesthetic gases or carbon monoxide. It is widely used in a number of countries (e.g. Australia), mostly for small birds and rodents, and it is suitable for animals of up to 3 kg.
- CO_2 is available commercially in cylinders (for food, medical, or industrial purposes). The compressed gas can be injected via a de-pressuriser and regulator into a special bag or container that contains a cage and the animal to be killed. Another source of CO_2 is 'dry ice' or solidified CO_2. When placed into a (covered) beaker of water inside the container it will release CO_2 gas (1 kg produces approximately 0.5 m³).

Closed chamber and gas cylinder used to euthanise invasive animals (rats, cats, mynas, etc.). Traps having captured animals can be put directly inside the chamber and the glass on its cover allows to control that the animal is dead.

- Animals lose consciousness very quickly after being placed into a chamber containing up to 70% CO_2. The gas has a narcotic effect on their brain, and prolonged intake leads to death due to lack of oxygen and direct depression of respiratory and cardiac functions. However, pure CO_2 can cause breathing difficulty and distress in conscious animals.

- It is preferable to put the trap or cage with the animal inside into the container pre-filled with CO_2 or in a special plasticised sack (sold for example with certain cages for mynas). Avoid removing the animal from the trap to place it in the container, as there is a risk the animal may escape, or injure the operator.

- A continuous inflow of CO_2 should be directed into the bottom of the sack or the container, and a constant level of CO_2 should be maintained for at least 3 minutes; anaesthesia will then occur within 60 seconds, and subsequent death within 5 minutes. An aperture is required at the top of the container/sack (i.e. bag loosely tied at the top) to allow air to be evacuated and a constant flow of gas during the required amount of time.

- Ensure sufficient exposure to high concentrations of CO_2 by completely filling the chamber with CO_2. Some animals may move to the upper part of the chamber to avoid the higher concentrations of gas (which is heavier than air) in the lower part of the chamber. Also limit the number of animals at any one time in the chamber.

INHALATION OF CARBON MONOXIDE (CO)

- Because carbon monoxide (CO) binds to haemoglobin in the red blood cells, with a much greater affinity than oxygen, it reduces oxygen supply to tissues. Concentrations of CO greater than 2% are enough to cause loss of consciousness within minutes, and at 4% to 6% death occurs due to respiration failure and cardiac arrest.

- Carbon monoxide is also extremely toxic to humans and difficult to detect. When it is inhaled (e.g. as a product of combustion) it can cause death or result in permanent nervous system damage. CO can be obtained in compressed cylinders (and in some countries, as combustible cartridges), but because it is so hazardous it cannot be recommended for routine euthanisation.

- Exhaust gases from idling petrol engines containing carbon monoxide have been used to euthanise animals (particularly mynas in Australia). This method is no longer considered acceptable by many authorities because modern cars may not produce the required high levels of CO. For example, modern car exhaust pipe converters can remove up to 99% of the CO produced. It is only when the engine first starts, when the converter is dormant during the engine warm-up, that sufficient CO may be obtained over a period of one or two minutes. In addition, exhaust gases contain irritating contaminants (e.g. ozone, nitrogen dioxide, nitric oxides) many of which are removed by the converter, but not when the engine starts. Exhaust gases also become rapidly unacceptably hot. However, scientific experiments have shown a better performance of CO obtained from cooled exhaust gases compared to CO_2 as a humane method for euthanising small birds (mynas, sparrows)[1].

INHALATION OF ANAESTHETIC GASES

- Anaesthetic gases such as halothane, isoflurane, or methoxyflurane may be used for small animals of less than 7 kg. However, they should only be administered by a veterinarian.
- The animal is placed in a closed container containing cotton or gauze soaked with an appropriate amount of an anaesthetic. The anaesthetic gas can also be vaporised into the container or a plastic bag placed around a cage containing the animal. The animal will become unconscious quickly and quietly, although some resistant behaviour may occur. It will die after 10-20 minutes.
- Once the animal is unconscious, it can also be removed from the cage and euthanized with a lethal injection.
- The use of chloroform and ether is not acceptable due to concerns about risks to humans and animal welfare. Chloroform may cause hepatotoxicity, renotoxicity and is a suspected carcinogen. Ether is highly explosive, induces a slow anaesthesia, causes irritation to skin and mucous membranes, and may cause excitement in some species.
- Anaesthetic gases are strictly restricted dangerous substances that should be kept under lock and key. Their supply, possession, use and storage must comply with national legislation.

LETHAL INJECTION

- Injection of a barbiturate overdose either by the intravenous (into a vein), intraperitoneal (into the body cavity) or intracardiac (into the heart) routes is one of the most humane methods of euthanasia.
- Intraperitoneal injections work much more slowly than intravenous injections, but are easier for an operator working alone. Intracardiac injection should only be used if the animal is heavily sedated, unconscious or anaesthetised because they can be traumatic to the animal.
- For an intraperitoneal injection, a non-irritant anaesthetic solution of barbiturate e.g. NembutalR (sodium pentobarbitone) can be used. Commercial 'euthanasia solutions' (e.g. Euthatal® or Lethabarb®) are very alkaline and may cause irritation and pain.
- Before injecting a barbiturate, animals need to be immobilised in a restraint cage or sack, or sedated by tranquillising or anaesthetic drugs (e.g. ketamine) or anaesthetic gases.
- Barbiturates are substances restricted to veterinary prescription and supply that must be kept under lock and key. Possession, use and storage of these barbiturates or sedative drugs must comply with national legislation.
- Barbiturates must be administered by a veterinarian or a well trained conservation officer.

STUNNING OR CONCUSSION

- 'Stunning' is achieved by using a hard instrument (e.g. metal pipe, heavy wooden stick) to deliver a single, strong and fatal blow to the central skull bones. Stunning alone may be acceptable for small or young animals with a soft skull such as rats, mice, birds, reptiles, amphibians or fish, usually weighing less than 250 g.
- Stunning provokes depression of the central nervous system and immediate unconsciousness. For a small animal, a sufficiently strong blow can be enough to cause death by destroying brain tissues. For middle size to large animals (weighing over 250 g), stunning may not be sufficient and must immediately be followed by another procedure such as cervical dislocation, or exsanguination (not suitable for birds). The latter consists in cutting the main blood vessels around the throat (carotid and jugular veins).
- 'Concussion', which can also be used for small animals, involves holding the animal in one hand by its body and legs, and swinging it in order to violently strike the back of its head against the edge of a hard surface (e.g. concrete, hard wood) to produce instant death.
- Stunning and concussion must be properly performed to be effective and humane. If not, some consciousness and pain can occur. Training of operators is therefore essential.

CERVICAL DISLOCATION

- This method can be used for small animals which can be easily handled, such as small to medium sized birds, rodents and rabbits (weighing less than 1 kg approximately). It involves separating the skull and the brain from the spinal cord. This damages the brain stem, which stops breathing and heart activity and leads to rapid death (brain activity persists for about 13 seconds in rats).
- Cervical dislocation instantly kills small animals but requires practice and is not aesthetically pleasant, as reflexes and violent muscular contraction may be evident for some time. It can be performed on small birds with special pliers or by breaking the neck (described below).
- To perform a cervical dislocation on a bird, take the legs in the left hand (for a right handed operator) and place the head between the first two fingers of the right hand with the thumb under the beak. A sharp jerk of each hand, pulling the head backward over the neck will break the spinal cord and carotid arteries. With a rat, hold the head in one hand and the body in the other, and separate your hands with a quick elongation and twist. Stunning or anaesthesia should be performed first on rats heavier than 150 g.
- To kill rabbits it is best to combine stunning and cervical dislocation (or exsanguination, see above). Stun the rabbit by a very strong blow with a stick (or closed fist) to the back of the skull, behind the ears, while it is held upside down by its two hind legs. The neck is then taken between two fingers and pushed down until it is stretched and the head moves backwards, until dislocation is felt.
- All cervical dislocation techniques must be performed by skilled operators.

SHOOTING AT CLOSE RANGE

- Shooting may be most appropriate when a trapped animal cannot be easily restrained or removed from the trap, and when there is a risk that it may escape or inflict injury to the operator.

- Shots must be fired at close range (e.g. 5 cm to 1-2 m using a shotgun or a handgun) and aimed at the brain (but the barrel should never touch the head), to ensure that the projectile penetrates and causes instant loss of consciousness. Chest shots (aimed at the heart) may be performed when the animal cannot be restrained. The shooter should approach the animal quietly, keeping unnecessary personnel away from the area to minimise stress.
- Smaller calibre rifles (minimum .22 magnum) can also be used to euthanise
- most species of animals at short range (< 5 m).
- Shooting needs to be performed only by skilled operators (army or police marksmen, trained rangers, etc.) having necessary experience with firearms and who hold the appropriate licences and accreditation. Storage and transportation of firearms and ammunition must receive previous official authorisations and comply with any legislative requirements.

Note: Drowning of animals is <u>not</u> considered to be a humane method of euthanisation. Nor is the freezing of animals to death[2, 3].

CONFIRMING THE DEATH

After an animal is euthanized, its death must be confirmed, and in the case of uncertainty, the euthanising procedure should be repeated, or exsanguination performed. Use the following symptoms to confirm death:
- *Absence of heart beat,* as determined by palpation or with a stethoscope. Neither absence of respiratory movement nor absence of pulse is a sufficient indicator of death in small animals.
- *Loss of corneal and palpebral reflexes,* as determined by touching the eyeball or stroking the eyelids, and finding that the eye remains open and the lid does not move.
- *Glazing of eyes,* as indicated by the cornea losing its clear moist appearance and becoming opaque, dry and somewhat wrinkled.
- *Loss of colour in mucous membranes* (such as interior of mouth) that become pale and mottled. When pressure is applied to the membrane, there will be no refill of blood, and the membrane will become dry and sticky. This method of assessment can be used only for large animals.

DISPOSING OF CARCASSES

Carcasses should not be discarded until death has been established. They need to be disposed of properly and in accordance with acceptable local practices and regulations, such as burying in a deep hole or incineration. Ensure that carcasses of poisoned animals or those that may contain residues of euthanising toxins cannot be consumed by other animals or by humans (this is especially important when dealing with chickens or rabbits).

References / Further reading
compiled and adapted from Sharp & Saunders, 2005, 2008; Reilly, 2001; and Close *et al.*, 1996.
1 Tidemann & King 2009; 2 Wildlife Ethics Committee 2013, 3 UFAW (undated).

Which management option to choose?

Chapter 5

Different natural habitats at Ile du Nord
(North Island), where cats and rats were
eradicated in 2003 and 2005 respectively.
Colin Bell

Four main management options

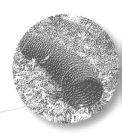

This chapter presents the different options available for the management of invasive species.

There are four main ways to fight against invasive alien species that have managed to establish themselves in an area[1,2]. These should be considered in the following order of priority:

Eradication consists in permanently eliminating all the individuals of a targeted population along with their propagules (seeds, larvae, eggs, etc.). Unless accomplished at a very early stage of colonisation, this is normally a challenging operation, hence the importance of early detection and rapid reaction.

Note: in medical terms, eradication has a different meaning and is used to designate the complete removal of a vector or the disappearance of a disease globally (akin to global extinction in conservation biology).

Containment is a form of control in which the invasive species is strictly confined to a particular area, by actively eliminating or eradicating individuals or propagules from surrounding buffer zones so as to prevent its spread from the area of confinement to non infested areas.

Control of an invasive species consists of reducing the size of its populations so that its negative impacts on the environment, human health or economic activities are reduced to an acceptable level. Control is a recurrent operation that needs to be conducted on a regularly basis.

Mitigation involves measures aimed at reducing the impact of the invasive species indirectly, without taking direct action against it, but rather by acting on the native species, habitats, goods or infrastructure affected, or at the ecosystem/agro ecosystem in general (e.g. predator fences or predator proof nest-boxes; practices that hamper the development of invasive species).

Managers should choose which option is best adapted to each case. This depends on a fair assessment of the requirements and the feasibility of the operation, taking into account the probability of success, cost effectiveness and any possible detrimental effects. Figure 17 presents the different questions and options of the decision making process.

MAKING THE RIGHT CHOICE

Eradication is usually the most desirable option, but normally also the most difficult to achieve – not only in terms of eliminating the target species, but also maintaining pest-free status in the long term (e.g. rats, ants). Hence it is important to consider what will be realistically achievable, not just technically and financially, but also taking into account the extent of the logistical, political or social support required.

It is necessary to assess both the feasibility of eradication and the long-term sustainability, as explained in more detail in the next section. For example, eradicating rats from an island requires not only a high level of preparation and technical *savoir-faire*

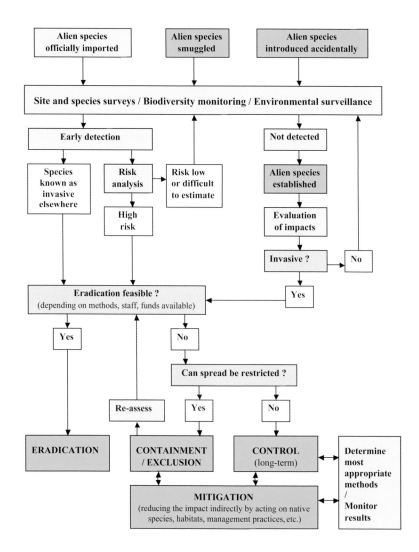

Figure 17: Decision tree on which management option to choose. Yellow boxes: areas where we can and should act. Green boxes: questions and important decision points. Pink boxes: areas out of our control.
Inspired from Wittenberg & Cock 2001, and Soubeyran 2010

during the eradication phase, but also a constant strict discipline in terms of control of supplies, transport procedures and other abatement measures over the long term to prevent reinvasion.

The eradication attempts of rats in 2000 for Curieuse and Denis Islands, where rats were found less than a year after the operation, may well have failed due to insufficient rigour in applying procedures to prevent reinvasion. If there is not sufficient determination and management capacity to control all landings and apply strict protocols over the long term – and indeed a legislative background to discourage uncontrolled landings – it may be wiser not to risk spending so much money and energy in attempting eradication. This is particularly important for the inhabited inner islands of Seychelles.

Compared to control, which implies never-ending management activities and costs, eradication is a one-off operation that is much more efficient in terms of results and more cost-effective over the long term. Nevertheless, continuous measures will be required after the eradication to prevent reinvasion (a constraint that should also be seen as an opportunity for job-creation and community stewardship). Eradications are normally expensive operations (see p. 162) requiring a major investment of money and human effort. The main disadvantage is that the investment may be wasted in the case of failure. In the case of rat eradications, the persistence of, or the reinvasion by, a couple of rats or a single pregnant female is enough to restart a population and ruin a costly operation.

Eradications are always a challenge and often present difficulties greater than anticipated[3]. Sometimes the feasibility of eradication is questionable due to uncertainty about the efficiency of available methods, the full commitment of stakeholders, the availability of resources to complete the operation, or the ability to later maintain the island pest-free over the long term[4]. In such cases, it may be wiser to choose containment or control rather than to risk a failure. Where the level of risk is minor or reasonably low, however, then eradication should be the preferred option, especially in the case of recent or geographically limited invasions for which the chances of success are higher.

Whatever management option is used, its application needs to be conducted in conformity with the laws and regulations of Seychelles. Some non-target plants and animals currently protected under the law may be affected, and safety issues per-

Although eradication is the most challenging management option, it is also normally the most efficient and cost-effective in the long term.
Gérard Rocamora

taining to the use of pesticides need to be respected. This is why we recommend that advice is sought from the appropriate government services: Environment Department (MEE), Seychelles Agricultural Agency (Plant Protection or Veterinary Services) or Ministry of Health before embarking on any operation against IAS, especially when using chemicals.

Large scale operations, such as island wide eradication of rats or bird species, require prior consultation and approval from the Environment Department (MEE).

Government services, environmental NGOs (see list of useful contacts p. 383), private companies or private experts can provide guidance to ensure that the necessary plans comply with existing Seychelles regulations and that non target species and human health are not put at risk.

To ascertain the success of invasive alien species management operations, indicators need to be identified by managers in advance of the operation[4]. What indicators are used will depend on the species and situation being dealt with. For example, to declare success in the eradication of a plant species, a metric normally used in Hawaii is that annual checks of the infestation site must reveal no plants for a period of times equivalent to 2-3 times the period that seeds remain viable in the soil[5]. Different indicators can be used for rat eradications on islands depending on the intensity of the trapping survey and the bioclimatic region where the eradication is taking place. No animals trapped and no sign of rats for two years is normally required before declaring rat-free status, especially in temperature and subantarctic regions. For Seychelles, and other tropical islands in general, our own experience suggests that one year with no sign of rats, as determined by an intensive trapping survey, should be enough for an eradication attempt to be considered a success (see rodent accounts, p. 192 and following). Similar metrics or benchmarks also can be used to assess the success of Containment, Control, and Mitigation activities.

References / Further reading
1 Wittenberg & Cock, 2001; 2 Soubeyran, 2010; 3 DWCT / MWF, 2010; 4 PII, 2011; 5 J. Beachy/Univ. of Hawaii, *pers. comm.*

Discussing the best technical options and preparing the operations in detail with all stakeholders are essential steps for the success of invasive species management.
Elvina Henriette

Eradication

When an alien species has invaded an area despite preventive measures, eradication is normally the preferred option. It is usually the more cost-effective rapid response to the early detection of a new non-native species[1]. Many different types of invasive species have been successfully eradicated around the world, particularly on islands. These include mammals, birds, reptiles or amphibians, invertebrates (insects, snails), many plants (broadleaf trees, palms, creepers, herbaceous weeds, etc.), and diseases for plants, animals and humans (fungal, bacterial or other). Such eradications have been motivated by agricultural, health and environmental considerations, including the desire to restore island ecosystems in order to create island refuges for endangered native species. The numbers of eradications conducted worldwide have increased spectacularly during the decade of the 2000s, and this trend continues[2]. By 2013, a total of 1,224 successful eradications of invasive plants and animals had been reported on 808 islands around the planet[3]. Most have been achieved against land vertebrates on small islands for conservation purposes.

This increase is also visible in Seychelles, where more than 40 successful eradications of 12 different animal species (six of them predatory), six mammals and six birds, have been conducted for nature conservation purposes and confirmed successful (see p. 61 for more details on animal eradications in Seychelles). Eradication was also attempted for the Melon fruit fly, an agricultural pest, but it was not successful (see case study p.114). Plant eradications have been less common and less documented; local eradications with variable success have been attempted for Albizia, Devil's ivy (Philodendron), Merremia and other creepers, Water lettuce and the native Bracken fern. More details on all these operations are provided in the species accounts and case studies.

PREPARING THE OPERATION

Comprehensive preparation and planning are critical for the success of an eradication. Some operations, e.g. the eradication of rodents or goats, can be technically and logistically very complex and require a long **pre-eradication phase** (more than a year if possible) during which a long list of pre-requisites will need to be thoroughly considered.

Important aspects that need to be taken into account to prepare an eradication are summarised in Figure 18.

Source: Courchamp F., Chapuis J.L. and Pascal M. 2003. Mammal invaders on islands: impact, control and control impact. *Biological Review* 78: 347-383.

In the case of rodent and cat eradications, a kit from the Pacific Invasives Initiative (Resource Kit for Rodents and Cat Eradications[4]; see p. 140) proposes a very well documented step by step process to select, design, implement and report on such eradications, which may also apply to eradications of other (animal) species.

The following recommendations, derived from lessons learnt during both successful and failed eradication attempts, both in Seychelles and abroad[1, 4, 5], will help increase the probability of success. Many of them also apply to other management options directed against invasive species, but they are particularly crucial

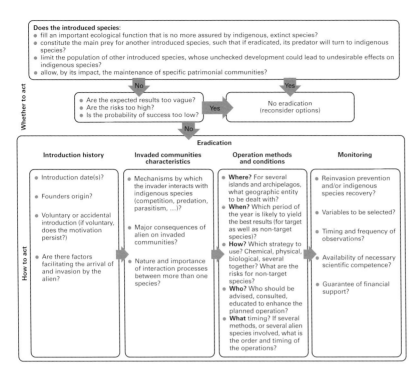

Figure 18: Decision diagram and important aspects to be taken into account when considering a mammal eradication.
Source: Courchamp *et al.* 2003

for eradications. They apply in particular to bird and mammal eradications which are complex operations. These recommendations can be further developed in future by organisations (government services, private islands, NGOs) and private experts as knowledge from local management experiences continue to build up in Seychelles.

KNOW YOUR TARGET SPECIES AND ITS ENVIRONMENT

An eradication is like a highly strategic battle (the title of this section could be *know your enemy and the battle field!*). It is also a race. For it to succeed, every member of the target species needs to be removed, before any of them can reproduce. So start by knowing your target species well, become familiar with it, and try to understand how individuals may react to the elimination methods planned[1]. Excellent knowledge of the field is also indispensable.

Identify your invasive species correctly using available identification guides, documents or databases accessible through the Internet, or get assistance from experts (biologists, agriculturalists, etc.).

Gather all available information on the status, biology and ecology of the targeted species, and on methods previously successful in eradicating it.

Gather maximum information on the island or property where eradication is to take place, and conduct visits to identify hotspots of infestation and areas with particular problems (e.g. difficult or restricted access).

Undertake additional observations or specific research to enhance available knowledge.

Understand how the targeted invasive species impacts the native species and how it interacts with the ecosystem, in particular trophic relationships with other invasive species that may be present (does the targeted invasive species help control other invasive species, or is it itself consumed by other invasive species?)[5].

THE PACIFIC INVASIVES INITIATIVE RESOURCE KIT FOR RODENT AND CAT ERADICATION

Pacific
INVASIVES
INITIATIVE

Driving principles

- Keep your eyes on changes that may affect sustainability
- Engage with stakeholders from the start
- Implement biosecurity measures as early as possible
- Monitor outcomes to demonstrate success
- The implementing agency must take responsibility
- Start easy and grow with experience
- Plan thoroughly
- Seek independent advice
- Allocate sufficient time for developing capacity and sharing lessons

The Process stages	Project documents required
Project selection	
Feasibility study	Feasibility study report
Project Design	Project Plan Biosecurity Plan Monitoring and Evaluation Plan
Operational planning	Operational Plan
Implementation	Project Updates* Adaptation Plans*
Sustaining the project	Final Project Report

* Suggested additions to the original table from PII 2011. Frequently, plans are modified during the implementation process, as data about the efforts are generated, and/or roadblocks are hit.

This kit is a capacity building tool for practitioners. It proposes a logical step by step process, providing detailed practical advice and essential supporting documents including guidelines and report templates for each step, and access to a range of information sources, including current knowledge and best practice. Although its focus is on rats and cats in the Pacific islands, many of the tools provided can be adapted to improve the effectiveness of other alien invasive species eradication projects worldwide. It provides guidelines are on the following topics:

- Rodent identification.
- Stakeholder engagement.
- Project selection.
- Feasibility study.
- Choosing the correct eradication technique.
- Consent and permits.
- Managing environmental effects.
- Guidance for project managers.
- Planning and managing the eradication operation.
- Bait and baiting.
- Rodent surveillance techniques.

- The use of Brodifacoum.
- Cat eradication and monitoring techniques.
- Non-target species.
- Biosecurity.
- Monitoring and evaluation.

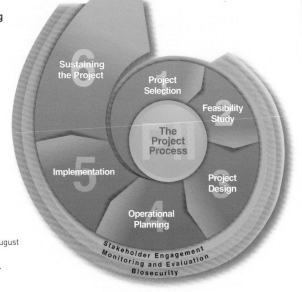

References / Further reading
- Pacific Invasives Initiative. 2011. Resource Kit for Rodent and Cat Eradication. http://rce. pacificinvasivesinitiative.org. Accessed on 7th August 2014.
- See also Boudjelas 2011; Boudjelas *et al.* 2011.

ASSESS FEASIBILITY

Feasibility assessment is an indispensable step in the eradication process. It weighs benefits and costs and provides evidence with which to decide whether the project is feasible or not, or with what conditions[1]. It helps to ensure that the project is scientifically and technically sound, secure in terms of finance and commitment from stakeholders, and that the chance of success is high in relation to the costs[6]. Feasibility assessment also identifies possible unforeseen problems and points that need to be resolved.

To this end, the following actions need to be taken:
- Verify that the eradication will be legally, politically, and socially acceptable. Pay attention to ethical and animal rights considerations.
- Determine the technical feasibility and long-term sustainability of the project.
- Verify that the project is environmentally acceptable. Some methods (i.e. toxic chemicals) may not be acceptable in the long term but can be done as a one off treatment.
- Determine how to source the required skilled staff, financial resources and equipment.
- Verify that the costs will be affordable and funding sufficient, even if the project takes longer than expected. Expect difficulty to increase towards the end of the programme. The last individuals are often the most difficult to eliminate!
- Identify any key issues that need to be resolved (e.g. toxicity trials for non-target species) before the project can proceed.
- Involve independent sources of expertise in a review of project documents (through science committees, external experts, etc.).

ELABORATE A STRATEGY

To elaborate a strategy for the eradication is like defining a *battle plan*; and the following need to be taken into account:

- Consider the various available best practice methods and protocols, compare their track-record in terms of success, and select the more efficient ones.
- Consider combining several methods of eradication. Individual members of the target species that may not have been eliminated by one method (e.g. trap-shy animals; plants resistant to a certain treatment) may be removed by another method.
- All individuals in the population must be at risk; not a single one should be missed. Make sure methods to detect individuals of the target species at low density are available[1].
- Start first by a method which has the potential to decimate (knock-down) the population, then treat remaining survivors with alternative methods[4].
- Start with passive and low disturbance techniques (e.g. poisoning) which are less likely to create aversion in the targeted species, then use more direct and visible methods (trapping, hunting).
- If necessary, test methods in the field and adapt them to become more effective.
- Choose the period of the year when the invasive species may be more easily targeted (e.g. when food availability is reduced so that baits are more attractive)[5].
- Evaluate potential risks to non-target species and adapt protocols to minimise them[7,8].
- Eradication methods should be as selective and as 'humane' as possible (see Box on ethics and animal rights p. 120). Persistence of chemical toxins in the ecosystem should be of short duration. Biological agents should be totally host-specific.

Eradications have been achieved using mechanical, chemical and biological methods (see p. 118). Plants are often best eradicated by combining mechanical and chemical methods. Animal eradications often combine chemical or biological methods to create an initial 'knock-down' in the population, and then follow using mechanical methods.

EVALUATE RISKS TO HUMANS AND NON-TARGET SPECIES

Some eradication methods (e.g. poisons, traps, shooting, use of helicopters, etc.) may be dangerous and present risks to human health. These risks need to be identified, so they can be minimised through the implementation of appropriate information campaigns, training, protecting equipment and operational safety measures. Particular attention must also be paid to reducing collateral casualties on non-target animals and plants. Despite all precautions taken, these are often inevitable[1].

In the case of common species, they may be considered as an acceptable cost that needs to be balanced by the long-term benefits of the eradication. For example, moorhens showed a significant decrease in numbers immediately after the 2005 rat eradication on Ile du Nord (North Island) due to rodenticide consumption; but a few years later numbers had grown well beyond their initial abundance[9] (see chapter 6).

In the case of rare endangered species, however, no risks should be taken. During the 2000 rodent eradication on Frégate, the entire Seychelles magpie-robin population and part of the Seychelles fody population were put into captivity for more than two months to avoid poisoning (direct or secondary) from the aerial spread of rodenticide[10]. In 2005, in preparation of the rat eradication on Conception, captive trials were conducted to confirm that Seychelles white-eyes were not attracted to the bait, and that invertebrates that could potentially consume the bait (e.g. ants, cockroaches, flies) were very few and unlikely to be consumed and a source of secondary poisoning to the birds[11].

ENVISAGE IMPACTS OF THE SPECIES REMOVAL ON THE ECOSYSTEM

Reconsider the eradication, or adapt the strategy and methods planned (e.g. make alternative plans; consider a multispecies eradication), if there is evidence that removing the invasive species will do more harm than good, as indicated in the following scenarios[5]:

- What consequences will removing the targeted species have on other species and on the ecosystem in general?
- Does the invasive species now fill an important ecological function no longer assumed by indigenous species? Would its removal negatively impact the maintenance of rare native biodiversity (e.g. rare endemics for which the main native host plants have disappeared and have been replaced by alien ones).
- Is the targeted species an important prey or predator for other invasive species? The elimination of one invasive species could increase predation on native species by other invasive species.

Acquiring in-depth knowledge on the species' biology and ecology, and on how the ecosystem functions may require several years of scientific research. This is not always possible. When time and funds are limited, eradications may sometimes have to be undertaken (after the different recommended steps are completed) making assumptions based on the best available information and lessons learnt from previous experiences.

Despite all the unknowns and the complexity of the task, efforts must be made to predict and avoid any negative impacts the eradication may have on a particular ecosystem[12, 13]. This implies a basic knowledge of trophic and competition relationships between species. In some cases, removal of an invasive species has led to unexpected population explosions of previously less dominant invasive species. This is a phenomenon referred to as 'surprise effects'[14, 15]. In Seychelles, the removal of Black rats on Bird Island in 1996 was followed by a possibly related explosion in Crazy ant numbers[16].

It follows that an 'ecosystem approach' that takes into account long-term restoration objectives and a range of possible interactions between species is preferable to a 'species-focused approach'[17]. Concern about 'surprise effects,' however, should not scuttle eradication projects. Rather, managers need to be aware of them and to plan for the unexpected. For example, after the Bird Island experience, anyone planning a rodent eradication on an island with invasive ant species present should prepare for a possible ant outbreak and plan ant control measures.

MULTIPLE SPECIES ERADICATIONS

Eradicating several species simultaneously or successively may reduce the risks of 'surprise effects'[17, 18]. The order of the eradications should be planned in a way that anticipates trophic chain reactions, in order to make the remaining pests most susceptible to control, hence maximising chances of success and minimising risks and costs[19, 20].

For example, it may be best to remove cats completely immediately after the eradication of rats, as suggested by past experience from Seychelles. In that way, cats are not only deprived of a food source but are subject to secondary poisoning from rodenticide, making their eventual eradication easier. In addition, cats can also help control rats. To eradicate cats first may result in an increase in rat damage to the ecosystem such as higher predation on rare endemic species or seabirds[12], although this is not the most likely scenario[21-23]. On the other hand, to only eradicate rodents may cause an increase in the predation pressure of cats on native birds and invertebrates.

However, rats may not necessarily increase after cat removal[24], and such decisions on the sequence of eradications need to be made on a case by case basis depending on the situations[25]. They should ideally be based on the assessment of the trophic relationships between the two predatory species, and on the current and predicted extent of ecosystem damage depending on the scenarios envisaged. In this approach, modelling can be a useful tool[26, 27]. But recent studies and empirical evidence suggest that even with more rats, environmental damage is often lower when rats are alone than when they are with cats[20, 26]. When all introduced predators cannot be eradicated together, eradicating the top predator usually generates a positive outcome[22].

Multiple species eradications, normally achieve success in a shorter period of time and require less effort than a series of independent single species operations[7].

PLAN THOROUGHLY AND BE ON THE SAFE SIDE

Planning an eradication operation well in advance and with plenty of 'marge de manoeuvre' is key to ensure that all pre-eradication requirements are met and that all is ready in time. The following actions are needed:

- The objectives of the programme and its expected outcomes clearly defined;
- The various stakeholders, their roles and responsibilities, and how the project will be managed (steering committee, etc.) clearly defined;
- Key requirements (human resources, equipment, bait, training, etc.) identified;
- A realistic budget, with at least 20% in excess for unforeseen changes or increases in prices (particularly important for projects over several years) prepared[1, 4];
- Realistic timelines, based on previous similar experiences and possible unexpected problems set;
- Changes and management risks (problems that may occur and how you plan to solve them) anticipated. Envisage different scenarios and expect the unexpected!

ENSURE STRICT BIOSECURITY PROCEDURES ARE IN PLACE

Prevention and incursion responses are essential to guarantee that the island or the natural area under consideration will remain free of the targeted species, and also of any new invasive species arriving from outside infested areas. To this end the following actions are needed:

- Efficient protocols to prevent reinvasion, and to contain and quickly eliminate invaders need to be identified as early as possible.
- Abatement measures for prevention, surveillance and incursion response need to be well understood and strictly applied before the eradication starts.
- Regular training of personnel, and audits to assess compliance with procedures need to be planned. Weaknesses and necessary improvements required must be identified.
- Concentrate on the species that present the greatest threat to the island (or the natural area) and its biodiversity, economic, health, cultural values, etc. Examples of such species include rats, invasive ants, lizards, weeds or creepers, etc.

The main pathways that need to be controlled on islands are boats, but also aircraft and the passengers themselves. Procedures for pest-proof rooms, loading at the port of departure, and unloading upon arrival are key to prevent reinvasion and to ensure the long-term success of the operation[28-30] (see the example of abatement measures for North Island, Annex).

COMMUNICATE

Support from stakeholders and the public is critical for the success of an eradication programme[1, 31], and can be obtained in the following ways:

- Communicate continuously with the authorities, local communities, stakeholders and the general public.
- Before the programme starts ensure that there is a good understanding of the problems in order to address any concerns and minimise chances of opposition by particular groups.
- Involve as many stakeholders as required, most importantly those who are affected in their day to day life by the target species, or will be by the management techniques.

PREPARE THE DETAILS OF YOUR OPERATION

It is essential that the timeframe and sequence of events required, as well as the responsibilities are well defined:

- Identify and describe in detail the methods and technical activities that need to be conducted for the entire eradication process to be completed, including pre-eradication; eradication; and post-eradication phases[4].
- Identify a motivated team that comprises good field workers, a project leader, and experienced field practitioners.
- Pay particular attention to key equipment and logistical requirements. This may include the use of specialised machinery or means of transportation (boats, helicopters, etc.), availability of fuel and maintenance aspects.
- List all items and equipment required for the operation, including spares and materials needed for logistical support (food, camping gear, tools, first aid kits, medicine box, documentation, satellite phone, laptops for data input, etc.).
- Use tick lists to make sure you have not forgotten anything.
- Prepare a timetable listing in chronological order all key activities that need to be conducted. Show who is responsible for each activity and indicate any particular precautions required (use for example a table with the six following columns:

When (date and time deadline)	What (activity)	Who (person in charge)	Conditions & Precautions	Done (tick when done)	Remarks

Logistics are crucial when working on islands, especially if these are very remote and far from the civilised world. The expedition that eradicated rats from several islands important for seabirds at Cosmoledo atoll (c. 1,200 km from Mahé) was conducted by ICS and IDC in November 2007, but took several months to prepare[32-34]. It involved a barge carrying over 15 tonnes of cargo with equipment, food, water, rodenticide and fuel, plus a helicopter. Twenty people worked continuously for three weeks, including eight crew members on the barge; a ground team of eight people under tents; a coordinator and medical assistance on Mahé contactable through satellite phone; and three helicopter crew members including a specialised mechanic. The Bell Ranger helicopter had to fly island-hopping to Cosmoledo, with plane escort from IDC planes during part of the trip, and was carrying an extra reserve of fuel as the distances to be crossed (380 km) were at the limit of its autonomy. It had to remain over a week in the southern islands, and fly back to Mahé at the end of the operation.

All key specialised equipment (e.g. spreader-bucket, helicopter spares) had to be duplicated in case of any breakdowns. Clearly, the long preparation, the meticulous organisation and the importance of the complex logistics set in place were decisive for the success of the operation.

DECIDE WHAT TO MONITOR

Plan what, how and when monitoring will be conducted to ascertain the success of the project, and to evaluate the impact of the eradication on the ecosystem.
• Select a number of indicators that can be used to measure success.
• Design a monitoring protocol for data collection.
• Describe how the data will be analysed, including which statistical tests may be used. The programme needs to be based on sound science from the start, with accurate and unbiased data, and impartial assessments. More details on monitoring can be found on page 166.

DURING THE ERADICATION PHASE

Incorporate the following guidelines when managing your team:
• Follow your operational plans insofar as possible. Take time to think and consult with experienced people before you decide on any changes (in a remote location, a satellite phone is an invaluable tool for this and obvious security reasons).
• In case of uncertainty, adopt a *precautionary principle* and make the choices that will increase the chances of success of the eradication. Put all the chances on your side!
• Keep members of the team motivated and focused on their work, with a high moral and a good team spirit.
• Assign team members a variety of tasks. This will provide training and development of multifaceted local capacity.

An eradication is all about team work. It requires a strong motivation, determination and tenacity, and a certain *'fighting spirit'* against adversity when operations become longer and more difficult than expected.

AFTER THE ERADICATION PHASE: ASCERTAINING SUCCESS AND GENERAL EVALUATION

Post eradication evaluation should include the following:
• Post-eradication monitoring conducted to ascertain success by means of a number of pre-defined indicators. For rat eradications, for example, no signs of rats detected during survey trapping after 12 months may be used as an indicator of success (as per our own experience on tropical islands, or after 24 months as in temperate or subantarctic bioclimates (see "Ascertaining success", p. 225). In Mexico, a survey method using a grid of wax tags spaced 50 to 200 m has proved its efficiency in validating eradication success within a few weeks or months after the operation[35-37].
• Continued monitoring for years to come in order to determine long-term impacts on the ecosystem and to demonstrate the benefits of the eradication. Of course, biosecurity measures will also need to be implemented.
• Production of a final report describing the operation, especially in terms of the results of the elimination of the targeted species; non-target species mortality; problems

encountered (methods and techniques, logistics, funding, management, etc.); an evaluation of the success or failure of the project; and what lessons were learnt[38].

There are different ways to take into account the above requirements when considering an eradication. This includes the production of an initial feasibility study; or a more complete proposal including aspects of feasibility and a general eradication plan. As the project progresses, logistics and implementation of the eradication techniques will need to be reviewed and changes integrated into the project.

References / Further reading
1Wittenberg & Cock, 2001; 2 Genovesi, 2011; 3 Glen *et al.* 2013; 4 PII, 2011; 5 Courchamp *et al.*, 2003; 6 Emerton & Howard, 2008; 7 Hoare & Hare, 2006; 8 Plentovich *et al.*, 2010b;9 Rocamora & Labiche, 2009c; 10 Merton *et al.*, 2002; 11 Rocamora, 2005; 12 Courchamp *et al.*, 1999; 13 Ringler *et al.*, 2014; 14 Caut *et al.*, 2007; 15 Courchamp *et al.*, 2011; 16 Hill, Holm *et al.*, 2003; 17 Zavaleta *et al.*, 2001; 18 Griffiths, 2011; 19 Innes & Saunders, 2011; 20 Russell, 2011; 21 Campbell *et al.*, 2011; 22 Nogales *et al.*, 2013; 23 Rayner *et al.*, 2007; 24 Bonnaud *et al.* 2011; 25 Le Corre, 2008; 26 Russell *et al.*, 2009a; 27 Nishijima *et al.*, 2014; 28 DoC (2003, reviewed 2010); 29 Harper 2014; 30 Rocamora, 2015; 31 Soubeyran, 2010; 32 Rocamora, 2007a; 33 Rocamora & Jean-Louis, 2009; 34 Rocamora, 2010a; 35 Samaniego *et al.* 2013; 36 Samaniego *et al.* 2015; 37 Aguirre *et al.*, in press; 38 see for example Climo & Rocamora, 2006.

Strict biosecurity measures must be already in place before the eradication of an invasive species, in order to prevent reinvasion.
Marc Stickler / North Island

Conservation dogs trained for rat detection to prevent island reinvasions in New Zealand.
James Russell

Containment and exclusion

"**Containment**" of an alien species involves restricting its spread beyond currently infested areas; while "**Exclusion**" entails restricting it from reaching a particular island or geographic area where it is currently absent. These approaches entail:

1. Creation of buffer zones in which numbers of individuals and propagules are eliminated and controlled in order to reduce chances of spread.
2. Abatement measures to prevent colonisation of non-infested areas.
3. Surveillance of non-infested areas to enable early detection and rapid eradication in the case of incursion.

Containment and exclusion can be technically difficult and require constant effort. The success of containment relies on the simultaneous implementation of efficient control measures in buffer zones to minimise the threat of spread; strict biosecurity protocols to exclude the species from non-infested areas; and effective monitoring methods to quickly detect and eradicate invaders. Success is more likely achieved at islands or in areas characterised by some kind of natural isolation, or by species that spread slowly over short distances or for which fecundity and dispersal can be limited. In such cases, containment can be highly effective to restrict the impact of the invasive species and to prevent damage to other parts of the country, and it can be an interesting alternative when eradication is not feasible[1].

The advantage of containment and exclusion is that areas or islands (such as biodiversity sanctuaries) can be kept pest-free over the long term, a strategy that has been in use in neighbouring La Réunion and in various other French overseas island territories[2, 3]. However, containment can be inconvenient in that it requires efficient and coordinated action on different fronts. Moreover, the threat of invasion will always be there, like a 'Damocles spade'. Hence, the success of containment usually requires coordinated efforts by various stakeholders in different places to contribute to surveillance and preventive measures. It follows that a good communication about the importance of containment is essential to secure widespread support and contribution from local communities and the public in general

The difficulties and challenges that containment or exclusion represent are exemplified by efforts to prevent rodents from spreading to the numerous small and middle-sized rat-free islands that exist in Seychelles. Failure to carry out any of the following measures is likely to, sooner or later, lead to recolonisation by rats[4, 5]:

- Effective control of the number of rats in a large exclusion area around piers, airports and storage rooms;
- Strict adherence to abatement protocols (such as chemical treatment and/or systematic checking of cargo, boat loading procedures on Mahé and Praslin, unloading, storage and checking procedures on the rat-free island);
- Immediate detection and elimination of invaders on a rat-free island (through rodenticide bait-stations, gnaw-sticks, tracking tunnels, survey trapping and permanent alert).

Another example of the importance of containment can be found in efforts to contain the Ringed-necked parakeet *Psittacula krameri* on the island of Mahé until it can be eradicated (see case study p. 276). The parakeet represents a serious threat to the endemic Seychelles black parrot should it reach Praslin, which hosts the entire breeding population of black parrots. Parakeets are also being excluded from Silhouette (one individual was shot in 2014), where there are plans to create, through translocation from Praslin, an alternative population of Seychelles black parrots to act as a native ecological replacement of the extinct endemic Seychelles parakeet *Psittacula wardii*[6-8].

Other species that need to be contained include: several invasive plants (such as Koster's curse, nowadays only present on Mahé and Silhouette); and ants (such as Crazy ant or Big-headed ant) that are still restricted only to a few islands. A list of species that need to be contained and require specific attention to avoid being spread to other islands is provided in Table 16, p. 184.

By opposition to containment, exclusion aims at protecting from invasive species areas of high biological value such as nature reserves or pristine habitats. Fencing has been successfully used as a conservation management technique to keep invasive animals (ungulates, but also predators such as rats, cats, or mongooses) out of biodiversity hotpots[9, 10]. Any incursion of the invasive species into the fenced area has to be immediately eliminated. This strategy can be applied in cases when eradication is not feasible at a large scale, but can be envisaged inside the exclusion area[1].The concept of 'mainland islands', i.e. pest-free biological islands within large infested 'mainland' or 'inland' areas where eradications (for example of rats or cats) cannot be envisaged[10-12], has originated in New Zealand, and could also be applied in Seychelles (see following box).

References / Further reading
1 Wittenberg & Cock, 2001; 2 Anonyme 2010; 3 Soubeyran, 2008; 4 DoC, 2003-2010; 5 Rocamora, 2015; 6 Rocamora & Laboudallon, 2009; 7 Rocamora & Laboudallon, 2013; 8 Reuleaux *et al.* 2014; 9 Micol & Jouventin 1995; 10 Courchamp *et al.*, 2003; 11 Burns *et al.* 2012; 12 Veitch *et al.* 2011.

Temporary beach fencing during landing of materials and other biosecurity measures aim at 'excluding' invasive rats from rat-free islands such as Aride.
Gérard Rocamora / ICS

PREDATOR FENCING: A SOPHISTICATED EXCLUSION TECHNIQUE THAT COULD BE EMPLOYED IN SEYCHELLES

On large islands (or continental areas) where eradication is not feasible or too costly, exclusion by physical fencing can be conducted[1]. Fencing to contain feral ungulates has been used successfully in many cases, for example with feral goats and pigs in Hawaii, or with cows in Ile Amsterdam, Terres Australes Antarctiques Françaises[2-4]. In New Zealand, Australia, and more recently in Hawaii, predator-proof fences up to several kilometres long have been erected to prevent access to rats, cats, and other invasive mammals[4, 5]. Their purpose is to create sanctuaries for native species over entire peninsulas or inland areas (contributing to the concept of 'mainland-islands' or 'biological islands').

However, predator-fences are very expensive to build, check and maintain. They need to prevent access below and under the ground to burrowing animals like rats; and their expected lifespan may only be about 20 to 25 years[4]. Vegetation needs to be regularly trimmed and cleared on both sides of the fence so that falling trees or branches will not break it or provide pathways for invaders[5]. Storms and other climatic accidents can also create serious damages. These might include exceptionally high waves, which is why these fences are often kept in the open near the coast. Even with the best available materials, corrosion will affect the structure. Therefore very regular inspections are required to implement biosecurity procedures, and detect and repair immediately any potential problem[6].

Intensive (and permanent) control is also necessary in a buffer zone by using a grid of bait stations and traps on the outside of the fence, in order to reduce to a minimum the density of the potentially invasive animals. A similar grid (including tracking tunnels and other detection devices) is also required inside the fence and needs to be monitored regularly[5, 6]. Despite the quality of these fences, incursions regularly occur and need to be dealt with immediately before the invaders inflict serious damages and start breeding.

On Kaena Point (Oahu, Hawaii), a predator-fence of 630 m has been operational since 2011. Its total cost was about $500,000 and cost-effectiveness (compared to the previous and less efficient control operations) is expected to be reached after 16 years. Incursions have occurred on average every seven months for cats and mongooses, every two months for rats, and even more often for mice[5, 6].

Predator-fences might be tried on the large islands of Seychelles to rehabilitate and create predator-free peninsulas or inland areas with high biodiversity value. On Mahé, for example, the peninsula of Cap Matoupa is probably the most appropriate location to consider; and a mainland island has also been suggested on Desroches. This technique could provide sanctuaries for native species of plants and animals in general, and opportunities to maintain or reintroduce to the large islands flagship species of threatened landbirds (e.g. Seychelles white-eye, Seychelles magpie robin, Seychelles warbler), and seabirds. Seychelles could join the very restricted club of nations that have so far erected predator fences. This could bring emblematic animal species, now tucked away in small private islands of difficult access, back to Mahé and other large islands where they once thrived.

Members of the public sometimes complain about the fact that the rarest emblematic wildlife species of Seychelles are out of their reach in private islands and remote nature reserves where they are mainly accessible to foreign visitors. It has been suggested in public meetings that government and conservationists should also work at bringing these species back to the large inhabited islands of Seychelles, closer to the human population. This would greatly benefit education programmes for the general public and school children that would get a direct access to these species. This is something predator-fences would be able to achieve.

However, predator-fencing is very demanding in terms of financial and human resources, and also in terms of biosecurity[4-6]. The management of these fences requires a level of monitoring, discipline and savoir-faire well above the routine abatement measures needed to keep a middle-sized island rat-free. Such projects would therefore need to ensure from the beginning that they can be made sustainable in the long term.

Predator fences against rodents and cats could be erected in certain favourable sites in Seychelles. However, this technique requires a higher level of maintenance, monitoring, technicity and biosecurity than keeping an island rat-free (Kaena point; Oahu, Hawaii islands).
Gérard Rocamora

A double gate system with a slide door (slightly inclinated to close automatically) controls the entrance of the fenced area. Note the shoe brushes next to the entrance point to avoid carrying alien seeds into the reserve.
Gérard Rocamora

References / Further reading

1 Courchamp *et al.*, 2003; 2 Micol & Jouventin, 1995; 3 Veitch *et al.* 2011; 4 Burns *et al.* 2012; 5 Young *et al.* 2013; 6 L. Young & E. VanderWerf, *pers. comm.*

Control

When eradication and containment have failed, are judged non feasible, too expensive, or too risky, long-term control measures become the preferred option. By combining impact mitigation (see next section) and control, it can be possible to 'live with the species' in the best possible way and keep its impacts below an acceptable level.

What level of abundance can be considered acceptable for an invasive species depends largely on the overall management objective. There is often a trade-off between the seriousness of the impacts of the species and the costs of its control[1, 2]. For very high-impact species, such as invasive rodents or predators, the abundance of the invasive species will need to be drastically reduced to avoid significant ecological, economic or public health damage. On the other hand, for low-impact or relatively harmless species, such as certain invasive trees or agricultural weeds, reptiles (e.g. Pacific gecko), or some invertebrates, relatively high densities may be tolerated.

In some cases, preliminary research may be required to assess the long-term impact of the target species and determine what density threshold could be acceptable[3]. Suppressing the targeted invasive species will generally lead to an increase in native biodiversity. Other harmful aliens that previously had been controlled by the target invasive may also benefit. However, this problem is unlikely to be as acute in a situation in which the targeted invasive is controlled as when it is eradicated.

Control is the most basic and popular form of management to fight against invasive species. In our everyday lives, just around our houses and gardens, we are confronted with the need to control an ever increasing number of invasive species such as rodents, creepers and weeds, other exotic plants and trees, insects or fungal diseases that destroy our fruits and vegetables.

Any kind of living organism can be controlled to a certain extent using one or several suitable available methods. The different type of methods (physical, chemical and biological) that can be used to control, contain or eradicate alien invasive species are presented in the previous chapter.

The best appropriate control method and protocol needs to be selected based on previous experience, although this is less crucial than for eradication[4]. For example, a given pest management method may be more effective at certain periods of the year, when stages of the targeted species will be more vulnerable. This may depend on biological cycles (e.g. adults or larvae for insect control), or on the availability of food (e.g. rodenticide being more efficient to control rats when little or no alternative sources of food are available). The species accounts presented in the second part of this book provide practical examples of recommended methods to control the most common invasive species in Seychelles, and more methods can be found from additional sources[5-8]. At La Réunion, control techniques have been trialled and detailed protocols have been established and recommended specifically for each of the most important invasive plants[9, 10]. In Seychelles, research is also being conducted to determine the best protocols to control effectively the most common invasive plants and trees (see p. 326 and 340).

Control will normally be achieved more easily in less infested areas (as shown at La Réunion with invasive plants[11], see p. 112). This allows native biodiversity to recover quicker. However, in certain cases, for example those involving rats, food resources in the habitat are scarce when the density of the targeted animals is high; therefore the attractiveness of bait (rodenticide, traps) and the efficiency of the treatment are increased.

Despite being much less intensive than eradication in the short-term, all control methods except classical forms of biological control require long-term funding and commitment[1,3]. One of the main problems with control is that its interruption or cessation – due in particular to funding problems or a change in priorities – will lead to an increase of the targeted invasive species. Levels may revert to those equal to –or even higher than – before control was started, and damage to native biodiversity may be irreversible.

The success of control may also depend on factors intrinsic to the ecosystem. These may include rainfall and hygrometry that may affect the efficiency of herbicide treatments; the availability of naturally available food in the case of rodents, etc.; or on the intensity of other restoration efforts conducted (e.g. rehabilitation of vegetation, elimination of animals dispersing invasive plants, etc.). Several types of control methods and mitigating measures can be combined to form an integrated pest management strategy[1,6] (see p. 157).

Control success will also depend on communication and public participation, especially when addressed at widespread invasive species that affect the whole country. In some instances, collective efforts may be hampered by the inappropriate attitudes or management habits of some. In the case of rats, for example, garbage bags deposited out of bins, bins not collected daily, insufficient cleanliness and absence of rat control at bin sites will generate proliferation of rats in a neighbourhood, and will hamper individual control at individual neighbouring houses. The same may apply to the control of mosquitoes or invasive pests like the Hairy caterpillars. Most importantly, we as members of the public, private companies, NGOs and civil society in general, should not expect one specific government agency that may be spearheading and coordinating at national level a particular control campaign, to be the sole stakeholder taking responsibility and to do it all. Fighting invasive species concerns us, and we all need to play our bit to make it work.

Control programmes can be effective in maintaining the densities and the nuissance created by invasive species below an acceptable level (Creeper control in Vallée de Mai).
Lucia Latorre/SIF

References / Further reading

1 Wittenberg & Cock, 2001; 2 Orueta & Ramos, 2001; 3 Courchamp *et al.*, 2003; 4 Soubeyran, 2010; 5 GISD 2014; 6 CABI 2014; 7 Orueta, 2007; 8 Soubeyran, 2008; 9 Triolo 2005; 10 GEIR 2014 www.especesinvasives.re; 11 Brondeau & Triolo, 2007.

PLANNING AND SETTING UP A CONTROL OPERATION

Control plans need to be defined and reviewed in an adaptative way, depending on results obtained and new knowledge acquired:

Once the type of control operation has been decided, an Action Plan (Control Plan) needs to be prepared. It should identify:

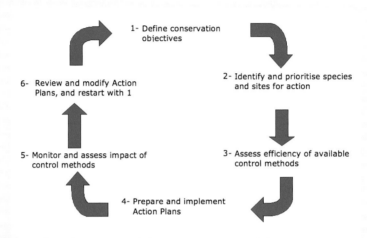

1- Define conservation objectives

2- Identify and prioritise species and sites for action

3- Assess efficiency of available control methods

4- Prepare and implement Action Plans

5- Monitor and assess impact of control methods

6- Review and modify Action Plans, and restart with 1

- Clear conservation objectives
- Priority invasive species targets and priority areas for intervention
- Choice and assessment of methods of control
- Technical, economical and logistical details
- Assessment of impact of the control (positive and negative consequences)
- Type of monitoring and data management

References / Further reading
Tu *et al.* 2001; Soubeyran 2010;

Adapted from Tu *et al.* 2001 and Soubeyran, 2010.

Mitigation of impacts and ecosystem management

Mitigation aims at reducing the impact of invasive species without taking direct action against them. The action is directed instead at the native species, habitats, goods or infrastructure affected, or at the ecosystem (or agro-ecosystem) in general.

Mitigation is sometimes seen as a method of last resort that results in 'living with' the invasive whilst minimising its impact after other management methods have failed or are deemed not applicable. Mitigation measures alone may be neither sufficiently effective nor provide a long-term fully satisfactory solution to the problem, but they will generally help to significantly reduce its extent until other alternatives become available, or when used combined to other management techniques[1].

For example, the problems caused by rats in households, hotels and residential areas in general can be significantly reduced by a clean environment and efficient garbage disposal, such that food sources available to the rats are minimised. Regularly trimming vegetation around buildings, blocking of accesses, installation of disks around electric and telephone wires (to prevent access of rats to roofs), daily collection of fruits that have fallen on the ground, etc. are simple mitigation measures that will help to reduce rat abundance around houses and backyards.

Mitigation has been commonly used for the conservation of threatened species and sites at various levels. Measures can be used to address particular problems of endangered endemic species, for example by providing predator-proof nest boxes to rescue a critically endangered bird from extinction. On Praslin, rat-proof nest boxes designed by Senior Conservation Officer Victorin Laboudallon (Department of Environment) in the 1980s have been successful in preventing rat damage to the nests of the Seychelles black parrot in the wild[5-7]. Translocating endangered endemic birds from infested to predator-free islands, a technique commonly used in New Zealand and Seychelles (see chapter 2), may be seen as an extreme form of mitigation against invasive predators. The same may also apply to the breeding and rearing in captivity of endangered birds and reptiles for reintroduction in predator-free or predator-controlled areas. This technique, also common in New Zealand, was conducted with great success in Mauritius and has prevented the extinction of vanishing endemic species such as the Mauritius pink pigeon *Nesoenas mayeri*, the Mauritian kestrel *Falco punctatus* and the Echo parakeet *Psittacula echo*[3].

Ecosystem management is also a form of mitigation applied at landscape level. It consists in managing the entire ecosystem in order to favour native species and inhibit introduced species in general; which prevents obvious advantages of cost-effectiveness and scale compared to single-species management[4]. This may be done for example by limiting the creation of roads and tracks in pristine conservation areas to prevent the penetration of new invasive plants. Measures to manage habitats or ecosystems may include practices that hamper the development of populations of certain invasive species (e.g. crop rotations in agriculture, fire in certain types of pastures), or that favour the development of populations of native predators or parasitoids that will counter the invasive species[1]. Another approach is to divert the impact of the invasive to less vulnerable species, for example by planting alternative native host plants to distract invasive predators or parasites.

In general, the best way to minimise the invasion and spread of invasive species, particularly of plants, is to conserve natural habitats in a pristine state, i.e. not altered by disturbances such as roads or construction[4]. Adequate ecosystem management can limit the number and the magnitude of invasions; in particular, pristine ecosystems are much less prone to invasions than human-modified ones. Mitigation and ecosystem management can be used efficiently in conjunction with control, containment or exclusion, but it is often labour intensive and costly[1].

References / Further reading

1 Wittenberg & Cock, 2001; 2 Soubeyran, 2010; 3 Jones & Merton 2012; 4 Simberloff, 2014; 5 Nevill, 2001b; 6 Rocamora & Laboudallon, 2009; 7 Rocamora & Laboudallon, 2013.

Erection of rat-proof nest-boxes especially designed for the Seychelles Black Parrot is a good example of mitigation measure.
Gérard Rocamora

Acting on multiple fronts: An integrated management strategy

An integrated management strategy involves combining, in an intelligent and cost-effective way, a variety of control techniques and actions that will most effectively reduce the impact of invasive species. This is inspired from Integrated Pest Management (IPM), or as it was first called 'integrated control' (*lutte intégrée* in French), a practice that originated in the 1960s to fight agricultural pests[1].

Integrated pest management was set up in reaction to observed negative impacts from the massive spraying of pesticides (that had prevailed during several decades) on non-target species including natural enemies of the targeted pests, which can some-times result in higher levels of the pest after the effects of the treatment has ceased. According to the UN's Food and Agriculture Organisation, IPM is defined as "the careful consideration of all available pest control techniques and subsequent integration of appropriate measures that discourage the development of pest populations and keep pesticides and other interventions to levels that are economically justified and reduce or minimise risks to human health and the environment". The principle is to make use of and favour the development of the natural enemies, antagonists, competitors or other biological agents already present in the habitat that can help control the pest. In contrast to classical chemical treatment that reduces pest populations to the bare minimum (in order to maximise immediate profits), integrated pest management attempts to minimise chemical treatments and to reduce the impacts of these pests to an acceptable level. This is normally achieved by limited timing and location of spraying, which minimises the risk that pests develop insecticide resistance, reduces the cost of treatment and ecosystem pollution. During the last few decades, integrated pest management and biological control have proved to be effective alternatives to control pests while reducing pesticide use[2]. However, IPM has been given nowadays too many different definitions and in many cases it has consisted in little else than using pesticides more judiciously, and acceptable levels of presence may be set so low that regular spraying of chemical remains the norm[3].

A parallel exists between integrated pest management and the fight against alien invasive species, where greater success is also generally achieved when a variety of methods can be efficiently combined. The main difference is that natural enemies of the aliens are normally not present and some may have to be introduced as part of biologi-cal control before more classical integrated pest management can be implemented[4].

When defining a management plan against an invasive species, especially when it is unlikely to be eradicated in the short term, it can be appropriate to use a variety of methods in combination, for example mechanical (e.g. trapping) and chemical treat-ments. These methods may be applied simultaneously to different areas, or to differ-ent categories of individuals, or in sequence depending on their respective efficiency and potential to create aversion in the pest, as seen in the sections on control and eradication. For example, cutting woody invasives or sisal by hand or with chainsaw may be efficiently combined with an application of herbicide to the stumps. Likewise, herbicides can be injected after the physical removal of creepers.

Ecosystem management (or habitat management) less favourable to the spread of the targeted invasive species can be added, as well as specific mitigating measures applicable to rare and endangered species in order to enhance the efficiency of the whole process[3]. Construction of exclosures and boardwalks may be combined with the removal of exotic plants in order to protect pristine stands of native vegetation from introduced herbivores and the trampling of visitors. This has been done in several high biodiversity areas of the Hawaiian Islands. However, it is difficult to provide general recommendations regarding how to conduct integrated control of invasives as this varies from one situation to another. Even if integrated management protocols may be replicated between different areas, they need to be tested and fine tuned for each specific location.

In the context of an integrated strategy against invasive alien species for Seychelles, it is important to fight not only on the ground, but also on the fronts of prevention, training, communication and networking. This principle is clearly mentioned in the Seychelles Biosecurity strategy (see p. 160), as it is in the biosecurity strategy on neighbouring island territories such as La Réunion and the Republic of Mauritius[6, 7]. There is also a need to prioritise the most appropriate

Integrated pest management and biological control have proved to be effective alternatives to control pests. Pheromone trap used against the Mediterranean fruit fly *Ceratitis capitata* in La Réunion.
Elvina Henriette

actions by conducting a risk and cost-benefit analysis, and preparing contingency plans that will facilitate rapid responses to IAS incursions[8]. This is needed at the national level, and also for particular islands. Emergency funds should always be available for a situation in which immediate action is required. A national Emergency Response Plan for for Pests, Diseases and Alien Invasive Species exists in Seychelles[5] (see website link p. 110).

Public education and communication are critically important long-term components of any strategy. Regular campaigns involving media, NGOs and the primary institutions that are active in the field of invasive alien species are required to further sensitise the general public, its political and economic leaders, and the scholars about the negative roles played by IAS and the need to take action against them. This can be done using examples of widespread invasives such as rats, creepers, the Spiralling white-fly, or the Hairy caterpillars *Euproctis* sp. (see case study p. 42) which affect each and everyone.

Coordination and collaboration amongst the various concerned stakeholders will be essential to meet the challenges ahead of us. The creation of a functional Seychelles Biosecurity Service (planned within the current Seychelles Agricultural Agency) is expected to play a leading role in these aspects. All these points are highlighted and developed in a report on the *Evaluation of threats of introduction and spread of IAS through production sector activities in Seychelles*[8].

References / Further reading
1 see for example Abrol & Shankar 2012; 2 Witt 2014; 3 Simberloff, 2014; 4 Wittenberg & Cock, 2001; 5 Government of Seychelles, 2014; 6 Anonyme, 2010; 7 Government of Mauritius 2010; 8 Dogley W., 2009.

SOME BEST PRACTICE GUIDELINES FOR ISLAND AND PROPERTY MANAGERS

Here are some management recommendations on biosecurity and the control / containment of invasive plants and animals:

- Regularly survey potential entry points (boat landing sites, helipad, airstrip, etc.) and paths used by visitors and island staff in search of new exotic plant species.
- Limit habitat alteration (e.g. construction, roads, fires, paths, etc.) that will favour penetration of invasive plants.
- Contain tourism infrastructure in defined areas and channel visits into natural habitats through well defined paths.
- Restrict access to areas with high biodiversity value, and implement a policy of visual inspection of shoes, socks and trousers to remove alien seeds from clothing at the starting point of such paths.
- Restrict access to areas with high biodiversity value.
- Use indigenous rather than exotic plants for landscaping. This will contribute to ecosystem restoration.
- Limit and control transfer of earth, sand, plants, etc. between infested and non-infested zones, and restrict vehicular movements between these areas.
- Avoid transporting green waste from one site to another– especially when it contains waste from invasive plants that reproduce by cuttings. Destroy such materials on site, or seal them for transportation. Clean machinery, clothes (especially socks), and shoes used during campaigns to eliminate invasive plants before moving to new areas.
- Keep areas around habitations clean and tidy, and trim vegetation around buildings to prevent access to rodents.
- Strictly control bins and do not leave any sources of food (scraps, fruits, etc.) that may attract rats or provide nourishment that will boost their reproduction. These actions will significantly reduce problems caused by rats.
- Undertake control of problematic species such as rodents or highly invasive plants.
- Communicate with other island managers, NGOs, private experts, etc. about invasive species, levels of infestation, identification, and methods of eradication, control or mitigation.
- Plan to eradicate major pests and prioritise what actions are needed.
- Remove artificial containers of rainwater that provide breeding sites for mosquitoes (e.g. old tyres, empty tins, flower containers, empty coconuts, etc.), both near houses and in natural habitats.

Further reading
Wittenberg & Cock, 2001; Merton *et al.*, 2002; DoC, 2003-2010; Soubeyran, 2010; PII, 2012; Harper, 2014; Rocamora, 2015. See also recommendations in species accounts for rats, Tiger mosquito, and invasive plants.

SEYCHELLES NATIONAL BIOSECURITY STRATEGY

The National Invasive Alien Species (Biosecurity) Strategy for Seychelles 2011-2015 (NBS) was developed and finalised through a process of national stakeholder consultation under the auspices of the GEF full-size project: *"Mainstreaming Prevention and Control Measures for Invasive Alien Species into Trade, Transport and Travel across the Production Landscape"*.

The National Biosecurity Strategy 2011-2015 was drafted in response to the existing Invasive Alien Species (IAS) problems, and to the growing threat IAS introduction posed to the biodiversity, production landscapes and economy of Seychelles by the rapid development of international trade, travel, transport and tourism.

The strategy is based upon best current international practices as set out under the International Plant Protection Convention, the Convention on Biological Diversity and the Global Invasive Species Programme. Its main aim is to put in place an overarching national strategic approach and effective legal, administrative and technical framework to address the prevention, control, mitigation and eradication of invasive alien species in the Seychelles archipelago.

The NBS sets out clear Vision and Mission statements supported by objectives and an action framework to attain the Mission during the five-year time span.

Vision

The impact of invasive alien species upon ecosystem services, biodiversity and economic sectors is prevented, minimised and where possible reversed.

Mission

A comprehensive, integrated and transparent Biosecurity framework is in place and functioning effectively to manage the spectrum of invasive alien species issues in Seychelles.

The objectives focus on the main management components (i.e. prevention, control and eradication, strengthening of the existing institutional framework) and also the need to raise awareness of invasive alien species issues. The NBS is structured as a framework strategy that identifies key areas for attention and proposes means and mechanisms to address them.

Objective 1: The introduction and establishment of invasive alien species in Seychelles are prevented and minimised.

Objective 2: The spread and impact of invasive alien species in Seychelles are prevented, minimised and effectively managed.

Objective 3: A comprehensive, empowered and transparent institutional and legislative framework is established.

Objective 4: Biosecurity issues are fully integrated across all sectors of the community.

Objective 5: A targeted invasive alien species communication strategy is developed and under implementation.

The Seychelles Agricultural Agency, under the portfolio of the Minister responsible for Agriculture, is responsible for the administration of the NBS through its Plant and Animal Health Section which will be legally empowered for this role under the pending Biosecurity Act[a]. The Act further mandates a representative National Biosecurity Committee to steer, manage and elaborate the NBS in line with the guidance provided under the document's Action Framework, and thereby provide the national authorities, partners and stakeholders a clear roadmap for the implementation of the Strategy.

. . . .

a   The National Biosecurity Committee has been functional since 2015 and includes representatives from a dozen of ministries, civil society institutions and parastatal agencies that are dealing with environment and energy, natural resources (agriculture and veterinary services), health, transport, tourism, finance, and media.

Further reading

Government of Seychelles, 2011. The National Invasive Alien Species (Biosecurity) Strategy for Seychelles 2011-2015. Ministry of Home Affairs, Environment and Transport & Ministry of Industry, Natural Resources and Industry, Seychelles.

Contributed by Didier Dogley (Minister of Environment, Energy and Climate Change) and John Nevill (Consultant, Mainstreaming 'Biosecurity' project).

Integrated pest management intends to minimise chemical treatments and promotes more environmentally friendly techniques such as biological control using pheromone trap (Anse Boileau agricultural research station, pheromone trap for Melon fruit fly and Mediterranean fruit fly).
Gérard Rocamora

Managing invasive species: How much does it cost?

Managing invasive species has a cost. Nevertheless, management is generally cost-effective in the short, middle or long term in view of the potential damages that invasive species can inflict. These costs need to be evaluated carefully and taken into account during the decision process of whether to control or eradicate a particular invasive alien species.

The initial cost of eradication is high; but post-eradication costs for ongoing surveillance and implementation of abatement measures to prevent reinvasion are lower. In the case of vertebrate pests, eradication includes substantial expenses for feasibility studies and planning during the pre-eradication period, and high operational costs. This is especially so when islands are remote and sophisticated techniques such as helicopter bait spreading are required. Post eradication costs such as intense survey trapping to ascertain the success of the eradication, biodiversity monitoring and operations to safeguard rare and threatened wildlife (e.g. endangered birds, Giant tortoises, etc.) are also substantial. In contrast, control programmes involve much lower preparation and continuous operational costs, but these need to be supported over the long term. When eradication is technically feasible, it is usually a better and cheaper long-term option than control, as long as the required rigorous discipline to prevent reinfestation can be maintained.

Vegetation restoration or rehabilitation requires complete or partial removal of invasive plants and replanting with native species. High costs of maintenance over a period of 2-3 years need to be taken into consideration.

Table 10 shows some rough indicative costs for different types of invasive species management operations in Seychelles, based on both published and unpublished cost estimates. For an eradication, costs increase with the size of the island and its remoteness from Mahé. Eradication costs may be shared and significantly reduced (up to 1/3) when several operations are combined – for example, eradicating the same species on several islands, or eradicating several species on the same island. It has been possible to conduct eradication operations that involve aerial baiting with a helicopter on an entirely local basis in Seychelles since 2007. This has helped to reduce costs significantly compared to earlier operations.

The cost of ecosystem monitoring should also be taken into account when considering an invasive species management programme, particularly in the case of eradications. Documenting ecosystem recovery has important implications for management recommendations, to help plan future operations, and to predict changes that will occur in the ecosystem. Unfortunately this has not been considered a priority in most past operations, both in Seychelles and abroad. So, the amount of information currently available on ecosystem recovery following eradication or control programmes is limited.

We recommend that a minimum basic ecosystem monitoring scheme addresses birds (seabirds and landbirds), reptiles, invertebrates and plants; and that it is repeated annually or every few years. If possible, pre-eradication surveys should span several years (see "Ecosystem recovery after the eradication of rats and cats", p. 169). It is difficult to provide general figures indicative of costs, given the diversity of situations that might be encountered (island size, distance to mainland, availability and cost of

Table 10: Indicative costs for different types of invasive species management operations in Seychelles

Invasive species management operation	Indicative costs (2012)
- Rat control on forested land (with rodenticide)	SR1,700 to 2,600 (€100 to 150)/ha/year
- Rat eradication on a small island (<30 ha; manual)	SR10,000 to 15,000 (€600 to 900)/ha
- Rat eradication on a larger inner island (aerial)	SR8,000 to 10,000 (€500 to 700)/ha
- Rat eradication on a larger outer island (aerial)	SR12,000 to 20,000 (€700 to 1,200/ha)
- Construction of a rat-proof room (10 to 100 m²)	SR5,000 to 8,000 (€300 to 500/m²) (with boat transport)
- Removal of exotic plants with replanting and manual maintenance (c. 1000 saplings/ha)	Up to SR250,000 (€15,000)/ha) (sloping lowland & accessible intermediate forest) Up to SR800,000 (€50,000)/ha (upland forest)
- Control (removal) of invasive coconut palms only (100 to 200 adults/ ha and regeneration) in a small granitic island.	SR16,000 to 35,000 (€1,000 to 2,000/ha; without replanting or maintenance) (sloping lowland & accessible forest)
- Gradual removal of invasive coconut palms (100 to 200 adults/ ha and regeneration), plus replanting and maintenance in a coralline island.	About SR65,000 (€4,000/ha) (lowland coconut plateau forest; 'Canopy' method).

expertise, volunteers, presence of threatened species, etc.). However, experience from the ICS FFEM project suggests that a minimum of SR50,000 to SR100,000 (c. € 3,000 to 6,000, excluding transportation and accommodation costs) should be budgeted each year for routine ecosystem monitoring. This will enable biodiversity data to be collected and at least stored in databases for detailed analysis when the opportunity arises (University student, project funding, etc.).

When preparing budgets for eradication operations, it is advised to always add a contingency of up to 20% to cover extra costs due to unforeseen difficulties.

Further reading
Beaver & Mougal 2009; Asconit & ICS 2010; Henri *et al.* 2004; Kueffer & Vos, 2004; PCA, 2009; Micol & Bernard 2010; Mwebaze *et al.* 2010; PII 2011; von Brandis, 2012; G. Rocamora & ICS, unpublished data.

An area replanted with native species after removal of exotic plants on Ile du Nord (left: Aug. 2006; right: Nov. 2011).
Katy Beaver / North Island

Improving knowledge and measuring change

Chapter **6**

Malaise traps are used to monitor
flying insects and measure changes
in invertebrate communities after rat
eradication (Conception Island).
Gérard Rocamora

Monitoring biodiversity

Our capacity to limit the impacts of invasive species and make informed decisions during the course of control or eradication operations relies on the continued collection of scientific data. Managing invasive species effectively requires a good knowledge of the initial distributions and abundances of the targeted species and an understanding of their biology. Preliminary studies are needed to collect the following basic information about invasive animals: preferred habitats, diet, breeding biology, reproductive output, etc. For invasive plants, information is also needed about timing and mode of reproduction, pollinators, abundance of seed production, dispersion etc. This will help to define the best appropriate protocols to eradicate, control, or mitigate the impacts of the targeted invasive species. This brief chapter provides some basic guidance about how biodiversity monitoring programmes may be set up to evaluate the impact of invasive species management on species and ecosystems, and examples of changes measured following the removal of rats and cats in some islands.

Knowing how best to monitor the success of management efforts can be challenging. In order to evaluate the degree of success of the management efforts, and whether or not the overall objectives of the programme are being met, a number of targets or indicators needs to be set from the beginning and a monitoring programme designed and implemented[1]. This will define the methods to be employed, the quantity of data to be collected, and how the data will be analysed so that statistically meaningful conclusions can be drawn[2, 3]. The approach typically used to investigate the success of management efforts and its impact on biodiversity is a 'time approach', whereby certain biodiversity parameters are monitored periodically on a same area or island where invasive species management protocols are taking place. This approach is also used to monitor the recovery of an ecosystem over time after the eradication of an invasive species. Alternatively, a 'space-for-time approach' can also be used by measuring simultaneously the same biodiversity parameters across areas or islands under different management regimes. The two approaches are complementary and their combined use is encouraged[33]. Similarly, when investigating the impact of invasive species, combining data of pre- and post-invasion stages with data from invaded and non-invaded sites has proved very useful.

   Ideally, as part of a usual 'time' approach, a baseline study should be conducted against which the efficiency of management operations and long-term changes observed in the ecosystem can be measured. Such a study should address the main taxonomic groups – i.e. plants, mammals, birds, reptiles, amphibians and invertebrates (for terrestrial ecosystems). Detailed studies were conducted to evaluate changes in invertebrate, plant and bird communities on islands where rats were eradicated during the ICS-FFEM project[4-6] (Ile du Nord, Conception, Cosmoledo; see case study p. 96 and the following section on ecosystem recovery). In a different study conducted with SNPA, changes in pollination webs on native vegetation present in granite outcrops have been closely monitored after the removal of exotic plants from these areas[7]. If possible, the baseline monitoring should begin several years before the start of the management operations in order to record the temporal variations in the distribution and abundance of both the targeted invasive species and the other species present. The data produced will help to differentiate between the impact of the management operations and other factors at play (such as interannual variations of ecological).

Unfortunately, such comprehensive baseline studies are rarely possible. Pre-eradication or pre-control data are often limited to a year or less prior to the management operation starts, as decisions to implement these programmes are generally not planned much longer in advance. Also, experience shows that securing funds to conduct baseline monitoring programmes is not easy, as island owners and site managers tend not to perceive such studies as essential to the success of the eradication or control operation. Biodiversity monitoring during eradication and control programmes should be given more importance, as a better understanding of the ecological processes involved will enhance the success of future operations[8].

Monitoring protocols may be set up specifically to determine how the targeted invasives impact native species and the ecosystem in general, and to evaluate the efficiency of the management protocol. For example, the abundance of rats and the breeding success of native landbirds may be measured before and after rats are either controlled or eradicated. Regular monitoring is also critical to ascertain the success of a particular eradication programme, to confirm the pest free status after a certain period, and to immediately detect any reinfestation (as determined when survey trapping is conducted to ascertain the continuous rat free status of an island; see p. 226). The setting up of standardised georeferenced databases is key to the success of long-term monitoringprogrammes. Modern tools such as Global Positioning System, Geographic Information System, or remote sensing are helpful to georeference the data and visualise changes in distribution and abundance. A routine surveillance monitoring that involves plant and animal surveys (e.g. once every 1-3 years on sites of high biodiversity value) will enable detection of any new invasives and prevent their establishment. Species with the highest potential risks of becoming invaders may be identified and monitored.

Partnership with universities, environmental NGOs and other research institutions should be sought by the managers of high biodiversity value islands or sites when setting up monitoring protocols. National biodiversity databases on flora and fauna already exist in Seychelles, such as the KBA biodiversity database[9, 10], and can help with invasive species monitoring. For example, the interactive Seychelles Plant Gallery[11] can provide an updated illustrated list of all exotic invasive species recorded in any particular island (and in the country as a whole); it is accessible online to the public, and contributors can send their observations accompanied with pictures to complete or update it.

Many different methods can be used to conduct biodiversity surveys and measure ecosystem change. General monitoring methodologies can be found in a variety of manuals, for example for birds[12, 13], reptiles and amphibians[14], invertebrates[15-17], or plants[18, 19]; some of which available from the internet. Standardised simple methods are described in local handbook publications for Seychelles (e.g. circular plots for seabird monitoring[20]; rapid biodiversity monitoring of plants, rodents and invertebrates[21]); and more detailed methodologies are provided in specific studies on Seychelles landbirds (e.g. using point counts[22, 23]); invertebrates (e.g. using pit-falls, malaise traps and leaf counts[5]; or investigating plant pollinators[23]); reptiles (e.g. point-counts, quadrats[4, 24]); and plants (trail step-point transects, vegetation plots, photo monitoring[25-29]) (see also next section on Ecosystem recovery). Additional references for biodiversity studies in Seychelles are provided in the Seychelles Biodiversity Metadatabase (gap analysis) as well as references for international best practice[30]. Modern techniques to collect data in the field through the use of CyberTracker are also being trialed in Seychelles[31]. Examples of methodologies used to monitor ecosystem response around the world can also be found in the proceedings of the conference 'Island Invasives: eradication and management', in the section Results and Outcomes[32].

References / Further reading

1 Wittenberg & Cock, 2001; 2 PII, 2011; 3 PII, 2012; 4 Rocamora & Labiche, 2009a, 2009b, 2009c; 5 Galman, 2011; 6 PCA, 2009; 7 Kaiser-Bunbury & Mougal, 2014; 8 Rocamora *et al.*, in prep.; 9 Senterre *et al.*, 2013; 10 Senterre & Kaiser-Bunbury, 2014; 11 Seychelles Natural History Museum & PCA, 2014 www.seychellesplantgallery. com; 12 Bibby *et al.*, 1998, 2000; 13 Gregory *et al.*, 2004; 14 Bennett, 1999; 15 McGavin, 1997; 16 Wheater *et al.*, 2003; 17 Drake *et al.*, 2007; 18 Rew *et al.*, 2006; 19 Stohlgren, 2007; 20 Burger & Lawrence, 2000; 21 Hill *et al.*, 2001; 22 Henriette & Rocamora, 2011a; 23 Cresswell *et al.*, 1997; 23 Kaiser-Bunbury *et al.* 2011; 24 Noble *et al.* 2011; 25 Fleischmann, 1997; 26 Fleischmann *et al.*, 2005; 27 Beaver *et al.*, 2007; 28 Senterre *et al.*, 2009; 29 Senterre *et al.*, 2012; 30 Senterre *et al.*, 2010a, 2010b; 31 Kaiser-Bunbury & Senterre, 2014; 32 Veitch *et al.*, 2011; 33 Thomaz *et al.*, 2012.

Vegetation monitoring is essential to understand the impact of invasive species on the ecosystem.
Gérard Rocamora

Ecosystem recovery after the eradication of rats and cats

The impact of introduced rats and cats on island ecosystems has been well documented[1-6]. However, data that assess the impact of eradications of rats and other introduced predators on native wildlife are limited, especially in the tropical world where long-term ecological consequences have been little studied[7-9]. Ecosystem recovery on islands following rat and/or cat eradication appears to have been poorly studied, especially in the long term[10]. Nevertheless, most of the reported changes following rat and cat eradications around the world have been positive[11]. Observed benefits for island biodiversity include increases in the abundance of native animals and plants, spontaneous colonisation by seabirds[12-13] and the opportunity to reintroduce rare endangered endemic species not tolerant to the presence of introduced predators. However, the imbalance created by the removal of rodents or/and cats and the complexity of ecological interactions between species may also lead to unforeseen negative effects, especially when other invasive species are present and may – directly or indirectly – benefit from the eradication of rats (see 'surprise effects' and "Multiple species eradications", p. 143)[14, 15].

For Seychelles, there are very few publications that document the impacts of rat eradications[16-21], and much of the information that exists is actually scattered in unpublished reports and newsletters, or as empirical observations made by managers and conservationists. Long-term biodiversity monitoring programmes to document the consequences of rat and cat eradications on other species and on the process of ecosystem recovery are very rare and not implemented with regularity. A review study[22] has been conducted to synthesise all available information on biodiversity monitoring and ecosystem recovery on islands where rats – and in some of them cats – have been eradicated in Seychelles.

In Seychelles, between 1996 and 2010, invasive rats were eradicated from 11 islands of 10 ha to 219 ha in size (6 granitic and 5 coralline), and cats from six of them (see Table 5, p. 66). In addition, cats alone had been already removed before the 1990s from three other islands not included in our review (Aride, Cousine, and Picard of Aldabra; see details p. 64). Monitoring consisted in 13 formal protocols combined with empirical wildlife observations. Table 11 summarises the major biodiversity changes observed at 10 of these islands. A minimum of 60 populations from 25 native vertebrates (five reptiles, 12 landbirds and eight seabirds) and four large invertebrates (snails, millipedes, crabs) benefited from these 20 operations[22]. When considering also islands where only cats were removed, no less than 67 populations from 26 native vertebrates (including the Aldabra rail *Dryolimnas aldabranus*) were beneficiaries.

Reptiles and landbirds typically showed increasing or stable trends, although some populations declined strongly at first, such as the Common moorhen *Gallinula chloropus* or the Seychelles skink *Trachylepis sechellensis*, but then recovered well beyond their initial abundances. Reptiles such as the Bronze-eyed gecko *Ailuronyx seychellensis* (on Anonyme and Grande Soeur), or the Burrowing skinks *Pamelascincus gardinieri* (on

Shearwaters are also species that can recolonise island rapidly after control or eradication of introduced predators (Wedge-tailed shearwater, Ste Anne).
Gérard Rocamora

Conception and Grande Soeur) have clearly benefited from the removal of Black rats. Giant tortoises *Aldabrachelys gigantea* have increased their breeding success and their numbers on Ile du Nord, Grande Soeur and Frégate. Resident landbirds often demonstrated clear increases after the eradications, such as the Seychelles magpie-robin *Copsychus sechellarum* on Frégate (where it also benefited from other conservation measures),or the Seychelles white-eye *Zosterops modestus* and

Table 11: Observed major biodiversity changes and ecosystem recovery in Seychelles islands (≥ 10 ha) where rats have been eradicated. Years indicate period when indicated changes were noticed. **Bold**: major changes observed or measured from a monitoring protocol. Normal characters: non-significant changes or empirical observations. *Italic*: changes not necessarily related to the impact of rat or cat eradication (from Rocamora *et al.* in prep.).

Island	Year of eradication	Vegetation	Invertebrates	Reptiles (lizards/geckos)	Landbirds	Seabirds
Bird Island (Ile aux Vaches)	1997	Expansion of *Pisonia grandis* despite being initially affected by Crazy ants. Pisonia had colonised after the abandonment of coconut plantation in the 1970's	**Expansion and explosion of density of invasive Crazy ant after the eradication (1997).** *Control of ants conducted in 1998-2000*		Strong increase of moorhens, colonisation by Seychelles blue pigeon, introduction and expansion of Seychelles sunbird	Sooty tern colony size stable (annual census 1972-2013). **Increase in Lesser noddies (2002-09) and White-tailed tropicbird breeding numbers (2006;** then fluctuating).
Frégate	2000	*(continued re-creation of c. 60ha of native woodland through coconut removal and replanting)*	**Strong declines of endemic snails (2002)** then recovery for *Pachnodus* **(2012); stability or minor decline of Giant scorpion (2002) and Giant beetle (2002, 2011)**		Increase in distribution and numbers of introduced Seychelles white-eye (2001-2010)	**Increase in ground-nesting White-tailed tropicbird (2006).** Increases in tree-nesting Lesser noddy and Fairy tern probably linked to habitat restoration (2006, 2010, 2013).
Denis	2002	*Expansion of Pisonia grandis (2 to 3.6 ha) (re-creation of c. 35ha of native dominated woodland through coconut removal and replanting)*			Strong increase of moorhens, Increase in distribution and numbers of introduced Seychelles endemics (Warbler, Fody, Magpie-robin, Flycatcher) 2004-2012.	**Increase of Wedge-tailed shearwater and Fairy tern;** decrease of Brown noddy in 2000-2002 (cutting of coconuts).
D'Arros	2003	*(re-creation of c. 11 ha of native woodland through coconut removal and replanting)*				**Colonisation and increase of Wedge-tailed shearwaters (2009, 2013). Colonisation of Lesser noddy;** increase also related to habitat restoration.
Anonyme	2003	*(removal of Cocoplum and replanting of native woodland eg. 0.5ha)*		Increased sightings of Bronze geckos (observed 2008-2009)		
North Island (Ile du Nord)	2005	Increased abundance of coconut seedlings in unmanaged areas (2008) *(+ habitat restoration of c.45ha of native woodland through coconut removal and replanting)*	**Decrease in total ground invertebrates, richness, diversity & evenness: spiders, earwigs, flies, hemipteran bugs and caterpillars on NI; homopteran bugs, ants, beetles, cockroaches and caterpillars on CI (2009). No significant variations on leafs** except decline of hopper bugs on NI. Decline of total leaf invertebrates, homopteran	**Immediate decline of** *Trachylepis sechellensis* **(2005)** then recovery beyond initial abundance (2009); Fluctuation/stability for Green gecko.	**Decline of moorhens followed by recovery (2007) and explosion in numbers (2009), recovery of Turtle-doves (2007), increase in Madagascar fodies (2007-2009);** increase in distribution and numbers of introduced Seychelles White-eye (2008-2013)	Increase in distribution and numbers of a small Wedge-tailed shearwater colony (2005-2013)
Conception	2007	Colonisation of coastal plateau by *Pisonia grandis* (c.1 ha) and 'explosion' of *Phoenicophorium borsigianum* palm seedlings (2010-2014)	bugs, ants, beetles, cockroaches and caterpillars; rediscovery of snail *Stylodonta* and Giant millepede on CI (2009)	**Decline of** *Trachylepis sechellensis* **(2008)** then recovery beyond initial abundance (2010-2014); Fluctuation/stability for Green gecko. Burrowing skinks observed more abundant (2010-2014)	**Increase of Seychelles White-eyes, Blue pigeons and Sunbirds; recovery of Madagascar turtle-doves, stability of Madagascar fodies, and extinction of Ground-doves (2008-2013);** Mynas stable or decreased.	First breeding of Wedge-tailed shearwater and establishment of a small colony (2010-2014).
Grande Ile (Cosmolédo)	2007	Very high density of tall grasses *Dactylotenium sp.* (observed in 2014; unclear if partly related to rat eradication)	Increase in total ground invertebrates, spiders, ants, beetles (2008). Decline of total leaf invertebrates, beetles, flies, ants (2008). Increase of Robber crab and Hermit crabs abundance (2014)		**Increase of Souimanga sunbird and of Madagascar cisticola (2008).**	Higher numbers of Fairy terns, Black-naped terns and Red-tailed tropicbirds (2014). Recolonisation by breeding Masked boobies
Grand Polyte (Cosmolédo)	2007				**Increase of Souimanga sunbird and stability of Madagascar cisticola (2008).**	after c. 50 years of absence on Gde Ile.
Grande Sœur & Petite Sœur	2010			Increased sightings of Bronze geckos on Gde Sœur (observed 2010-2014)	Strong increase of moorhens (2012-2014),	**Establishment of a small Wedge-tailed shearwater colony on ˜Pte Sœur (2011-2012)**

the Seychelles blue pigeon *Alectroenas pulcherrima* on Conception. The latter species colonised and established naturally a breeding population on Bird Island since the rats were removed[23].

Five species of seabirds: Wedge-tailed shearwater *Puffinus pacificus* (five), Fairy tern *Gygis alba* (two), Lesser noddy *Anous tenuirostris* (one), White-tailed tropicbird *Phaethon lepturus* (one) and Masked booby *Sula dactylatra* (one) spontaneously (re) established a total of nine new breeding populations across seven of these islands (all within their current or historical range). Nine species of seabirds (all except the Sooty tern *Onychoprion fuscatus*) have benefitted from rat (and cat) eradications. Seabird recovery was observed in all the islands except one (Anonyme). Population increases of already existing populations were observed in six islands; and at least 22 seabird populations were beneficiaries of the rat (and cat) eradications: 11 recolonisations and 11 populations increases of already existing populations. Only 32% of these population beneficiaries were in the six granitic islands, and 68% were in the five coralline islands. Shearwaters have also increased in numbers and range on islands such as Desroches, Alphonse and Ste Anne, where rat and cat control have been conducted. Seabirds appear to play an important role in the recovery process, by inducing vegetation changes and bringing in nutrients through their excrements. This has also been shown in other countries[2a, 2b]. Vegetation changes include the rapid development of native *Pisonia grandis* trees (see pictures p. 361), the sticky seeds of which are dispersed by seabirds, at three islands (Bird, Denis and Conception), and the increase of endemic palm *Phoenicophorium borsigianum* seedlings at Conception. At Cosmoledo atoll, it is yet unclear whether the observed spectacular spread of several invasive grasses is related to the eradication of rats.

The response of invertebrates to the eradication of rodents has been complex and diverse. Changes in invertebrate populations associated with the eradication of rats were reported on eight islands. These included changes in abundance and distribution for 10 species of insects, arachnids, snails, myriapods and crustaceans at five different islands. On Conception Island, two years after the eradication, two species

The Masked booby has recolonised Grande Ile and increased in numbers on Grand Polyte since the eradication of rats and cats.
Gérard Rocamora

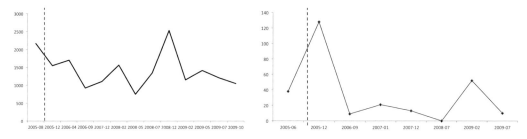

Figure 19 (left):
Variation in time of invertebrate density (number of invertebrates for a total of 32 permanent samples) counted during leaf point counts on Ile du Nord. The vertical dashed line represents the date of the rat eradication (Galman 2011).

Figure 20 (right): Abundance variation (in number of individuals/50 pitfalls) of *Subulina octona* (Gasteropoda: Subulinidae) caught in pitfall traps on Ile du Nord. The vertical dashed line represents the date of the rat eradication (Galman, 2011).

previously thought to be locally extinct were found. These included living specimens of the endemic snail *Stylodonta unidentata* and of the endemic Giant millipede *Sechelleptus sechellarum*, both of which have been continuing to recolonise the island[21]. On Cosmoledo atoll, seven years after the rat eradication, an increase in crustaceans (Robber crabs *Birgus latro* and Hermit crabs *Coenobita sp.*) was recorded at both Grand Ile and Grand Polyte Islands[24]. On Frégate, Giant scorpion *Chiromachus ochropus* abundance remained stable during the rat eradication; and the Giant tenebrionid beetle *Polposipes herculeanus* populations have maintained an overall stability[19, 25]. The endemic Enid snail *Pachnodus fregatensis* and Streptaxid snail *Conturbatia crenata* suffered strong declines during the eradication process; but the former has since returned to its initial levels[20, 26].

On high (granitic) islands, overall invertebrate abundance and diversity significantly decreased, both on the ground and on the surfaces of leaves, during the four years following the rat eradications. This contrasts with the situation on the flat (coralline) islands where, a year after the eradication, abundance had significantly increased on the ground[21]. At Conception, invertebrate groups that showed significant declines on the ground in pitfall traps (i.e., ants, snails, thrips, lacewings and springtails) differed from those found in pitfalls at Ile du Nord (i.e., spiders, earwigs, flies, hemipteran bugs and caterpillars). At Ile du Nord snails and springtails also decreased, but not significantly. Groups that showed significant declines on leaf surfaces at both Conception and Ile du Nord include Homopteran bugs, particularly Auchenorrhyncha (hoppers), heteropteran bugs, caterpillars, beetles, cockroaches, spiders and lacewings

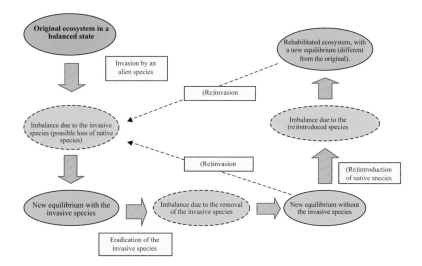

Figure 21: Possible evolution of an ecosystem following invasion by an alien species and restoration actions such as the eradication of this species and the (re) introduction of native ones (adapted from Galman, 2011).

and sometimes Sternorrhyncha (i.e., coccids, aphids, whiteflies). At Cosmoledo atoll numbers of beetles, flies and ants declined[21].

The general decrease in invertebrate populations observed after rat eradications may be explained, at least in part by interactions with other species: on the ground by the observed increase of skink populations and ground feeding birds; and on the leaves by the increase in the abundance of arboreal insectivorous birds. It is more difficult to explain the significant increase in invertebrates (particularly spiders, beetles and ants) observed on the ground at Cosmoledo[27]. This exemplifies the difficulty of adequately understanding ecosystem processes, even in apparently very simple, small closed island ecosystems[28].

Post-eradication invertebrate abundance appears to be unpredictable and difficult to interpret in view of the complex ecosystem interactions and changes taking place in trophic webs, and through habitat modifications, and species reintroductions. Rats had been significant consumers and predators that impacted a great diversity of plants and animals. So, their removal makes resources available to other species and shifts the balance that existed prior to eradication. Increased densities of landbirds (not necessarily related to rat suppression) are often accompanied by significant reductions in the abundance of certain arthropod groups, such as spiders[29]. This is evidence of cascading effects in trophic webs[30]. However, it is also possible that some of these invertebrate variations may be due to stochastic events.

In Seychelles the variations that have been observed in overall invertebrate abundance after rat (and cat) eradications suggest that existing balance between prey and predators, and more generally, between interacting species, have been significantly altered. The up and down variations in populations of the snail *Subulina octona* on Ile du Nord (Fig. 20), seem to correspond to the process of establishment of a new balance within the invertebrate community and the ecosystem as a whole[21]. However, because of the long-term damage to the ecosystem caused by rodents, it may not be possible to return to the original pre-invasion equilibrium on rat-freed islands. There may have been local extinction of some native species, severe habitat changes that may be irreversible, the presence of other invasive species and the (re)introduction of rare native species. Figure 21 summarises the possible evolution of an ecosystem following invasion by an invasive species and restoration actions such as the eradication of this species and the conservation (re)introduction of native species.

On balance, rat and cat eradications appear to have consistently benefited island ecosystems in every case. Most importantly, seven island translocations of five globally threatened birds (listed below), and one of a rare non-globally threatened reptile

Vegetation changes in restored areas can be monitored using photography (North Island 2009-2012). See also on p. 361.

The picture on the right was taken from the same spot as that on the left three years later, but with a different lens, showing the three small native trees in the foreground that have grown (and the same three coconut tree crowns in the background).
Katy Beaver/PCA

(Seychelles Black-mud terrapin *Pelusios subniger parietalis*) were conducted to three de-ratted islands (Denis, Ile du Nord and Frégate Islands). These species are the six threatened vertebrates listed below. In addition, the endemic Seychelles sunbird *Cynniris dussumieri* was also translocated to rat-free Bird Island. If we consider the islands from which cats alone were removed; an additional non-threatened endemic bird, the Aldabra rail (that was reintroduced to Picard Island), is also a beneficiary and the total number of successful island translocations to predator-free islands (covering both reintroductions and conservation introductions) allowed by the eradications rises to 16 (seven reintroductions and nine conservation introductions; see Table 6 and details p. 75). In total, 17 populations of six globally threatened species have benefited from 21 rat or cat island eradications conducted in Seychelles[22]. For four of the species: Seychelles magpie-robin (EN), Seychelles white-eye (EN), Seychelles warbler (VU), and Seychelles fody (NT formerly VU), this has contributed to the downgrading of their IUCN threat status. The other two threatened species beneficiaries are the Seychelles black paradise flycatcher (CR) and the Aldabra Giant tortoise (VU). Among the few negative effects reported was the spread of two invasive animals, the Crazy ant and the Common mouse, on Bird Island.

Numerous indices relating to both flora and fauna indicate that ecosystem recovery is taking place at all islands where rats and cats have been removed in Seychelles, although it has been difficult to conclusively quantify the impacts of these eradications.

Figure 22: Abundance variation (in number of reptiles/point) of the endemic Seychelles skink *Trachylepis sechellensis* on Ile du Nord after the 2005 Black rat eradication (decline, then steep increase). The sign 'ERAD.' indicates the eradication period (Rocamora & Labiche, 2009c).

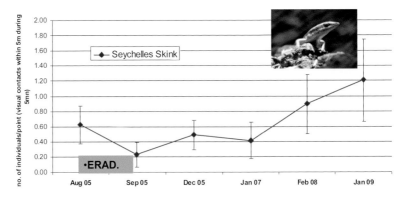

Figure 23: Abundance variation (in number of birds/point-count) of several native bird species on Ile du Nord after the 2005 Black rat eradication: Common moorhen *Gallinula chloropus* (increases), Striated heron *Butorides striatus* (decreases; non stat. significant) and Seychelles sunbird *Cynniris dussumieri* (fluctuates); (Rocamora & Labiche, 2009c).

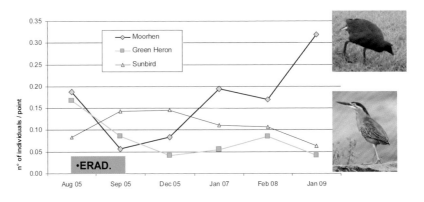

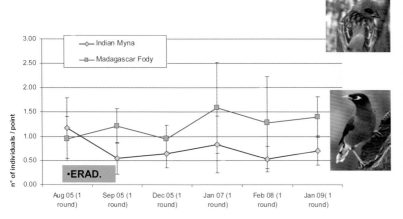

Figure 24: Abundance variation (in number of birds/point count) of introduced Common mynas *Acridotheres tristis* (decline first; then stability) and Madagascar fodies *Foudia madagascariensis* (stability; then increase) on Ile du Nord after rat eradication (Rocamora & Labiche, 2009c).

Ecosystem recovery is a process that may be slow if not combined with additional conservation measures that involve ecosystem rehabilitation (habitat restoration, eradication of other invasive species, and reintroductions).

In the past, pre-eradication baseline surveys were often absent or insufficient and long-term monitoring protocols were not regularly implemented. Unfortunately, some island managers/owners were either insufficiently motivated or simply could not cover the full costs of long-term ecosystem monitoring. This is not specific to Seychelles and explains the relative scarcity of data on ecosystem recovery after rat eradications around the globe. Documenting ecosystem recovery needs to be given higher priority in the future.

Further reading

1 Towns *et al.*, 2006; 2a Mulder *et al.*, 2009; 2b Russell, 2011; 3 St Clair *et al.*, 2011; 4 Towns *et al.*, 2011; 5 Medina *et al.*, 2011; 6 Harper & Bunbury 2015; 7 Courchamp *et al.*, 2003; 8 Lorvelec & Pascal, 2005; 9 Varnham, 2010, 10 Russell & Holmes, 2015; 11 Veitch *et al.*, 2011; 12 Jones, 2010; 13 Le Corre *et al.*, 2015; 14 Courchamp *et al.*, 2011;15 Ringler *et al.*, 2015; 16 Merton *et al.* 2002; 17 Hill, Holm *et al.*, 2003; 18 Thorsen *et al.*, 2000; 19 Gerlach 2005a, 20 Gerlach 2005b; 21 Galman, 2011; 22 Rocamora *et al.*, in prep; 23 A. Skerrett/SBRC *pers. comm.*;24 G. Rocamora/ICS, *pers. obs.*; 25 Canning, 2011; 26 Raposo & Canning, 2012; 27 Galman *et al.*, in prep.; 28 Watari *et al.*, 2011; 29Gruner, 2004; 30 Recher & Majer, 2006; 31 Thomaz *et al.*, 2012.

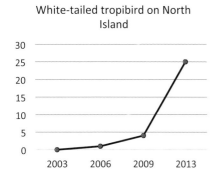

White-tailed tropibird on North Island

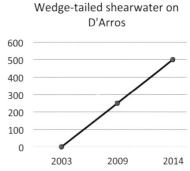

Wedge-tailed shearwater on D'Arros

Figure 24bis: Increase of White-tailed tropicbird and Wedge-tailed shearwater numbers of pairs on North Island and D'Arros Island respectively after the rat and cat eradication.
Data: North Island & D'Arros Research Center

Which are the species that require particular attention for being or becoming invasive?

Chapter **7**

Among the many ornemental plants introduced
in Seychelles, some could become invasive.
Gérard Rocamora

This chapter aims at identifying the species known to be invasive and problematic in Seychelles, and those not yet confirmed as having a significant impact but that require particular attention. It provides a series of species lists that go beyond the most problematic 44 invasive species that are treated through case studies or species accounts in the second part of the book.

The following tables list the most important alien plants and animals reported to be invasive – or presenting a confirmed risk of invasion – in the environment of Seychelles (in the broad sense, i.e. not only in natural ecosystems but also in areas degraded or modified by humans.) Species affecting only agriculture or health are not covered here, but the particular case of native invasive species is presented.

The lists for plants have been compiled from a variety of sources, including previous publications: *Case Studies on the Status of Invasive Woody Plant Species in the Western Indian Ocean*[1,2]; the *National Invasive Alien Species Baseline* report[3]; a review on invasive creepers: *Invasion risk from climbing and creeping plant species in Seychelles*[4]; and the *Review of IAS control and eradication programmes in Seychelles*[5]. Priority species listed during a national workshop held on 12 October 2010 in Victoria under the UNDP/GEF Biosecurity project, and the opinions of various botanists, naturalists, and members of the Plant Conservation Action Group consulted were also taken into account. The report *Evaluation of threats of introduction and spread of IAS through production sector activities in Seychelles*[6] was also useful. These lists are not based on hard scientific evidence obtained from rigorous research, but rather on concordant observations obtained from different knowledgeable individuals, although in some cases there is no consensus on invasive status.

Table 12 lists plant species that are already considered to be fully invasive, i.e. that have spread over vast areas and which already have a confirmed significant negative impact on the ecosystems of Seychelles. These can be considered as priority species for action.

Table 13 lists those species only locally invasive or with a limited degree of invasiveness in Seychelles but that are sometimes invasive in other countries, and which could spread and become widely invasive. For many of these, we do not have sufficient information yet to say whether these plants are able to spread over wider areas. The fact that they have spread in one or a few places is not sufficient to classify them as fully invasive as their spread may result from specific conditions at a particular time. However, they do deserve special attention.

Table 14 lists the most important invasive terrestrial animals having an environmental impact in Seychelles; this includes a few species that so far appear to only have a limited impact.

Table 15 lists a number of alien weeds and creepers that can be considered naturalised, that invade mainly agricultural and other open habitats created or modified by humans.

Table 16 details the island distribution of the main plant or animal aliens – invasive or showing signs of invasiveness – that require special attention to prevent their spread to other islands in Seychelles. Members of the public must be warned not to transport them to other islands. Evidence of spread needs to be reported to local and national authorities, and efforts should be made to eradicate them as soon as they are seen at a new island.

Table 12: Problematic invasive plants and fungi having a confirmed significant impact on Seychelles environment and landscapes.

Bold: most problematic across main and/or many small islands; Blue: priority sp. treated in the book; Italic: no consensus on invasive status; +: identified as priority by stakeholder workshop; *: possibly or probably native; §: mainly for coralline islands; (Ag): problematic also for Agriculture; ^: proposed creole name. All these species are also highly invasive in other island territories and countries around the world.

Compiled from Kueffer & Vos 2004; Beaver & Mougal 2009; Nevill 2009; Senterre 2009; Biosecurity workshop 12 Oct 2010 and *pers. comm.* from B. Senterre, K. Beaver, L. Chong-Seng, C. Kueffer & C. Kaiser-Bunbury.

English name	Kreol name	Latin name
Woody species		
Cinnamon	Kannel	*Cinnamomum verum*
Devil tree +	Bwa zonn (Bwazonn)	*Alstonia macrophylla*
Albizia	Albizya	*Falcataria moluccana*
Cocoplum	Prin-de-frans (Prindefrans)	*Chrysobalanus icaco*
Chinese guava +	Gouyav-de-Sin (Gouyavdesin)	*Psidium cattleianum*
Koster's curse +	Fo watouk	*Clidemia hirta*
White cedar	Kalis-di-pap (Kalisdipap)	*Tabebuia pallida*
Red sandalwood/Coralwood *	Lagati	*Adenanthera pavonina*
Indian laurel	Bwa zozo (Bwazozo)	*Litsea glutinosa*
White leadtree	Kasi	*Leucaena leucocephala*
Casuarina *	Sed	*Casuarina equisetifolia*
Rose apple +	Zanbroza	*Syzygium jambos*
West indian lantana +	Vyey fiy (Vyeyfiy)	*Lantana camara*
Australian holly / Christmas berry+	Larb de Nwel	*Ardisia crenata*
Pawpaw / Papaya §	Papay	*Carica papaya*
Butter tree	Bwa ber (Bwaber)	*Pentadesma butyracea*
Coconut tree *	*Kokotye / Koko*	*Cocos nucifera*
Creepers		
Merremia * (Ag)	Lyanndarzan	*Merremia peltata*
Devil's ivy or Golden pothos +	(Filodendron)	*Epipremnum pinnatum*
Japanese climbing fern +	Fouzer zaponnen / Fouzer file §	*Lygodium japonicum*
Cat-claw creeper	Lalyann grif sat ^	*Macfadyena unguis-cati*
Bengal trumpet/Blue trumpet vine (Ag)	Lalyann tronpet ble ^	*Thunbergia grandiflora*
Hiptage	Lalyann papiyon	*Hiptage benghalensis*
Arrowhead vine	Lalyann feyaz ^	*Syngonium podophyllum*
Rangoon creeper/Chinese honeysuckle	Santonin	*Quisqualis indica*
Coral vine/creeper; Bride's tears (Ag)	Lantigonn	*Antigonon leptopus*
Aquatic weeds		
Water lettuce +	Leti lanmar	*Pistia stratiotes*
Other non woody plants		
Sisal + § (Ag)	Lalwa or lalwes	*Agave sisalana*
Mauritius hemp/Green-aloe	Lalwa or lalwes	*Furcraea foetida*
Snakeweed / Porter weed §	Zepible	*Stachytarpheta jamaicensis/ urticifolia*
Fungal diseases		
Takamaka wilt	Maladi takamaka	*Leptographium calophylli*

Table 13: Plants locally invasive or with a limited degree of invasiveness in Seychelles, sometimes invasive in other countries and requiring special attention.

Bold: most problematic across main and/or many small islands; Blue: priority sp. treated in the book; Italic: no consensus on invasive status; *: possibly native; §: mainly for coralline islands; (Ag): problematic also for Agriculture; $: affects introduced Sandragon *Pterocarpus indicus* but also the overall ecosystem and could also affect natives.
Preliminary list compiled from Kueffer & Vos 2004; Beaver & Mougal 2009; Nevill 2009; Senterre 2009; Biosecurity workshop 12.11.10 and *pers. comm.* from B. Senterre, K. Beaver, L. Chong-Seng, C. Kueffer & C. Kaiser-Bunbury). Those species with small distributions (e.g. *Acacia mangium*, Water hyacinth) should be eradicated before they become highly invasive.

English name	Kreol name	Latin name
Woody species		
Shoebutton ardisia	(no name)	*Ardisia elliptica*
Santol	Santol	*Sandoricum koetjape*
Rubber tree	Kaoutsou	*Hevea brasiliensis*
Blue strawberry flowers	Bwa demon (Bwademon)	*Memecylon caeruleum*
Shrubby dillenia	Bwa rouz blan (Bwarouz blan)	*Dillenia suffruticosa*
African tulip tree / Fountain tree	Pis pis (Pispis)	*Spathodea campanulata*
Black plum	Bwa mozambik (Bwamozambik)	*Vitex doniana*
Clove tree	*Zerof*	*Syzygium aromaticum*
Bird's eye bush	*Bwa kok (Bwakok) (also Bwademon)*	*Ochna kirkii*
Mickey mouse plant *	*Bwa bouke (Bwabouke) (Bwa mang)*	*Ochna ciliata*
Black wattle / Brown salwood	Akasya gran fey	*Acacia mangium*
Soap pod	Akasya pikan	*Acacia concinna*
Cashew	Kazou	*Anacardium occidentale*
Ylang-ylang	Ilangilang	*Cananga odorata*
Liberian/Robusta coffee	Kafe	*Coffea liberica/canephora*
Star apple	(no name)	*Chrysophyllum cainito*
Willow-leaved justicia *	Lapsouli	*Justicia gendarussa*
Pink peppercard	Bwa lansans	*Schinus terebinthifolius*
Creepers		
Heartleaf philodendron	(Filodendron)	*Philodendron hederaceum*
White convolvulus creeper	(Lalyann)	*Merremia dissecta*
Puero/Tropical kudzu (Ag)	(no name)	*Pueraria phaseoloides*
Passion fruit	Fri lapasyon	*Passiflora edulis*
Wild potato yam (Ag)	*Mortora / Ponm Edwar maron*	*Dioscorea bulbifera*
Rosary pea vine / Crab's eye *	*Reglis*	*Abrus precatorius*
Asparagus fern	Fouzer	*Asparagus setaceus*
Japanese honeysuckle	(Chèvrefeuille du Japon)	*Lonicera japonica*
Aquatic weeds		
Water hyacinth	Lisdo anvaisan	*Eichhornia crassipes*
White Waterlily	Lisdo	*Nymphaea lotus*
Minute duckweed	Lantir ver	*Lemna perpusilla*
Other non woody plants		
Upright elephant ear/Giant taro	Vya	*Alocasia macrorrhiza*
Castor oil plant §	Tantan	*Ricinus communis*
Prickly chaff flower	Lerb serzan	*Achyranthes aspera*
Wild pineapple	Zannannan moustikenn	*Ananas comosus*
Elephant's ear	Kaladyonm	*Caladium sp.*
Mother-in-law plant / Dumbcane	Vya tang	*Dieffenbachia seguine*
Black Taro	*Sonz*	*Colocasia esculenta*
Golden Bamboo	Banbou	*Bambusa vulgaris*
Fungal diseases		
Sandragon or Angsana wilt $	Maladi Sandragon	*Fusarium oxysporum*

Table 14: Examples of naturalised alien weeds and creepers invading mainly agricultural and other open habitats created or modified by humans.

Certain species such as Guinea grass and Devil's pumpkin are known to have also invaded natural habitats in other countries (e.g. Hawaii). *Italic*: no consensus on invasive status.

English name	Kreol name	Latin name
Creeping beggarweed/Spanish clover	Gro tref	*Desmodium incanum*
Guinea grass	Fatak	*Panicum maximum*
Common asystasia	Manztou	*Asystasia gangetica*
Yellow alder / Sundrops	Koket	*Turnera ulmifolia* var. *angustifolia*
Coffee senna	Kaspyant	*Senna occidentalis* (*Cassia occidentalis*)
Glorybower / Bleeding heart	Ker de Zezi	*Clerodendrum thomsonae*
Indian acalypha / Indian nettle	Lerb sat	*Acalypha indica*
Stonebreaker	Kiranneli	*Phyllanthus amarus*
Corky-stem passionflower	Lepeka, Lagrenn lank	*Passiflora suberosa*
Wild passion fruit	Bonbon plim	*Passiflora foetida*

The lists may be incomplete and are intended mainly to provide guidance for prioritisation. They remain open to future additions – and hopefully deletions too if successful control and eradication programmes can be developed for some of them (see also p. 111 regarding the need to establish national priorities for the management of invasive plants and animals in Seychelles).

Most of the species listed are exotic and invasive, but for a few species (indicated) it is unclear whether they are native, invasive or both. Among these is the Coconut tree, which is not an invasive in the strict sense but covers large areas on medium/small granitic islands and on outer islands where it was planted and exploited extensively during the 19th and 20th century. The legal status of the Coconut tree is problematic given that while it is subject to ongoing control programmes on small islands where ecosystem rehabilitation is underway, in 1996 it was declared a protected species. Other species like the Devil's tree (Bwa zonn) are still on the Protected Species list because of their value for timber, hence permission must be sought from the Environment Department before undertaking their removal, as is the case for Coconut trees. Whether the creeper *Merremia peltata* is native or exotic is unclear; and this is also the case for Coralwood (Lagati) *Adenanthera pavonina*. Some authors consider these species to be native to Seychelles while others consider them to be introduced. On the other hand, some species considered to be native may actually be recent arrivals to Seychelles having arrived with the help of human transportation (for example, in the case of certain species of mosquito). Seychelles paleoecology and palynology (which includes the study of pollen, spores and other microscopic organisms) will be able to help distinguish between native, doubtfully natives, and invasive species[7, 8], and DNA analysis is also likely to prove useful (as in the case of mosquitoes[9]).

Table 15. Invasive terrestrial animals with a significant environmental impact in Seychelles. **Bold**: most problematic across main and/or many small islands. Blue: priority sp. treated in the book. *Italic*: no consensus on invasive status; +: identified as priority by stakeholder workshop; ++: limited or potential impact only; *: hybridises with endemic fodies; **: hybrid introduced *S. p. picturata* with local race *S. p. rostrata*; ^: proposed creole names; $: also reported to affect some non-cultivated plants.

English name	Kreol name	Latin name
Mammals		
Black rat +	Lera lakelong ^	*Rattus rattus*
Brown rat +	Lera later ^	*Rattus norvegicus*
House mouse	Souri	*Mus musculus*
Feral cat +	Sat maron	*Felis catus domesticus*
Tenrec	Tang	*Tenrec ecaudatus*
Feral goat	Kabri	*Capra hirtus domesticus*
Feral rabbit	Lapen	*Oryctolagus cuniculus*
Black-naped hare ++	Liev ^	*Lepus nigricans*
Birds		
Common myna +	Marten	*Acridotheres tristis*
Barn owl +	Ibou	*Tyto alba affinis*
Indian crow +	Korbo	*Corvus splendens*
Ringed-necked parakeet +	Kato ver	*Psittacula krameri*
Red-whiskered bulbul	Zwazo konde	*Pycnonotus jocosus*
House sparrow+	Mwano	*Passer domesticus*
Feral chicken	Poul maron	*Gallus gallus domesticus*
Madagascar turtle-dove**	Tourtrel dezil	*Streptopelia madagascariensis*
Madagascar Fody*	Kardinal, Sren	*Foudia madagascariensis*
Reptiles		
Crested-tree lizard +	Lezar lakret	*Calotes versicolor+*
Red-eared terrapin +	Torti latanp rouz	*Trachemys scripta elegans+*
Invertebrates		
Tiger mosquito	Moustik tig ^	*Aedes albopictus*
Crazy ant +	Fourmi maldiv	*Anoplolepis gracilipes*
Big-headed ant +	Fourmi grolatet ^	*Pheidole megacephala*
Spiralling white-fly	Bigay blan / Mous blan	*Aleurodicus dispersus*
Coconut white-fly	Bigay blan koko / Mous blan koko ^	*Aleurotrachelus atratus*
Papaya mealy bug $	Bigay blan papay ^	*Paracoccus marginatus*
Giant african snail	Kourpa zean (lalevros & lalevpal) ^	*Achatina fulica/immaculata*
Rosy wolf snail ++	Kourpa moustas	*Euglandina rosea*
(No common name)	Kourpa strie ^	*Subulina striatella*
American cockroach	Kankrela ameriken ^	*Periplaneta americana*
Coccid sp.	Lipou sp.	*Icerya sechellarum*
White-footed ant +	Fourmi lipyeblan ^	*Technomyrmex albipes*
(also alien scale insects and mealy bugs in general)		

THE CASE OF NATIVE INVASIVE SPECIES

Cases of native species having become invasive are not uncommon in the literature. In Seychelles, the creepers *Merremia peltata* and Coralwood (Lagati) *Adenanthera pavonina* are considered native (or 'subtantially native) by some authors based on early reports (1840 and 1768 respectively, the latter prior to the first human settlement), their regional distribution; and/or their presence in well preserved native ecosystems where they do not spread[4, 10]. Native invasiveness is generally associated with changes in landscape management that shift the natural balance so that it benefits particular species. Such changes can be caused by widespread fires, development or abandonment of agricultural or fallow land, residential or urbanised areas, etc. But they can also result from unexplained spontaneous changes in the ecology of certain species. This phenomenon of invasiveness by native species is not restricted to plants and can also be observed in animals. Such changes may sometimes be attributed to the introduction (or spontaneous arrival) of foreign populations or subspecies of the same species that are clearly invasive and will interbreed with the local populations. For example, the introduced grey-headed subspecies of Madagascar turtle dove

The creeper *Merremia peltata*, considered native in Seychelles, may have become invasive as a result of the introduction of – or hybridisation with – exotic populations of the same species.
Gérard Rocamora

Nesoenas picturata picturata, originally from Madagascar, has invaded most of the inner islands and the Amirantes and interbred with the Seychelles red-headed form *N. p. rostrata*. The phenotypes of the latter have now almost disappeared. In this case, the difference in morphology between the local and the foreign birds make the phenomenon visible. But, similar changes may also occur unnoticed in other animals or plants species for which there are little of no visible differences in morphology between genetically differentiated local and introduced populations. It may for example be the case with *Merremia peltata*; but only molecular genetic analysis could reveal such a phenomenon (see box on hybridisation p. 49).

Table 14 provides examples of the many naturalised weeds and creepers that tend to invade agricultural and other man-made habitats in Seychelles, but that are not considered to cause any significant environmental impact. Although reported in the past to be invasive in plantations, after fires or in woodlands modified by humans, the native Bracken Fern *Dicranopteris linearis* (locally known as Fouzer grif lyon), is also not causing any environmental problems and thus has not been included in any of the lists.

References/further reading
1 Kueffer & Vos 2004; 2 Kueffer *et al.* 2004; 3 Nevill 2009; 4 Senterre 2009; 5 Beaver & Mougal 2009; 6 Dogley W. 2009; 7 Coffey *et al.* 2011; 8 Van Leeuwen *et al.* 2008; 9 Le Goff *et al.* 2012; 10 Gerlach, 2006.

Table 16: Island distribution of alien species – invasive or showing signs of invasiveness – that require public attention to avoid spread to other islands. These invasive species should be reported to Environment Department, concerned island managers and book authors when discovered on new islands.
Source: Dogley W. 2009; Nevill 2009, Senterre 2009, and present work

Scientific name	English name	Distribution (2014)
Woody plants		
Ardisia crenata	Australian holly, Coral ardisia	Mahé.
Dillenia suffruticosa	Shrubby dillenia	Mahé.
Litsea glutinosa	Indian laurel	Mahé, Anonyme, Ile au Cerf, Conception & Ste Anne; Frégate.
Clidemia hirta	Koster's curse	Mahé; Silhouette, Ile du Nord; few individuals eliminated on Praslin.
Memecylon caeruleum	Blue strawberry flowers	Mahé, Praslin, Ile au Cerf
Creepers		
Hiptage benghalensis	Hiptage	Mahé.
Merremia peltata	Merremia	Mahé, Silhouette, Ile Denis.
Lygodium japonicum	Japanese climbing fern	Mahé.
Thunbergia grandiflora	Bengal trumpet/Blue trumpet vine	Mahé
Epipremnum pinnatum	Devil's ivy	Mahé & Ile au Cerf; Praslin, La Digue & Félicité; Silhouette & Ile du Nord; Frégate; Ile Denis.
Macfadyena unguis-cati	Cat-claw creeper	Mahé, Ste Anne.
Quiscalis indica	Chinese honey-suckle, Rangoon creeper	Mahé; Praslin, La Digue, Cousin & Curieuse; Silhouette.
Syngonium podophyllum	Arrowhead vine	Mahé; Praslin, La Digue & Curieuse; Silhouette; Frégate.
Antigonon leptopus	Bride's tears, Coral creeper, Coral vine	Mahé; Silhouette & Ile du Nord; Frégate.
Dioscorea bulbifera	Wild potato yam	Mahé, Silhouette.
Aquatic weeds		
Pistia stratiotes	Water lettuce	Mahé, Praslin, La Digue; formerly in artificial ponds at Frégate.
Eichornia crassipes	Water hyacinth	Mahé, Praslin, La Digue.
Nymphaea lotus	White waterlily	Mahé; Praslin & La Digue.
Mammals		
Rattus norvegicus	Brown rat	Mahé & Ste Anne; Praslin & La Digue.
Rattus rattus	Black rat or Ship rat	Mahé, Ste Anne group & Thérèse; Praslin, La Digue, Curieuse, Félicité, Marianne & Ile Ronde (Praslin); Silhouette. All outer islands except Bancs Africains, D'Arros, St Joseph, Marie Louise, Desnoeufs, Boudeuse & Etoile, St François & Bijoutier (Alphonse group), Goëlettes and Bancs de Sable (Farquhar), and all Cosmoledo Atoll minus Menai (infested).
Mus musculus	House mouse, Common mouse	Mahé, Praslin, Aride, La Digue, Bird; D'Arros, Desnoeufs, Marie-Louise.
Reptiles		
Calotes versicolor	Crested tree lizard; Oriental garden lizard	Ste Anne. One observed on Eden Island in Aug. 2012, captured in 2014. One observed and captured in Victoria in 2014.

Scientific name	English name	Distribution (2014)
Hemidactylus frenatus	Common/Asian house gecko	Mahé, Praslin, Silhouette, Bird; Platte, Desroches, Alphonse, Bancs africains, D'Arros, Desnoeufs, Marie-Louise, Poivre, Remire.
Trachemys scripta	Red-eared slider	Mahé.
Birds		
Acridotheres tristis	Common myna, Indian myna	All inner islands except Cousin, Cousine, Aride, Ile aux Récifs, Frégate, and Denis (to be confirmed). Absent from all outer islands.
Corvus splendens	(Indian) House crow	Mahé (occasional since eradicated).
Passer domesticus	House sparrow	Desroches, D'Arros, Desnoeufs, Marie-Louise, Poivre, Rémire, St. Joseph; Alphonse, St Francois & Bijoutier. Occasional on Mahé since eradicated.
Psittacula krameri	Ring-necked parakeet	Mahé. One eliminated on Silhouette in June 2014.
Pycnonotus jocosus	Red-whiskered bulbul	Formerly on Assomption (eradication in 2014); and Aldabra (one individual eliminated).
Tyto alba (race affinis)	Barn owl	Mahé & Ste Anne group, Praslin, La Digue, Curieuse & Félicité; Silhouette. Occasionally on Anonyme, Conception; Aride, Cousin, Cousine, Grande Soeur, Ile du Nord; Frégate & Bird.
Fish		
Oreochromis mossambicus	Tilapia	Mahé, Praslin, La Digue, Silhouette.
Poecilia immaculata/ reticulata	Guppy, Milionfish, Rainbowfish	Mahé, Praslin.
Invertebrates		
Bactrocera cucurbitae	Melon fruit fly	Mahé, Praslin, La Digue.
Pheidole megacephala	Big-headed ant	Mahé, Île Ronde (Mahé); Praslin, Aride, Cousin, Cousine, La Digue, Marianne; Desroches, Coëtivy, Farquhar.
Anoplolepis gracilipes	Crazy ant	Mahé, Ste Anne, Anonyme; Praslin, Cousin, La Digue, Petite Soeur, Félicité, Marianne; Bird.
Achatina immaculata	Pink-liped Giant African snail	Mahé, Ile au Cerf, Ile Ronde, Ste Anne & Thérèse; Praslin, Curieuse, Félicité, Grande Soeur and La Digue. Only empty shells have been found on Ile aux Récifs.
Lissachatina fulica	Pale-liped Giant African snail	Mahé, Anonyme, Conception, Ile au Cerf, Ile aux Vaches Marines, Ronde, Longue, L'Islette, Ste Anne & Thérèse; Praslin, Curieuse, Félicité, Grande Soeur & La Digue; Silhouette; Frégate & Bird; D'Arros & Desroches.
Stegomyia albopicta (Aedes albopictus)	Tiger mosquitoe	Confirmed on Mahé, Anonyme, Conception, Ile au Cerf, Longue, Moyenne, Ronde, Ste Anne & Thérèse; Praslin, La Digue & Félicité; Silhouette & Ile du Nord. Also on Bird, Denis, Plate, Desroches and probably D'Arros.
Pathogens		
Leptographium calophylli (also Verticillium calophylli)	Takamaka wilt disease or Takamaka verticillium wilt	Mahé, Ste Anne, Ile au Cerf, Longue & Thérèse; Praslin, La Digue, Curieuse, Félicité & Grande Soeur; Silhouette & Ile du Nord; Denis. Possibly on Marianne and Anonyme.

Invasive species – How can you help?

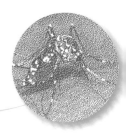

As a citizen, you can contribute significantly to reduce the impact of Alien Invasive Species in Seychelles, limit their spread and prevent new infestations:

- **Don't pack a pest!** Do not bring any plants (including vegetables and fruits), or animals (including pets) and related unprocessed products into Seychelles – by plane, boat or any other means of transport. To do so is a serious offence. If you have inadvertently brought some, you should declare it immediately upon arrival into the country.
- **Don't bring in a pest!** Do not transport plants and animals between islands without taking precautions and obtaining previous authorisation from government authorities. You may spread dangerous alien invasive species, including new parasites and diseases.
- **Be aware of hitchhikers!** Before travelling to a nature reserve or national park, and to any area or island with high biodiversity such as a pristine native forest or wetland, carefully clean and inspect your clothes, socks, shoes, hiking boots, camping gear and other equipment so as not to transport any alien seeds or wildlife diseases.
- **Don't transmit plant diseases!** When moving equipment between islands, disinfect all instruments that have been used to cut vegetation (machetes, saws, axes, chainsaws, etc.). Use alcohol or a flame to disinfect in order to avoid spreading plant diseases.
- **Keep pets at home!** Never set your pets free in the wild. They will have a very negative impact on native species. There is also a good chance that your pet will suffer or die when having to fend for itself. Call veterinary services to find alternative solutions.
- **Don't release a pest!** Do not discharge old water from your aquarium into rivers and wetlands.
- **Fight rodents and weeds!** Control rodents and plant pests around your house and properties. Eliminate or trim vegetation near your house that could provide a route of travel for rodents to enter your house or climb onto your roof.
- **Fight the bite!** Eliminate all sources of standing water in your property and around your house that could provide habitat for the larvae of mosquitoes that can transmit diseases like Chickungunya and Dengue fever. Keep grass short and remove bushes and tall grasses immediately adjacent to buildings.
- **Go native!** Use native plants as ornamentals in your garden for landscaping.
- **Get involved!** Participate as a volunteer on campaigns working to control invasive plants.
- **Report a pest!** Report any sighting of a suspected new alien to the Department of Environment by phoning the Green line 2 722 111
- **Be clean!** Keep your bin sites and public bins always clean and tidy.

Together, we can make a real difference!

Visitors and scientists travelling to Round Island Nature Reserve, in Mauritius, strickly following biosecurity rules and carrying in sealed barrels with screw-on lids items previously checked and certified free of invasive pests.
Gabby Salazar/MWF

Left: Anything that retains water such as tyres, plant pots, boxes, empty tins and coconuts, leaf palms, and other potential mosquitoe larving sites must be removed, particularly around houses, a preferred habitat for the invasive Tiger mosquito.
Gérard Rocamora

Right: Keeping bin sites and backyards clean, and food and garbage are out of reach of rodents, is critical to prevent their proliferation and to optimise the effectiveness of control operations.
Gérard Rocamora

WEBSITE LINKS ON INVASIVE ALIEN SPECIES

- CABI – Invasive Species Compendium: http://www.cabi.org/isc
- Convention of Biological Diversity: http://www.cbd.int/island/invasive.shtml
- www.cbd.int/invasive/
- Global Island Partnership (GLISPA): http//glispa.org
- Island Conservation www.islandconservation.org
- Global Island Vertebrate Eradication Database http://diise.islandconservation.org/ (or http://eradicationsdb.fos.auckland.ac.nz/)
- IUCN/SSC Invasive Species Specialist Group: http://www.issg.org/
- Global Invasive Species Programme Database: http://www.issg.org/database/welcome/
- Cooperative Islands Initiative on Invasive Alien Species on Islands http://www.issg.org/cii/islands_and_invasives.html
- Pacific Invasives Initiative http://www.pacificinvasivesinitiative.org/
- Pl@ntInvasion http://community.plantnet-project.org/pg/groups/516/plntinvasion/
- Groupe Espèces Invasives de La Réunion. www.especesinvasives.re/
- Site du comité français de l'Union Internationale pour la Conservation de la Nature.
- Initiative sur les Espèces Exotiques Envahissantes en outre-mer http://www.especes-envahissantes-outremer.fr/
- Biofouling solutions (identification sheets, general information and management recommendations about biofouling and invasive marine species) www.biofoulingsolutions.com.au

SOME INVASIVE ALIEN SPECIES GUIDES AND TOOLKITS

- Clout, M.N. & Williams, P.A. (eds.) 2009. **Invasive species management: a handbook of principles and techniques. – Oxford University Press**, Oxford, UK, 308pp.
- Dogley, W. 2004. **Plant pest and disease management – A manual for farmers and gardeners.** Department of Natural Resources. Ministry of Environment and Natural Resources, Victoria, Seychelles. 108 pp.
- Emerton L. and Howard G. 2008. **A Toolkit for the Economic Analysis of Invasive Species.** Global Invasive Species Programme, Nairobi. 113pp.
- McNeely J.A., Mooney H.A., Neville L.E., Schei P.J. and Waage J.K. (eds.). 2001. **Global strategy on invasive alien species.** IUCN, Cambridge, U.K., in collaboration with the Global Invasive Species Programme. X + 50 pp.
- Pacific Invasives Initiative. 2011. **Resource Kit for Rodent and Cat Eradication.** http://rce.pacificinvasivesinitiative.org. Accessed on 7th August 2014.
- Pacific Invasives Initiative. 2012. **Resource Kit for Invasive Plant Management.**
- http://ipm.pacificinvasivesinitiative.org. Accessed on 7th September 2014
- Shine C. 2008. **A Toolkit for Developing Legal and Institutional Frameworks for Invasive Alien Species.** Global Invasive Species Programme. Nairobi.
- Soubeyran Y. (coord.). 2010. **Gestion des espèces exotiques envahissantes. Guide pratique et stratégique pour les collectivités françaises d'outre-mer.** Comité français de l'UICN, Paris.

- Soubeyran Y., Caceres S. and Chevassus N. 2011. **Les vertébrés terrestres introduits en outre-mer et leurs impacts. Guide illustré des principales espèces envahissantes.** Comité français de l'UICN et ONCFS, Paris.
- Tye A. 2009. **Guidelines for invasive species management in the Pacific: a Pacific strategy for managing pests, weeds and other invasive species.** SPREP, Samoa.
- Wittenberg R. and Cock M.J.W. (eds.). 2001. **Invasive Alien Species: A Toolkit of Best Prevention and Management Practices.** CAB International, Wallingford, Oxon, UK, xii - 228.

OTHER PUBLICATIONS OF INTEREST

- Mooney, H.A., Mack, R.N., McNeely, J.A., Neville, L.E., Schei, P.J. and Waage, J.K. (eds) 2005. **Invasive Alien Species: A New Synthesis.** Island Press, Washington, DC, 368 p.
- Simberloff, D. & Rejmánek, M. (eds.) 2011. **Encyclopedia of biological invasions:** University of California Press. Berkeley.
- Simberloff D. 2013. **Invasive species. What everyone needs to know.** Oxford University Press.
- Tassin, J. 2014. **La grande invasion. Qui a peur des espèces invasives ?** Editions Odile Jacob. Paris, 216 p.

Gérard Rocamora

Part 2
Identification and
management of
priority species

Adult Black rat
(*Rattus rattus*).
James Russell

Species accounts are presented in the alphabetical order of their common English names within each chapter – Mammals – Birds – Reptiles – Invertebrates – Broadleaf trees – Creepers – Other plants – Fungal disease.

Names of species appear in the following order: English, French, Creole and scientific name.

In the distribution section of the species accounts, islands are normally listed in the following order: Inner islands: Mahé and satellites (by alphabetical order), Praslin/La Digue and satellites, Silhouette and Ile du Nord; Frégate, Ile au Récifs and Mamelles; Bird and Denis. Outer islands: Plate, Coétivy; Desroches and other North Amirantes islands (by alphabetical order); South Amirantes islands (Boudeuse/Etoile, Desnoeufs and Marie-Louise) and Alphonse group; Farquhar atoll (individual islands by alphabetical order), Providence and St Pierre; Cosmoledo atoll, Astove, Assomption and Aldabra atoll.

Mammals

ONE OF
WORLD'S WORST
100
INVASIVE
SPECIES

J. Russell

Black (or Ship) rat

Rat noir

Lera lakelong

Rattus rattus

IDENTIFICATION AND BIOLOGY

The Black rat is a relatively large blackish or brownish-grey rodent, with a body length of up to 23 cm and a very long tail. Compared to the Brown rat, its body is more slender (not sturdy)[1,2]. Its head is characterised by a pointed (not slanted) snout, with larger ears (19-26 mm) normally covering (or just touching[3]) the eye when folded forward. It has larger eyes, and a tail that is longer than the body (see drawing below)[4,5]. Belly coloration is variable, greyish-brown to whitish, often paler than rest of body. The hindfoot is smaller (28-38 mm) and of uniform, usually dark colour compared to the Brown rat. In Seychelles, adult males normally weigh less than 220 g, and females no more than 200 g[6,7]. On the dry inselbergs of Morne Seychellois National Park, rats were significantly smaller with shorter tails and body lengths, and weighed less than those caught from intermediate or montane forests[8]. Analysis of stomach contents revealed a predominantly invertebrate diet with a minority of plant parts for Black rats and a clear preference for plant matter for Brown rats, hence an interesting resource partitioning. For both species, female stomachs had significantly more plant matter compared to males[8].

Black rats have acute hearing that is sensitive to ultra sound, and a highly developed olfactory sense. They are extremely agile climbers, mainly nocturnal and very arboreal, whereas the Brown rat is mainly terrestrial.

Black rats occupy all terrestrial habitats types including mangrove trees where they can build bulky nests, as on Aldabra[4, 10], and they are less dependent on human habitations and the proximity of water. These ecological differences reduce the competition between the two species[9, 11, 12]. On Denis Island, a radio-tracking experiment on eight live-trapped rats showed that they all regularly climbed and stayed in the crown of coconut trees, and would come down to the ground at least every night, although some would not venture more than 15 m from their tree[13]. Omnivorous and opportunistic, this rat eats all sorts of plant material (leaves, stems, moss, roots, buds, fruits, flowers, seeds, bark) and a large variety of both invertebrates (insects, crabs, arachnids, etc.) and vertebrates including eggs, nestlings and adults of landbirds or seabirds, hatchlings and young of reptiles (lizards, geckos, turtles and tortoises)[2, 4], and probably amphibians (caecilians and frogs) too. They also consume all kinds of human food (grain, biscuits, bread, waste etc.) and agricultural crops. Extremely prolific, Black rats breed all year round in Seychelles, although more during the wetter season[6, 10, 14]. Each year a female produces up to 6 litters of 3 to 10 young each. Females usually have 10 nipples (but some may have 12) and reach maturity at c. 3 months. The gestation period is of 20-22 days, and weaning occurs at 21-28 days. Black rats are extremely adaptable and able to prosper in new environments. Longevity is 2 years or less[2-4].

ORIGIN

A human commensal for thousands of years on the Indian sub-continent, the Black rat had already reached western Europe between 2,500 and 1,500 years ago, and was later spread around much of the globe by ship (hence its name 'Ship rat') during colonial times[2, 4]. It is one of the three most wide-spread species of rats, having been introduced to the majority of archipelagos of the world; and it is considered the most successful colonising mammal after humans together with the Brown rat and the House mouse. First reported in 1773 on Mahé, it was probably already present in Seychelles before that date[15]. It is reported to have colonised Bird Island in 1967 from Praslin in a consignment of thatching leaves[16]. Black rats in Madagascar and Comoros appear to have originated either from the native area on the Indian sub-continent or from the anciently colonised Arabian peninsula, probably brought by merchants during the 8th-10th centuries or before[17]. Genetic studies to determine the origin of Seychelles rats are in progress.

DISTRIBUTION & ABUNDANCE IN SEYCHELLES

Black rats are present on Mahé, the Ste Anne group and Ile Thérèse; Praslin and La Digue, Curieuse, Félicité, Ile Ronde (Praslin) and Marianne; and Silhouette[6, 18, 19]. They never colonised Conception; Aride, Cousin, Cousine and Ile aux Fous; Frégate; Ile aux Récifs and Mamelles[16, 19] and were eradicated from Bird, Denis, Grande Soeur, Petite Soeur, Ile du Nord, Anonyme (although individuals may reinvade and survive for short periods) and Ile aux rats[20, 21]. They occur on all the outer islands, including the whole of Aldabra atoll, except Bancs Africains, D'Arros and St. Joseph atoll, Boudeuse and Etoile, Desnoeufs and Marie Louise islands; St François and Bijoutier (Alphonse group); Bancs de Sable, Goëlettes, Déposés, Ile du Milieu and Lapins (Farquhar); and all Cosmoledo islands but Menai; all of which they never colonised[3, 16]. They were eradicated from Grand Polyte, Petit Polyte and Grand Ile (Cosmoledo)[21].

The Black rat is the commonest species of rat, being much more abundant and omnipresent than the Brown rat on large islands such as Mahé[23, 24]. It occurs at high densities in natural and human-modified habitats, as well as around houses and buildings to which they are drawn for shelter and foo. Rats can reach extremely high densities in Seychelles and on tropical islands in general[25], where abundance of water and food, absence of predators, scarcity of competitors, and absence of severe climatic con-straints allow continuous successful breeding all year-round. Available abundance indices are not always directly comparable as inconsistent trapping protocols have been used. For a description of how abundance indices are calculated refer to "Index-trapping", p. 218. Recorded abundance indices, based on lines of single traps spaced at 25 m intervals, were the following: on Mahé in June 2011, 12.1 rats per 100 trap-nights (corrected) at La Misère residential areas (where rat control was taking place), 44.4 at Helvetia and 55.7 at Pointe Larue-Anse aux Pins; 93.2 on Ste Anne in the woodlands[25]; and 64.7 on Grande Soeur in March 2010[26].

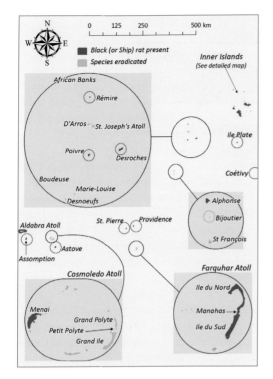

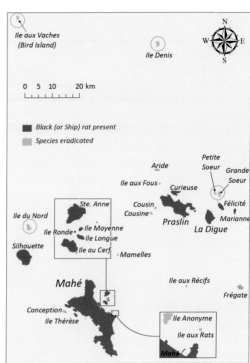

Within the Morne Seychellois National Park, in July-August 2006, abundance indices based on lines of paired traps spaced at 25 m intervals in different habitat types averaged 26.0 (15.7 to 36.6) Black rats per 100 trap-nights (corrected)[8] in inselberg vegetation. Higher abundances of 32.3 rats (27.1 to 43.2) were recorded in montane forest (500 to 900 m a.s.l) and 45.8 (37.1 to 60.4) in intermediate forest (200-400 m a.s.l.), although these abundance also include a small percentage of Brown rats. Important fluctuations in abundance may occur between seasons[6, 10]. On eight small inner islands (both granitic and coralline) in 1999-2000, abundance indices varied seasonally. During the South-East monsoon they ranged from 17.5 to 99.5 rats per 100 trap-nights (corrected), and during the North-West season from 3.8 to 63.2 rats[22]. Inter-annual variations of abundance may also occur. On Bird Island, the abundance index based on single traps spaced at 50 m intervals, was 26.5 rats/100 trap-nights (corrected) in November 1995, but increased to 141 in November 1996[14, 20]. In contrast, the same index remained very stable on Curieuse (82.2 in July 1998 and 84 in July 2000) and Denis (35.6 in July 1998 and 35.4 in May/June 2000)[14, 27].

On Grand Polyte (Cosmoledo), a very high index (single traps spaced at 25 m intervals) of 250 rats/100 trap-nights (corrected) was found in October 2005

compared to only 60.0 in November 2007. On nearby Grand Ile this index was 81.5 rats/100 trap-nights in October 2005 and only 25.6 in November 2007[21,28]. On Ile du Nord, the same index was 175 rats per 100 trap nights (corrected) in August 2005, compared to only 23 rats per 100 trap-nights in August 2003[29, 30].

During population outbreaks, up to 2 to 3 rats may be found in one single cage trap! The above-mentioned very high index on Cosmoledo corresponds for example to 25 rats trapped in a single line of 20 traps in one night, i.e. an average of 1.25 rats caught per trap. At Pouhou islet (Mayotte), measuring 0.6 ha, 55 rats trapped with 25 cage-traps over two nights gave an abundance index of 110 rats/100 trap-nights (uncorrected) or 220 rats/100 trap-nights (corrected). During the rat eradication with cage-traps that immediately followed (December 2004-January 2005), 117 rats were captured, giving an exact density of **195 rats/ha**, possibly a world record![31]. At Juan de Nova and Europa (Iles Eparses, Mozambique Channel), rat densities attained 26/ha and 65/ha respectively in forests; and home ranges varied between 0.35 and 0.84 ha in forest and from 1.3 to 3.1 ha in grassland[32, 33]. In Seychelles, density estimates and home ranges have only been measured on Aldabra, and ranged in 2013 between 22 and 32 rats/ha in three different habitat types[25]. In the granitic islands, in view of the

Black rats have characteristic long ears, big eyes, and a tail longer than body. An individual on Mahé.
Gérard Rocamora

extremely high abundance indices reached during outbreaks, of similar values to those recorded on Pouhou (Mayotte), densities are likely to reach and exceed 100 rats/ha during such periods[29].

IMPACTS AND THREATS

The Black rat is one of the worst invasive animals on earth[34]. It not only has a huge negative ecological impact on island ecosystems, but also on agriculture, public health and the economy of human societies in general[1,2]. Its negative ecological impact is worse than the Brown rat as it has easy access to both ground and arboreal resources, and both rats are preying upon a large diversity of native plants and animals and threatening the rarest and more fragile ones[4,25]. The Black rat restricts the regeneration of native plant species by eating their seeds and seedlings[35-37], and has the potential to devastate and wipe out colonies of both ground and tree-nesting seabirds[38-40], and entire populations of small reptiles and landbirds[42]. It has caused or contributed to catastrophic declines, range contractions and global extinctions of many native species and island endemics around the world, including plants, rare invertebrates and small verte-brates (probably including the extinct Aldabra brush warbler)[43-45]. In Seychelles, a number of species have become excluded from islands invaded by Black rats (see p. 58). Among these are endemic arboreal pas-serines like the Seychelles fody *Foudia sechellarum*, the Seychelles white-eye *Zosterops modestus*, and many breeding seabird species (e.g. breeding terns restricted to rat-free islands or islets). The many species of seabirds, landbirds and reptiles showing signs of recovery after Black rat eradication illustrate the very strong negative impact this species has on the ecosystem (see next section).

Black rats and Brown rats together serve as the reservoirs and primary vectors of the Leptospirosis bacteria, Hepatitis (E) viruses and other rat-trans-mittable diseases[46]. (This is currently under study as part of regional research projects led by CRVOI-IRD). The economic impact of the Black rat is considerable due to the damage it causes to agriculture crops, commercial goods in warehouses and households, structural equipment (telephone and electricity wires, etc.), and public health (see p. 33). Damage to commercial coconut plantations (mainly caused by eating young fruits) was probably considerable in Seychelles as in other Indian Ocean island countries (e.g. over 40% of harvest reduction due to rat damage in Maldives coconut plantations[48,49]).

ERADICATION AND CONTROL PROGRAMMES
Eradication

Twelve attempts have been made to eradicate Black rats from islands of 10 ha or more in Seychelles (see Table 17 below). In addition, two tiny islets of less than 1 ha have also been successfully freed of rats. All of these have been reported to be successful from an operational point of view. However, as for Brown rat eradications, the success rate in terms of the final outcome of these operations is only 75% (excluding the two islets; see also p 62) as rats were found again on some islands within a year after the opera-tions (Denis in 2000, Curieuse in 2000, Ile du Nord in 2003)[14,29]. These failed outcomes were explained as reinvasion facilitated by insufficient biosecurity measures and uncontrolled boat landings. However,

Contrary to Brown rats, Black rats have tails that are longer than their bodies.
Gérard Rocamora

in the absence of DNA samples collected prior to the attempts, it is not possible to prove that these rats were reinvaders and not survivors[47]. This is particularly the case for Curieuse, the largest and highest of the medium-sized granitic islands (5th granitic island in size), which has c. 10 ha of mangrove where arboreal rats may have survived. No specific measures were taken to compensate for the fact that aerially spread baits must have been rapidly destroyed by high tides after each drop[14, 50]. The repeated eradication attempts on Denis Island (a flat coralline island) and on Ile du Nord (a high granitic island) were also challenging operations for the following reasons. Both islands had small hotels and resident populations of over 100 persons, and were covered with large extensions of coconut trees[29]. In the case of Denis Island, large densities of hermit crabs also consumed the bait[30]. The eradication of Black rats from two islands of uninhabited Cosmoledo atoll, located 1,200 km from Mahé, used a helicopter and was also a challenging operation due to its remoteness and the complex logistics involved[51]. The permanent grid of 66 bait stations set up across the small island of Anonyme (9.8 ha) is an original and efficient protocol designed to first eradicate rats, and then to prevent the re-establishment of reinvaders that may swim or be carried over by floating debris from neighbouring Mahé (only c. 500 m away).

Before the first Black rat eradication attempt was conducted (on Bird Island), eight different bait types were tested for acceptance by rats, durability in the environment and attractiveness to non-target species. Pestoff Rodent Bait 20R (with Brodifacoum 20 ppm) cereal pellets of 12 mm and Pestoff Talon wax blocks (with Brodifacoum 50 ppm), both commercialised in New Zealand, were chosen and used successfully[14, 20]. During the following eradication operations in Seychelles, similar bait types have been used, mainly Pestoff Rodent Blocks (waxed blocks with 20 ppm brodifacoum); and 10 to 12 mm Pestoff Rodent Bait 20R cereal pellets, dyed green (to track bait ingestion by rats, and found to be less attractive to birds in New Zealand). Flavour (normally coconut) has often been added to the bait (in pellets or in blocks) to try and increase its attractiveness to rodents, although its effectiveness is not proven (at least in Seychelles)[53]. Out of the 14 Black rat eradications conducted in Seychelles, four operations were ground applications, and the other 10 were aerial applications done with helicopter. Bait rates varied between 6 and 12 kg/ha for applications with bait stations, and between 15 to 25 kg/ha for hand-spread pellets. For aerial applications, rates varied between 23 to 40 kg/ha in 2 to 4 drops (pulses) spaced 6-11, 8-16, and 12-19 days each. On four occasions, differential GPS (DGPS) was used to guide the helicopter during the drops to ensure optimal coverage.

Eradication of Black rats has been attempted in subantarctic, temperate, and tropical climates[72-75, 56]. Over a total of 336 operations of sufficient data quality reported in DIISE until September 2014, 258 were successful but reinvasion occurred in 75 islands. In tropical islands only, out of a total of 166 operations, 138 were successful but reinvasion occurred in 33 cases[56]. Chances of failure were described as low (8%)[57], but current figures show higher failure rates of 11.0 % and up to 33.3% when reinvasions are added, and 17 to 37% for tropical islands only. In Mauritius, Black rats were successfully eradicated from Ile aux Aigrettes (20 ha) in 1987, Gabriel Island (15 ha) in 1995 and Ile Platte (253 ha) in 1998, through hand broadcasting of brodifacoum (20 ppm) - and bromadiolone (50 ppm) for the latter two islands[58]. The 2006 attempt on Eagle Island (Chagos; 240 ha, coconut dominated) through two hand broadcasts of wax Brodifacoum 50 ppm blocks on a 30 m grid failed[59]. Successful eradication of Black rats has taken place on the southern Indian Ocean subantarctic territories of France (TAAF) (e.g. on Ile St Paul, c. 800 ha) and on several small tropical islands (e.g. Surprise Island, New Caledonia)[69-71]. In

2014, Macquarie Island (off Tasmania, Australia), measuring 12,872 ha; is the largest island on earth from which Black rats have been eradicated, and Hermite Island (Australia), measuring 1,020 ha, is the largest in the tropical world[60].

Control

The ubiquitous Black rat is the main target for rat control in Seychelles, except in Victoria and around villages where the Brown rat dominates. Past methods consisted in using traditional traps ('Lasonmwar'), and later maize and sugar waxed blocks with anticoagulants (such as warfarin after the 1950s)[20, 61]. Current control methods include bait stations providing blocks or cereal pellets with anticoagulant (Brodifacoum, Diphenacoum, Flocoumafen, etc.), hand spread pellets in case of serious outbreaks, and trapping using various trap models (cage-traps, snap traps, and occasionnally the non-humane glue traps) (see General recommendations section). Biological control was attempted in 1949-52 through the introduction of the Barn owl (*Tyto alba affinis*), with limited success in controlling rats and disastrous consequences for native species such as seabirds (see p. 248). Regular de-ratting operations are conducted on the three large granitic islands (Mahé, Praslin, La Digue) in selected sites around Victoria (e.g. market, public buildings, harbours, bin sites) by government agencies and pest control companies under the authority of the Department of Public Health of MoH. Permanent control is also organised by many owners of private houses and businesses. On government-owned islands managed by IDC (e.g. on Desroches, Alphonse, Silhouette), ongoing control is organised by IDC, hotel and ICS staff around habitations and areas of high biodiversity value. On Aldabra, SIF conducts rat control around the research station, campsites and on lagoon islets that are important for breeding seabirds.

Black rats represent 80 to 90% of rats trapped around La Misère and Barbarons residential areas and woodlands[3]. There, rat control is conducted using a grid of 50 m spaced home-made bait stations installed by ICS to protect the endangered Seychelles white-eye. This method has also been used since 2006 at the two properties of the President of UAE (over c. 25 ha)[63], and in 2011-2012 around La Misère and Souvenir villages, and Grande Anse forestry station (c. 25 ha in total)[64]. In Seychelles, traps are normally baited with fresh or roasted coconut, traditionally considered by Seychellois to be the most effective bait for rats. In bait stations, Pestoff Rodent Blocks (waxed blocks with Brodifacoum 20 ppm) are used, and rat trapping conducted during 2006-2008 indicated that rat abundance had been reduced by 85 to 95%[63]. Since 2011, on Ste Anne Island, the Beachcomber tourist resort has adopted the same technique and extended rat control to natural habitats around the hotel with a grid of c. 140 bait stations spaced at 40 m intervals. These cover c. 25 ha and include a small recently established Wedge-tailed shearwater colony (see picture of grid bait stations on Sainte-Anne p. 89).

Ecosystem monitoring and recovery

Between 1996 and 2010, a minimum of 24 native vertebrates (5 reptiles, 11 landbirds and 8 seabirds) and several large invertebrates (snails, millipedes, crabs) have benefited from the eradication of Black rats on nine islands of 10 to 201 ha in Seychelles[65]. On Ile du Nord, 3 years after the eradication, the endemic Seychelles blue pigeon *Alectroenas pulcherrima* significantly increased, and Moorhens *Gallinula chloropus* and Seychelles skinks *Trachylepis seychellensis* doubled their initial abundance; significant changes within invertebrate communities were also recorded (see "Ecosystem recovery after the eradication of rats and cats", p. 169). Seabirds also greatly benefited from Black rat eradications. In total, five species (Wedge-tailed shearwater *Puffinus pacificus*, Fairy tern *Gygis alba*, Lesser noddy *Anous tenuirostris*, White-tailed tropicbird *Phaethon lepturus*; and Masked booby *Sula dactylatra* on Cosmoledo atoll) have spontaneously established a total of seven new breeding populations across five islands where Black rats were removed. Following the Black rat eradication on Ile du Nord in 2005 and on Petite Soeur in 2010, several nests of White-tailed tropicbirds and a small colony of Wedge-tailed shearwaters were discovered on each island, and since then these colonies have been developing. It is likely that some of these seabirds used to nest sporadically on the same islands before the eradication, and were able to increase their numbers after rats were removed. Not many changes in vegetation have been detected after Black rat eradications. However, the *Pisonia grandis* forest did significantly expand in both Denis and Bird islands, and an increase in coconut regeneration has been reported on Ile du Nord.

Table 17: Summary of Black rat eradications conducted in Seychelles. More details may be found in the references given as well as in Beaver & Mougal 2009.

Island (Size in ha)	Date of eradication attempt	Method (No. of applications)	Bait	Outcome
Bird[14] (101)	Oct 1996 – April 1997	Ground application. Bait station (bottles / pipes) grid every 25 m along 50 m spaced lines, refilled after a week then monthly with blocks (7) + hand spread pellets (2)	Wax bait blocks with Brodifacoum 50 ppm; + 12 mm cereal pellets 20 ppm (total 11.6 kg/ha).	Successful outcome. 30-70% of 3 introduced birds (Indian myna, Ground dove, Turtle dove hybrid) and some Turnstones died.
Curieuse[14] (286)	July 2000	Helicopter (2 drops). DGPS navigational guidance.	Brodifacoum 20 ppm cereal pellets (total 23 kg/ha).	Failed outcome. Rats (many) present in Aug. 2001. The eradication may have succeeded, but measures to prevent reinvasion lax or absent, and rats could have survived in the mangrove. 60% of Giant tortoises penned. 10-40% of above introduced birds, Madagascar fody and a few Turnstones died.
Denis[54] (143)	July 2000	Helicopter (2 drops). DGPS navigational guidance.	Brodifacoum 20 ppm cereal pellets (total 23.6 kg/ha).	Failed outcome. Rats still present in Aug. 2001 (but mice eradicated). The eradication may have succeeded but reinvasion likely following poor implementation of protocols. 5 Giant tortoises penned. Mortality reported for above introduced birds
Denis (143)	2002	Ground application (4). Hand spread along a 25 m grid.	Brodifacoum 20 ppm cereal pellets (minimum 25.4 kg/ha).	Successful outcome. No details available.
Ile du Nord[30] (201 ha)	August 2003	Helicopter (3 drops). DGPS navigational guidance.	Brodifacoum 20 ppm cereal pellets (total 31 kg/ha).	Failed outcome. Success of eradication operation unclear as rats were reported in Dec. 03 and captured in March 2004, possibly new invaders or survivors. Some mortality recorded amongst introduced mynas, Barn owls and Madagascar turtle doves.
Anonyme[52] (10 ha)	July to Oct. 2003	Ground application. 66 bait station grid spaced c. 35 m apart, plus intensive cage trapping.	Brodifacoum 20 ppm wax blocks (total c. 6 kg/ha), trap bait = roasted coconut.	Successful outcome. 1/3 of stations not refilled for the first month due to negligence. Intensive trapping started in Oct. after some rats suspected to intake sub-lethal dosis. No mortality of non-target species.
Anonyme[21] (10 ha)	Nov. 2006	Intensive cage trapping, plus ground application of rodenticide resumed (66 bait station grid spaced c. 35 m apart).	Brodifacoum 20 ppm wax blocks.	Successful outcome. Follows reinvasion and establishment of a small breeding population on plateau after interruption of bait station refilling in July 2006 (island sold).
Ile aux rats (near Mahé)[21] (<1 ha)	Nov. 2006	Ground application, hand spread (1).	Brodifacoum 20 ppm cereal pellets (total 15 kg/ha).	Successful outcome. No mortality of non target species recorded.

Ile du Nord[28] (201 ha)	August-Sept. 2005	Helicopter (4 drops). Visual navigation with mapped fly lines; DGPS navigational guidance only for 2nd drop. 50 m bait station grid on high risk area	Brodifacoum 20 ppm cereal pellets (total 39.9 kg/ha) + 20 ppm wax blocks.	Successful outcome. Grid stations on plateau (162) refilled for 7 months, plus 31 permanent bait stations. High mortality of mynas (53% of population), Madagascar turtles doves (84%) and Common moorhens (60%); Barn owls eradicated, and minor mortality noted on other introduced birds.
Cosmoledo: Grand Ile[51] (143 ha) Grand Polyte (21 ha) & Petit Polyte (<1 ha)	Nov. 2007	Helicopter (2 drops). Visual navigation with mapped flight lines, plus field marks with 4m white palls with colour flags over a 100m grid. No DGPS.	Brodifacoum 20ppm cereal pellets (total 29.7 kg/ha). 'Collars' of 20 ppm bait waxed blocks in small area of mangrove.	Successful outcome. Declared rat free in Nov. 2008 after 660 trap-nights on Grand Ile, 330 on Grand Polite and 30 on Petit Polyte. Rat free status confirmed in March 2014. No mortality recorded for non-target species. C. 20 permanent bait stations installed but not refilled between Nov. 2008 and March 2014 (no access allowed due to piracy risks).
Grande Soeur[55] (85 ha)	August-Sept. 2010	Helicopter (3 drops). Visual navigation with mapped fly lines No DGPS.	Brodifacoum 20 ppm cereal pellets (total 35.9 kg/ha).	Successful outcome. Declared rat free in Nov. 2011 after intensive survey trapping, continued until Nov. 2012. Giant tortoises all penned during operation. Important mortality recorded for Mad. Turtle-doves and mynas (probably over 80 and 40% respectively). 20 permanent bait stations.
Petite Soeur[55] (35 ha)	August-Sept. 2010	Helicopter (3 drops). Visual navigation with mapped fly lines No DGPS.	Brodifacoum 20 ppm cereal pellets (total 32.1 kg/ha).	Successful outcome. Declared rat free at the same time as Grande Soeur in Nov. 2011 after intensive survey trapping. No mortality recorded for non-target species. 7 permanent bait stations.

MANAGEMENT RECOMMENDATIONS

Read "General recommendations and considerations for the control and eradication of rodents", p. 214. The following are particularly relevant to the Black rat.

Control and eradication protocol

- Control can be conducted using trapping or chemical treatment (anticoagulants). Island eradications always require the use of anticoagulants, except for very small islands (less than a few hectares) where trapping alone has been used successfully.
- Cleanliness and tidiness are essential requirements for success in controlling Black rats in gardens and residential areas. Trim vegetation regularly around your house and other buildings to prevent contact between branches and roof. Remove or prevent easy access to any food and fresh water: food waste in bins, fallen fruits, pet food, etc.

- Do not only place lines of traps or bait stations immediately adjacent to buildings (or bird nests); but also place them 25-50 m away, in order to create a buffer zone where rats are trapped or poisoned before they reach the sensitive areas

Collars with bait blocks were placed around branches in mangrove trees during the Black rat eradication at Grande Ile (Cosmoledo atoll).
Gérard Rocamora

Densities of rats can be very high in tropical islands.
Gideon Climo/North Island

from which you want to exclude them. This is particularly important for Black rats, which live mainly in woodland. Traps/bait stations around buildings should only serve as a last resort form of protection.

Chemical treatment

- Mangroves are one of the biggest challenges in the eradication of Black rats from tropical islands, because the poison bait can be washed away at high tide. Place bait stations or simple 'collars' of waxed bait blocks around branches well above the maximum sea water level (one around each tree in low density areas, or one every 15-20 m in areas of dense mangrove). (See p. 199; trialed in 2007 on the small extensions of mangroves of Grande Ile, Cosmoledo). A system with 'bolas' made of rodenticide, designed in Micronesia, has also been trialed on Aldabra on a larger scale[10].
- Whether conducting a ground based or an aerial application of rodenticide, ensure a very regular and thorough distribution of rodenticide during eradication attempts, especially in mature forests and areas dominated by coconut trees, a favourite habitat for Black rats. Increased availability of bait is necessary in order to compensate for the alternative food sources (fruits, invertebrates) abundant almost year-round in wet tropical climates. This problem is particularly acute for Black rats, which have access to above-ground resources.
- Bait stations may be spaced 25 to 50 m from each other depending on rat density. For an eradication attempt, spacing of 25 to 35 m is recommended (e.g. grid for eradication on Anonyme Island, 2003), while 40 to 50 m should suffice for control (e.g. Ste Anne control grid).

Trapping

- Avoid placing traps in open areas (e.g. bare ground or mowed grass). Rats will feel more comfortable and safe approaching while relatively hidden (e.g. near a rock, at the bottom of a tree, along a wall, etc.). Rats avoid venturing into wide open areas, especially Black rats which are arboreal and feel more vulnerable on the ground.
- Certain individuals may learn to avoid traps ('trap shyness'). This is why trapping alone is normally not recommended for eradication.
- Coconut (fresh or roasted) appears to be the most practical and efficient bait for rat trapping in Seychelles.

See p. 127 recommended methods to euthanase animals captured alive.

There are no sophisticated biological control methods (e.g. genetically modified animals capable of producing only male offspring; use of viruses to disseminate immunocontraception) that can be recommended yet against rats, but research is being conducted in that direction (see p. 125)[67, 68].

References / Further reading

1 CAB International, 2014; 2 GISD, 2014; 3 G. Rocamora, *pers. obs.*; 4, 5 Grant *et al.*, 2003; 6 Hill *et al.*, 2003c; 7 G. Rocamora & E. Lagadec / CRVOI unpublished; 8 Fanchette& Kerridge, 2009; 9 Foster *et al.*, 2011; 10 Harper *et al.* 2015; 11 Grant, 1972; 12 Taylor, 1978; 13 Gideon Climo, *pers. comm.*; 14 Merton *et al.*, 2002; 15 Cheke, 2010; 16 Racey & Nicoll, 1984); 17 Tollenaere *et al.*, 2010; 18 Rocamora & François, 2000; 19 Nevill, 2009; 20 Beaver & Mougal, 2009; 21 Rocamora & Jean-Louis, 2009; 22 Hill 2002; 23 Ministry of Health, 1994;24 G. Rocamora & E. Lagadec / CRVOI, unpublished; 25 Harper & Bunbury, 2015; 26 G. Rocamora, unpublished; 27 Merton, 2001; 28 Climo & Rocamora 2006b; 29 Climo & Rocamora 2006a; 30 Climo, 2004a; 31 Rocamora et Said, 2005; 32 Russell *et al.*, 2011; 33, 34 Lowe *et al.*, 2000; 35 Drake, 2007; 36, 37, 38 Jones *et al.*, 2008; 39 Mulder *et al.*, 2011; 40, 41 Ruffino *et al.* 2015; 42 Atkinson, 1985; 43 Courchamp *et al.*, 2003; 44 St Clair, 2011; 45 Towns *et al.*, 2006; 46 Bovet *et al.* 2013; 47 Russell, Miller *et al.*, 2009; 48 Fiedler, 1984; 49 Dolbeer *et al.*, 1988; 50 G. Rocamora, 2005 *in litt.* to Don Merton; 51 Rocamora 2007a, 2007b, 2010a; 52 Rocamora, 2008; 53 Bill Simmons/ Animal Control Products (NZ); *pers. comm.*; 54 Climo, 2004b; 55 G. Rocamora / Sisters Pty Ltd, unpublished data; 56 DIISE, 2014; Successful and failed eradications only, excludes reinvasions, whole island eradications only, satisfactory or good data quality only; 57 Howald *et al.*, 2007; 58 Bell 2002; 59 Daltry *et al.*2007; 60 Burbidge 2011; 61 Lionnet, G. 1966; 62 Buckle & Smith, 1994; 63 Labiche & Rocamora, 2009; 64 Henriette, 2013; 65 Rocamora *et al.*, in prep.; 66 Rocamora G. and Labiche A. 2009c.; 67 Dobson, 1988; 68 Campbell *et al.*, 2015; 69 Lorvelec & Pascal, 2005; 70 Soubeyran, 2008; 71 Soubeyran *et al.* 2011; 72 Thomas & Taylor, 2002; 73 Galván *et al.*, 2005; 74 Howald *et al.* 2007; 75 Towns & Broome, 2003.

Brown rat.
Jack Jeffrey.

ONE OF
WORLD'S WORST
100
INVASIVE
SPECIES

Brown rat

Rat brun ou Rat surmulot

Lera later

Rattus norvegicus

IDENTIFICATION AND BIOLOGY

The Brown rat is a relatively large brownish grey rodent, with a body length of up to 25 cm and a tail of similar size or less. Compared to the Black rat, its body is heavier and sturdier (not slender) and its head has a more slanted (not pointed) snout and eyes. Its ears are smaller (14-22 mm) and normally do <u>not</u> cover the eyes when folded; while its tail is shorter than the body (see drawing)[1, 2]. Belly coloration is variable, often more pale (whitish or greyish) than in the Black rat. The hindfoot is large (30-41 mm) and whitish / pale on top. In Seychelles, adults normally weigh over 200 g, males up to 385 g, females up to 280 g[3].

It has acute hearing that is sensitive to ultra sound, and a highly developed olfactory sense. Brown rats are omnivorous and opportunistic, and eat all kinds of plant and animal (both invertebrates and vertebrates) material, much like the Black rat; but its prey tends to be more terrestrial and non-arboreal. This may include the following: large and small ground-dwelling insects; the eggs or young of lizards, marine turtles and tortoises; eggs, nestlings or adults of both landbirds and seabirds especially those nesting or foraging on the ground; grain, fruits and all sorts of agricultural crops, as well as human waste[1, 2, 4]. In the forests of the Morne Seychellois National Park, stomach content analysis revealed a clear preference for plant matter for Brown rats (whereas Black rats had a predominantly invertebrate diet)[5]. Brown rats are mainly nocturnal and seldom forage above ground. They are often associated with human habitations (especially when Black rats are present), and prefer habitats where fresh water is available, although they can survive without[3, 4]. Although not good climbers like Black rats[17], Brown rats are quite strong swimmers. These ecological differences and subsequent niche partitioning reduces competition between the two species[18, 19]. Extremely prolific, Brown rats breed all year round in Seychelles, although more during the wetter season[3, 6]. They may produce up to 6 litters per year, with 6-12 young in each. Females have 12 nipples and reach maturity at 50-60 days. The gestation period is 21-24 days, and weaning occurs at 28 days[1, 2].

ORIGIN

probably North China and/or Mongolia. The Brown rat has been a human commensal for thousands of years, having colonised Europe during medieval times, and spread during the 18th century to all other continents except Antarctica. Now it is one of the three most widespread species of rats, having been introduced to almost all archipelagos around the world[1, 2]. It appears to be a more recent coloniser of the islands of the Western Indian Ocean[7, 8], particularly in Seychelles where it was first reported in 1994 on Mahé[9, 10]. The Brown rat was formally identified on Conception and D'Arros only in 1997[11], and 2002 respectively[12]. Nevertheless, it was probably already present on Mahé prior to 1994, given that the presence of (presumably Brown) rats was reported on nearby Conception as early as 1965 by a former worker[13], and on D'Arros in 1944 in an official report[14]. The subjects of these early reports are unlikely to have been Black rats, unless the latter were subsequently out-competed when Brown rats later colonised these two islands. Brown rats

colonized Frégate Island in 1995 during the construction of the island resort[15].

DISTRIBUTION & ABUNDANCE IN SEYCHELLES

Brown rats are present on Mahé, Ste Anne, Longue andlle aux Cerf[35, 37], Praslin[9] and apparently La Digue[16], these being the only islands where they coexist with the Black rat. Brown rats were formerly present on Frégate (1995-2000), D'Arros (until 2003) and Conception (until 2007) until their eradication from these sites. The Brown rat was the only species of rat ever reported from these three islands.

On Mahé, the Brown rat is less abundant and omnipresent than the Black rat, and occurs in relatively low abundances in natural habitats[20], particularly in damp and wet localities of intermediate and montane forests[5], and also in residential areas where rats are controlled, e.g. 10.0 Brown rats/100 trap-nights (corrected) in June 2011 at La Misère residential areas[20]. For a description of how abundance indices are calculated refer to "Index-trapping", p. 218.

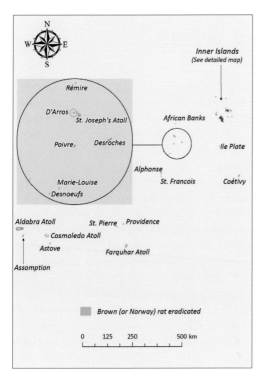

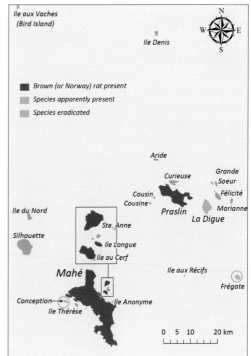

Table 18: Summary of Brown rat eradications conducted in Seychelles[22, 44-49].

Island (Size in ha)	Eradication attempt	Method (No. of applications)	Bait	Outcome
Frégate (219)	Nov. 1995 to June 1996	Bait station grid on plateau (c. 25 m) and snap trap lines.	Flocoumafen / Difenacoum until March. Brodifacoum 50 ppm wax blocks + 20 ppm pellets	Failed outcome. Interrupted in June 96 after several Seychelles magpie-robins were killed by traps and secondary poisoning.
Frégate[22, 46] (219)	June-July 2000	Helicopter (3 drops). DGPS navigational guidance.	Brodifacoum 20 ppm cereal pellets (total 35 kg/ha)	Successful outcome. Endemic birds (39 magpie-robins, 330 Seychelles fodies) and Giant tortoises held captive for 11 weeks. Confirmed rat free in June 2002 but Seychelles white-eyes already transferred in Oct/Nov. 2001. Mortality on introduced and migrant birds only.
D'Arros[47] (140)	August-Sept. 2003	Ground application (3). Hand spread along 50 m spaced tracks	Brodifacoum 20 ppm cereal pellets (total 17 kg/ha).	Successful outcome. About 50% of Seychelles Fodies, and Giant tortoises taken into captivity, but no significant mortality, as with other birds. Declared rat-free in 2005.
Conception[23] (69)	August-Sept. 2007	Helicopter (2 drops). Flight lines on map only. No (D)GPS.	Brodifacoum 20 ppm cereal pellets (total 26.7 kg/ha).	Successful outcome. Captive trial conducted in 2005 to assess poison risks to Seychelles White-eye. No mortality of non-target species observed. Declared rat-free in Oct. 2008 after intensive survey trapping of over 1,800 trap-nights.

However, Brown rats can reach high abundance in areas with insufficient control and/or inadequate waste management, or in natural habitats on islands where Black rats are absent. Available abundance indices are not always directly comparable as a variety of trapping protocols have been used: 220 rats per 100 trap-nights (corrected) on Conception in May 1998, and 27.2 to 38.8 (corrected) on Frégate in May 1998/June 2000 (single traps spaced 50 m)[21, 22]; 60 to 89 in 1999-2000 (traps in pairs spaced 25 m)[6]; and 121 rats /100 trap-nights in August 2007 (single traps spaced 25 m) on Conception[23]. In certain years, at the end of the South-East monsoon (driest) season, the abundance of rats on Conception appeared to be very low and coincided with outbreaks of (skin) disease[24]. The abundance index on D'Arros was only 20 rats/100 trap-nights (single traps spaced 25 m) when the species was first formally reported in May 2002[12]; and even in 1944, rats were also reported to be 'few' on D'Arros island[14]. No density values are available for Brown rats in Seychelles.

IMPACTS AND THREATS

The Brown rat is one of the worst invasive animals on earth[25]. Alike other introduced rats, it has a very negative ecological impact on island ecosystems and also on agriculture, public health and the economy of human societies in general[1, 2]. It preys upon a large diversity of native plants and animals and threatens the rarest and more fragile species[26]. It has caused or contributed to the decline, range reduction or global extinction of many native species and island endemics around the world, including rare ground dwelling invertebrates and small vertebrates (mammals, birds, reptiles)[27-30]. This species restricts the regeneration of native plant species by eating their seeds and seedlings[31], and has the potential to devastate and wipe out colonies of ground nesting seabirds[32, 33] (e.g. terns or shearwaters) and entire populations of small reptiles and landbirds[34]. On Conception, during severe droughts, this species was reported 'chewing' the stems of the endemic Latanier feuille *Phoenicophorium borsigianum* leafs to obtain sap, resulting in a high proportion of palm leafs being cut or damaged[35, 50]. Damage by rats (unknown species) has also been observed on Mahé on the endemic 'Vacoa de montagne' *Pandanus multispicatus*)[36]

Between 1995 and 2000, the species invaded Frégate island, and populations of the ground-foraging endemic Seychelles magpie-robin *Copsychus sechellarum* (the fledglings of which are easy prey) and other endemic species restricted to Frégate (e.g. Giant Tenebrionid beetle *Polposipus herculeanus*, Giant scorpions *Chiromachus ochropus* and endemic snails *Pachnodus fregatensis* and *Conturbatia crenata*) were at risk[38]. The rediscovery on Conception, two years after the Brown rat eradication, of two rare endemic invertebrates thought to be locally extinct during the period when rats were present on the island, illustrates the strong impact of this rat on certain large invertebrates (see next section)[39]. Brown rats occasionally climb trees and may plunder easily accessible nests, and capture low perched arboreal adult passerines at night, as observed for Seychelles white-eyes *Zosterops modestus* during a captive trial on Conception island in 2005[40]. Nevertheless, the Seychelles white-eye and the Seychelles fody *Foudia sechellarum*, another endangered endemic landbird, were able to co-exist with Brown rats and form healthy populations on Conception and D'Arros, respectively, before the rats were eradicated[11-12].

Together with the Black rat, this species is the reservoir and main vector of Leptospirosis bacteria, Hepatitis (E) viruses and other rat-transmittable diseases[41]. The plague *Yersinia pestis*, an IAS bacterium transmitted via fleas, caused millions of human deaths after this rat colonised Europe, and it is still present in Madagascar[42]. Economic impacts of this rat are also considerable due to damage to agricultural crops, to commercial goods in warehouses and households, and to structural equipment (telephone and electricity wires, etc.) (see p. 33).

ERADICATION AND CONTROL PROGRAMMES
Eradication

Four attempts have been made to eradicate Brown rats from islands in Seychelles (see table below) including two ground-based and two aerial operations with helicopter. In terms of the final outcome of these eradications, a success rate of 75% has been obtained, similar to that for Black rat eradications

Above and following page: Physical caracteristics of the Brown rat include a thick tail shorter than the body, a slanted snout, small ears, relatively small eyes and a sturdy body shape. The Brown rat is mainly terrestrial (by opposition to Black rat that is mainly arboreal) and likes the proximity of humans and water.
Jack Jeffrey

(see also p. 62). The only failed attempt was the ground based attempt conducted on Frégate during the colonising phase in 1995, where extreme 'neophobia' (avoidance of new objects such as traps and bait stations) was reported[38].

During the first (and ground-based) attempt on Frégate, baits with Flocoumafen and Difenacoum were first used, and then wax coated cereal blocks with Brodifacoum 50 ppm[44]. In the following Brown rat eradication operations in Seychelles, bait types similar to those used for Black rat operations were used. Pestoff Rodent Bait 20R cereal pellets (with Brodifacoum 20 ppm), green dyed (to track bait ingestion by rats, and tested as less attractive to birds in New Zealand) were used for the aerial applications or to be hand spread, and Pestoff Rodent Blocks (waxed blocks with Brodifacoum 20 ppm) for the ground bait stations (both commercialised in New Zealand). Flavour (normally coconut) was added to the bait (in pellets or in blocks) to try and increase its attractiveness to rodents, although its effectiveness has never been proven[45]. Bait rates for the applications varied between 17 and 35 kg/ha in 2 to 3 drops (pulses) spaced 5-11 to 11-24 days each. On Frégate Island, differential GPS (DGPS) was used to guide the helicopter during the drops to ensure optimal coverage[22, 46-49].

Eradication of Brown rats has been attempted in subantarctic, temperate, and tropical climates[51-54, 73]. Over a total of 191 operations of sufficient data quality reported in DIISE until September 2014, 184 were successful but reinvasion occurred in 53 islands. In tropical islands, out of a total of only 17 operations, 16

were successful and one failed (the 1995-96 Frégate one)[54]. Chances of failure were described as low (5%[50]; current figures from DIISE show failure rates of 3.7% but up to 31.4% when reinvasions are added, and of 6% up to 12% for tropical islands only. In Mauritius, Brown rats were successfully eradicated from one island, Coin de Mire (also called Gunners Quoin, 15 ha) in 1995 through hand broadcast of brodifacoum (20 ppm) and bromadiolone (50 ppm) on a 25 m grid[55]. Successful eradication of Brown rats also took place on Tromelin Island (Iles Eparses-TAAF, to France)[56-58]. In 2014, Campbell Island (New Zealand), measuring 11,300 ha, became the largest island on earth where Brown rats had been eradicated, and Frégate Island Seychelles, measuring 219 ha, the largest in the tropical world[54, 73].

Control

The Brown rat tends to be targeted during control programmes conducted around towns, villages and residential areas, where this species is mainly present – and sometimes dominant[4, 59]. Control methods consist of bait stations providing blocks or cereal pellets with anticoagulant (Brodifacoum, Diphenacoum, Flocoumafen, etc.), hand spreading of pellets in cases of serious outbreaks, and trapping (cage-traps, snap traps)(see "General recommendations and considerations for the control and eradication of rodents", p. 214). Regular de-ratting operations are conducted on the three large granitic islands (Mahé, Praslin, La Digue) in selected sites such as those around Victoria (e.g. market, public buildings, harbours, bin sites) by government agencies and pest control companies under the authority of the Department of Public Health of MoH. Permanent control is also organised by many owners of private houses and businesses. Brown rats usually represent 5% to 20% of the rats trapped around La Misère and Barbarons residential areas and woodlands[3, 20]. There, rat control is conducted by means of a grid of 50m spaced home-made bait stations installed by ICS to protect the endangered Seychelles white-eye. This has been done since 2006 at the two properties of the President of UAE (c. 25 ha)[63], and in 2011-12 around La Misère and Souvenir villages, and Grande Anse forestry station (c. 25 ha in total)[64]. Traps are normally baited with fresh or roasted coconut in Seychelles. See Black rat (p. 192) and Barn owl (p. 248) texts regarding past unfortunate attempts for biological control through the introduction of the latter.

Ecosystem monitoring and recovery

Since the eradication of Brown rats from Conception in 2007, numbers of the endangered Seychelles white-eye and the Seychelles blue pigeon *Alectroenas pulcherrima* have increased[65]. In addition, two rare endemic species, the land snail *Stylodonta unidentata* and the Giant millipede *Sechelleptus sechellarum*, were rediscovered in 2009[66], and a small breeding population of Wedge-tailed shearwaters has become established[67]. Also, vegetation changes include an increase in endemic palm seedlings and the rapid development of native *Pisonia grandis* trees (see "Ecosystem recovery after the eradication of rats and cats", p. 169). On D'Arros Island, the eradication of Brown rats has been followed by the establishment and expansion of a breeding population of Wedge-tailed shearwaters (c. 250 pairs in 2011)[68]. On Frégate island, the eradication has been followed by a major increase in numbers of Seychelles magpie-robins and also Moorhens *Gallinula chloropus*; and two other globally threatened species, the Seychelles white-eye and the Seychelles warbler *Acrocephalus seychellensis* have been successfully introduced[65]. In addition, rare endemic invertebrates such as the snail *Pachnodus fregatensis* and the Giant tenebrionid beetle *Polposipes herculeanus* have also recovered to pre-invasion levels[60-62, 69, 70]. On Coin de Mire (Mauritius) after the eradication, two native skinks increased in numbers, strong seeding of endemic *Dracaena*, *Pandanus* and *Latania* sp. was observed, and a species of night gecko was rediscovered[1, 55].

MANAGEMENT RECOMMENDATIONS

Read "General recommendations and considerations for the control and eradication of rodents", p. 214 and other text references for practical advice on trapping and chemical treatments. The following are particularly relevant to the Brown rat.

Control and eradication protocol

- Control can be conducted using trapping or chemical treatment (anticoagulants). Island eradications always require the use of anticoagulants, except on very small islands (less than a few hectares) where intensive trapping may be used alone.
- Cleanliness is <u>essential</u> to prevent proliferation of rats around residential areas, especially Brown rats that are particularly attracted to human habitations and water sources. Remove or prevent access to any food or fresh water easily accessible from ground level around houses, gardens and buildings, including food waste in bins, fallen fruits, pet food, sewage conduits, etc.
- Brown rats seem to be easier to control than Black rats because they live mainly on the ground where bait stations or traps operate. However, we have noted that certain adult Brown rats may become 'clever' and learn to avoid traps and sometimes bait stations, hence the importance of alternating control techniques.

Chemical treatment

- Bait stations may be spaced 40 to 50 m from each other, although shorter distances – down to 25 m

Rattus rattus

longer than rest of body	slender	big	big	pointed
Tail	Body shape	Ears	Eyes	Snout
shorter than rest of body	sturdy	small	small	slanted

Rattus norvegicus DR

Brown rat tails (upper) are thicker and shorter compared to Black rat (lower) for a similar body length.
Gérard Rocamora

– may be required in residential habitats or areas where abundant food resources are available and home ranges may subsequently be reduced. This is especially important if eradication is sought.

Trapping

- Make sure that traps are stable and well placed in order to reduce the chances of traps being set off prematurely. Brown rats are particularly wary, easily frightened and able to learn from a bad experience, and therefore more likely to become 'trap-shy'. Hence trapping alone is not recommended for eradication as it generally fails to eliminate all individuals.
- Coconut (fresh or roasted) appears to be the most practical and efficient bait for rat trapping in Seychelles.

See p. 127 recommended methods to euthanase animals captured alive.

There are no sophisticated biological control methods (e.g. genetically modified animals capable of producing only male offspring; use of viruses to disseminate immunocontraception) that can be recommended yet against rats, but research is being conducted in that direction (see p. 125)[71, 72].

References / Further reading
1 CAB International, 2014; 2 GISD, 2014; 3 G. Rocamora, unpublished; 4 Grant et al., 2003; 5 Fanchette & Kerridge, 2009; 6 Hill et al., 2003c; 7 Cheke, 2010; 8 Harper & Bunbury, 2015; 9 Ministry of Health, 1994; 10 Meyer, 1994; 11 Rocamora 1997a, 1997b; 12 Rocamora & Matyot, 2002; 13 Herman Bristol, pers. comm. to Victorin Laboudallon; 14 Christianson E., 1944; 15 Thorsen & Shorten, 1997; 16 Perley Constance, pers. comm.; 17 Foster et al., 2011; 18 Grant, 1972; 19 Taylor, 1978; 20 G. Rocamora & E. Lagadec/CRVOI-IRD, unpublished; 21 Merton 1999, 22 Merton et al. 2002; 23 Rocamora & Labiche, 2009d, and unpublished; 24 G. Rocamora, pers. obs.; 25 Lowe et al., 2000; 26 Harper & Bunbury, 2015; 27Atkinson, 1985; 28 Courchamp et al., 2003; 29 Towns et al., 2006; 30 St Clair, 2011; 31 Drake, 2007; 32Jones et al., 2008; 33 Mulder et al., 2011; 34 Ruffino et al. 2015; 35 Rocamora & François, 2000; 36 Pat Matyot in Fanchette & Kerridge, pers. comm.; 37 Nevill, 2009; 38 Thorsen et al., 2000; 39 Galman, 2011; 40 Rocamora, 2005; Rocamora, 2005; 41Bovet et al. 2013; 42 Duplantier et al., 2002; 43 Beaver & Mougal, 2009; 44 Thorsen & Shorten, 1997; 45 Bill Simmons; pers. comm.; 46 Merton, 2001; 47 Climo, 2004b; 48 Rocamora & Labiche, 2009d; 49 Rocamora & Jean-Louis, 2008, 2009;50 Rocamora, 2007d; 51 Thomas & Taylor, 2002; 52 Galván et al., 2005; 53 Howald et al. 2007; 54 DIISE, 2014; Successful and failed eradications only, excludes reinvasions, whole island eradications only, satisfactory or good data quality only; 55 Bell, 2002; 56 Lorvelec & Pascal, 2005; 57 Soubeyran et al., 2011; 58 Le Corre et al., 2015; 59 Buckle & Smith, 1994; 60 Gerlach, 2005a; 61 Gerlach, 2005b; 62 Gerlach et al., 2005; 63 Labiche & Rocamora, 2009; 64 Henriette, 2013; 65 Rocamora et al., in prep.; 66 Galman 2011; 67 G. Rocamora, pers. obs.; 68 Kappes et al. 2013; 69 Raposo & Canning, 2012; 70 Canning 2011a; 71Dobson, 1988; 72 Campbell et al., 2015; 73 Towns & Broome, 2003.

The original 'Rat motel' station (Rowley Taylor, Tawhitinui Island, NZ). The inner box can take a snap trap, rodenticide or a tracking card, and has sawdust added for homeliness shelter.
James Russell

ONE OF
WORLD'S WORST
100
INVASIVE
SPECIES

Mice eat a lot of plant material
(Aride Island).
James Russell

House (or Common) mouse

Souris domestique

Souri

Mus musculus

IDENTIFICATION AND BIOLOGY

The House mouse is a small, grey-brown rodent, measuring 65-95 mm and weighing 12-30 g. Its long tail is of similar length to its body. It has a relatively shorter tail, and larger eyes and ears than do young Black rats[1], with comparatively smaller feet and head[40]. Its biology has been poorly studied in the tropics, including Seychelles. It is omnivorous, and eats all sorts of plant parts (leaves, roots, buds, fruits, flowers, seeds) and a large variety of invertebrates in natural habitats, as well as human food (grain, biscuits, bread, soap, scrap etc.)[2]. In the Southern Ocean islands and polar climates their diet consists of plant material and invertebrates[4, 5, 34], but they have been documented predating on landbird and seabirds chicks on Gough Island (South Atlantic)[6, 7, 8], something that does not seem to happen in Seychelles (see Impacts and threats). They are good climbers, primarily nocturnal, and often linked to human habitations. Mice are extremely prolific; they breed all year round, although more during the wetter season, producing up to 10-13 litters a year usually comprising 5-6 young (gestation period: 18-21 days). Females reach maturity at 5 weeks. The ecological and behavioural plasticity of the species allows it to adapt and prosper in new environments[2].

ORIGIN

Indian sub-continent and South-East Asia. The House mouse has been a human commensal for at least 8,000 years. It has spread around much of the globe and now is the mammal with the largest distribution on Earth[1, 31]. It may have been introduced to Seychelles by early navigators from the Arabian peninsula (as is the case in Madagascar[9]) or by the first settlers coming from the Mascarenes in the 18th century[10]. It was reported to be present

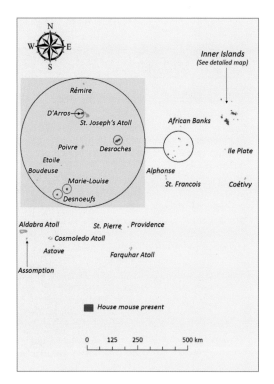

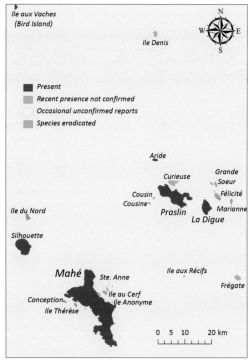

on Bird Island by the early 1900s and on Aride before 1918[11].

DISTRIBUTION & ABUNDANCE IN SEYCHELLES

The species has been recorded at Mahé, Praslin, La Digue, Aride[11], Curieuse and Silhouette, and also at Bird, D'Arros, Desroches, Desnoeufs and Marie-Louise[12, 13]. There have been occasional unconfirmed reports from Ste Anne and other islands where mice may have been mistaken for young rats. Mice have been eradicated from Denis and Frégate[14, 15]. On Curieuse, the species was reported to be present by residents in 2001[13], after the rodent eradication took place, but its presence has not been confirmed with certainty in recent times. The species appears to be concentrated mainly around habitations and buildings where numbers may at times erupt, but seems to only occur at low densities in natural habitats[17]. When rats are present mice are harder to detect as their abundance is likely to be reduced; but they may also be underestimated due to competition (for both food and traps), and a less conspicuous behaviour[16].

IMPACTS AND THREATS

In Seychelles, few studies have been carried out to assess the specific impact of the House mouse on ecosystems and the economy. At Aride, stomach contents of c. 30 individual mice found in the forest were analysed and found to contain almost exclusively seeds and fruit, and very few invertebrate remains (ants)[3]. There was no evidence of seabird carrion, despite being in the South-East monsoon, when corpses are abundant[3]. At subantarctic islands, mice have direct impacts on plants, invertebrates, and birds[20, 30]; they are responsible for local extinctions of invertebrates[4], feed on endemic plants and seeds, and prey upon eggs and chicks of endangered land-bird and seabirds, including bleeding large Albatros chicks to death[6, 7, 8, 17]. There, mice pose a significant conservation threat on islands where they are the only introduced mammal present, but their impacts appear much reduced where this is not the case[8, 20]. There is no proof one way or the other whether they have a significant negative environmental impact in Seychelles, including where they are the only mammal present. It is possible that they could affect the breeding success of seabird colonies, threaten endangered plants through excessive seed consumption, and compete with other animals for food; but this has yet to be demonstrated. Mice may also be responsible for indirect effects on native wildlife through competition for food with small reptiles and landbirds[22, 23], interactions with other invasive

Table 19: Summary of House mouse eradications conducted in Seychelles[22, 44-49].

Island (Size in ha)	Eradication attempt	Primary target	Method (No. of applications)	Bait	Outcome
Bird (101)[14, 15, 18]	Oct.-Nov. 1996	Black rat	50 m grid of bait stations (2)	Brodifacoum 50 ppm cereal wax blocks (11.6 kg/ha)	Black rats & rabbits eradicated, but mice still present (Jan. 1998; survivors or reinvasion).
Frégate (219)[14, 15, 18]	June-July 2000	Brown rat	Helicopter (3 drops), with DGPS.	Brodifacoum 20 ppm cereal pellets (35 kg/ha)	Success. Both Brown rats and mice eradicated.
Denis (143)[14, 15, 18]	June-2000	Black rat	Helicopter (2 drops) with DGPS, ground baiting	Brodifacoum 20 ppm cereal pellets (23.6 kg/ha)	Failed to eradicate both Black rats and mice (reinvasion or survivors).
Denis (143)[18]	June-July 2002	Black rat	hand spread along 25 m tracks (4)	Brodifacoum 20 ppm cereal pellets (25.4 kg/ha)	Success. Both Black rats and mice eradicated. Rodent-free status still confirmed[24].
Curieuse (286)[15, 18]	July 2000	Black rat	Helicopter (2 drops), with DGPS.	Brodifacoum 20 ppm cereal pellets (23 kg/ha)	Failed to eradicate both Black rats and mice (survivors or reinvaders found within a year)Current status of mice unclear[25].
D'Arros (140)	August-Sept. 2003	Brown rat	hand spread along 50 m spaced tracks (3)	Brodifacoum 20 ppm cereal pellets (17 kg/ha)	Brown rat eradicated, but mice still present (probably survived)[26].

species (e.g. trophic competition with rats; serving as an alternative prey for cats)[16, 21], or by inducing long-term vegetation and habitat changes, among others. Mice may also have a significant impact on Seychelles'economy[19]. They are a pest on agriculture crops and stored food in homes, and are a potential reservoir (although relatively unimportant vectors) of infectious diseases including salmonella, plague or leptospirosis. They are clearly a nuisance for the tourism industry.

ERADICATION AND CONTROL PROGRAMMES

Six rodent eradication attempts were conducted on islands with mice in Seychelles. Although the primary targets of these eradications were rats and rabbits, rather than mice, it was hoped that mice could also be eradicated in the process. Only two operations (33%) succeeded in removing the mice. These included the aerial eradication conducted on Frégate in 2000, and the ground operation conducted on Denis in 2002[15, 18] (See Table 5, p. 66). More details on the various operations can be found in the texts for Black and Brown rats.

Over a total of 81 operations of sufficient data quality reported in DIISE until September 2014, 55 were successful but reinvasion occurred in 4 islands. In tropical islands only, out of a total of 24 operations, 16

(66%) were successful and no reinvasion occurred[27]. Chances of failure were described as high (38%, 17 out of 45 attempts[28]); current figures show a failure rate of 32.1%, and 33.3 % for tropical islands only[27]. In Mauritius, mice were eradicated in 1995 with cereal Brodifacoum pellets 20 ppm from Ile Cocos (15 ha) and Ile aux Sables (8 ha) using a 10 m grid of plastic bait stations, and in 1998 from Ile Plate (also known as Flat Island, 253 ha) using a 25 m grid of bait stations raised 15 cm above ground[29]. In 2005, House mouse was eradicated from an islet of Rangiroa atoll in French Polynesia; and in 2006 from Surprise Island, in New Caledonia, at the same time as rats[35, 36]. As of 2014, Macquarie Island (subantarctic, Australia, in 2011), at 12,873 ha, was the largest island on Earth where mice had been eradicated, followed by Rangitoto Island (New Zealand, in 2009), at 2,432 ha[27]. In the tropical world the largest island where mice have been eradicated is Ile Plate (Mauritius, in 1998), at 253 ha, followed by Frégate Island (Seychelles, in 2000) at 219 ha[27].

Control

Regular control of mice has been taking place around houses on several small inhabited islands infested with mice. These include Bird, Curieuse, Aride, D'Arros and Marie-Louise, as well as the larger islands of Mahé,

Praslin and La Digue. Kill traps are most commonly used to trap mice. On Aride, traps are placed in plastic protective boxes to avoid damage to non-target species of birds and skinks. Control has also been conducted on some islands with rodenticide pellets which are available from local outlets.

MANAGEMENT RECOMMENDATIONS

The section "General recommendations and considerations for the control and eradication of rodents", p. 214 applies; but the following recommendations are particularly relevant to the Common mouse. More details can be found in the *Pacific Invasives Initiative toolkit for rodent and cat eradications*[37] and other text references.

Control and eradication protocol

- Try to eradicate mice and rats at the same time, as the former may increase dramatically in numbers on islands or at areas where rats have been removed (a phenomenon called competitor release effect[16]). This is probably because more resources become available to mice along with possible release from rat predation. The same may apply to cats[32].
- In tropical climates with abundant food resources all year round, rodents can access alternative sources of food (fruits, invertebrates) that may deter them from eating sufficient rodenticide. The small body size of mice exacerbates this problem[28].
- Prior testing of attractiveness and palatability of possible non-toxic baits on mice populations is recommended before finalizing protocols, especially for eradication.

Be very strict with food and garbage management so as to prevent mice to access it, especially around habitations[40, 41]. This key factor is even more important than for rats for protocols to be effective and succeed in controlling or eradicating mice.

Chemical treatment[28]

- Anticoagulants are the most widely used method worldwide to control rodents. All successful mouse eradications have involved use of anticoagulants[27].
- Brodifacoum has been used in 80% of eradication attempts, but other anticoagulants also used successfully include pindone, bromadialone, and flocoumafen.

- Rodenticide may be distributed either using bait stations, hand spreading or aerial drops by helicopter (all of which show similar rates of success of c. 50% for mice).
- Bait stations should not be spaced at intervals of more than 20 m for mouse eradication. On a New Zealand island, some mice had home ranges as small as 0.15 ha[33]; but in tropical climates with abundant food resources (such as in Seychelles), home ranges are likely to be even smaller, thus decreasing the chances of mice encountering bait.
- The most likely reasons for failure include insufficient bait coverage to put every single mouse at risk. This may be caused by bait stations being spaced too far apart (in certain islands spacing had to be 5 m), or spreader bucket malfunctions, especially in habitats with heavy vegetation cover.
- Other possible reasons for failure include bait not being sufficiently attractive, or development of an aversion or resistance to bait following long term rodenticide control.
- Combining several methods, such as aerial eradication and bait stations with manual spreading of bait in high risk areas (infestation hotspots), or possibly using several toxicants is likely to increase chances of success.
- Consider the impacts of other species – e.g. rats, rabbits, cats, tenrecs or crabs. How will they interact with mice, and may they diminish the chances of mice accessing the bait?
- Mouse eradications remain a challenge and require a great deal of attention during the planning and feasibility assessment stage, and later during implementation. It is particularly important to consider the attractiveness of the bait provided, perfect bait delivery, and treatment of hotspots.

Trapping

- Trapping is recommended as an effective method for occasional or routine mouse control, especially in homes and around buildings, including farms[38, 39]. Many different models of kill traps (snap traps) or cage traps can be used (see pictures).
- Trapping alone will not eradicate mice, but can be used in high density areas, combined with poisoning, to catch animals that may not have taken enough rodenticide.

Essential factors that increase chances of success in mice eradications (modified from MacKay *et al.* 2007).

1. Will the mice eat the bait? Conduct trials to test the attractiveness and palatability to mice of different types of bait. Try various flavours and toxicant concentrations.

2. Will the chosen method of distribution allow every mouse on the island to have access to rodenticide?

3. Consider the effects of non-target species, including any other mammals present. Will they compete with mice and prevent them accessing the poison?

4. Are there areas that may support high mouse numbers which require extra rodenticide, such as human habitations, buildings, dumpsites and surroundings, or dense grassland?

References / Further reading
1 GISD, 2014; 2 CAB International, 2014; 3 J. Gerlach, *pers. comm.*; 4 Le Roux *et al.*, 2002; 5 Smith *et al.*, 2002; 6 Cuthbert and Hilton, 2004; 7 Wanless *et al.*, 2007; 8 Jones & Ryan 2010; 9 Duplantier *et al.*, 2002; 10 Cheke 2010; 11 Warman & Todd 1984; 12 Racey & Nicoll, 1984; 13 Hill, 2002; 14 Merton, 2001; 15 Merton *et al.*, 2002; 16 Caut *et al.*, 2007; 17 Jones *et al.*, 2003; 18 Beaver & Mougal, 2009; 19 Nevill, 2009; 20 Angel *et al.*, 2009; 21 Russell, 2011; 22 Newman, 1994; 23 Mills *et al.*, 2002; 24 Arjan de Groene/GIF, *pers. comm.*; 25 Allen Cedras/SNPA, *pers. comm.*; 26 R. von Brandis, *pers. comm.*; 27 DIISE, 2014 - Successful and failed eradications only, excludes reinvasions, whole island eradications only, satisfactory or good data quality only; 28 MacKay *et al.*, 2007; 29 Bell 2002; 30 Howald *et al.*, 2007; 31 Lowe *et al.*, 2000; 32 Van Aarde *et al.*, 1996; 33 MacKay *et al.*, 2011; 34 Russell, 2012; 35 Soubeyran *et al.*, 2011; 36 Courchamp *et al.* 2011; 37 PII, 2011a; 38 Buckle & Smith, 1994; 39 Balbaa & Zakaria, 1983; 40 Orueta, 2003; 41 Orueta, 2007

Different traps for mice (left to right): scented (also baitable) spring trap, live capture trap, T-rex spring (snap) trap.
Vikash Tatayah / MWF

PROCESSING RATS AND MICE

- Handling of rats can be risky and requires precautions to ensure people are not exposed to infectious diseases. Considering the ubiquitous presence of rats and the high prevalence of leptospirosis and other rat-transmitted diseases in Seychelles (with dozens of cases every year – see Fig. 1, p. 34 - and a mortality rate higher than that caused by AIDS), the Seychelles Public Health Authority warns members of the public and practitioners to be very cautious when handling rats, and to adopt the safest possible available methods to euthanize them.

- Immobilised rodents captured with glue traps can be killed instantly by delivering a strong blow to the animal's head with a solid stick. Although we do not recommend this method as it is non selective and considered inhumane, glue traps can be bought in shops and their use is authorised in Seychelles. In the case of live traps, the rodent can be transferred into a solid bag made of cloth or thick plastic, and restrained by an operator wearing thick gloves to avoid being bitten through the bag or if the rodent escapes. The operator should either deliver a **strong fatal blow** to the animal's head, or if sufficiently experienced perform a **cervical dislocation** by holding the head and body of the rat with the two separated hands and making a quick elongation and twist (for animals under 150 g only unless the operator is very experienced). However, these methods may not be suitable for untrained members of the public and they may not succeed in killing the rat instantly. Moreover, there is a risk of injury and contamination should the rat escape during the process, particularly when two rats are caught together in the same trap.

- For professional control operations, **euthanization by using carbon dioxide from commercial cylinders** is recommended. This procedure requires a trained operator and specially designed bags or containers in which carbon dioxide must reach a concentration of over 70% in the chamber before it becomes functional. This method induces a rapid anaesthesia and subsequent death within 5 minutes. Carbon monoxide may be used for euthanasia (only a 6% concentration is required) but it is a very toxic gas and it is not easily available commercially. Rats in cage traps can be put into a large plastic bag or container with a small aperture (to let the residual air escape) and can be euthanized with Carbon monoxide gas from the exhaust pipe of a cold started petrol (not diesel) car engine. However, this method is controversial and hence cannot be recommended (see "Euthanising invasive animals", p. 127).

Rodents in live traps can be euthanised humanely with a CO_2 chamber.
Fabrice Boulet / www.stopinsectes.com

- Although submersion in water for 5 minutes can kill rats in a relatively quick and safe manner (average time to death 2.6 minutes), drowning is not considered a humane method as per international animal welfare standards and therefore cannot be recommended. Trapped animals put in the sun to die from overheating suffer a long and painful death (10 to 15 minutes or more), as do animals that are burnt alive with flammable liquids. Both of these methods have been reported on occasion, but are unacceptable acts of cruelty to animals and should never be used.

References / further reading
Close *et al.* 1996, 1997; Rural Development Service, 2006; Sharp & Saunders 2005; Reilly, 2001; UFAW 2009; Yamamoto *et al.*, 1983.

General recommendations and considerations for the control and eradication of rodents, including lessons learnt from Seychelles

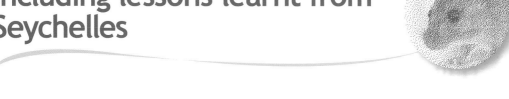

CONTROL

Rodent control can be most effectively conducted by chemical treatment through anticoagulant **rodenticides** and through **trapping** (see boxes below). Use of anticoagulants is the most widespread method used worldwide to control rodents; it is also the easiest and most efficient method[1-3]. Other possible methods that have been tested in Seychelles, such as repellents or the emission of ultra sound, appear to lose their effectiveness after some months[4,5]. Biological control methods, such as introduction of diseases or parasites specific to rats, are being investigated abroad but these are hampered by the tremendous breeding potential of rats and their capacity to develop resistance[6]. Introduction of cats to most inhabited islands, or the introduction of the African barn owl to control rats had disastrous consequences (see p. 248). Contraceptives administered through bait consumption are still experimental but with promising potential. Development and release of genetically modified rats that will produce exclusively male offspring is also a possibility currently being investigated by several research teams around the world to control rodents[7].

ELIMINATE FOOD, SHELTER AND ACCESS AVAILABLE TO RATS

Eliminating food and shelter available to rats is critical to prevent rat proliferation and to optimise the effectiveness of any control operation. Food, water, and shelter attract rats, facilitate their multiplication, and undermine control efforts[1-3]. Well fed females produce more offspring and alternative available food sources discourage rats from eating rodenticide and from entering traps. Actions required include:

- Strict management of waste, with <u>closed</u>, solid bins preventing access to rodents.
- Removing fallen fruits from the ground.
- Removing any possible alternate food (fruits, waste, coconuts, etc.) and shelter sources, particularly near human habitation (dumping sites, green waste piles, etc.) through mulching or burning. Food waste may be buried (more than 1m deep), incinerated or macerated.
- Removing food leftovers from pets' plates before night, cleaning and placing plates beyond the reach of rats. This will also prevent pet contamination by rat transmitted diseases.
- Keeping gardens and grounds adjacent to buildings clean and tidy.
- Trimming vegetation near buildings to prevent contact between branches and the roof, as these will provide an easy access to rodents.
- Closing off or filling holes that provide access to buildings, and to shelters where rats may breed.

ELIMINATE RATS BEFORE THEY REACH SENSITIVE AREAS

- Create a buffer zone where rats are eliminated <u>before</u> they reach the sensitive areas you want to exclude them from. Do not just place lines of traps or bait stations immediately adjacent to the buildings (or bird nests) that you want to protect. This will not be sufficient. It is essential to also conduct control 25 to 100 m away in the surrounding vegetation. Traps or bait stations immediately near sensitive areas should only serve as a last resort protection and may not even be required in an effective control programme.
- Alternate control techniques (trapping/poisoning) to eliminate individuals that have either learned to avoid traps or are insufficiently attracted by rodenticide when other food sources are abundant.

CONDUCT RAT CONTROL PERMANENTLY AROUND BIN SITES AND OTHER POTENTIAL HOTSPOTS OF INFESTATION

- Pay special attention to sites that attract rats in search of shelter and food. This includes bin sites, areas around kitchens, and also areas where green waste has accumulated.
- Checking and refilling bait stations every two weeks at these sites is the easiest recommended method but trapping can also be used in conjunction at regular intervals.

SET UP PERMANENT RODENT CONTROL OVER LARGE AREAS THROUGH A GRID OF BAIT STATIONS

- Permanent grids of bait stations spaced at 40-50 m and baited with waxed cereal blocks are effective for rodent control and to reduce rat abundances up to 95%. Such grids have been functioning on Mahé since 2006 (Seychelles white-eye sites at Barbarons and La Misère) and Ste Anne (in and around the properties of the Ste Anne Resort & Spa since 2011) (see Black rat text).
- Permanent grids of bait stations may be used efficiently to create 'biological islands' or 'mainland islands' where rodents are controlled over large areas, as is the case in other countries to protect relictual passerine populations from rodents (e.g. New Zealand, Mauritius[78]).
- Number each bait station, and cut and maintain tracks to facilitate access. Install them on site 1-2 weeks prior to baiting them in order to avoid 'neophobia' (shyness to new objects).
- Refilling of bait stations can be conducted every 1 to 4 weeks depending on consumption.
- Traps baited with coconut can be used to further increase the efficiency of the grid of bait stations. Traps can be distributed along transects or uniformly across the grid (e.g. one for each grid square), and checked every day.
- Trapping removes rats immediately and eliminates rodents that may not be attracted to waxed cereal blocks in bait stations. However, it requires a considerable extra man-power effort. When cereal blocks in bait stations are attractive and palatable to rats, it is likely that most of the rats trapped would have died anyway within a few days from rodenticide consumption.

ERADICATION

Rodent eradications on tropical islands are delicate operations which require a high level of discipline and motivation; and guidance/supervision from persons with previous experience in such operations is essential (especially for aerial broadcasting and large islands). The recommendations and considerations below do not pretend to guide the readers to conduct a rodent eradication entirely by theirselves but rather to highlight the most important best practices and key points to keep in mind for the various types of operations, and to improve their understanding of all the complex different issues involved. These apply to the three species of rodents present in Seychelles: Black rat, Brown rat and House mouse. More detailed information is available from the list of references, including in a Tropical Rat Eradication special issue of Biological Conservation[24] and the **Pacific Invasives Initiative toolkit for Rodent and Cat eradication**[11] www.pacificinvasivesinitiative.org/rk/index.html (see p. 140). The latter provides practical advice for all the stages of an eradication project (i.e. Project Selection – Feasibility Study – Project Design – Operational Planning – Implementation – Sustaining the Project) with specific guidelines for each step. See also general considerations in the 'Eradication' section in Part 1 of this guide (p. 138).

As of 2014, a total of 650 attempts to eradicate rats from islands had been undertaken around the world, out of which 69 (11%) failed and 134 (21%) ended by reinvasion. This represents a global success rate of 87%, that is lower in tropical islands (81%) compared to non-tropical regions (92%)[24, 25]. In Seychelles, 16 rodent eradication attempts had been conducted as of 2014, out of which 12 were successful in permanently removing rats (although two attempts failed to eradicate mice). In some cases, rats appeared to be eradicated but reinvaded the island within months.

- All successful eradications of rats from islands have included use of anticoagulants (except for some very small islands where intensive trapping was used; e.g. Pouhou islet, 0.6 ha, Mayotte).
- The best season for conducting rat eradication in Seychelles is during the driest months, i.e. July to September in the inner islands, and July to November in the southern outer islands. This coincides with the period when natural food is

RODENT TRAPPING

- Many different models of kill traps, glue traps (banned in some countries) or cage traps (simple or multiple capture) can be used for routine rat or mouse control[1-3, 8-11] (see some illustrated examples).

- Snap-traps (kill-traps) should be used with a plastic, mesh or sheet metal cover to limit risks to non-target species.

- Traps are usually placed in lines, at a fixed distance from each other (usually 15 to 25 m for control purposes), with one or two traps per point (see also Index-trapping section[12]).

- Insofar as possible, place traps in the shade and/or check them early morning before the sun is too harsh so that non-target species (and trapped rats) are not affected by the heat of the day.

- Cage traps (live-traps) present no risks for non-target species and are generally very effective when properly baited. However, captured rodents must be properly euthanized (see p. 213).

- Grilled (roasted) coconut appears to be the most practical and preferred bait traditionally used in Seychelles, but fresh (non-grilled) coconut is also effective. However, there is no clear evidence that rats prefer it roasted, this has yet to be scientifically tested.

- Sherman traps (or the similar Elliot traps) are foldable and ideal for situations in which large numbers of traps are needed, as they occupy very little space. However, they are less effective than cage traps.

- Bait can be changed systematically every night or simply when its attractiveness has been reduced by rain, mould, heat or partial consumption by invertebrates.

- Avoid placing traps in open areas (e.g. on bare ground or mowed grass at a distance of several meters from vegetation). Instead place them where rodents can approach while remaining relatively hidden: near a rock, at the bottom of a tree, along a wall etc.

- Remove vegetation, dead leaves, etc. in the immediate vicinity (i.e. within c. 30 cm of the trap) so that it does not interfere with the closing mechanism.

- Make sure that traps are stable and do not shift position when pressure is applied to any of the corners as such movement can traumatise a rat. Some rats are easily frightened and can become trap-shy after a bad experience.

- When both rats and mice are present, or you do not know which of the two are present, you should set traps for both species at each site.

- Trapping can be used to complement anticoagulant rodenticides thereby reducing the amount of chemicals needed. Traps may also catch individuals that are not sufficiently attracted to rodenticide bait.

- Trapping on its own (without complementary use of rodenticides) is <u>not</u> recommended to eradicate rats, although trapping alone has successfully eradicated rats from some very small islands (e.g. Pouhou islet, 0.6 ha, Mayotte, in 2005[13]).

Cage traps prevent harm to non-target species, but trapped animals then need to be euthanased.
Gérard Rocamora

Foldable live traps (Sherman-type) facilitate transportation, but may be less efficient than cage-traps.
Nik Cole / MWF

Snap (kill) traps are practical and effective but may represent a danger for certain non-target species.
Nik Cole / MWF

Traps need to be strong enough to resist dammage from certain non-target species (e.g. Robber crab *Birgus latro*). Gérard Rocamora

minimal (except when seabird colonies are present), and dry weather will ensure maximum longevity of the bait.

- Certain rodent eradications can be very challenging and require a lot of attention during the planning and feasibility assessment, and later during implementation, particularly with regards to attractiveness and quantity of bait provided, perfect bait delivery, and treatment of hotspots[11, 22].

- The success of a rodent eradication depends as much - or even more - on the quality of the preparatory phase than on the eradication phase itself, during which well established existing methods to eradicate rodents can be applied with proper guidance[21].

- The most likely reason for failure is insufficient bait coverage. **Every single rodent must be put at risk**, and this will not happen if bait stations are too far apart, hand broadcasting is too irregular or if there is a malfunction of the spreader bucket, especially in hilly habitats with dense vegetation cover.

- Heavy consumption by non-target species such as land crabs, hermit crabs, and the presence of dense coconut (*Cocos nucifera*) forest are factors associated with eradication failure[26].

- Reasons for failure – or failed outcome – in Seychelles may include survival of a few rodents due to too much alternative food being present (e.g. in hotspots around residential areas, green waste piles) or insufficient bait being available (e.g. in mangroves or swamps where water destroys it rapidly). This includes in particular breeding females and young emerging after bait is no longer widely available[21, 26, 30]. Reinvasion due to insufficient discipline in the implementation of rodent prevention protocols is also a likely major reason for failure[31, 32].

CHOOSING THE ANTICOAGULANT AND DISTRIBUTION TECHNIQUE

- Over 70% of successful rodent eradications worldwide have used Brodifacoum, which appears to be the most widespread and effective toxin[28, 29]. All rodent eradications attempts in Seychelles have used Brodifacoum, mainly as 10 or 12 mm cereal pellets with 20 ppm (Pestoff Rodent Bait 20R made in New Zealand), green dyed to repel birds and detect consumption by rodents (see monitoring section below).

- Importation of rodenticide into Seychelles must receive the prior approval of the Pesticide Board (Public Health Department, Ministry of Health).

- Bait should be stored in a dry, well ventilated place, away from any solvents or contaminants that may affect its palatability, and should never get wet. Do not leave the bait bags in the sun for long periods as condensation inside the bag may affect the bait's quality. Temperature differences must also be avoided when unloading and storing bait[7].

- Do not use bait older than 3-6 months for an eradication, although bait quality will depend on storage conditions. Moisture and development of mould will affect the quality and palatability of the bait for rodents and will make it unsuitable for an eradication.

- Development of aversion or resistance to bait (or loss of bait attractiveness) following long term rodenticide control could be a possible reason for failure. In such cases, it is preferable to use a different active ingredient for the eradication[27].

- Brodifacoum belongs to the second generation of anticoagulants. It is mainly toxic to mammals, but also (and to a lesser degree) to birds and reptiles in decreasing order[33]. It has no effect on plants, and little effect if any on invertebrates (see "Non-target species and species interactions", p. 222) although impact on snails has been reported or claimed[34, 35].

- Despite an extremely low solubility in water, Brodifacoum residues can affect aquatic wildlife including marine fishes[23].

- Brodifacoum pellets may not break down for up to c. 3 months in New Zealand, but only 4-6 weeks in Seychelles, a period during which the anticoagulant remains active. This actually depends on rainfall and climatic conditions: break-down is quicker in the open, slower in forests[7].

- Brodifacoum may present a relatively high risk to non-target species if broadcasted as pellets. If such risks are considered too high, alternative toxins such as Diphacinone can be used although rats will need to feed on more bait before getting a lethal dose[36].

- The 3 different ways of distributing rodenticide for eradication are: **aerial broadcast** (helicopter with spreader bucket), **hand broadcast** and **bait stations** along lines or grid. With bait stations, rodents need to come to the bait whereas with aerial or hand

INDEX-TRAPPING

- Index-trapping is used to calculate an index of abundance in order to compare the density of rodents at different islands, sites or habitats, or in different seasons at the same site.
- Set traps (usually 20 to 30) along a line at a fixed distance. Traps can be laid out in pairs with 1m between each of the pair. Traps are checked the following morning, then rebaited and reset in the afternoon for the next night.
- For each trap, record if sprung or unsprung (closed/open), whether bait has been eaten or is still present (bait gone/bait OK), and if a rat/mouse or another animal has been captured. For example: Site 1. Trap 1: Closed, bait gone; Trap 2: 1 rat / Site 2: Trap 1: 1 Hermit crab; Trap 2 Open, bait OK; etc.

You can calculate:

- the *uncorrected index* I_{UN} = total number of rats caught divided by the total number of uncorrected trap-nights (number of traps × number of nights they have been in activity, irrespective of whether some were found closed, open, with or without bait left) × 100 (expressed as the number of captures per 100 trap-nights).
- a *corrected index* I_{COR} = number of rats per corrected trap-night × 100 (recommended) using Cunningham & Moors (1996) method taking into account the traps that have been set off during the night (whether empty or occupied by a rat or another animal).
- a *simplified index* I_S = total number of rats caught divided by total number of traps that either caught a rat or were found open (excludes traps that closed with nothing inside, or traps that caught another animal) × 100.

Corrected number of trap-nights = total number of trap-nights – ½ (number of traps having captured a rat or another animal + number of closed empty traps).

This correction is based on the assumption that sprung (closed) traps have remained unsprung and capable of catching a rodent during half of the night on average. No correction is done for unsprung traps with the bait removed as they were still theoretically capable of catching a rodent. Example: 50 traps set during 3 nights that caught 7 rats, plus 13 sprung, empty traps. Total No. of trap-nights: 50 × 3 = 150. No. of trap-nights lost: (7+13)/2 =10. Corrected No. of trap-nights = 150-10=140. I_{COR} = 7 × 100 / 140 = 5.0 captures per 100 trap-nights. By comparison I_{UN} = 4.7 and I_S = 5.1

- Index lines are usually run for several nights, or until no rats are caught (as during an eradication). A minimum of 100 corrected trap-nights in each habitat is recommended.
- Bait type should never be changed when conducting index trapping, so that all results are directly comparable regardless of site or season.
- Setting two traps at each site, at c. 1 m from each other, is recommended by some authors, as this increases the potential capture rate (particularly useful in high density areas where 100% of traps may catch one or more rats if only one trap is placed at each site). However, the habitat area surveyed is then divided by two compared to one trap per site with the same number of traps.
- Recommended distance between trapping sites is 25 m (max. 50 m) but shorter distances (e.g. 15 m) may be chosen in areas with very high densities if only one trap is placed per site.
- Change bait (coconut) every day for all traps so that all keep the same chances of catching a rodent.

Adapted from Cunningham & Moors 1996.

broadcast, it is the bait that comes to the rodent.

- Choose the distribution technique after taking into account local parameters and the results of the feasibility study[11]. Considerations include the following:
    - Aerial broadcasting is the method least likely to fail but tends to be costly, especially if you need to bring to and keep a helicopter on remote islands.
    - Hand spreading requires the involvement of more people (team members must be selected carefully) and also the cutting of tracks which may have an impact on native vegetation.
    - Bait stations require significantly less bait and are associated with lower risk to non-target species. However, the length of the operation tends to be longer and the risk of failure greater.
- Combining several baiting methods, such as bait stations or manual spreading in high risk areas (infestation hotspots, around habitations, etc.), in addition to aerial broadcasting will increase chances of success during challenging operations.
- Aerial broadcast is recommended to eradicate rats from islands over 30 ha (heavily forested, especially if access is difficult) or 300 ha (open ground), whereas hand-broadcast and bait stations are normally more appropriate for islands below these limits[11].
- Special attention should be given to mangrove trees, where Black rats can live and breed in large bulky 'nests' made of vegetation, as observed on Aldabra and Cosmoledo[52, 30]. Bait blocks may be tied to mangrove trees (as in Cayo Centro, Chinchorro bank, Mexico[53, 54]), or hanged using 'colars' (as on Cosmoledo[52]) or 'bolas' (as in Micronesia[30]). Densely forested areas, particularly those largely covered with coconut trees, where rats can find an abundant source of food and shelter, also deserve much attention (see also Black rat text p. 192; and also p. 86).

PLANNING AND PREPARATION
(see also p. 138)

- Prepare feasibility studies and the eradication plan well in advance (at least two years and one year, respectively, for large scale or logistically complex operations). This should include an analysis of climatic data for previous years in order to tune the operation with the driest period of the year[50], an assessment of environmental impacts, an inventory of non-target species and a basic understanding of how trophic webs function on the island[21, 32, 50].
- Identify all key requirements – human resources, equipment, bait, training, etc. – based on previous similar eradications[11].
- Assess the risks to project implementation and success depending on factors such as meteorological conditions, environmental conditions on the island including availability of natural food for rodents, obtention of permits, implementation of food and waste management plan and biosecurity measures to prevent reinvasion, status and condition of the rodent population, etc. and be ready to react and modify your plans accordingly in case of changes[20, 21].
- Bait availability trials within study plots are recommended to calculate the density of bait required so that sufficient bait is available for consumption by rats for a minimum of four nights, and also by non-target species[21-23]. Alternatively, bait rates may be approximately determined from previous operations conducted in similar conditions.
- Non-toxic bait dyed green or with another biomarker (e.g. fluorescence) can be used to determine the species other than rodents that may take and consume the bait[7, 49].
- Prepare a realistic budget, with at least 20% in excess for unforeseen changes such as increases in prices or possible unexpected problems. Anticipate problems that may occur and how you plan to solve them, and envisage different scenarios[21].
- Transportation and logistics, excellent planning and organisation are key to the success of rodent eradications, especially when conducted in challenging environments such as the remote outer islands of Seychelles[37].
- Remove insofar as possible alternate food sources (fruits, human waste, etc.), especially in possible infestation hotspots (e.g. near human habitation). Eliminate dumping sites, green waste or coconut piles, etc. using machinery, mulching or burning. Eliminate food waste by deep burial (more than 1 m deep), incineration or maceration[20].
- In wet tropical or equatorial climates where food resources are abundant all year round, it is difficult to prevent rodents from accessing alternative sources of food (fruits, invertebrates) that may discourage

RODENTICIDE TREATMENT

Various rodenticides may be used to control or eradicate rodents, but the most commonly used are Brodifacoum, Diphacinone, Diphenacoum, Pindone, Bromadiolone, Cholecalciferol or Flocoumafen (2nd generation anticoagulants)[11, 14]. These inhibit the synthesis of vitamin-K-dependent clotting factors in the liver, which ultimately provokes death by internal haemorrhage (normally within 3-10 days)[1, 2]. Rodenticides are usually distributed by any of the following three methods: bait stations, hand broadcasting, and aerial broadcasting.

Bait stations

- Bait stations are used to distribute rodenticides and protect bait from weather conditions. They come in a variety of shapes and materials (see some illustrated examples). Some are available commercially, but effective ones can also be home-made using PVC pipes (e.g. models used by ICS, or MWAF in Mauritius[15]).

- Do <u>not</u> position bait stations in the open, but rather close to a wall, a tree or vegetation cover. (See instructions for traps, below). Clean them regularly if they become dirty or smelly to ensure that they remain attractive to rats; and keep vegetation cleared within c. 30 cm of them.

- Do <u>not</u> use strong poisons that will kill rats too quickly as this is likely to create an immediate aversion and wariness among rats to any sort of bait. It normally takes 3 to 4 days for a rat to die after eating a lethal dose of anticoagulant, so they do not associate consumption of rodenticide with danger.

- Wax blocks can remain attractive and palatable to rodents for 2 to 4 weeks in the granitic islands depending on weather conditions [16] (exceptionally up to 6 to 8 weeks during dry periods or in the drier outer islands).

- Refilling can be done every two to four weeks depending on consumption.

- Bait station grids baited with wax rodenticide blocks or pellets can also be used for eradication. On Anonyme Island (10 ha), Black rats were eradicated in 2003 with a 35 m grid of 66 stations kept permanently to prevent rats swimming ashore from Mahé (500m away) to become re-established[17].

Hand broadcasting

- Hand broadcasting of rodenticide is used to control rats during high infestations or in particular hotspots, or to eradicate rats from relatively flat islands where access by foot can be easily organised everywhere on the island by cutting tracks[11].

- This method enables a much better spread and therefore better availability of bait to rodents compared to bait stations (as the rats do not need to go to the bait station). However, bait effectiveness will be much reduced in time as its durability and palatability will be affected by weather.

- Bait can be hand-spread along parallel lines spaced 15 to 50 m apart, at sites (or stations) separated by similar intervals along the line, depending on the rodent species present and on the island characteristics. Black rats require higher density of bait than do Brown rats; and the House mouse requires the highest density bait distribution. Islands with high vegetation or hilly topography require a relatively denser geographical coverage of bait.

25 m

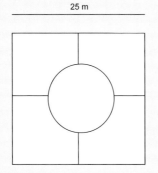
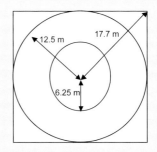

Rodenticide (Pestoff) pellets conditionned in a waterproof plastic bucket for transportation and storage. Gérard Rocamora

A rigorous method is required when spreading rodenticide by hand for an eradication. Gérard Rocamora

- The bait quantity (or dose) to be hand-spread at each site (or station) during each application is calculated from the total quantity required per hectare (in kilograms/ha), which depends on the density of rats and non-target species (e.g. crabs, snails, insects.) eating the bait[11, 23].

- Divide the intended rate per ha (1 ha = 10,000 m²) by the total area (in m²) to be treated between each baiting station (e.g. 25 m × 20m= 500 m² for lines spaced 25 m apart and stations at 20 m intervals along the line) to determine the dose to be hand spread at each station. For example 12 kg/ha = (12,000/10,000) × 500 = 600 g

- Each dose is normally divided into five equal parts using a calibrated scoop (e.g. 600/5 = 120 g). Four are hand-spread at an angle of 45° from the line in perpendicular directions, and the remaining 20% distributed within a radius of a few meters around the point (see drawing below)[11].

Aerial broadcasting

- Aerial broadcasting allows a better, more regular spread, making the bait more available to rodents than can be achieved by either hand spreading or bait stations. It is the preferred method for eradication of rodents in large hilly islands with dense tropical vegetation[11, 21]. This method requires an experienced helicopter pilot with a full understanding of the application process.

- A spreader-bucket slung from a helicopter ensures the most uniform and best delivery of pellets (which normally measure 10-12 mm); with an overlap in coverage of 50%, controlled by DGPS (Differential Global Positioning System) printouts to avoid any gaps.

- DGPS is highly recommended but not indispensable for islands of less than 200 ha, with repeated applications and a 50% overlap in coverage[21, 22], if relatively precise navigation can be undertaken by the pilot, using fly paths drawn on an orthophoto and natural landmarks (large trees, rocks, etc.), or gridded ground marks (e.g. flagged poles) for guidance[18, 19].

Rat eradication in remote islands require camps with large teams for several weeks (ICS on Grand Ile, Cosmoledo, Nov. 2007). Gérard Rocamora

Rat eradication can be conducted using rodenticide pellets spread by helicopter (ICS team on Grand Ile, Cosmoledo Nov. 2007). Gérard Rocamora

- Flying lines can also be marked on flat islands characterized by low vegetation (less than 3-4 m high) by using colour flags stuck at the top of 4-5 m plastic poles positioned every 100 m. This method was used in 2007 during the eradication of rats from Grande Ile and Grand Polyte, on Cosmoledo atoll[19].
- During an eradication two to four successive broadcasts of pellets may be required[20]. Aerial applications should be undertaken from different angles (30 to 90 degrees) to minimise chances that some areas will not to be covered.
- First fly the contours of the island to ensure that all coastal areas (including beaches, offshore rocks, cliffs, etc.) are well covered and receive sufficient bait to compensate for the presence of bait-eating crabs.
- Check the average swath width of spreader-buckets with the envisaged apertures and the chosen bait prior to the eradication, even though some buckets already have tables providing this kind of information[22]. This is done by spreading bait over a relatively flat, open surface with little or no vegetation (e.g. a large rocky area), and measuring every 10 to 20 m the distances reached by the bait perpendicular to the fly line. Check the correct functioning of all equipment (spreader-bucket, GPS) well in advance.
- Always keep a spare spreader-bucket available in case of a break down, especially if the operation is being carried out far from any logistical base, as interruption of an operation and helicopter transport is very costly. Good quality equipment and proper functioning of a spreader bucket are essential.

Bait usually spread for eradication is cereal pellets (normally of 10-12 mm diameter, and 20-25 mm long) containing Brodifacoum (or another active ingredient) at a concentration of 20 ppm, although higher concentrations may also be used. In Seychelles, all pellets used for this purpose were Pestoff 20R Brodifacoum cereal pellets (made in NZ).

The attractiveness and palatability of bait pellets is key to the success of a rat eradication. Gérard Rocamora

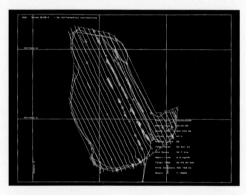

GPS devices can determine precisely the coverage of the helicopter. Gérard Rocamora

them from eating sufficient rodenticide.
- Assess rat abundance, body condition and reproductive status, and natural resource availability during the feasibility study, then closer to the date of the planned eradication[21]. If conditions are unfavourable (low density of rats, abundance of natural food, good body condition and high proportion of females breeding) then the operation may be delayed. Conditions are optimal when food is scarce, and rats numerous and hungry.
- Prior to eradication, take c. 20 samples of the rat population (e.g. foot flesh in 90% alcohol) to allow distinction between survivors and reinvaders

through genetic analysis in case the operation fails and rodents are still present after the operation is completed[38].

NON-TARGET SPECIES AND SPECIES INTERACTIONS

- Negative consequences to non-target species are almost inevitable during eradications. However, these effects can be minimised by taking precautions; they tend to be short term and are rapidly compensated for by the long term benefits of the eradications[22, 32, 78] (see also Ecosystem recovery section).

- Fatalities on non-target species after rat eradications (mainly birds and reptiles, but also fish and possibly snails) have been widely documented[39-42, 51], including in Seychelles[20, 32, 34, 35, 43-45].
- The persistence of rodenticide residues in trophic webs (invertebrates, reptiles, fish, birds) for weeks and months has also been documented around the world[41, 46-49].
- Impacts to non-target organisms have generally not been considered of concern as precautions were taken in case of significant risks, but exposure to bait may persist for months. Minimising the quantity of toxicant applied and long-term monitoring of potentially affected species (including toxicant residues) should be encouraged[21].
- Before the operation, identify all potential non-target species (birds, mammals, domestic or feral mammals, reptiles, crabs and other invertebrates, etc.) and assess how they may be put at risk, especially native species, and how they may interfere with the operation.
- Consider risks of both direct poisoning and secondary poisoning through food webs before the pellets break down (e.g. landbirds feeding on non-affected invertebrates that have eaten the bait).
- Take maximum precautions to minimise mortality of non-target species (e.g. provide covers for kill traps, check traps early morning, restrict access to bait stations, use green dyed rodenticide pellets less attractive to birds, etc.). This may include specific studies and captivity during the campaign for rare native species (e.g. birds, reptiles or invertebrates)[32, 34, 49].
- Insofar as possible keep all livestock and sensitive species (e.g. Giant tortoises) penned for 2-3 months depending on bait breakdown time. Pellets need to be removed from the enclosures of captive non-target species during aerial broadcasts, and whenever possible pools of water need to be covered and pellets (that will float on the surface) removed[20, 32].
- People are unlikely to be at risk from Brodifacoum during rodent eradications. Even at highest used concentration (50 ppm) a 15 kg child would need to eat 75 g of bait to have a 50% chances to die. Nevertheless, always keep Vitamin K1 antidote available during such operations, which can be administered by injection or orally in case of accidental injestion

Hermit crabs can consume large quantities of bait (with no harmful effect to them). Gérard Rocamora

of bait (this can also be administered to livestock or sensitive species)[11].
- Try to eradicate rats first and cats immediately after when both species are present. Experience shows that cats can be greatly affected by secondary poisoning from consuming rats, and sometimes through direct consumption of rodenticide cereal pellets (see Feral cat text p. 230). However, if cats appear to be a bigger problem than rats (e.g. for seabird colonies) and if their localised control is not possible, they may need to be eradicated before the rats.
- Other invasive species that had originally been controlled by rats through predation (e.g. invasive ants [76]) or competition (e.g. mice) may also benefit from the eradication and may increase in number, provoking new immediate impacts to native species that may require additional management[72-74] (see also 'surprise effects' in Part 1, p. 143).
- In the presence of several invasive species, it is best to eradicate them all at the same time (i.e. conduct multiple species eradications)[75]. See House mouse text (p. 208) regarding possible interactions between mice and rats, and betweeen mice and cats.

BAIT APPLICATIONS (RATES, NUMBERS, INTERVALS)

- Grids for ground-based applications (bait stations or hand broadcast) can be positioned using a measuring tape or rope of known length, a GPS, or a hip chain and clearly marked with very visible colour ribbons in vegetation, or coloured pegs planted in the soil. It is generally easier to start grids from a line crossing the island over its maximum length.
- Bait applications must be conducted during dry

periods (no significant rains for 2-3 days prior to the application) and should be followed by at least 3 to 4 nights without significant rainfall. Rainfall greater than 15-25 mm is likely to irreversibly affect bait consistency and palatability[20, 21].

- The quantity of bait to be hand-spread at each point (dose) needs to be calculated taking into account the estimated consumption by rodents and non-target species (from field trials), the planned number of applications, and experience from previous similar eradications.

- Consider the effects of other species (rabbits, cats, tenrecs, crabs, etc.); how will they interact with rodents; will this diminish chances for the rodents to access bait?[11]

- Bait density should be increased in areas where non-target species are likely to consume significant quantities of bait, typically land crabs and hermit crabs in Seychelles, hence higher application rates are required in coastal zones[21].

- Plan additional baiting (by hand) in all areas where aerial distribution is not effective or may be insufficient. These may include sites identified as potential infestation hotspots, areas where rodents may have easier access to alternative food sources (around human habitation, dump sites, and areas where green waste accumulates), sites that do not retain pellets (cliffs), or which do not benefit from aerial distribution (caves, under roofs, inside buildings)[20].

- Bait rates used for successful operations varied from 6 kg to 40 kg/ha in Seychelles. Average rates include 9 kg/ha for Black rats using bait station grids and Brodifacoum wax blocks (n = 2); 25 kg/ha for Brown rats using Brodifacoum 20 ppm cereal pellets (n = 3; two aerial & one hand broadcast), and 31.3 kg/ha for Black rats using the same pellets (n = 6; five aerial & one hand broadcast).

- Number of applications, intervals between applications and rates for each application depend on the amounts consumed by rodents and non target species, and on palatability and durability (longevity) of bait (as a result of rainfall, temperature, sunshine, etc.). This can be monitored with density quadrats (see next section).

- Average number of applications for successful operations in Seychelles have been the following: 2.7 (n = 3) and 3.0 (n = 6) for Brown rat and Black rat eradications, respectively; and 3.5 (n = 5) in the inner islands and 2.0 (n = 2) at Cosmoledo for Black rats.

- Intervals between applications are normally 7 to 10 days, but may be shorter in case of rainfall immediately following an application, or longer when bait consumption is low and if bait quality and durability remains high.

- Intervals used during past successful operations in Seychelles have been 5-11 days between 1st and 2nd application, 8-24 days between 2nd and 3rd applications, 12-19 days between 3rd and 4th applications, resulting in periods of 30-50 days with good quality bait available to rodents.

- During the aerial drops, keep continuous control of the amount of bait broadcast in relation to the area approximately covered; if necessary, reduce or increase the diameter of the bucket's aperture to adjust the density of the bait spread.

MONITORING (RAINFALL, DENSITY QUADRATS, RATS, WILDLIFE OBSERVATIONS AND MORTALITY)

- Closely monitor rainfall and weather predictions before and after each application.

- Randomly select 6 to 10 permanent bait density quadrats of 10 m × 10 m each divided into four (5 m × 5 m) sub-squares before the applications, and mark them with pegs and string, in order to check bait densities later, and to monitor bait consumption and durability[20].

- Check bait availability and quality by visiting density quadrats immediately after each application, and every following day. This will allow adjustment of application rates for the subsequent applications depending on bait quantity, consistency and palatability[21, 23].

- In each of the quadrats, search for baits in a systematic way (circling from the outside to the inside of each sub-square) and record bait numbers (full baits and pellet pieces may be distinguished). Place all baits found at the centre of each subsquare.

- On the following days, note the numbers still present and their quality, rating from fresh bait with good consistency and palatability (1) to that which is extremely degraded and unpalatable (5). Visit quadrats every day for the first few days, then less often after no changes are recorded[20].

- Record daily the mortality of any non-target species

Density quadrats are used to check the quantity and uniformity of bait spread by helicopter (here in rocky habitat; Petite Sœur). Gérard Rocamora

Pellets from two different drops gathered in a density quadrat, some freshly spread and others no longer palatable. Gérard Rocamora

(date, location, species identity, number). Likewise, record any rodents found dead[20].

- Record wildlife observations, especially concentrations, species of special interest, and interactions with other species (e.g. species eating bait or dead rodents).
- Monitor the abundance of rats daily along several trap lines covering the main habitat types until no rats are caught for c. 5-7 days (see index-trapping box), then conduct survey trapping (random trapping distributed all over the island) to detect any possible areas with surviving rats[20].
- Verify pellet consumption by rodents by checking for the presence of green dye in faeces (e.g. in cage-traps), or by dissecting euthanised animals and checking the colour of the intestines and stomach. Other biomarkers, such as fluorescent pyranine, can also be used[21-23].

ASCERTAINING SUCCESS

- In temperate areas two years without any sign of rodents is considered the standard before success can be declared (irrespective of any trapping conducted). However, experience from Seychelles and elsewhere shows that this period may be reduced to **one year** if intensive survey trapping is conducted during that period (e.g. a total of c. 1,000 trap-nights or more with a minimum of 250 trap-nights at the end of the period).
- The use of two independent detection methods (e.g. gnaw sticks, traps, tracking ink tunnels, etc.) has been suggested to confirm success within a year in tropical environments[57]. In Mexico, an innovating survey method using a grid of wax tags spaced 50 to 200 m and based on modelling has proved its efficiency in validating eradication success within a few weeks or months after the operation[54-56].
- In tropical humid ecosystems, rodents have more resources available, breed all year round (c. every three months) and multiply more quickly than those in

The green dye of the bait allows to tracks its consumption by rats (here, in faeces). Gérard Rocamora

temperate and subantarctic climates; hence surviving rats can be detected much quicker, within months.

- In all three unsuccessful Black rat eradication attempts in Seychelles, rats (either reinvaders or survivors) were detected within a year (or much earlier) of the end of the operation (see Black rat text). On Eagle Island (Chagos, 240 ha), where a Black rat eradication was attempted in 2006, evidence that rats had either survived or reinvaded was found well under a year[58].
- A year after the rat eradication conducted under the ICS FFEM project, the following four islands were declared rat free based on intensive survey trapping: Ile du Nord (North Island; Black rats) in Oct. 2006 after 8,700 trap-nights[20]; Grand Ile and Grand Polyte (Cosmoledo, Black rats) after 900 and 390 trap-nights respectively[19, 59]; and Conception (Brown rats) after 1,800 trap-nights[18, 59]. In 2001, Frégate Island was considered free of Brown rats (which can be detected more easily than Black rats at low density) by DoE a year after the eradication, and Seychelles white-eyes were transferred there 16 months after the operation[60, 61].
- However, we recommend a two year period before any reintroduction can be conducted, in order to test the effectiveness of the rat abatement measures used to prevent reinfestation.

PREVENTING REINFESTATION: PROTOCOLS, SURVEILLANCE AND EARLY RESPONSE

- After a successful eradication programme on an island, rodent protocols to prevent recolonisation are indispensable for the long-term success of the operation[22, 32].

- Abatement measures need to be in place well before the eradication starts so that there is time to test the new rules and procedures and for all people on the island to get used to them.

- Preventing reinvasion by rodents is often more challenging than the eradication itself as it requires a **strong management discipline** that needs to last forever, whereas the eradication phase – even when challenging – only lasts for a limited amount of time.

- The most likely ways for a rat-free island to be invaded or reinvaded in Seychelles are from any of the following: infested cargo brought by a supply boat (or aircraft), an uncontrolled rat-infested visiting vessel landed on the island shores, or by a rat swimming ashore from a nearby infested island or boat moored a few hundred meters away from the island beaches.

- Prevention strategy consists in placing multiple barriers in the most likely introduction pathways of rodents onto the island. This includes preventing rodents from getting onto boats that regularly supply or visit the island (boat loading protocols), and preventing rodents from getting off the boats onto the island (unloading protocols, rodent-proof room, traps, bait stations)[62-65] (see also p. 103).

- Permanent bait stations need to be installed adjacent to the main beaches and landing points; and these need to be checked and regularly refilled every 2-4 weeks depending on weather conditions[62, 65]. However, bait stations with rodenticide may represent a risk for certain sensitive species (e.g. Seychelles magpie-robin could be secondary poisoned by eating coackroaches having consumed rodenticide bait[43, 44]).

- Building a **rodent-proof room** containing bait stations and traps where all goods and materials are unpacked and checked for rodents and other pests is a key requirement. Likewise, bait stations and gnaw-sticks need to be kept on board regularly visiting (supply) vessels[20, 65, 66].

- Prior checking of cargo and certifying it rodent-free before it is loaded and shipped to the island is the first priority step to prevent invasion (see details of a **boat loading protocol**, in Annex).

- Avoid mooring supply boats at night on rat-infested peers or jetties to prevent rats boarding vessels prior to shipment. Maintain regular control of rats by using bait stations and traps. Keep piers/jetties free of any food, rubbish and rodent shelters. Use line-guards or half plastic bottles to prevent rats from climbing on board along mooring ropes. Strong lights at night will discourage rats[11, 65].

- Some islands with small harbours have built rodent-proof fences, but these need to be constantly checked and maintained to be really effective[32].

- Prevention measures exist for each of the islands of Seychelles where rodents have been eradicated. Examples of loading procedures on Mahé, unloading procedures after reaching the island, and rat proof room protocols are given in Annex.

- Use **check-lists** and conduct regular audits of protocols for rodent prevention, surveillance and early response.

- When importing large amounts of cargo/supplies to a rat-free island (e.g. for construction), place all the cargo into a container prior to shipment, and have it fumigated by a pest control company with Methyl bromide or Aluminium phosphide (for 3 to 5 days). This is the safest method to guarantee that no rat or other animal pest will invade or reinvade the island[62, 65]. Spraying with ordinary insecticides is effective against invertebrates but not against rats.

- When unloading cargo/supplies that do not fit into the rat-proof (biosecurity) room, set up a temporary fence on part of the landing beach to help contain potential invasive species. Display boxes containing live-traps (e.g. Sherman/Eliott traps) for rodents to hide and become trapped. Visually check large materials as they are unloaded[65].

- Regular supplies (particularly food) for rat-free islands are best transported in transparent hermetically closed **plastic containers** (boxes or tubes[66, 67]) that can be checked before being loaded onto the supply boat. This method can also be used for islands with small hotels and a substantial human population.

- Avoid or **ban cardboard boxes** for carrying supplies, which rodents can easily chew their way into and which can easily harbour invertebrate IAS. Use plastic boxes or dry bags instead. Cardboard boxes

also facilitate the invasion of smaller species such as invasive ants[62, 65].

- Strict control of all arriving boats is key to protecting an island from rodent invasion. However, unless the island is surrounded by a marine park, or is a legally declared nature reserve, access to visiting boats cannot be easily restricted in Seychelles. Public access to beaches is a constitutional right, and biosecurity requirements need to be solved through the implementation of protocols . These should be included in the Biosecurity Act (see p. 38, 84, 106) so that visitors can access those beaches in concertation with the island management.

Plastic drums are recommended as a systematic procedure to transport checked items to high biodiversity islands / areas.
V. Tatayah / MWF

Chewsticks and chewcubes are useful to detect the presence of rats.
Nik Cole / MWF

A rat-proof trailer transports cargo items from a barge to the pest-proof room.
C.J. Haveman / North Island

- Aircraft bringing passengers or cargo also need to be regularly inspected and fumigated.
- Keep the island tidy and clean, with a minimum of food resources available so that in case of reinvasion, rodents will be more easily attracted to bait stations or traps. Replace dumpsites by incinerators or composters for treatment of organic waste (particularly food waste). Export all solid non degradable items, and mulch or burn green waste (do not pile it up[20]).
- Regularly implement the following **surveillance techniques** to detect any invasion early and to eliminate the invader rapidly. These include the use of traps, checking for tooth marks on blocks in bait stations and on gnaw-sticks soaked in coconut oil or with peanut butter, and tracking tunnels with ink cards where rat foot prints can be seen[11, 63].
- On rarely visited islands (e.g. remote outer islands), permanently deploy 10-20 tunnels that can be used for trapping and baitingj, or use tracking cards during visits.
- In view of the capacity of rodents to swim or to raft aboard tree trunks or debris, islands located less than 600 m from islands infested by Black-rats or 1,500 m from those infested by Brown rats are probably at significant risk from natural reinvasion.

Beyond these distances, existing surveillance and early response techniques to eliminate rodents are considered sufficient to mitigate the reinvasion risk[11].

- Staff awareness and an understanding by all island inhabitants, crew of vessels and visitors is essential for effective prevention and surveillance. On islands with a large human population (e.g. those with hotels) and a relatively large staff turn-over, regular training, briefings and updates (e.g. quarterly presentations, posters, leaflets) are essential [11].
- Any report of infestation should be taken seriously. Interview the person to confirm and (if possible) further substantiate the report. Determine what was seen by asking open questions – i.e. colour, size, where, how close, how long etc. Produce a written report on any observations and incidents.
- If information is deemed to be reliable, proceed immediately with an **incursion response**. Act immediately to remove the invader before it can reproduce and establish a small breeding population.
- Take photos, keep any samples of evidence such as droppings, partly-eaten bait, teeth marks, fur or hair, etc. that will help to determine the identity and number of invaders[20].
- When rats invade a rat-free island they tend to travel extensively (probably in search of their own kind). This has been shown by radio-tracking experiments in New Zealand and by local experience in

Seychelles. On two consecutive nights, a single invading Brown rat (recognisable by physical body and tail characteristics) was photographed by camera-traps at two sites c. 1 km apart on Ile du Nord in February 2010, giving the false impression that a small population had already established [17].

- Check for signs of rats in areas to which rodents are attracted or from which they may originate (landing sites, houses, kitchens, dump sites, coconut piles/groves, etc.).
- Prepare a ***rodent incursion response kit*** that contains the following items: biosecurity instructions, traps, tracking cards, identification guidelines, maps, camera and batteries, rodenticide baits, expert contacts, dissecting equipment, ethanol, flagging tape, etc.
- Keep key stakeholders (management, advising experts, partners, etc.) well informed about status, progress and outcomes of any incursion response [66].
- Sophisticated night vision equipment including automated night wildlife camera are extremely useful for identifying and tracking invading rodents [20].
- After a prolonged period with no signs of invaders (bait intake, sightings, captures), response measures can be reduced and standard surveillance may resume. However, a single invading rodent can stay for months on a large island without being noticed.
- Methods to detect rodent invasions and to eliminate invaders rapidly need to become more efficient. Sometimes invading rodents cannot be detected for months if they avoid traps, bait stations, or poisoned food [68]. This leaves ample time for a single invading pregnant female to produce a founding population.

ECOSYSTEM MONITORING AND RECOVERY

Regular monitoring of the main components of an ecosystem (i.e. plants, invertebrates, birds, reptiles) should be considered a priority before, during and after any rodent eradication [21, 22]. Ideally monitoring should start 2-3 years before the eradication in order to establish a solid baseline of data, and be conducted for a few years (if possible) every 2-6 months after the eradication in order to identify short and medium term changes while taking into account seasonality. After that, monitoring should be conducted annually or at least once every few years, depending on opportunities to investigate long term impacts (see "Monitoring biodiversity", p. 166).

The removal of a rodent species is likely to have impacts on other species in the island ecosystem. Some information has been collected in Seychelles about how the eradication of rodents from islands affects the fauna and flora. Between 2005 and 2009, pre- & post-eradication standardised monitoring schemes were conducted on Conception, Ile du Nord and Grand Ile/Grande Polyte (Cosmoledo) as part of the ICS FFEM project [59]. After a rodent eradication, native species often showed rapid signs of recovery. Some of these were spectacular (e.g. for certain landbirds, reptiles and invertebrates), while others recovered at a slower long-term pace (e.g. seabirds) [69, 70]. This has been observed not only in Seychelles (see "Ecosystem recovery after the eradication of rats and cats", p. 169), but also in other parts of the world (New Zealand, Australia, Mariana Islands, etc.) [24, 71].

Invertebrate communities may also show unexpected patterns of recovery, whereby certain groups tend to become more abundant while others become more rare. By removing a single species which had a high impact on others, and often in a dominant position in the ecosystem, a rodent eradication may suddenly modify the balance amongst the remaining species. It may take several months or even years before the interactions between the various species stabilise and a new balance is found [77].

References / Further reading

1 Buckle & Smith, 1994; 2 Grant *et al.*, 2003; 3 Rural Development Service 2006; 4 Bill Simmons & G. Rocamora, *pers. obs.*; 5 CAB International, 2014; 6 Dobson, 1988; 7 Campbell *et al.*, 2015; 8 Orueta, 2003; 9 Orueta, 2007; 10 Orueta & Ramos, 2001; 11 Pacific Invasive Initiative, 2011a; 12 Cunningham & Moors, 1996; 13 Rocamora & Said, 2005; 14 Donlan *et al.*, 2003; 15 Tatayah, Haverson *et al.*, 2007a; 16 Beaver & Mougal, 2009; 17 G. Rocamora, *pers. comm.*; 18 Rocamora 2009d; 19 Rocamora 2007b; 20 Climo & Rocamora, 2006; 21 Keitt *et al.*, 2015; 22 Broome *et al.*, 2014; 23 Pott *et al.*, 2015; 24 Russell & Holmes, 2015; 25 DIISE, 2014 (516 successful and failed eradications only, excludes reinvasions, whole island eradications only, satisfactory or good data quality only); 26 Holmes *et al.*, 2015; 27 Greaves, 1994; 28 Howald *et al.*, 2007; 29 DIISE, 2014; 30 Harper *et al.*, 2015; 31 Merton, 2001; 32 Merton *et al.*, 2002; 33 Hoare & Hare, 2006; 34 Gerlach, 2005a, 2005b; 35 Gerlach & Florens, 2000; 36 Parkes *et al.*, 2011; 37 Rocamora, 2010a; 38 Russell *et al.*, 2009b; 39 Courchamp *et al.*, 2003; 40 Bell, 2002; 41 Pitt *et al.*, 2015; 42 Pain *et al.*,. 2000; 43 Thorsen & Shorten, 1997; 44 Thorsen *et al.*, 2000; 45 Merton, 2001; 46 Cole & Singleton, 1996; 47 Booth *et al.*, 2001; 48 Walker & Elliott, 1997; 49 Rocamora, 2005; 50 Ringlet *et al.*, 2014; 51 Erickson *et al.*, 2002; 52 Rocamora 2007a, 2007b; 53 Conservacion de Islas, 2015; 54 Samaniego

Home-made model of rat bait station used in Mauritius by MWF. Vikash Tatayah

Home-made bait station used in Seychelles by ICS & IBC-UniSey. Gérard Rocamora

Home-made Novacoil bait station on ground (New Zealand). Lesley Baigent

Home-made Novacoil bait station in a tree. Bill Simmons / Pestoff

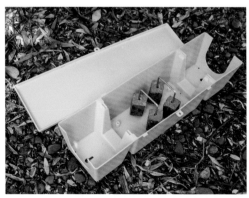

Commercial bait station 'Departure Lounge'.
Bill Simmons / Pestoff

et al. 2015; 55 Samaniego *et al.* 2013; 56 Aguirre *et al.*, 2015, in press; 57 Russell *et al.*, 2008; 58 Daltry *et al.*, 2007; 59 Rocamora & Jean-Louis, 2009; 60 Rocamora, 2001; 61 Rocamora & Henriette-Payet, 2008; 62 DoC, 2010; 63 Merton *et al.*, 1989; 64 Broome 2007; 65 Rocamora, 2015; 66 Harper, 2014; 67 Tatayah, Birch *et al.*, 2007; 68 Russell *et al.*, 2005; 69 Jones, 2010; 70 Rocamora *et al.*, in prep.; 71 Veitch *et al.*, 2011; 72 Caut *et al.*, 2007;73 Courchamp *et al.* 1999; 74 Russell *et al.*, 2009a; 75 Ines & Saunders, 2012; 76 Feare, C.1999; 77 Galman, 2005; 78 Maggs *et al.* 2015; 79 Millett *et al.*, 2001.

Commercial bait station 'Dead Rat Caffé'. Bill Simmons / Pestoff

S. Caceres & J.N. Jasmin

Feral cat

Chat haret

Sat maron

Felis catus

IDENTIFICATION AND BIOLOGY

The Feral cat is a medium sized predator, normally c. 35-55 cm long without the tail, weighing normally 1.5 to 3.5 kg (but exceptionally up to 6 kg), and rarely taller than 30 cm at the shoulders, with highly variable fur colourations (including grey, black, brown, and some with white patches)[1]. They are adaptable to all sorts of habitats and climates, from subantarctic to dry semi-desert, and humid equatorial climates[2]. They are very well adapted to hunting and to nocturnal life. Strictly carnivorous, they feed mainly on birds (small passerines, seabirds)[3, 4], reptiles (lizards, geckos, sea turtle and tortoise hatchlings), small mammals (mice, rats), and large insects[1, 5, 6]. Cats are quite prolific, breeding twice a year and producing litters of up to 4 to 6 kittens (gestation period is 63-65 days) and reaching sexual maturity at 7-12 months[2]. Their calls are characteristic and easily recognisable, as are their footprints.

ORIGIN

Wild cats originate in Africa and the Middle East, the first evidence of domestication dating back 9,000 years ago in Cyprus[2]. They have been spread by people throughout Europe, and later to other continents and many islands around the world. Domestic cats were most likely brought to Seychelles from Mauritius and La Réunion by the first settlers, and introduced to virtually every island with a human settlement to help control rats, and also as pets. Cats often form feral populations, particularly after settlements are abandoned.

DISTRIBUTION & ABUNDANCE IN SEYCHELLES

Feral cats are present in the inner islands on Mahé, Ste Anne, Ile au Cerf, Longue and Thérèse (last two

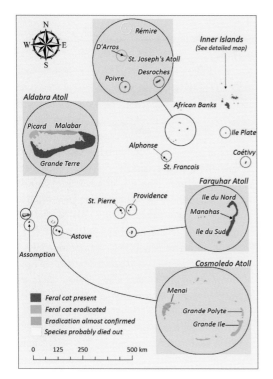

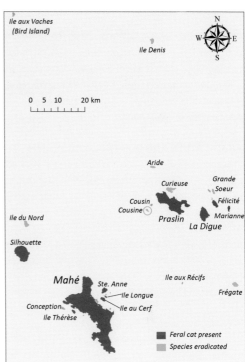

not checked recently)[7-9]; Praslin, La Digue, Félicité, and Marianne[10, 11]; Silhouette; and in the outer islands at Plate, Coëtivy, Desroches, Poivre, Alphonse, Farquhar (Ile du Nord & Ile du Sud), Providence, and St. Pierre (to be confirmed); Menai (Cosmoledo), Astove, Assomption, and Grande Terre (Aldabra)[3, 12-15, 28]. They probably died out on Rémire given an absence of recent reports. Feral cats have been eradicated from a total of 10 islands including Aride, Cousine, Frégate, Curieuse, Denis, Ile du Nord, D'Arros, and died out naturally on Picard (Aldabra atoll)[16]; Grande Soeur, Grand Polyte and probably Grand Ile (Cosmoledo atoll). At D'Arros, only a handful of neutered domesticated cats remain[17]. Feral cats form populations of variable densities (0.1 to 3 ind./ha) depending on the resources available[18]. The cat eradication at Cousine Island (26 ha), was conducted exclusively by trapping, and resulted in the capture of 77 cats (i.e. 3.0 cats /ha)[19].

IMPACTS AND THREATS

Feral cats are amongst the worse threats to island biodiversity[20]. They can severely damage the native fauna of islands, and seabirds in particular[6, 21], as reported from a minimum of 120 islands where they have been introduced worldwide[22]. They are directly responsible for the global extinction on islands of 33 birds, mammals and reptiles species (14% of total), and remain a threat to at least 175 other native vertebrates (123 birds, 27 mammals and 25 reptiles) worldwide, including 38 critically endangered species (8% of total)[22]. Populations of rare and endangered species of endemic landbirds, seabirds or reptiles, are at particular risk. In the 1980s, on Frégate, they were found to be responsible for the near-extinction of the Seychelles magpie-robin, which had disappeared from Alphonse after cats were introduced[23]. On Cousine Island, stomach contents collected from Feral cats consistently showed the remains of seabirds[19]. On Aldabra, Feral cats have been shown to prey primarily on turtle hatchlings[24], cockroaches, and rats, as well as tortoise hatchlings, geckos, skinks, crustaceans, other insects, birds and fish[3, 5]. The occurrence of the endemic flightless Aldabra rail is restricted to cat-free islands, suggesting cats are responsible for its local extinction on other islands. On Grand Ile (Cosmoledo), cats preyed upon large numbers of Sooty tern adult and chicks[12]; and on Juan de Nova (Iles Eparses, France), they are estimated to kill an average of six birds per day during the Sooty tern season[4]. Indirect negative impacts include reduction of native species breeding success due to behavioural changes, habitat changes due to disruption of seed dispersion, reduction of nutrients additions and vegetation disturbance from seabirds, and others.[25, 26]. Elimination of Feral cats usually leads to a substantial increase in populations of seabirds and other cat prey[18, 27] but it can also lead to some negative effects such as increases in other invasive species (mice, rats, rabbits)[23].

ERADICATION AND CONTROL PROGRAMMES

In Seychelles, cats are known to have been eradicated, directly or indirectly, from at least nine islands (ten with Grand Ile) (see p. 65 and p. 87 for island areas).

Aride

Feral cats were eradicated using dogs and young boys to catch them during the 1930s. They are reported to have been introduced there in 1918, when the island was exploited for coconuts, seabirds and agriculture[29].

Frégate

In 1960, 86 cats were poisoned or trapped, and a further 22 were trapped in 1979-1980 to protect the Magpie-robin population[16]. In 1981-82, a team from Forestry & Conservation (Department of Agriculture) with a technical expert from the New Zealand Department of Conservation (DoC) conducted an intense operation, during which 56 cats were poisoned or trapped, and eradication was subsequently confirmed. Sixty gin leg-hold traps using a variety of baits were set across the island at night and closed during the day. In addition, Pestoff 1080 poison (Sodium Fluoroacetate) was injected into small cubes of fresh fish and placed on the ground (on a leaf or a small rock) at c. 20 m intervals along paths at dusk. Any cubes that were not eaten during the night were collected before dawn the following morning[23]. This programme was supported by the International Council for Bird Preservation (now BirdLife International).

Cousine

The cat eradication programme took place between April 1983 and June 1985, during a series of one-week visits by a DoE employee (Victorin Laboudallon). In all there were 43 days of intensive trapping. A total of 73 cats (33 male and 40 female; including 23 juveniles and 50 adults) were trapped, using a combination of leg-hold Gin traps and Victor traps (both as 'walk-in' and 'walk-through'). Each day, 16 to 28 traps were set across the island[19]. Cats had only been introduced to Cousine in 1971-72[16].

Curieuse

The July 2000 rat eradication attempt significantly reduced Feral cat numbers before the cat poisoning campaign had even started. The early cat mortality (which led to some cats being found dead and others disappeared) was caused by a combination of primary poisoning when cats consumed the Brodifacoum cereal pellets, plus secondary poisoning when cats ate poisoned rats[30]. This was immediately followed by the cat poisoning campaign during which 70 bait stations (modified "Philproof" from Animal Control Products, New Zealand) were deployed around habitations and across the island. Pelleted chicken cat food proved ineffective and was replaced by canned tuna oil. One week of pre-feeding with non-toxic bait was followed by the addition of toxic cat bait at 0.1% 1080 (Pestoff 1080; active ingredient Sodium Fluoroacetate). Baits were changed daily over a period of 2-3 weeks. A week after poisoning started, 90 leg-hold traps (Lanes Ace and Victor 1.5) were set at 100 m intervals along tracks and baited with canned tuna. The last cat was trapped in February 2001 and control trapping continued into 2001[31]. This programme was supported by the Dutch Trust Fund (DTF).

Denis

Cat eradication started in June 2000, shortly after the rat eradication (during which one cat was found dead), as part as the same DoE-DTF programme, and using the same methodology as for Curieuse[30, 31]. The 52 bait stations were baited with canned tuna oil, and leg-hold traps were also used. The last cat was trapped in September 2001.

Ile du Nord

Cat eradication started after the rat eradication attempt of September 2003 (during which one cat was found dead). Fifty Sentry cat bait stations were deployed across a network of paths cut through the island, around human habitation, and at known cat feeding areas and dumping sites . Stations were pre-baited for 7 days with a mixture of rice, canned tuna and vegetable oil. Then poison 1080 gel at 0.1% was applied for 5 days, during which the toxic bait was changed daily. Seventy-two Victor leg-hold traps (mainly n°2) were then deployed along tracks at 150-180 m intervals, baited with canned tuna oil and rebaited daily, during c. 3 weeks. Sixteen cats were found dead from poison, and five more were trapped during the first two trapping weeks. No cats were caught during a second week of trapping a month later[32].

D'Arros
Cat eradication started immediately after the rat eradication campaign in August 2003, and combined poisoning and trapping with leg-hold traps, with similar methods than on Ile du Nord (North Island). A total of 21 feral cats were killed, and a few domestic cats were neutered[16, 33].

Grand Polyte & Grand Ile (Cosmoledo)
Feral cat presence was recorded on both islands between 1968 and 2003[12]. In November 2005, cats were observed on Grand Ile, and old scat noted on Grand Polyte[34]. After the two aerial drops of rodenticide (Brodifacoum 20 ppm pellets) on both islands in November 2007 to eradicate rats, the number of cat observations was significantly reduced on Grand Ile during survey rat trapping[13]. Remaining cats on Grand Ile were estimated to number a dozen or less in November 2008[35]. No cats were subsequently observed on Grand Polyte, possibly having died as a result of secondary poisoning and reduced food resources. Follow-up assessment of the cat eradication could not be conducted due to logistical problems and piracy in 2009 (when the boat Indian Ocean Explorer was captured by pirates). In March 2014, cat tracks were not observed on the beaches where cats used to search for turtle hatchlings in the past, nor around four cat feeding tables and sand patches (see picture) deployed during four days and nights on both islands (lagoon side only on Grand Ile)[36]. In November-December 2014, feeding tables were set up again on each island (three for three nights on the NE seaward side on Grand Ile; five for two nights around Grand Polyte); Grand Ile was extensively surveyed during day time as in March, with no sign of cats observed[14]. One last visit in 2015 to survey Grand Ile at night and set up cat feeding tables at its southern end is expected to confirm the total absence of cats on Grand Ile.

Grande Soeur
The presence of various feral cats had been occasionally recorded[10] until the rat eradication took place in September 2010 when the last cat (a visibly sick individual) was observed. No specific poisoning or trapping was necessary and cats are no longer there. The last cats appear to have died during the rat eradication from direct consumption of rodenticide and/or secondary poisoning from consumption of rats.

Cat control operations are taking place more or less regularly on at least five other islands. On islands managed by IDC, control sessions targeting feral cats are organised several times a year (every 2 to 6 months) by IDC pest control services, in collaboration with ICS conservation staff, and pest control activities in the hotels present in these islands. These normally involved trapping or poisoning sessions (active ingredient Methomyl mixed with canned tuna or injected into fresh fish)[38].

Félicité
A total of c. 35 cats were trapped between 2010 and 2012 by the island management.

Alphonse and Desroches
Cat control has been conducted on a relatively regular basis, particularly around the shearwater colonies, since at least 2009. Different kinds of traps have been used with varying success, but overall results are considered inadequate on Alphonse since cat predation continues to happen there. Poisoning sessions are now conducted on a regular basis in these islands and cat eradication is envisaged. But, the number of animals eliminated has been no more than a few individuals per control session[38, 40].

Silhouette
Cats appear to have been subject to irregular control in the vicinity of La Passe. Because cats are a threat to the Critically Endangered Sheath-tailed bat Coleura seychellensis and to seabirds, cat control needs to be conducted more systematically in the area of the bat roosts, and extended to the whole plateau. Shearwaters trying to dig nesting burrows have been reported in several occasions, but some have been preyed upon (presumably by cats or Barn owls)[41].

Farquhar
Control is being conducted at irregular intervals around human settlements on Ile du Nord[38], but Feral cats occur in relatively high numbers on both Ile du Nord and Ile du Sud[37, 42].

Aldabra

Feral cats occur on Grande Terre (the largest island), where they have been subjected to occasional control by cage trapping and shooting[16, 43]. Leg-hold traps cannot be used due to the presence of many non-target species (i.e. Robber crabs, etc.), and eradication is envisageable but very challenging, and some recommendations have been made[5]. Cats appear to have died out naturally from Picard Island during the 1990s.

Extensive literature is available on cat eradications around the world[18, 27]. By 2014, up to 75 successful, and 12 failed cat eradication campaigns have been undertaken, with an overall success rate of 86%[62]. In Mauritius, cats were eradicated from Ile Plate (Flat Island, 253 ha) using leg-hold traps[44]. In Iles Eparses and French Antarctic territories, cat control is being undertaken through trapping and hunting and eradication is considered[45, 46]. More details and references on cat control and eradication programmes conducted outside of Seychelles can be found through the Global Invasive Species Database www.issg.org/database/welcome and the Invasive Species Compendium database www.cabi.org/isc/.

MANAGEMENT RECOMMENDATIONS

The following recommendations are summarised from experience obtained in Seychelles, and abroad[47, 55-57]. More detailed information is available from text references, in particular the *Pacific Invasives Initiative toolkit for Rodent and Cat eradication (2011)* http://rce.pacificinvasivesinitiative.org. which provides practical advice for all the stages of an eradication project (i.e. Project Selection – Feasibility Study – Project Design – Operational Planning – Implementation – Sustaining the Project) and has specific *Guidelines on Cat Eradication and Monitoring Techniques.*

Control and eradication protocol
- Cats can be controlled by several methods, mainly trapping, poisoning and shooting.
- Feral cats have high population turnovers, and may recover 70% of their initial numbers within 6 months, hence some authors recommend bi-annual sessions of control[55].
- Cat eradications are delicate technical operations that require motivated staff and guidance from persons with good knowledge of Feral cat behaviour and ecology. This is particularly important when trying to eliminate the last survivors through trapping.
- Rat eradications can generate a significant knock-down in numbers of cats through both direct consumption of bait and secondary poisoning (up to 50-100%). Engaging in cat eradication immediately after rodents have been removed maximizes this pre-campaign reduction.
- Cats may participate (although not necessarily in a significant way) in the control of mice or rats. In some cases it may be preferable not to eradicate cats first, so that rodents do not increase and generate more detrimental effects on native wildlife, as has been documented in some cases[48, 27, 22]. However, rats may not necessarily increase after cat removal[49], and the best timing to remove cats versus rats may depend on each particular circumstances[50]. Recent studies and empirical evidence suggest that even with more rats, environmental damage is often lower when rats are alone than when both rats and cats are present[25, 51]. When all introduced predators cannot be eradicated together, eradicating the top predator usually generates a positive outcome[22].
- Inhabited islands or those that can be visited regularly are usually declared cat free within six months of no sightings or recent signs of cats.

- Use a combination of several methods, typically with an initial knock-down phase to decimate the population, and a follow-up phase to eliminate the survivors.
- Plan carefully the sequence in which the activities are planned, especially for eradication.
- First use methods that kill the greatest numbers of cats with limited risk of creating aversion in the cats. Start with poisoning, followed by trapping and then shooting or the use of trained dogs saved until last. This strategy is particularly important for islands which are large and/or with difficult terrain.
- Target periods when available food is seasonally minimal, so that cats will more easily take the bait, for example outside the seabird breeding season and/or immediately after a rat eradication attempt. Poor timing is one of the primary causes of failure along with inappropriate choice of eradication methods.
- Separate control periods by intervals of several

weeks of no control in order to avoid the development of aversion behaviour in the cats; but avoid excessively long periods without control that could lead to an increase in cat numbers.

- Make sure food scrap from picnics, restaurants or dumpsites is properly disposed of, buried or incinerated, and is not available to cats. This will help to limit the population growth as any source of readily available food will enhance cat survival and breeding success.

Poisoning

- Sodium Fluoroacetate (Pestoff 1080) is the most recommended toxicant to poison cats. It has been used successfully in many cat eradications in Seychelles and around the world. It is absorbed and transformed into a complex substance which interferes with normal cell functioning, and is transmitted via the central nervous system to the heart and lungs.
- 1080 is extremely toxic to mammals including humans, and birds are also at high risk. Hence extreme caution must be exercised to ensure that non-target native species do not have access to the bait or to contaminated equipment. A promising new toxin, PAPP, which is more specific than 1080, has been successfully tested in New Zealand[52].

- Use protective gloves and clothing when handling concentrated solutions of 1080, as well as bait and contaminated equipment, as skin absorption of very small quantities may cause poisoning. As of 2014, despite some positive trials conducted in the past to develop an antidote 53, 54, and the successful use of acetamide by veterinarians to save poisoned dogs, there is no commercially available antidote.
- Place baits at intervals of 25 to 150 m depending on cat density and the protocol chosen. Place commercial cat bait stations directly on the ground, or use home-made feeding tables to limit access by crabs and other non target species.
- Plan to spend a few days trying a variety of baits. These might include small cubes of fresh fish or a mixture of canned tuna oil with rice. Then habituate the cats by pre-baiting using the preferred (non poisoned) bait for at least a week. Add 1080 and conduct poisoning during sessions lasting 5 to 10 days.

Open feeding tables can be used only at night, when native Pied crow nor other non-target species can take the poisoned bait. Gérard Rocamora

- Use 0.05% 1080 gel at a dosage of c. 0.5 g/10 g of bait. Mix the toxin with the preferred bait using a special syringe that allows quick and precise refill of exact doses. Alphachloralose can also be used for control, but it is not recommended for eradication.
- Place the bait at dusk and collect it before dawn to avoid any poisoning of non-target species such as birds. All uneaten poisoned bait should be removed after each session and disposed (buried or incinerated) so that it is not available to other wildlife or humans.
- Keep 1080 gel bottles in a protective metal case during transport, and preferably store in a cool room.
- Collect and keep records of all dead carcases and dispose of them by burying or incineration.

Trapping

- Trapping can complement chemical methods on large islands, but may also achieve eradication if pursued with sufficient intensity on islands that are small and with easily accessible terrain.
- Cage traps (for single or multiple capture), and paw-traps or leg-hold traps (when no non-target species are at risk), are available commercially from New Zealand, Australia and Europe. Many different types of traps are available[55-57].
- Recommended traps are Victor size 1½ leg-hold traps (see p. 121). Soft-jaw models have a rubber insert to prevent open injuries to the animal's leg. However, these more humane traps (compared to hard-jaw models) are less effective for eradications, as animals have a greater chance of escaping.
- Where ground non-target species are at risk, Victor

traps can also be used grouped by three into a wooden box of c. 1 m^2 with a chimney that allows access to cats only.

- Leg-hold traps (Gin, Victor, etc.), hard or soft jawed, can be installed in any of the following ways: individually as a baited walk-in set (also called 'blind' or 'cubby'); with the cat entrance restricted to one side, bait nailed at c. 25-30 cm above ground and 15-25 cm horizontal distance from trap); or as a double walk-through set (two traps at a distance of c. 1 m from each other that cats can enter from two opposite directions). Traps are camouflaged with a light sprinkling of dry soil or leaf litter; with triggers positioned on the sides.
- Cage traps have been used with mixed success. They can be useful when other traps cannot be used because of the presence of non target species, or in inhabited areas where animals need to be captured alive. The Havahart 1089 is a recommended model.
- In low density areas, or when trying to capture the last surviving individuals, trapping efficiency can be increased by using captive animals (cats or potential prey), lures (e.g. cat call device, cat urine or faeces conserved in glycerine), or devices that stimulate cat curiosity.

- Create clean paths of c.1 m wide across the whole island, and 'track pads' (patches of sand or soil of c. 1 m^2) every 150 m to position the feeding (poisoning) tables and/or traps, and read tracks (see picture).
- Include in your trapping protocols a pre-feeding period to test the effectiveness of various baits (tuna oil, fresh fish, industrial cat food pellets, raw fish, fried fish, etc.) and possible difficulties likely to be encountered with non-target species. Use a variety of bait(s) other than the ones that you use for poisoning in case some surviving cats have developed an aversion to the initial baits.
- Pay most attention to maintaining and setting traps, as any animals escaping from poorly maintained or set traps are likely to become shy and put in jeopardy or considerably extend the duration of the operation.
- Selection of trapping sites and the setting of traps are both critical. This must be done, at least initially, by a person with previous experience. Traps are normally placed along maintained tracks, open areas and habitat borders (ecotones), and attached with two strings.

- Check traps every morning and evening. Do not stress trapped cats unnecessarily; do not treat trapped cats cruelly or leave them in harsh conditions.
- Wear gloves when handling traps and equipment to minimise human scent; rinse traps with water and put them out for at least a week before use.
- Keep records of weight, age, sex and reproductive status of all animals caught, and GPS geographical references of all traps, routes and any interesting observations.
- Systematically record information about the impact on non-target species (date, numbers, locations, etc.).

Euthanasia

- Feral cats caught in traps need to be handled by experienced persons. Those in leg-holds can be shot in the head from a short distance with a silenced rifle (e.g. .22), and their carotids cut to ensure death[59]. Trapped cats, including those in cage-traps, can be eliminated humanely by initial injection of an anaesthetic, e.g. cocktail of Medetor (medetomidine chlorhydrate) and Zolétil (tilétamine + zolazépam), or Xylazine/ketamine using a 'peashooter' (blow pipe or sarbacane) or injection pole, followed by a lethal injection of barbiturates such as sodium pentobarbitone (at 150 mg/kg)[60]. This should be done by a veterinarian or a trained conservation officer; amd these controlled drugs need to be kept under lock and key.
- Carbon dioxide (CO_2 at more than 60% of volume) from gas cylinders can also be used in an appropriate chamber. The cage containing the cat can be put into the chamber or into a large plastified bag with an opening at the top to let residual air escape. This method induces a rapid anaesthesia and subsequent death within 5-8 minutes, but this time may be extended to15-20 minutes for young cats[61] (see "Euthanising invasive animals", p. 127).
- Alternatively, trapped cats can be held with an animal control pole (a cable noose mounted on an 80 cm long stick), taken out of the cage-trap, stunned by a heavy blow to the head with a stick, and then exsanguinated.

Shooting

This method has been employed in many eradication around the world, to complement efficiently poisoning or trapping.

- Do not employ shooting as a primary method of eradication (as hunted animals may rapidly become very shy); but use it instead to eliminate the last individuals.
- Cats may be attracted to an open area by food bait, cat calls, or another captive cat, where they can be shot with a rifle (e.g. .303 or .22 magnum).
- Hunting at night using lower powered spotlights can be effective. Rifles should have silencers and telescopic sights, and ideally each shot should kill.

A weather resistant "Philproof Mini" cat feeding bait station. Originally adapted for dispensing 0.1% 1080 fish/polymer baits in New Zealand, it was also successfully used on Curieuse Island in combination with Victor 1.5 soft catch leghold traps. Bill Simmons / Pestoff

Biological control

Viruses can be used to control cats although eradication is unlikely to be obtained. At least 5% of the population needs to be contaminated[55, 57]. The Feline Panleucopenia Virus (FPV), also known as Feline infectious enteritis, feline ataxia, or cat plague has been used successfully on Marion Island (South Africa). Feline immunodeficiency virus (FIV) is another lentivirus that affects cats worldwide. Both viruses can be used simultaneously and may show a synergic effect[58]. This technique can be used to 'knock-down' large populations in islands large and/or of difficult access; it could be recommended on the remote islands of the Aldabra group, but probably not for Silhouette due to the danger of spreading the disease to domestic cats on nearby Mahé.

DETECTION METHODS TO CONFIRM ERADICATION

Methods to detect any remaining cats may include lack of captures in traps, but also involve searches for scat and cat latrines, footprints in sand patches or track pads, prey remains, spotlighting at night or during crepuscular hours, use of lures, and reports from staff and public. Trained dogs may also be used.

PREVENTING (RE)INFESTATION

- Cats do not normally swim across open water and are very unlikely to recolonise islands unless assisted by humans.
- Cat pets should not be allowed on inhabited islands where cats have been eradicated. On D'Arros, a few domestic cats that were present prior to the eradication were exceptionally authorised after being neutered, marked and registered.

References / Further reading

1 GISD, 2014; 2 CAB International, 2014; 3 Seabrook, 1987; 4 Peck et al., 2008; 5 Seabrook, 1990; 6 Bonnaud et al., 2011; 7 Rocamora & François, 2000; 8 Merton 1999; 9 G. Rocamora & P. Matyot pers. comm. 2014; 10 Hill, 2002; 11 Senterre et al., 2003 (KBA database);12 Rocamora et al., 2003; 13 Rocamora & Jean-Louis, 2009; 14 Martin & Pinchart, 2015; 15 M. Betts, D. Brown, A. Mazarin Constance, J. Mortimer & G. Rocamora/ ICS-IDC, pers. obs.; 16 Beaver & Mougal, 2009; 17 R. von Brandis, pers. comm.; 18 Nogales et al., 2004; 19Laboudallon, 1987; 20 Lowe et al., 2000; 21 Medina et al., 2011; 22 Nogales et al., 2013; 23 Watson et al., 1992; 24 Seabrook 1989; 25 Russell, 2011; 26 Medina et al., 2014; 27 Campbell et al., 2011; 28 Racey & Nicoll 1984; 29 Warman & Todd, 1984; 30 Merton, 2001; 31 Merton et al., 2002; 32 Climo, 2004a; 33 Climo 2004b (report not available); 34 Climo & Rocamora, 2006; 35 Roland Nolin, pers. comm.; 36 G. Rocamora/ICS-Pangaea expedition, pers. obs.; 37 G. Rocamora pers.comm.; 38 David Brown/IDC, pers. comm.; 39 Steve Hill pers.comm.; 40Aurélie Duhec, Sam Balderson & Pep Nogués/ICS pers. comm.; 41 Angela Street/ICS pers. comm.; 42 A. Duhec & Richard Jeanne/ICS, pers. comm.; 43 Wanless et al., 2002; 44 Bell, 2002; 45 Soubeyran et al., 2011; 46 Chapuis et al., 1994; 47 PII 2011b; 48 Rayner et al., 2007; 49 Bonnaud et al. 2010; 50 Le Corre, 2008; 51 Russell et al., 2009; 52 Murphy et al. 2011; 53 Cook et al., 2001; 54 Biomarkers http://www.scoop.co.nz/stories/GE0204/S00020.htm; 55 Orueta & Ramos 2001; 56 Orueta 2003; 57 Orueta 2007; 58 Courchamp & Fugihara. 1999; 59 Close et al. 1997; 60 Reilly 2001; 61 Sharp & Saunders 2005

ONE OF
WORLD'S WORST
100
INVASIVE
SPECIES

Feral (European) rabbit.
Florent Pouzet/Biotope

Black-naped (Indian) hare.
Jorge Orueta

Feral (European) rabbit & Black-naped (Indian) hare

Lapin sauvage & Lièvre à collier noir (indien)

Lapen maron & Lyev maron

Oryctologus cuniculus & Lepus nigricans

IDENTIFICATION AND BIOLOGY

Feral rabbits and Black-naped hares are medium size mammals that measure 30-50 cm and 40-70 cm in length, and weigh up to 2.5 and 7 kg respectively. They have characteristically long ears, large eyes, long back legs and a short tail[1,2]. The hare is larger, with a slimmer silhouette, longer ears and grey-brown fur with a black patch on its neck, whilst the feral rabbit has variable fur (white, brown, grey, black)[3,4]. Both species are nocturnal but can also be seen during the day feeding mainly on grasses and herbs and also on young shoots of bushes and germinating seeds. Rabbits are extremely prolific, they may have up to 5-7 litters (of up to12), and 25 to 30 offspring per year, and they can breed when aged 3 to 4 months (gestation period of 29-35 days)[1] The Black-naped is much less prolific; it has a sexual maturity of c. 1 year, and1-4 offspring per litter (gestation period 41-47 days)[2]. Normally silent, they produce tiny sharp calls, especially when inside their underground burrows.

ORIGIN

The Feral rabbit originates from Southern Europe & North Africa where it was domesticated[5,6]. It was probably introduced to Seychelles by early settlers as a food source (as was done worldwide on more than 800 islands), and was intentionally released on certain islands or escaped from captivity[7]. The Black-naped hare originates from India and Sri-Lanka and was introduced to Cousin, via Mauritius and Java, in the 1920-1930s by Mauritian workers, also as a food

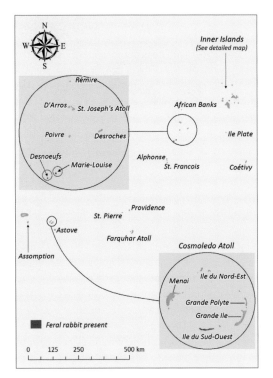

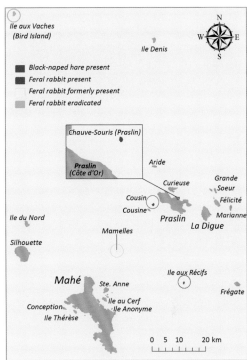

source[7, 8]. It has also been introduced to some other islands in South-East Asia and the Pacific.

DISTRIBUTION & ABUNDANCE IN SEYCHELLES

In the inner islands, Feral rabbits are found on Ile aux Récifs, Chauve-Souris (Côte d'Or, Praslin)[3, 4], and used to be present on Bird where they were eradicated in 1996[9], and on Mamelles where they died out (sometime between 1984 and 2000)[10, 11]. In the outer islands they occur on Ile du Sud-Ouest (Cosmoledo Atoll), and Desnoeufs & Marie-Louise in high densities[12-14]. They had also been released onto other outer islands (including Assomption[15]) where they apparently did not successfully establish. The Black-naped hare occurs only on Cousin, in low numbers and may have declined in the last decades (max. 170 individuals for 27 ha in the 1980s, down to c. 30 in 2001) with the increase in woodland cover and subsequent reduction in open grassy areas[10, 16, 17].

IMPACTS AND THREATS

The European rabbit is considered one of the worst known invasive species on earth[26]. Rabbits, through excessive browsing, can inflict severe damage to native vegetation and associated ecosystems, agricultural crops, and also forestry. By destroying plant cover, they can cause extensive erosion, prevent regeneration, and impact the habitat of native animals including ground invertebrates and soil microfauna[5]. The rabbit has caused extensive ecological damage and is considered a threat to biodiversity on many islands and in countries where it has been introduced, such as Australia, New Zealand, New Caledonia, Kerguelen (Sub-Antarctic French Territories), Ascension Island (South Atlantic, UK), Laysan Island (Hawaii) and Mauritius[5, 6]. Impact of rabbits has not yet been quantified in Seychelles but on islands where its density is high it is likely to significantly impact certain native or endemic plants and habitats, and associated microfauna[10]. Black-naped hares may cause damage to crops and vegetation, as reported from Mauritius where they are considered an agricultural pest. They also impact the ecosystem in general by eating young trees and bushes when other food sources are scarce. On Cousin, they affect the regeneration of *Casuarina equisetifolia* but their overall impact on vegetation appears to be limited[16, 17].

ERADICATION AND CONTROL PROGRAMMES
Bird Island

This is the only island where there has been a programme targeting eradication of rabbits in Seychelles[12]. This occurred at the same time as the rat and mouse eradications in October/November 1996 (see Black rat text, p. 192). The rabbit eradication was attempted through two hand spread applications of cereal based pellets with Brodifacoum at 20 ppm (Pestoff Rodent Bait 20R), separated by 10 days. Transect lines were spaced at 25 m intervals and bait was spread at a density of 4-5 kg/ ha (100 g scattered around the operator's feet and in four perpendicular directions every 25 m along each transect). Open areas occupied mainly by rabbits received a reduced unspecified dosage. No more rabbits were observed after 3 weeks and the programme was confirmed successful in 1998[9].

Ile aux Récifs
The Environment Department of the MEECC has been repeatedly conducting shooting, trapping and poisoning of rabbits between 2010 and 2013 during long seasonal visits with army guards. However, this was not enough to eradicate the population[27]. A proper eradication plan and follow up is required.

In Mauritius, rabbits were eradicated from Ile Ronde (Round Island; 151 ha) in 1986 by hand spreading of Brodifacoum cereal pellets at 20 ppm[18]. Black-naped hares were eradicated from Coin de Mire (Gunners Quoin, 8 ha) in 1995 (while targeting Brown rats), using brodifacoum cereal pellets at 20 ppm plus some with bromadiolone at 50 ppm, distributed over a 25 m grid and near where animals were seen. However, rabbits were released by people shortly after the eradication of hares and had in turn to be eradicated using similar brodifacoum hand-broadcasted cereal baits pellets in 1998[6,19]. Cereals baited with chlorofacinone (another anti-coagulant) 50 ppm were also used successfully against rabbits on Ile Verte (148 ha, Kerguelen, TAAF)[6].

Larger islands where rabbits have been eradicated with cereal Brodifacoum baits include Enderby (710 ha, New Zealand, 1993), St Paul (800 ha, TAAF/Sub-Antarctic French Southern Territories, 2001) and Macquarie Island (12,872 ha, 2011)[20]. In Santa Clara (Mexico), brodifacoum pellets were also used as well as burrow fumigation with aluminium phosphorus. Other rabbit control/ eradication programmes around the world were based on biological methods, such as spread of virulent laboratory strains of viruses (myxomatosis or calicivirus; or rabbit haemorrhagic

disease RHD), with fleas or mosquitoes used as the main vectors (e.g. Australia and New Zealand)[6]. On Phillip Island (200 ha, Australia), eradication started in 1979 with Myxoma virus with fleas as vectors, but following interruption was only completed in 1986 through an intensive poisoning program with 1080 (sodium monofluoroacetate) supplemented by trapping, shooting and fumigating[6].

Shooting and trapping were also necessary to eliminate the last remaining animals and achieve eradication on Coin de Mire, Ile Ronde, Ile Verte and Santa Clara (using calibre 12 or .22 rifles). In the Canary Islands, capture trials showed greater effectiveness of home-made multiple capture traps, compared to simple capture commercial traps. Various baits such as carrots, apples, bread, cereals or other fruits and vegetables have been used successfully for trapping, with pre-feeding trials to determine bait preference and attract rabbits[21]. The unfortunate introduction of alien predators such as foxes or mustelids (ferrets, stoats or weasels *Mustela* spp.) to reduce rabbit densities had dramatic consequences for the ecosystems in Australia and New Zealand. At Isla Grossa (14 ha; Columbretes, Catalonia-Spain), eradication was achieved in 100 days after destruction of 175 rabbits mainly by shooting, plus use of nooses, bow and arrows, and various traditional trapping methods[21,22,23].

Rabbit eradication is normally followed by a spectacular recovery and change in vegetation composition, as evidenced in the numerous case studies around the world.

More details and references on rabbit control and eradication programmes conducted elsewhere can be found through the Global Invasive Species Database www.issg.org/database/welcome and the Invasive Species Compendium database www.cabi.org/isc/. No control/eradication information is available there for the Black-naped hare.

MANAGEMENT RECOMMENDATIONS
Control and eradication protocol
- Eradication plans should start with an effective chemical or biological 'knock-down' method to decimate the whole population, and if necessary

Multiple-capture rabbit trap.
Jorge Orueta

continue with various methods to catch surviving rabbits, such as trapping, shooting or use of trained dogs. This strategy is particularly important for islands that are large and/or have difficult terrain.

- Cereal pellets with Brodifacoum at 20 ppm, spread by hand or by helicopter, which can be used at the same time to eliminate rodents, are preferable to the use of haemorrhagic viruses or other biological agents that may present a risk to rabbit farming.

- Trapping protocols should include a pre-feeding period to test the effectiveness of possible baits (bread, fruits, vegetables)[21]. Cage traps (for single or multiple capture) and paw-traps (used when no non-target species are at risk) with a rubber protection to prevent open injuries to the animal's legs are available commercially from Australia or Europe.

- Trapping, netting at burrows, and shooting can complement chemical and biological methods on large islands; but these may also achieve eradication if used sufficiently intensively on islands that are small and with easily accessible terrain. Cage traps should be preferred when non-target species are at risk from other methods[21, 22, 23].

- Trained dogs can also be used to capture rabbits, or simply to locate burrows from which rabbits can then be eliminated through netting or fumigation.

- When complete eradication is not the objective, effective control can be achieved by combining the previous measures (trapping, shooting, use of trained dogs).

- Shooters should aim at the centre of the head, between the eyes (in front view), or at a point between the eye and the base of the ear directed towards the opposite eye (in side view).

- Biological control with viruses or RHD has not proved satisfactory and is not recommended, especially when eradication is attempted[6, 21], and nor is obviously the introduction of any alien predator as this would have dramatic consequences against native species[22, 23].

Euthanasia

- Trapped rabbits can be humanely killed easily by stunning with simultaneous (or subsequent) cervical dislocation to ensure immediate loss of consciousness and quick death[24]. This involves a very strong blow with a stick or closed fist to the back of the skull, behind the ears, while the rabbit is held upside down by its two hind legs (to be done by an experienced operator). Ex-sanguination by cutting the main blood vessels around the throat (carotid and jugular veins) may then be done. This is especially appropriate for animals suitable for consumption. Chemical methods (carbon dioxide, lethal injection; see "Euthanising invasive animals", p. 127) are also possible but rarely used in the field[25]. Animals should not be consumed if they have been treated with poisonous substances.

PREVENTING (RE)INFESTATION (SMALL ISLANDS)

- Rabbit farming should not be allowed on islands where they may escape and spread, especially where the species has been eradicated. Existing farming facilities should be discontinued and dismantled.

- The main concern about the presence of the Black-naped hare on Cousin is the risk that some animals may be transferred to other islands, although this is unlikely to happen on a Special Reserve.

References / Further reading
1 Macdonald & Barrett, 1993; 2 Ludrigan & Foote 2003; 3 Bowler, 2006; 4 Hill & Currie, 2007; 5 CAB International, 2011; 6 GISD 2014; 7 Racey & Nicoll, 1984; 8 Kirk & Bathe, 1994; 9 Merton *et al.*, 2002; 10 Nevill, 2009; 11 confirmed by G. Rocamora, *pers. obs.* 2012; 12 Beaver & Mougal, 2009; 13 Rocamora *et al.* 2003; 14 ICS, *pers. comm.* 2014; 15 J. Mortimer/ICS, *pers. comm.*; 16 Kirk & Racey, 1992; 17 Dunlop *et al.*, 2005; 18 Merton 1987; 19 Bell, 2002; 20 DIISE, 2014; 21 Orueta & Ramos, 2001; 22 Orueta, 2003; 23 Orueta, 2007; 24 Sharp & Saunders, 2005; 25Reilly, 2001; 26Lowe *et al.* 2006; 27 R. Fanchette, pers. comm (no reports available).

Adult tenrec.
Gérard Rocamora

Common tenrec

Tangue, Tenrec, Hérisson malgache

Tang

Tenrec ecaudatus

IDENTIFICATION AND BIOLOGY

The Common tenrec is a fat, largely nocturnal, medium-sized ground mammal (up to 2 kg) with the appearance of a hedge-hog, having uniform light brown hair and a long snout[1]. It is the only species of tenrec present in Seychelles. The young have a darker fur with five parallel rows of light coloured quills on their backs. The females are significantly more nocturnal than males and young. Tenrecs are harmless and defenceless except for the presence of strong white spines arranged longitudinally on the backs of young, replaced by a crest of rigid hairs in adults[2]. Tenrecs forage in woodland litter, from lowland plateaux to mist forests, also in gardens, agricultural fields and residential areas. Omnivorous, they feed mainly on invertebrates, small ground vertebrates and fallen fruits[2]. They have an acute sense of smell but poor eyesight. Although usually silent, they may produce rustling and plaintive calls[3,4]. In young animals, the dense quills of the central row act as a stridulating organ that produces a sound varying in frequency from 2 to 20 kHz[2]. The species aestivates during the driest periods, usually from June to September. It is the most prolific mammal on earth, with litters of up to 32 per female[1] although in Seychelles these are normally smaller (15 to 19)[5]. Females have up to 29 teats and the gestation period is 50 to 64 days[12].

ORIGIN

Madagascar. Common tenrecs were introduced to Seychelles probably as a game animal around 1880 via the Mascarenes[1,2]. It is still hunted there and in the Comoros archipelago for food, but this is no longer the case in Seychelles. This species has only been introduced in the Comores, La Réunion, Mauritius and Seychelles.

DISTRIBUTION & ABUNDANCE IN SEYCHELLES

Common tenrecs are found on Mahé, Thérèse, Praslin[6, 7], and also on Anonyme[8]. They are commonly observed at night, but only occasionally during the day-time, alone or in family groups. They are abundant in wooded habitats, including uphill forests and residential gardens; but they are uncommon on the lowland sandy plateaux[2].

IMPACTS AND THREATS

The impact of Common tenrecs has not been quantified, but it is likely to be significant with regards to native and particularly endemic ground invertebrate species, such as the Giant millipede (absent from all islands where tenrecs are present). They also impact small native vertebrates such as lizards, geckos and frogs, in particular the endemic caecilians and burrowing skinks which are more vulnerable. They are considered a pest in gardens as they destroy seedlings while foraging.

ERADICATION AND CONTROL PROGRAMMES

There have been no documented programmes to control or eradicate them from islands in Seychelles. However, tenrecs are regularly found trapped in rat cage-traps baited with coconut and they do feed on rodenticide cereal pellets or blocks, as observed on Anonyme. Where rats are controlled using bait stations tenrecs have access to, tenrec densities appear to have been reduced. This is the case at the Seychelles white-eye main breeding areas on Mahé between La Misère and Barbarons, and at Anonyme Island where rodenticide bait stations are used to prevent recolonisation by rats. Dogs help to control tenrec populations in residential and agricultural areas by killing large numbers of them. Young may be taken by cats and Barn owls.

At La Réunion (e.g. at Réserve Naturelle de la Roche Ecrite) and in the Comoros archipelago where the species is considered a delicacy, Common tenrecs are killed by local hunters, and are probably controlled at least in the most intensively hunted places where the species has become rare (e.g. in Mayotte[9]). In Mauritius, where residents also hunt them for food, Common tenrecs are a regular non-target catch for cat and mongoose traps set up by Mauritian Wildlife Foundation in predator controlled areas. However,

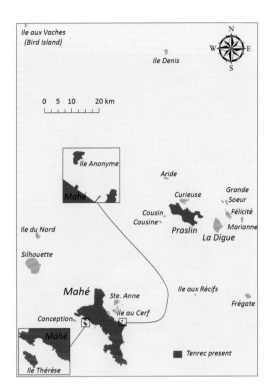

because tenrecs tend to be 'trap-happy' and are caught repeatedly after being liberated, they need to be controlled to maintain a high rate of trapping efficiency[10]. No control or eradication programmes are documented for this species on the Global Invasive Species Database.

MANAGEMENT RECOMMENDATIONS
Control and eradication protocol
- Cage traps baited with fruits or roasted coconut may be used to capture tenrecs and reduce their densities in sensitive areas important for native invertebrates and small ground vertebrates (lizards, caecilians, frogs).
- There are no specific traps for tenrecs but they can be caught with a variety of cage traps (rat traps; wooden box traps and tomahawk traps for cats).
- Home-made bait stations with a large entrance (PVC pipe of 5 inches of diameter; see Rat sections) baited with waxed cereal brodifacoum blocks at 20 ppm to control rats will help to reduce tenrec densities.

Euthanasia
- Trapped tenrecs can be humanely killed in the same way as rats, through physical methods such as stunning (along with ex-sanguination in neighbouring countries where they are a source of food)

A litter of young tenrecs.
Sarah Caceres & JN Jasmin

or chemical despatch (inhalation of carbon dioxide from commercial cylinders in a special chamber, or lethal injection of sodium pentobarbitone; see "Euthanising invasive animals", p. 127)[11].

PREVENTING (RE)INFESTATION (SMALL ISLANDS)

Tenrecs are not good colonisers and are unlikely to establish themselves on new islands without assistance. Intentional introduction to islands where they do not occur must be strongly discouraged. Application of contingency measures for rodents will suffice to prevent tenrec infestation in all currently rat-free islands.

References / Further reading

1 Vololomboahangy & Goodman, 2008.; 2 Racey & Nicoll, 1984; 3 Bowler, 2006; 4 Rocamora & Solé 2000; 5 Nicoll & Racey 1985; 6 Hill & Currie, 2007; 7 Nevill, 2009; 8 Rocamora *et al.*, 2005; 9 G. Rocamora & Soufou Said, *pers. obs.* 2002-2012; 10 Vikash Tatayah, *pers. comm.*; 11 Sharp & Saunders, 2005; 12 Nicoll, 1984.

Tenrec trap.
Vikash Tatayah / Mauritius Wildlife Foundation

Contributed by **Wilna Accouche** & **Nancy Bunbury** - Seychelles Islands Foundation.

ONE OF WORLD'S WORST 100 INVASIVE SPECIES

Adult goats on Aldabra. SIF

The feral goat (*Capra hircus*) has been described as one of the most destructive herbivores introduced to islands of the world[1]. Goats were introduced to Aldabra prior to 1878 and by 1929 'thousands' were reported on the atoll[2]. During the Royal Society expedition in 1967-1968, goats were reported on all four major islands of Aldabra (Grande Terre, Picard, Malabar and Polymnie) and on a smaller island in the lagoon (Ile Esprit)[3]. Goat numbers were estimated at 500–600 individuals in 1976-77, and at c.1300 in 1985[4]. Goats were no longer present on the smaller islands of Polymnie and Ile Esprit by 1976-77[5], presumably due to localised control. Concern that goats could potentially threaten the endemic biota of Aldabra prompted UNESCO and the Seychelles Islands Foundation to conduct a control program in January–March 1987 and in 1988[4]. It was found that goats were altering species composition and slowing regeneration of natural vegetation as well as reducing shade cover and forage for the Aldabra giant tortoises (*Aldabrachelys gigantea*)[5, 6, 7]. A total of 883 goats were eliminated from Picard, Malabar and Grande Terre, possibly up to 70% of the initial population[4]. This was followed by opportunistic hunting of goats by SIF staff.

An SIF ranger shooting at a goat.
Phillip Haubt

A dart containing an anaesthetic is used to capture goats.
Andy Gouffé

A Judas goat is equiped with a radio-transmitting collar.
SIF

In 1993, an eradication program using the 'Judas goat technique'[8], whereby 28 goats were captured and equipped with a radio transmitter, was employed and by 1995, a total of 832 goats were shot (using 0.243 calibre rifles with 2-7× telescopic sights). As a result, goats were eradicated from all islands except Grande Terre where it was estimated that 84 remained in 1995[9].

A follow-up effort was conducted in 1997 where a further 106 goats were culled on Grande Terre[9]. However, a few goats still remained on eastern Grande Terre as staff reported goat sightings in this area shortly after the eradication campaign. Between 2000 and 2005 c. 250 goats were shot on an opportunistic basis using conventional hunting methods and by 2005, it was estimated that c. 100-200 goats remained on eastern Grande Terre[10].

Given the low density of the remaining goat population, a final eradication programme, applying recent advances in Judas goat techniques, was planned in 2007[10] . Judas goats were captured, sterilised, fitted with radio-transmitting collars and released. Females were injected with hormone implants to make them more attractive to other males. As goats are gregarious, Judas goats seek out and associate with other goats. They can then be tracked and located and any associated non-Judas goats are removed. This technique increases the efficiency of eradication programs by reducing search time for hunters and locating remnant herds. A total of 202 goats were eliminated during the first phase of this eradication program between August and December 2007 by a team of experienced hunters and a wildlife veterinarian[10]. The team spent 1,153 hours walking 2158 km in the field and used GPS/GIS technology to ensure complete coverage of the island. Twelve Judas goats were released during this period but four of these were later eliminated to induce more effective searching behaviour in the others.

Since the end of this intensive phase in late 2007, monitoring was conducted on a monthly basis by SIF staff on Aldabra to locate any remaining goats[12]. Judas goats were located using radio telemetry, all non-Judas goats were eliminated and Judas goat associations closely monitored. Other potential areas were also searched for signs of goats. The last non-Judas goats were shot in March 2010, and the remaining Judas goats had to be recaptured for their radio transmitters to be replaced so that they could continue to be monitored[12]. With funding from the

European Union, the Seychelles Islands Foundation was able to finalise this important project and confirm eradication. Intensive searches were conducted and Judas goats regularly monitored. In December 2011, after more than one year since the last non-Judas goat was sighted, an aerial helicopter survey located only four Judas goats[12]. In January 2012, two professional hunters spent two months on Aldabra, covering over 1,000 km on foot, and found only the Judas goats, which were all shot except one, which was left alive in a final attempt for it to seek other goats. The last Judas goat was still alone by July 2012 and was shot on 3rd August 2012, marking the end of 25 years of eradication efforts during which 2,344 goats were culled on Aldabra[12]. It was a long-

Using the telemetry equipment to locate Judas goats.
SIF

term and expensive project, and commitment to follow-up monitoring and intensive training of local staff have been key ingredients to its success. Finally, the threat posed by feral goats to Aldabra's ecosystem has been completely removed.

References / Further reading

1 GISD, 2014; 2 Stoddart, 1981; 3 Stoddart, 1971; 4 Coblentz et al.,1990; 5 Gould & Swingland, 1980; 6 Coblentz & Van Vuren, 1987; 7 Bourn et al., 1999; 8 Taylor & Katahira, 1988; 9 Rainbolt & Coblentz, 1999; 10 von Brandis, 2007; 11 Beaver & Mougal, 2009; 12 Bunbury et al., 2013; 13 Jeanne Mortimer/ICS, pers. obs.; 14 Mike Betts/ICS, pers. obs.; 15 crew of boat 'Serenity', pers. comm. to V. Laboudallon; 16 Pigott, 1961; 17 G. Rocamora & ICS-CORDIO expedition, pers. obs.; 18 Nevill, 2009.

Map of Aldabra atoll and its main islands.
SIF

FERAL GOATS AND OTHER LARGE UNGULATES IN SEYCHELLES

Feral goats still occur on at least one or two islands in Seychelles. They were still present on Astove (March 2015)[13, 14] and possibly on Saint Pierre (March 2009)[15]. Goats were reported last time on Cosmoledo (Ile du Nord-Est) in 1961[16] and are no longer present[13, 17]. The only other large feral mammal present in Seychelles is the Feral pig, still present on Astove (March 2015)[14]. Feral pigs were also present on Menai (Cosmoledo) until 1999[14]; but had died out by 2002[17]. Small populations of large feral or semi-feral domestic ungulates (cows, pigs) also occured in some inner islands (e.g. Frégate, Ile du Nord; Cousine)[18], but were removed or died out (see p. 62).

Barn owl captured at Sainte Anne Island.
Andre Dufrenne

Barn owl

Effraie des clochers

Ibou blan

Tyto alba affinis

IDENTIFICATION AND BIOLOGY

The Barn owl is a large nocturnal or crepuscular bird of prey with a light brownish plumage, a white facial disc and pale underparts. Juveniles have a similar plumage, with darker upperparts and slightly golden brown underparts[1]. It nests in tree cavities, old buildings and rocky cliffs (mainly March-August on Aride). Loud shrills are uttered in flight, while young produce typical strong hisses 'Pssst'[2]. It hunts birds, rodents, and small reptiles in forest openings, gardens and alleys with large flying paths.

ORIGIN

Sub-Saharan Africa. It was introduced from East Africa to Plate Island as a trial (3 birds) in 1949; then to Mahé in 1951-1952 (27 birds) by Government agriculture services to help control rats[1,3].

DISTRIBUTION & ABUNDANCE
IN SEYCHELLES

Barn owls breed on the four main granitic islands where they are common (Mahé, Praslin, Silhouette and La Digue)[1]. They are also found Ste Anne, Ile aux Cerf, and Curieuse (possibly breeding); and they are reported as visitors on the smaller islands of the Ste Anne group, Anonyme, Conception, Thérèse; Ile du Nord; Aride, Cousin, Cousine, Félicité, Grande Soeur and Marianne; and very occasionally Frégate and Bird[3-5, 18, 19]. Highest densities may occur on islands with seabird colonies, such as Aride (73 ha) where 20 birds were eliminated in 1996[6]; the species abundance is normally much lower in rat infested islands and residential areas.

IMPACTS AND THREATS

The Barn owl preys on native seabirds, particularly Fairy terns and other white terns. Although an isolated individual that reaches a seabird island may not have a significant impact on abundant species, its impact on rare or declining seabirds such as the Roseate tern *Sterna dougalii*, can be significant. In 2006 and 2007, over 50 and 20 Roseate terns, respectively, were killed and partly eaten in the colony at Aride[7] (the owls prefer to eat the head and breast muscles, and often discard the rest of the carcass). Barn owls are believed to capture Bronze geckos *Ailuronyx seychellensis*, and also pose a threat to rare endemic landbirds such as the Seychelles magpie-robin *Copsychus sechellarum* or the Seychelles white-eye *Zosterops modestus*[8].

ERADICATION AND CONTROL PROGRAMMES
Aride Island

A breeding population estimated at 20 birds (min. 8 pairs) was eliminated in 1996 when 16 were killed or found dead[6]. Another 26 were killed between 1996 and 2002, when the last suspected nesting attempt was recorded[7]. During this period, one nestling was reared and kept captive as a decoy until it died in 2001. This decoy was very efficient at attracting incoming birds that were then shot. Despite regular reinvasion of birds coming alone or in small groups from Praslin, re-establishment of breeders has been prevented through regular monitoring of the 18 known former nests or roost sites. Between 2003 and 2009, this resulted in another 20 owls killed or found dead[7]; and between 2010 and 2013, another 14 were killed or found dead (1, 2, 7 and 4 in 2010, 2011, 2012 and 2013, respectively[9-12]).

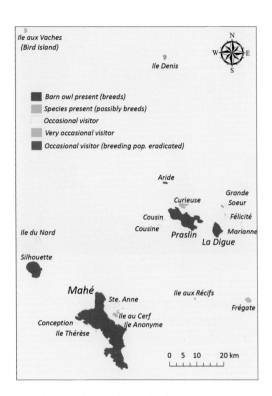

Cousin, Cousine, Frégate, Bird

Like Aride, these islands which also host seabird colonies have ongoing protocols to eliminate any invading Barn owls[3,4]. Every year, at least some invaders are killed by shooting or trapping on Cousine and Cousin, but incursions are less frequent on Bird and Frégate that are more remote.

Ile du Nord

Barn owls disappeared after the rat eradication attempts of 2003 and 2005, apparently having died (several corpses were found) as a result of secondary poisoning or starvation due to scarcity of prey. Single birds were found in 2006 and 2009 (considered as incursions), dead or dying[13-15].

Grande Soeur

Resident Barn owls appear to have vanished after the rat eradication of September 2010. One owl was found dead; and another was briefly heard calling in 2011. The latter may have been visiting from neighbouring Félicité, and may have either gone back or died.

No control or eradication programmes are documented for this species on the Global Invasive Species Database www.issg.org/database/welcome/

A noose carpet (bal-chatri type) trap with a skink as a live bait. Gérard Rocamora/ICS

MANAGEMENT RECOMMENDATIONS
Control and eradication protocol
Following are recommended protocols for small islands:
- Survey the island during the day and search for evidence of roosting or nesting activities at potential territorial sites which may include pellets, feathers, or concentrated guano areas.
- At night or at dawn, play recordings to attract owls for a maximum of 20 minutes of tape-luring at each site. Use a torch and tape-recorder or CD player to survey potential hunting habitat.
- Priority must be given to sites with rare breeding seabirds (e.g. Roseate tern colony).
- Use noose carpets (loops of nylon fishing line tied to a wire mesh – also called Bal-chatri) placed on top of regular perching palls or roosting sites, or over a box containing a live prey (e.g. a mouse) to capture the owls. Larsen traps (a cage with two or more compartments including a decoy bird to attract other individuals of the same species) can also be used.
- Use a .22 rifle – preferably not an air gun – to shoot at any Barn owls located, and destroy any nests found. Aim for the brain at short range (<20 m), or at the chest from a distance.
- A government bounty has existed since 1969 (currently SR50 per Barn owl, equivalent to €3 per bird). But the small numbers of owls killed annually make it an inefficient control measure for populations on large islands. Additionally, the bounty puts at risk the two endemic raptors, Seychelles scops owl and Seychelles kestrel, both of which have been mistakenly killed in the past.

Euthanasia
To euthanize Barn owls, the following protocols are recommended:
- Carbon dioxide from a gas cylinder can be directed into a special chamber containing the owl. This method induces a rapid anaesthesia and death within 5-8 minutes.
- Barbiturates, such as sodium pentobarbitone, can be injected by a Veterinarian or a specially trained officer.
- Shoot captured birds at close range.
- Induce concussion by striking a heavy blow to the back of the head by means of a hard stick (4-5 cm in diameter).
- See also "Euthanising invasive animals", p. 127 and selected references[16, 17] for more details.

Preventing reinfestation
To prevent reinfestation of small islands, the following protocols are recommended:
- Eliminate new invaders immediately to prevent establishment of a breeding population, using rifle, noose carpets, mist-netting or tape-luring.
- Keep a police-licensed .22 rifle in good working condition to be used only by licensed conservation staff or police marksmen.
- Strategically place perches in open habitat where incoming owls can perch and be easily seen and shot.
- Painting perching spots in white or a light colour may be attractive to the owls.
- Raise a captive Barn owl chick and keep it as a decoy to help attract any new incomers.
- Map all previously known and new roosting or nesting sites, and keep casual records of owl activity.
- Visit all known potential territories, including former nesting and roosting sites, at least once a year.

References / Further reading
1 Skerrett *et al.* 2001; 2 Rocamora & Solé 2000; 3 Nevill 2009; 4 SBRC 2014; 5 Present work; 6 Malcom Nicoll in Betts 1996; 7 Yeandle 2009; 8 Rocamora & Henriette-Payet 2009; 9 Sutcliffe *et al.* 2011; 10 Calabrese & Maggs 2012; 11 ICS 2013; 12 ICS 2014; 13 Climo & Rocamora 2006; 14 Rocamora & Jean-Louis 2009; 15 Beaver & Mougal 2009; 16 Reilly 2001; 17 Sharp & Saunders 2005, 2008; 18 Rocamora & François, 2000; 19 V. Laboudallon *pers. comm.*

ONE OF WORLD'S WORST **100** INVASIVE SPECIES

Common myna on an African tulip tree. S. Caceres & J.-N. Jasmin

Common myna

Martin triste

Marten ordiner

Acridotheres tristis

IDENTIFICATION AND BIOLOGY

The Common myna is a medium size blackish bird with yellow bill and eye patch and legs, and a white wing patch in flight. Juveniles have a paler and duller plumage compared to the adult, with brown throat and breast, and a dark-brownish head[1,4]. It produces loud whistles and rattled calls from before dawn to dusk, especially when concentrated in large noisy flocks; and it may imitate other birds[2]. Common mynas nest in pairs in buildings, coconut crowns and tree cavities, mainly between October and April, producing broods of 2-4 fledglings. This generalist and opportunistic bird is found in all habitats between sea level and high altitude secondary forests, but mainly around houses, coconut plantations and open farmland[1,3]. It is omnivorous (feeding on insects, spiders, seashore invertebrates, fruits, berries, grain, pet food, bread, cooked rice and human food waste, etc.), occasionally takes eggs and small vertebrates such as birds and lizards, and scavenges at rubbish dumps.

ORIGIN

Indian sub-continent, China & South-East Asia[4,5]. It was introduced to Mahé from Mauritius, probably at the end of the 18th century, apparently to help control large insects impacting crops[6]. It has been introduced to many tropical, subtropical and temperate regions of all continents except Antarctica and South-America, and many islands in the South Atlantic, Pacific Ocean and Indian Ocean[4,5].

DISTRIBUTION AND ABUNDANCE IN SEYCHELLES

Common mynas breed on all four main granitic islands (Mahé, Praslin, Silhouette and La Digue) where they

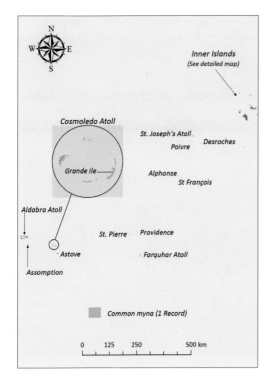

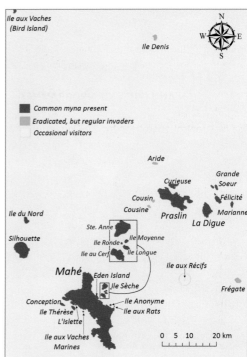

are very abundant, and are also common on most inner islands of more than 10 ha[7]. On Frégate, of 219 ha, a total of 745 birds were eliminated during the 9 months of the eradication (see next case study) hence an average density of 3.4 birds/ha. On Denis Island, of 143 ha, numbers were estimated at about 1,000 birds; hence an average approximate density of 7 birds/ha[40]. On the larger islands, evening roosts can host thousands of birds. Only occasional individuals are encountered at Aride, Cousin, Cousine, Frégate (following their eradication from these islands), and on Ile aux Récifs. They also breed on the coralline islands of Bird, and eradication was confirmed in July 2015 on Denis. There has been only a single record in the outer islands, at Cosmoledo, in November 2008[8].

IMPACTS AND THREATS

Common mynas damage fruit crops, prey on eggs and nestlings of native and introduced birds[4, 5], and compete with native cavity nesters, especially the Seychelles magpie-robin *Copsychus sechellarum*, the Seychelles kestrel *Falco araea*, the Seychelles scops-owl *Otus insularis*. Their impact on these rare and globally threatened species (including the Sey-chelles white-eye *Zosterops modestus*) can be high[9]. They also disturb nesting seabirds and take the eggs of some species. Fiercely territorial, mynas have

Head wound on a Seychelles warbler on Denis Island attributed to Common mynas.

Jildou van der Woude / University of Groningen

been reported on Frégate Island to chase Seychelles magpie-robins from their nest boxes and sometimes kill them. On Denis Island, up to 5% of the introduced Seychelles warblers *Acrocephallus sechellensis* were found with serious head wounds; and observations suggested these were caused by myna attacks (se picture)[10, 41]. Mynas are considered a nuisance by the tourism industry for stealing food in restaurants, and because of their droppings and associated risks of spreading diseases to humans (including mites, der-matitis, asthma, and severe skin irritations). Mynas are reported to carry avian malaria, Psittacosis and

Ornithosis (Chlamydophila), Salmonellosis and arbo-viruses (currently under investigation in Seychelles)[4].

ERADICATION AND CONTROL PROGRAMMES
Frégate Island
The first control operations for the Common myna at Frégate were initiated by BirdLife International between 1993 and 1997, when 326 birds were shot by police marksmen. Larsen traps, mist nets and use of alphachloralose had been previously tested with very little success. Between 1998 and 2002 conservation staff shot another 394 birds, and also used nest trapping to kill 12 pairs, leaving possibly no more than 7-8 individuals[11]. Numbers rose again to hundreds of pairs, but between June 2010 and February 2011, Frégate Island Ecology department managed to eradicate the species[12], using MiniMyna® Australian traps (724 birds caught), walk-in traps (5 birds), shooting (4 birds), and nest-box traps (3 adults, 9 chicks and 42 eggs destroyed) (see following case study). As of 2014, this was the only registered case (worldwide) of complete eradication of a large established island population of several hundreds of birds. Since then, several re-invasions have been successfully thwarted by immediately deploying baited Mini-Myna traps in the areas of activity - particularly on open areas like airstrips and lawns after mowing. There was an indication that a single exceptionally human- and trap-shy individual persisted in the wild (September 2014)[13].

Aride, Cousin and Cousine
In 1993-94, all but one of 17 mynas present on Aride were shot by police marksmen, then two more in 2001. In 2000-2002, a breeding population of 6-10 mynas was also eliminated on Cousin using a combination of a .22 airgun and nest-box traps[11]. After a period of irregular control, a small breeding population of 3 pairs was still present on Cousine in 2001[11] but was later eliminated by shooting[14, 15, 16]. Despite repeated re-invasions by small groups (1-3 birds), these three islands have managed to remain myna-free, and their management is authorised to keep air rifles to eliminate occasional invaders.

Denis Island
Eradication of mynas was first attempted at Denis in 2000-2001 under a GEF BirdLife Seychelles project, using Starlicide DRC1339 (2.5 mg/kg of bait), which

An efficient type of trap in use at Ste Anne Resort, outside view. Gérard Rocamora

produced drastic reductions in flock size of some 90% (i.e. 150 birds reduced to only 6-14) after four days of application. This was followed by shooting over a 12 day period resulting in another 26 birds killed. An estimated 70-90% of the entire population was killed and 40-60 individuals remained at the end of the operation in 2001[11]. By 2009 there were once again several hundreds of mynas, and c. 1,000 early 2010[40], and later in 2010 another eradication attempt was started by WildWings Bird Management, funded by a GEF/GIF grant. Funnel traps and mist nets succeeded in catching only 2 birds each, so their use was discontinued after two weeks. Decoy traps were highly successful and became the technique of choice. From May to August 2010 (weeks 1-14), 640 mynas were caught and between November 2010 and March 2011 a further 277 were culled. A total of 917 birds (90% of the original population) were removed from the island[17, 18]. Eradication efforts then lapsed until 2014, when the myna population had increased to over 200 birds (number estimated in August 2013)[42]. Between May and October 2014 another 129 birds have been caught with decoy traps, drop traps and ladder traps. Shooting by an experienced conservation hunter resulted in 66 mynas shot in a 4 week period, leaving by April 2015 only one known myna on the island that was shot in July 2015[19]. This is the second successful registered case worldwide (after Frégate Island) of complete eradication of a large established island population of hundreds of mynas.

Ile du Nord

During the rat eradication conducted at Ile du Nord under the ICS FFEM project in August-September 2005, mynas consumed small pieces of rodenticide cereal pellets (Brodifacoum 20 ppm). This reduced their numbers by over 50% in one month, from an estimated 890 to 470 birds (using the Distance Sampling method)[20]. A pre-existing myna eradication plan was updated and immediately implemented[21]. In 2006 and 2007, mynas were poisoned with Starlicide DRC 1339 (2.5 mg/kg of bait). Out of c. 400 mynas observed feeding, 52 were found dead. This further reduced their population to less than 100 birds[22], when risks to non target species and bait shyness became too high, and the operation ceased. Myna numbers increased again to over 250 birds after permission to use an air rifle was denied by police authorities for about 2 years. Meanwhile, nest box traps were used with little success (only 11 young and 9 eggs were destroyed in 22 months)[23]. Shooting finally took place between October 2008 to December 2009, killing 227 birds. Unfortunately, intensity and efficiency were insufficient to overcome the ability of the mynas to multiply their numbers[14]. In October 2012 WildWings Bird Management began another eradication attempt under a GEF/GIF grant; and by January 2013 over 700 mynas had been killed[18]. Unfortunately, the eradication efforts have since lapsed but may resume in 2015.

Ste Anne

A myna control operation was conducted around Ste Anne restaurant 'L'Abondance' by DoE in January 2006, using large walk-in traps. 52 birds were captured in 4 one day sessions of trapping, each separated by 4 to 8 days interval[24]. Since 2008, a couple of efficient home-built cage traps baited with bread have been capturing, on an ongoing basis, an average of 70-80 birds per month with no apparent development of trap aversion[25]. Extreme unexplained variations of the number of birds trapped per day have been recorded, such as 1 to 54 birds/day in June 2013. Between 2010 and 2013, there has been a peak of captures every year between June and August; possibly due to a scarcity in natural food and abundance in emancipated juveniles looking for food at this time of year[25]. Constant control efforts are required, as eradication is not possible due to proximity to Mahé and other infested nearby islands.

Grande Soeur

The rat eradication conducted at Grande Soeur in September 2010 reduced the myna numbers from an initial population of over 100 birds to 30-40 birds[26]. By June 2011, these had increased to 60-80 birds. In September 2011, c.4 days of poisoning with starlicide DRC 1339 reduced the population by 50%. This followed ineffective attempts using 6 MiniMyna traps, and installation of 10 nest-boxes with noose traps, and walk-in traps. Unfortunately, the operation was interrupted in November 2011, with an estimated 15-20 birds remaining on the island[26]. However, numbers did not increase in following years beyond an estimated 50-60 birds, possibly due to the competition from moorhens *Gallinula chloropus*, another ground feeding which experienced a spectacular recovery after the rat eradication (up to 30-50 birds in total). In June-July 2014, 33 mynas were captured and eliminated using a walk-in trap, and in October 2014 the population was estimated at less than 20 birds[27]. Total eradication could be achieved but re-invasion from neighbouring infested islands will require constant vigilance.

Petite Soeur

The September 2010 rat eradication at Petite Soeur reduced the dozen birds initially present by half. A year later, in September 2011, only 4-6 birds remained, one of which was trapped in a rat cage-trap; a MiniMyna trap was used without success to try and capture the others[26].

Details on control and eradication programmes conducted elsewhere around the world can be found in the Global Invasive Species Database and in the Invasive Species Compendium database[4]. Myna control is widely conducted in countries where the species has been introduced, such as Australia, New Zealand, Singapore, Israel, and islands in the Pacific (Fiji, Moturoa, Hawaii) and in the Atlantic (St Helena & Ascension) oceans. These efforts mainly involve trapping, netting and chemical control. Tidemann traps have been used successfully in Australia, and Tindall traps in New Zealand; whilst decoy traps, Kadavu traps, Larsen traps, Rat snap-traps and other foraging traps were less successful. Small invading myna populations (<50 birds) have been eradicated using decoy and funnel traps in the Canary (Tenerife, Gran Canaria) and Balearic Islands (Mallorca)[28]. These traps were also effective on St Helena. Large nets have been used successfully in Singapore to eliminate roosts of hundreds of individuals. In NSW Australia, booklets have been produced to promote myna control and provide advice and guidance to members of the public and institutions willing to participate in preventive and trapping activities[29, 30].

MANAGEMENT RECOMMENDATIONS
Control and eradication
The following methods have been used with success in Seychelles. However, what worked on one island or in a particular situation may not necessarily work in at a different one. When designing a control or an eradication plan for mynas, it is important to:

- Survey the island and identify hotspots where mynas concentrate to feed, and also to roost in the evening. Wherever possible, conduct a proper island census using point counts or transects (e.g. Distance sampling), and regular counts in high density areas.
- Consider a range of different methods that can be combined so that birds not eliminated by one method can be destroyed by another.
- Plan to use the most aggressive or stress inducing methods always at the end of the operation, when numbers have already been significantly reduced. The order in which the various methods are employed is important, especially when eradication is attempted. For example, shooting should never be used from the onset as it may make mynas shy and deter them from approaching bait food, traps or anything associated with humans. Knock-down methods likely to kill great numbers of birds with little or no risk of producing aversion, such as certain types of poisoning, may be used first. Traps can be used next; and shooting and nest-boxes with noose traps should be left for the end. Alternative scenarios with trapping as primary method have also proved efficient.
- Always plan to spend 1-3 weeks testing trap methods and bait types. For example, rice with grated coconut was used successfully on Denis Island as bait for mynas, but it did not work well on Ile du Nord in 2006, where birds preferred canteen food (fish, meat, etc.). Meanwhile, papaya which was not effective on Ile du Nord worked well on Frégate; and on Grande Soeur, bread was preferred over rice, but rice provided a better substrate for the starlicide. It is best to use food that mynas are accustomed to eat.
- Separate control sessions by pauses in effort of 1 week or more. This will minimise the development of aversion by some birds for a particular method. Mynas are very wary and learn quickly.
- Avoid long lapses in control effort, especially during the breeding season (October-May), to avoid a build up in myna numbers (small populations may double during one breeding season).
- Make sure that there are no human presences or other forms of disturbance during control sessions. Mynas are particularly active early morning (06:00 to 08:00), and late afternoon (16:00 to 18:00), when they are more likely to form feeding concentrations.
- Make sure food scrap from picnics, restaurants or dumpsites is properly disposed of, buried or incinerated, and is not available to mynas. Such sources of nutrition will enhance population growth by increasing breeding success or juvenile survival.

Poisoning
The following pertain to use of poison:

- DRC 1339, is very effective at drastically reducing numbers of mynas on islands with large populations, but it **cannot be used where non-target native species** (e.g. Moorhen, Giant tortoises), **especially ground-feeding threatened endemics** (e.g. Seychelles magpie-robin), **are at risk**.
- DRC 1339 is absorbed into the bird's bloodstream and impairs the liver and kidney functions. It causes birds to die from uremic poisoning 1-4 days after intake. Although its toxicity to humans and animals other than certain birds is low, it must be used with great caution.
- Wear protective gloves and face mask to prevent contacting or inhaling the powder.
- Place the poison in trays, directly on the ground or on special tables, in open areas with little vegetation.
- Conduct pre-baiting (using non poisoned bait) for a minimum of one week to attract the target birds.
- Starlicide DRC 1339 can be used at 2.5 mg (1 dose)/kg of bait, and mixed with rice, pounded dried bread, domestic food, or sprinkled on papaya fruits. Turmeric and sugar can be used to mask the flavour and bitter taste of starlicide.
- Exercise permanent surveillance of the poisoning stations from a hide to prevent non-target species from accessing the poison. Record how many mynas have (or may have) eaten the poisoned bait.

Drop traps can be efficient to capture mynas. Arjan de Gröene/GIF

Detail of a drop trap from Denis Island. Arjan de Gröene/GIF

- Keep DRC 1339 doses in a sealed envelope, out of direct light. The powder may lose its effectiveness after 12 months. For baits placed in full sun, a 50-70% loss of active content can be expected after 2 days, depending on the bait medium.
- Use of other, stronger and less selective toxins, is not recommended.
- Alphachloralose avicide is reported to not be effec-

tive enough in warm climates and also induces aversion.
- All uneaten poisoned bait should be removed after each session and disposed of (buried or incinerated) so that it is not available to other wildlife.
- Collect and keep records of any dead carcases found, and dispose of them (by burying or incinerating); but most will not be found.

Additionnal information in the use of poison to eliminate mynas can be found in the selected references[22, 31-33].

Trapping

The following protocols are recommended:
- For funnel or MiniMyna traps, select areas where foraging mynas concentrate, preferably in open areas with little vegetation, where traps can be installed directly on the ground or on tables.
- Start pre-baiting for a few days to attract mynas, after removing other food sources from the area (this applies to all trap types, except decoy traps).
- Set the trap, but leave it open for at least 5 days so that birds can enter, eat and come out freely. Then close the trap and start trapping (this applies to all trap types, except decoy traps).
- Stay away from your traps and observe with binoculars from a distance. Avoid accessing traps when mynas are around so that they do not associate traps with humans. Mynas start foraging at first light, hence it can be better to bait the traps the night before.
- Cage traps such as MiniMyna tend to be more efficient when they already contain one (unstressed) myna, and when several traps are placed a few meters from each other.
- Decoy traps have a special compartment to host a captive myna that will help attract others. These can also be set on territories. They do not require pre-baiting or to be left open prior to use.
- Decoy or funnel traps have proved in certain cases to be highly effective at reducing myna numbers. They have achieved complete eradication for island populations up to 745 birds (Frégate Island).
- MiniMyna traps are designed as walk-in tunnels and valves with spikes that prevent the return of mynas from the holding cage to the entrance cage. Check regularly that valves have all their spikes to

prevent birds from returning to the entrance cage and escaping.

- Check cage traps every morning and evening, and also during the day in hot conditions. Release all trapped native birds and make sure that shade, food and clean water are available for both decoy and any trapped birds.
- Try walk-in traps with one side that closes by pulling a string from a distance to catch birds that cannot be captured with cage traps.
- Try noose carpets (loops of nylon fishing line tied to a wire mesh) placed on top of perching, feeding or roosting sites, to capture mynas reluctant to enter cage traps (see Barn owl, p. 248).
- Place nest boxes with a hinged top and noose traps at the entrance at 5-6 m high on trees to capture the last breeding pairs. Any nests found must be destroyed along with eggs and nestlings.
- Do not stress trapped birds unnecessarily, approach the trap only when ready to collect and euthanize them. Do not treat trapped birds cruelly or leave them in harsh conditions.
- Remove mynas from cage traps by grasping them from the top with index and middle fingers round the neck and holding both wings firmly so that they cannot struggle and injure themselves. Put them into a breathable bag (e.g. a pillow case) and roll them so that they cannot move and injure themselves.
- Wear gloves when handling mynas and/or wash hands thoroughly after contact with birds, traps, food and water bowls and bags in which birds have been kept.

Additionnal information in myna trapping can be found in selected references[12, 24, 29, 30, 33-35].

This model of cage trap (four lateral trapping compartments and one central compartment for a decoy), can capture up to four birds. Arjan de Gröene/GIF

A cage trap with a decoy bird in a central compartment (Larsen trap type) on Denis Island. Arjan de Gröene/GIF

Euthanasia

The following protocols describe how best to euthanize mynas:

- Mynas, as any other invasive animal, must be humanely euthanized. Inhalation of carbon dioxide from available gas cylinders at high concentration in an appropriate chamber is the preferred recommended method for euthanization (widely used in Australia). MiniMyna® traps come with a special bag into which the entire trap can fit to perform this operation. Bring the carbon dioxide with a pipe from the cylinder down to the bottom of the bag, loosely tie the bag at the top to allow air to be evacuated and provide a constant flow of gas for 5 minutes.
- Carbone monoxide from the exhaust of a petrol engine has also been used in Seychelles[12]. The engine was cold when started and run for 6 to 8 minutes, with a pipe sufficiently long to ensure that exhaust gases were completely cooled when reaching the bag. Although this method is very practical and efficient[36], it cannot be recommended as it is no longer considered humane in a number of countries (see

"Euthanising invasive animals", p. 127).

- Physical alternative methods include concussion, which consists in holding the bird in one hand (legs, lower wings and body), and violently hitting the back of its head against a hard wood (or concrete) surface; and cervical dislocation (with special pliers or by breaking the neck), which kills small and medium sized birds instantly but requires a certain practice and is not aesthetically pleasant as reflexes remain for some time. The bird's legs are taken in your left hand (if you are right handed) and the head placed between the first two fingers of the right hand with the thumb under the beak. A sharp jerk with each hand, pulling the head backward over the neck will break the spinal cord and carotid arteries.
- Drowning is <u>not</u> considered a humane method for euthanizing animals.

See also "Euthanising invasive animals", p. 127 and selected references[37-39] for more details.

Shooting

Shooting can be very effective, but it needs to be done very consistently and by a skilled marksman with a good understanding of myna behaviour.

- Early morning and late afternoon are good times to shoot (but see following precautions).
- Use a .22 air rifle, equipped with silencer and telescope (7x or 8x).
- Never shoot at birds in a group, always at isolated individuals or pairs of birds to prevent survivors from becoming gun-shy.
- Shoot in different places each time, and preferably under cover. Use of tripod or some sort of support will greatly improve success of long distance shots.

Additionnal information in myna trapping can be found in selected references[11, 33, 34].

Hunter in camouflage suit with a myna bird caller at Denis Island. P.A. Åhlén / GIF

Preventing reinfestation (small islands)

Small islands located near islands with large myna populations are likely to be reinvaded regularly, and a constant vigilance needs to be maintained, according to the following protocols.

- Eliminate invaders immediately to prevent the establishment of a breeding population. Birds are easiest to target immediately after they arrive on the island, before they become established.
- Keep a police-licensed rifle (air rifle or .22 rifle) in good working condition on the island for use only by trained conservation staff or police marksmen.
- Keep a variety of traps in working condition and ready to use. These may include: MiniMyna traps, walk-in traps and nest-boxes with noose traps.
- Keep records of (re)invasion occurrences, noting the numbers of individuals trapped or poisoned.
- Make sure that potential food sources especially scrap from picnics or restaurants are properly disposed of, buried or incinerated, and are not made available to mynas.

References / Further information

1 Skerrett et al. 2001; 2 Rocamora & Solé 2000; 3 Present work; 4 GISD www.issg.org/database/welcome and ISC www.cabi.org/isc/ (accessed 10.09.14); 5 Lowe et al. 2000; 6 Guy Lionnet, pers. comm.; 7 Nevill 2009; 8 Roland Nolin & André Labiche, pers. obs.; 9 Safford & Hawkins 2013; 10 Markus Ultsch-Unrath pers. comm.; 11 Millett et al. 2005; 12 Canning 2011; 13 Dane Marx, pers. comm.; 14 Beaver & Mougal 2009; 15 Dunlop et al. 2005; 16 Samways et al. 2010; 17 Feare 2010a; 18 Chris Feare, pers. comm.; 19 Arjan de Groene & A. Labiche, pers. comm. & http://www.greenislandsfoundation.blogspot.com/2015/07/green-islands-foundation-completes.html; 20 Climo & Rocamora 2006; 21 Climo et al. 2006; 22 Bristol 2006; 23 Vanherck & Rocamora. 2009; 24 Fanchette 2006; 25 G. Rocamora / St. Anne resort, unpublished data; 26 G. Rocamora & Roland Nolin, unpublished data; 27 Gilles Saout, pers. obs.; 28 Saavedra 2010; 29 Anonymous (undated); 30 Pham & van Son 2009; 31 Feare 2010b; 32 ACVM 2002, 33 Orueta 2007; 34 Orueta & Ramos 2001; 35 www.mynamagnet.com; 36 Tideman & King 2009; 37 Reilly 2001; 38 Sharp & Saunders 2005; 39 Close et al. 2007; 40 van der Woude & Ploegaert, 2010; 41 van der Woude & Wolfs, 2009; 42 van der Woude et al., in prep.

CASE STUDY The eradication of Common mynas from Frégate Island

Contributed by Gregory Canning (Conservation Manager, Frégate Island Private)

The eradication of mynas from Frégate Island had been attempted previously[1]; but these efforts were unsuccessful, due in part to a lack of constant effort during the attempts, as well as to the use of inefficient methods. By 2002 the numbers of mynas had been significantly reduced as a result of shooting. However, with incomplete eradication and no follow up the population was allowed to increase to a substantial size.

In June 2010 a renewed eradication attempt was initiated and complete eradication was achieved in February 2011[2]. The principal method employed to catch the birds was the use of a commercially available trap – the MiniMyna trap manufactured by Myna Magnet Australia Pty. Ltd.[3]. Supplementary capture methods included shooting, nest box trapping and the use of a walk in trap.

MiniMyna traps were very successful during the myna eradication on Frégate. Gérard Rocamora

Chemical control was not considered as an option due to its non-selectivity and the possibility of impacting the Seychelles magpie-robin (*Copsychus sechellarum*) population.

MiniMyna traps were set up in areas where birds concentrated. Cages were baited with papaya and left open for approximately one week during the initial phase of the programme to allow the mynas to become habituated to the traps. Experimentation determined that cages placed within close proximity to one another (between one to two metres apart) were most effective at catching birds. Caller birds were kept in cages to attract other mynas to the cages. These birds were left overnight to attract mynas the following morning; and they were watered and fed to prevent stress. It was better to keep two caller birds in a cage rather than one, as two mynas appeared less stressed than one on its own.

The majority of mynas were trapped at the island dump site where the birds concentrated early in the morning and again in the afternoon when food waste was dumped. This method accounted for 92% of the 745 total birds caught and killed[2]. Once the birds had been trapped, the cage was placed in a non-permeable bag and carbon monoxide was used to kill the birds using exhaust gases from a petrol engine[4].

Shooting was not considered a viable primary method of eradication due to the size of the myna population and the high concentration of mynas in particular areas. Any shooting attempts in these areas would likely have produced an aversion to guns within a very short period of time. Some members of the population may already have been gun shy as a result of shooting in the past. Shooting was used only on individual birds, not on pairs or groups. Only 4 birds were killed by this method.

Nesting boxes that had been provided for the Seychelles magpie-robin were largely utilized by the mynas as nesting sites. In an effort to trap mynas, three nooses were placed over the entrance holes of the nesting boxes, hanging from above. They overlapped one another to ensure successful capture. Two or three further nooses were also placed in the central section of the box. This method was only used at one site after other methods failed to capture birds there; but it could have killed more individuals had it been used more regularly. Three birds were captured using this method.

Walk-in traps were the second most successful method during the myna eradication on Frégate. Greg Canning/Frégate Island

Aversion to MiniMyna traps was inherent, or developed in some individuals and pairs. Individuals that could not be eliminated by shooting and nest trapping, due to their nests being inaccessible, were targeted using a walk-in trap. This trap was designed and built on the island and baited with papaya. Birds would enter the trap and a door would be closed behind them. Once trapped, the bird was killed using an air rifle. Any non-target species trapped at the same time were released after the myna had been shot to prevent unintentional escape of mynas. This method eliminated 5 individuals.

Nesting boxes provided for Seychelles magpie-robins were regularly used by mynas. In the course of monitoring these boxes any myna pulli or eggs that were found were destroyed along with their nests. Myna nests with eggs and pulli found in the eaves of roofs were also destroyed. This method was used to destroy 9 pulli and 42 eggs.

Non-target species were sometimes caught in the cages and the walk-in trap. The vast majority of these individuals were released unharmed from the cage traps and all individuals were released unharmed from the walk-in trap.

Eradication of the Common myna has been correlated with a noticeable increase in the number of Seychelles magpie-robins that have fledged, as well as an increase in the use of nesting boxes by the latter. In areas where there had previously been high concentrations of mynas, Seychelles magpie-robins dispersed into those areas.

Traps are subsequently avoided by mynas that manage to escape, indicating a degree of learned response. Different capture methods ensure that this learned response is countered. The continuous and regular setting of traps and targeting of these individual birds ensured that eradication was successful. These birds must not be given an opportunity to increase in numbers once they have been reduced to ensure complete eradication.

Frégate experienced unusually low rainfall during the time that the programme was implemented and it is possible that a shortage of other readily available food encouraged the birds to enter the traps for easily available food. Birds that were habituated to the presence of humans entered the traps readily and were not dissuaded from entering the traps even when they observed the setting up and removal of traps.

This appears to have been the first ever successful eradication of a well established population of hundreds of mynas from Seychelles or anywhere else in the world[5, 6]. We hope that this successful experience can soon be followed by others.

References / Further reading

1 Millett et al. 2005; 2 www.mynamagnet.com; 3 Canning 2011; 4 Tidemann & King 2009; 5 Feare 2010a; 6 Saavedra 2010.

Feral chicken

Poulet domestique

Poul

Gallus gallus domesticus

IDENTIFICATION AND BIOLOGY

The Feral chicken is a large well known reddish brown bird. Its plumage colour is variable as a result of domestication. The male generally has long green glossy ornamental feathers on its tail, long golden-reddish neck hackles along its neck and back, large red crest and facial lobes, strong beak and spurs[1]. The female is smaller with very simple adornments. The call of the male, typically before dawn but also sometimes during the day, is characteristic. Feral chickens feed on the ground, mainly on seeds and invertebrates, but also on fruit, food scrap and occasionally small vertebrates (mice, reptiles, amphibians). Wild hens build a very rudimentary nest on the ground, and may incubate up to a dozen eggs per brood.

ORIGIN

Indian subcontinent & Southern Asia[1]. There it was domesticated thousands of years ago and introduced all over the world. It probably arrived to Seychelles from Mauritius and La Réunion around 1770 with the first settlers.

DISTRIBUTION & ABUNDANCE IN SEYCHELLES

Feral chickens are known to have established wild, self sustaining populations on a number of granitic and coralline islands[2], such as Cousine (until 1996)[3], Ile du Nord (until 2003)[4] and Desnoeufs (until 2007-08). Feral chickens are still common on Plate, D'Arros (100-200 individuals)[6], Marie-Louise and Alphonse (100-200)[7], and semi-feral populations (i.e. part of the population still being supplemented by humans) occur on Marianne and Rémire[5], and Desroches (including c. 20 totally feral in 2014)[8]. Semi-feral free-ranging chickens also breed in the wild on Bird Island[9], and Denis Island (about 100 individuals[10]). Free-ranging

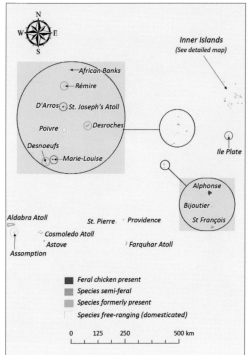

domesticated chickens occur on Mahé, Praslin and La Digue; some were also present on Assomption in 2013[9], and on Poivre in 2014[8].

IMPACTS AND THREATS

Feral chickens have a detrimental effect on forest-floor invertebrate communities, directly by feeding and indirectly through raking, and on the regeneration of native plants and trees through consumption of seeds and seedlings. Chickens compete with native ground-feeding landbirds (turtle-doves, Seychelles magpie-robin) or waterbirds (moorhens *Gallinula chloropus*, whimbrels *Numenius phaeopus*, turnstones *Arenaria interpres*). They also pose a threat to rare native invertebrates, reptiles (endemic skinks and chameleons) and amphibians (caecilians and small endemic frogs)[2]. Chickens may also provide a reservoir for avian transmittable diseases, including Asian lineage NHV1/H5N1 Bird Flu that can be very hazardous to humans[1]. This virus has never been recorded in Seychelles but this could happen if chicken eggs, feathers or meat are imported from countries such as China, Vietnam, Laos, Cambodia or Egypt[9]. Seychelles chickens are potential carriers of *Salmonella*, *Campylobacter* and endoparasites all of which pose health risks to humans; and these may already be present in some domestic poultry.

Late night or very early morning calls of the males can be a nuisance for hotel clients. The presence of large numbers of chickens is also a hygiene concern.

ERADICATION AND CONTROL PROGRAMMES

Little documentation is available on past eradications and current control programmes in Seychelles.

Cousine

Feral chickens were apparently present on Cousine in small numbers (no more than a few dozens) in the 1980s[11]. These were progressively eradicated during the early 1990s by capturing adults and destroying any nests found[3]. In June/July 1996, a few remaining birds and nests were captured and destroyed opportunistically during a BirdLife International/DoE bird-ringing course, and all chickens had disappeared by the end of the year[12].

Ile du Nord

The wild population of chickens present on Ile du Nord appears to have been eradicated through shooting and nest destruction in 2003[13].

Desnoeufs

Feral chickens were recorded on Desnoeufs in 1995-96 but had died out by 2007-08, probably as a result of regular capture for consumption by local workers[14]. Although not a proper eradication attempt, this was apparently enough to eliminate this feral population.

Alphonse
Ongoing opportunistic trapping of chickens has been conducted at Alphonse by IDC and hotel staff for their own consumption. During 2011 to 2013, ICS has been trapping up to 7 chickens per month using home-made and more effective automatic walk-in spring traps baited with ripe pawpaw[7]. A plan is in preparation for complete eradication, based on an initial knockdown of most of the population through a well planned poisoning phase. Feeding trials concluded that cooked rice would be effective bait.

D'Arros, Marie-Louise & Rémire
Resident staff members control feral chicken populations on these islands by capturing them with home-made traps for their own consumption[6,7,15]. However, there are plans to eventually eradicate feral chickens from these islands. Significant predation by Grey herons on the chicks has been noted on D'Arros[6].

The Global Invasive Species Database[1] reports control of feral chickens on Bermuda and Lord Howe Island (Australia) through poisoning, the Cayman Islands through shooting or trapping, and Diego Garcia (Chagos). It is however likely that there are other unreported operations taking place elsewhere around the world.

MANAGEMENT RECOMMENDATIONS
Trapping can effectively control feral chickens, or eradicate small populations of less than 50 birds. To eliminate larger, well established populations may require use of chemicals capable of reducing population numbers quickly, given that extensive trapping over a long period may prompt an increasing fraction of the survivors to become wary of traps. Intensive trapping can follow a poisoning regime in order to capture all animals that could not be poisoned. Shooting should be used only as a last resort to kill a few trap-shy isolated individuals; and should be performed by a skilled conservation officer or a professional police/army marksman. A trained dog can also be used to search and eliminate adults and nests.

Control and eradication protocol (small islands)
The following protocol is recommended:

Poisoning
- Conduct pre-feeding over a minimum of two weeks on points spaced across the chicken's distribution (e.g. every 150-200 m) to determine bait preference and to accustom chickens to come and feed.
- Follow with a poisoning period of 1 week using starlicide/avicide DRC 1339 at a concentration of 2.5 g/kg of bait or a 10% alphachloralose paste for application to slices of bread[16]. However, the use of this avian poison is not recommended on islands where endemic or rare ground-feeding birds (and even reptiles) are present, unless poisoned bait is made available in trays that are permanently watched from a hide (as is done for poisoning Common mynas).

Trapping and nest destruction
- Encourage chickens that have become wild or semi-feral to come and feed in enclosures to re-accustom them to semi-captivity and human dependence, and to facilitate their capture.
- Use homemade walk-in traps triggered by pulling a string from a distance, or more sophisticated automatic walk-in spring traps, or funnel traps.
- Survey the island in search of nests and destroy eggs and hens (if possible) when found. Trained dogs may also be used to locate and capture remaining individuals.
- Use a .22 rifle with a silencer – preferably not an air gun (not powerful enough) – to shoot at chickens. Aim at the brain at close range (<20 m), or at the chest from greater distances.

Euthanasia
- Chickens can be humanely dispatched by stunning with a strong blow to the head, that is immediately followed by exsanguination. Cervical dislocation is possible (with suitable equipment and trained staff) for birds of less than 3 kg but we do not recommend it.
  See also "Euthanising invasive animals", p. 127 and selected references[17-19] for more details.
  Note: Chickens treated with poisonous products must not be eaten by people, pets (dogs, cats) or non-target species.

A hand-made trap to catch chickens and cats (Alphonse Island). The animal is attracted into the trap with a bait suspended on a wood stick. When the animal touches the bait, it frees the small piece of wood , the spring (coconut leave) relaxes and the noose closes on the animal. Chicken are trapped by one leg and cats more often caught by one of the back legs or by the tail.
Aurélie Duhec/ICS

Juvenile semi-feral chickens on Desroches.
Pep Noguès

• Alternative methods of euthanasia include the following:
  – Use carbon dioxide from an available gas cylinder in a special chamber or bag (carbon monoxide from the exhaust pipe of an idling, cold-started, petrol engine is no longer recommended as it is not considered humane in a number of countries).
  – Barbiturates such as sodium pentobarbitone injected peritoneally or at the foramen magnum at base of skull by a veterinarian or a trained officer.
  – Shooting at short distance.

PREVENTING REINFESTATION (SMALL ISLANDS)
• Do not rear free-range chickens. Keep chickens in outdoor large enclosures and recapture or eliminate any escapees.
• Use traps or trained dogs to eliminate groups of birds that have become wild, as shooting is likely to rapidly cause shyness and compromise the elimination of survivors.

References / Further reading
1 GISD www.issg.org/database/welcome (accessed on16.10.14); 2 Nevill 2009; 3 Samways *et al.* 2010; 4 Hill 2002; 5 G. Rocamora *pers. obs.* 2013; 6 Rainer von Brandis/DRC *pers. comm.*; 7 Aurélie Duhec/ICS *pers. comm.*; 8 Pep Noguès/ICS, *pers. comm.*; 9 Chris Feare, *pers. comm.*; 10 A. de Groene/GIF, *pers. comm.*; 11 V. Laboudallon, *pers. comm.*; 12 Peter Hitchins, *pers. comm.*; 13 Bruce Simpson/ North Island, *pers. comm.*; 14 Roland Nolin, *pers. comm.*; 15 IDC management, *pers. comm.*; 16 Bill Simmons/ACP, *pers. comm.*; 17 Reilly 2001; 18 Sharp & Saunders 2005; 19 Close *et al.* 1997.

A male House sparrow.
Pep Nogués/ICS

House sparrow

Moineau domestique

Mwano

Passer domesticus **subsp.** *indicus*

IDENTIFICATION AND BIOLOGY
The House sparrow is a small brownish bird. The male has the following colouration: a grey cap and chestnut brown nape; black bill, throat and upper breast; whitish cheeks; a black striped brown mantle and tail; and greyish underparts. The females are duller, with a distinctive large pale 'eyebrow', blackish streaked upperparts, pale brown underparts and a brown bill. Juveniles are similar to adult females and have paler bills than adults[1]. House sparrows nest in the cavities of buildings and roof frames, mainly between October and April. The call is a loud monotonous 'chirrup'. It is an opportunistic, social and gregarious species that prefers buildings and gardens, but is also found in surrounding open habitats, feeding mainly on seeds, invertebrates, flower buds and food scrap.

ORIGIN
It probably originated in South-West Asia[2]; but it is now a cosmopolitan species having become a human commensal that has colonised all continents except Antarctica. The race present in Seychelles occurs from East Africa to the Indian subcontinent. Its colonisation of the Amirantes was probably ship-assisted from India, Tanzania or via Mauritius where it was introduced[1] (the exact origins could be determined through DNA analysis).

DISTRIBUTION & ABUNDANCE IN SEYCHELLES
The House sparrow is common on most islands of the Amirantes including Desroches, D'Arros and St Joseph Atoll (Ressource, Vars & St Joseph islands), Poivre and Rémire; Desnoeufs, Marie-Louise and the Alphonse group (including Bijoutier & St François islands)[1]. In view of the numbers that have been culled only on Desroches (756 in 2013), the total population of the Amirantes probably comprises several thou-

A female House sparrow. Pep Noguès / ICS

IMPACTS AND THREATS

The House sparrow is an aggressive competitor that can displace native passerines[2, 3]. On D'Arros, it excludes the native Seychelles fody and the introduced Madagascar fody around human habitation. In the Amirantes, House sparrows create a nuisance for the tourism industry and all human residents by breeding inside verandas and bungalows, stealing food and spoiling tables and kitchens of residences and restaurants with their droppings, and possibly risking the transmission of diseases. There is a permanent threat that a breeding population will become established in the inner islands. Should this occur, there would be a significant risk (that should be investigated) of transmitting avian diseases to native birds[2], in particular rare and globally threatened Seychelles endemics. House sparrows may also do considerable damage to fruit and grain crops.

sand birds. Single birds or small groups have been recorded repeatedly on the granitic islands in the following sites and years: on Mahé in 1965 (a group of c. 20), in 2000 (several birds), and in 2007 (1 bird, probably the same seen until mid-2008); on Aride in 1998 (1 bird), and in 2002 (2 birds); on Ile du Nord in 2007 (2 birds); and possibly on La Digue in 1999[4]. Near the Victoria harbour in 2002, a small breeding population of less than 20 birds was found and rapidly eradicated. There has been no other confirmed records since then (March 2015)[5, 17].

ERADICATION AND CONTROL PROGRAMMES
Mahé Island
In 2002, a newly established breeding population of less than 20 birds discovered next to Victoria harbour by the Division of Environment staff was eradicated

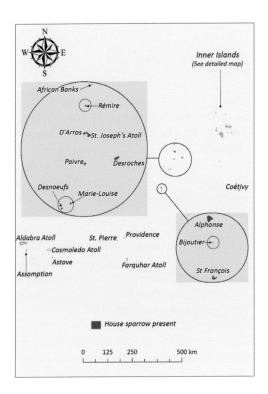

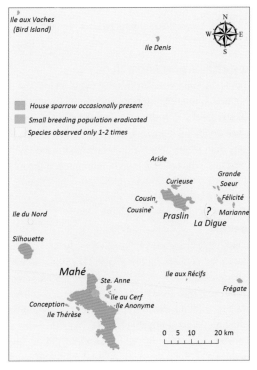

A home-made projected-net used to capture Madagascar fodies and House sparrows. Gérard Rocamora/ICS

within a year. Most birds were eliminated through a combination of different methods. These included: mist-netting (1 bird caught), rat glue on a cardboard baited with coconut (2 caught), feeding traps made from propped-up nets or crates (5 juveniles caught), and a specially designed wire and net trap measuring 15-30 cm that was placed at the entrance to nests located in the regulator holes of fuel storage tanks (8 adults and 2 juveniles)[6]. The female of the last surviving pair was shot a few months later, and the remaining male is assumed to have died. Since then, new arrivals have been reported on occasion, and these have been similarly dealt with[7].

Desroches and Alphonse Islands

Before 2009, limited control sessions had been undertaken on both Desroches and Alphonse islands around buildings, and hotel and canteen areas by IDC using chemical poisoning[8]. This involved the active ingredient tricosene metomyl co-formulants (Lanate®), mixed with jam and expired corn-flakes and sprinkled around the hotel area. Up to 10-15 sparrows were killed at a time using this method. This approach was later stopped because it was causing mortality to hermit

crabs and some introduced but non targeted bird species (turtle-doves and fodies)[9]. Poisoning was replaced by a combination of trapping and occasional nest destruction. The traps used were similar to cage rat-traps, and the frequency of control sessions had to be increased as trapping was found to be less effective than poisoning[10]. At Desroches, in 2010, a few dozen sparrows were eliminated as non-target species during a cat control poisoning operation[9].

At Desroches, in 2013, two control operations were conducted by DoE and IDC, and 756 sparrows were eliminated. These included 311 birds (38.2% in male plumage) between 29 March and 06 April 2013, and 445 birds (49.7% in male plumage) during 04-26 October 2013[11]. Most (c. 80%) of these birds were trapped at the Indian construction camp and c. 20% near the Hotel. Birds were captured primarily using mist-nests (c. 90%) and also with traps made of chicken mesh. Two types of traps were constructed. One was 0.85 × 0.25 × 0.50 m with funnel entrances on each side and a small door on top to get the birds out; and the other was a large walk-in trap (see p. 260) that measured 3 × 2 × 1 m with a door on the side to enter the cage and capture the birds with a small hand net or bag. Trapping was usually conducted during the first three hours after sunrise, and during the last two hours before sunset. Traps and nets were baited with small pieces of bread[12].

At Alphonse, in 2010, House sparrow control was occasionally undertaken using traps in and around the restaurant. This resulted in the elimination of 27 birds[10]. Since 2011, no organised attempts to eliminate sparrows have taken place in the Alphonse group. However, some expatriate workers from the Philippines occasionally kill sparrows around the staff canteen (no records available)[12]. Control sessions using traps and mist-nets (as on Desroches) are planned for the end of 2014.

Aride / Ile du Nord

In 1998, at Aride and Ile du Nord, a few sparrows were observed on occasion[4] but disappeared after a short period without having to be trapped or shot.

Details on control and eradication programmes conducted elsewhere can be found on the Global Invasive Species Database and the Invasive Species Compendium database[2]. Sparrows have been successfully controlled in many large cities in Europe using Avitrol® (active toxin 4-Aminopyridine) mixed with grain as bait. An eradication attempt took place on Ile Ronde (Mauritius) in 2008-2009, during which 320 sparrows were captured and killed primarily by using mist-nets, and glue on perching sticks[13].

MANAGEMENT RECOMMENDATIONS
Control and eradication protocols

- In the Amirantes, control of House sparrow populations should focus on preventing access to favourable nesting or roosting sites inside buildings, by blocking entrances larger than 2 cm and eliminating resting places. In addition, new buildings should be designed so that they do not provide suitable nesting places for sparrows.
- Destroy any nests found and block entrances to prevent re-use.
- Trapping or poisoning (using 4-Aminopyridine) may be used occasionally where sparrow densities are very high.
- At small inner islands managed for conservation, any sparrows that arrive can be eliminated using a .22 air rifle. On Mahé, where human activity is high, shooting should be done by police marksmen.

- In the case of already established populations of sparrows, a combination of methods should be first tested. These may include: feeding traps, nest traps, glue traps, mist-nets and shooting. Poisoning can be considered as an initial knock-down method only if there is no risk to non-target species or if they can be prevented from taking the bait.
- Plan the order in which the various methods will be employed during an eradication attempt. First try knock-down methods such as poisoning that are likely to kill great numbers of sparrows with little or no risk of developing an aversion. Nets or traps can be used later, and shooting left for the end to kill the last few birds.
- Proper disposal of food and leftovers and cleanliness at picnic sites and restaurants will prevent access by sparrows. Reducing their food supply will limit their population growth.

Euthanasia

- Captured birds should be euthanized in a humane way. This can be done chemically through the use of carbon dioxide or physically by concussion or cervical dislocation.

Using carbon monoxide from the exhaust pipe of an idling, cold-started, petrol engine is no longer recommended as it is not considered humane in a number of countries.

See "Euthanising invasive animals", p. 127 and selected references[14-16] for more details.

PREVENTING REINFESTATION
(INNER ISLANDS)

- High vigilance is required, as there is a permanent risk of reinvasion from passing boats.
- Keep the public informed through press articles and television about the risks of reinvasion, how to identify House sparrows, and the need to report sightings to Environment Department or local NGOs.
- Conduct regular surveys of the harbour and adjacent areas looking for House sparrows or their nests.
- Eliminate invaders immediately to prevent the establishment of a breeding population, using rifles, mist-netting (trained personnel only) or trapping.
- Keep a police-licensed rifle in good working condition (.22 air rifle) on small islands managed for nature conservation, to be used by licensed staff or police marksmen.

- Ensure that staff is trained in the use of traps that are known to have been effective in eliminating sparrows in the past including mist-nets, special nest and feeding traps.
- Keep records of (re)invasion occurrences and inform the Seychelles Bird Records Committee. Also record the number of individuals shot, trapped or poisoned.

A nest of House sparrow. Pep Noguès / ICS

References / Further reading
1 Skerrett *et al.* 2001; 2 GISD www.issg.org/database/welcome and ISC www.cabi.org/isc/ (accessed on 10.03.14); 3 Nevill 2009; 4 Skerrett *et al.* 2007; 5 Skerrett *et al.* 2011; 6 Fanchette 2003; 7 Beaver & Mougal 2009; 8 Betts 2009; 9 Jupiter 2011; 10 Nahaboo 2011; 11 Brown 2011, 12 Aurélie Duhec & Sam Balderson/ICS, *pers. comm.*; 13 Bednarczuk *et al.* 2010; 14 Reilly 2001; 15 Sharp & Saunders 2005; 16 Close *et al.* 1997; 17 A. Skerrett/SBRC, *pers. comm.*

House crows are often seen foraging on roadsides and garbage dumping sites. Jacques de Spéville

(Indian) House crow

Corbeau indien

Korbo endyen

Corvus splendens

IDENTIFICATION AND BIOLOGY

The Indian house crow is a medium-sized bird that is almost entirely black, sometimes with a deep purple gloss[1]. It has a greyish patch above from the nape and upper mantle, often extending below from the breast and upper belly, which is darker on juveniles. Crows nest high up in tree forks or in buildings, primarily from October to April. They are highly vocal, uttering a deep characteristic 'kwah-kwah'[1]. Crows are social and gregarious, and roost in groups, mainly around towns, in residential areas, open fields and rubbish dumpsites. Very opportunistic and omnivorous, they eat almost anything, including fruits, seeds, invertebrates, rodents, bird eggs and chicks, and food scraps. They are often aggressive to other animals and humans when nesting. They can travel considerably on board ships, where they feed on left overs and may be fed by crew members.

ORIGIN

South Asia from Iran to South China, including Sri-Lanka, Maldives and Laccadives[2]. House crows have expanded their range partly through ship-assisted introductions to the coasts of the Arabian Peninsula, the African coast from the Red Sea south to South Africa, Australia, Malaysia and Singapore. The species is also established in Mauritius and a small colonising population was reported in 2014 at Toamasina (East Madagascar)[14]. It is said to have arrived on Mahé aboard an Indian cargo vessel in 1977[3].

DISTRIBUTION & ABUNDANCE IN SEYCHELLES

A small population of House crows was first recorded on Mahé in 1977 (5 birds on Mahé, 1 on Ste Anne). Numbers increased until there were 25 birds during the mid 1980s. The birds were observed mainly between

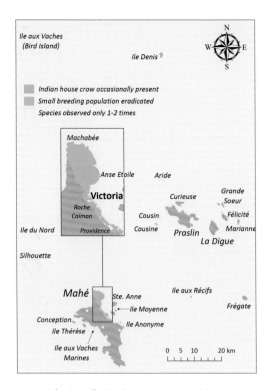

Machabée, Anse Etoile, Victoria, and a rubbish dumpsite (now closed) near Roche Caiman. Breeding occurred until at least 1981 in trees opposite the former dumpsite at Anse Etoile, but control efforts prevented population growth. Eradication was thought to have been reached by the end of the 1980's but two birds were observed at Glacis in 1992. The last pair was observed and shot around Victoria in 1994. Since then, 9 sightings of single birds and two sightings of pairs have been reported on Mahé between 1998 and 2009 (East coast and Glacis). This includes one bird in March 1998 (shot) and some others after 2000 that left by themselves[4], including one at Anse-aux-Pins (June 2002), one at Providence (June-July 2005), one at Victoria (Mars 2007) and one again at Providence (Dec. 2008). In May 2009, one was seen at Glacis (shot) and two around Victoria yacht basin in July 2009 (one of which was shot)[5]. On November 2012, two more individuals (also shot) were observed between the Mahé Gateway jetty to Ste Anne and Ile Persévérance[6]. One House crow was observed at Grand Anse in November 2014[7].

In 1981, a pair was observed building a nest on Silhouette, but no successful nesting was recorded[1]. In 2005, one bird was also reported on Silhouette and Ile du Nord, before it was shot[4]. Single birds were observed on Bird in 1978, on Praslin (l'Amitié) and Cousin, Moyenne and Ile aux Vaches Marines in the early 1980's, and at Aride in April 1992[1, 8, 13].

IMPACTS AND THREATS

House crows could have a serious impact as a predator on the native fauna of Seychelles[4], especially threatened endemic species of landbirds, seabirds, as well as small reptiles such as the Tiger chameleon *Archaius tigris*, amphibians, and rare invertebrates. For example, crows were observed destroying a Seychelles sunbird *Cinnyris dussumieri* nest[1]. Crows can also damage chicken farms by preying on the chicks. The omnivorous crows can significantly impact fruits and other agricultural crops. In addition, they have the potential to transmit infectious diseases (such as *Salmonella*, *Plesiomonas*, *Escherichia coli*, *Shigella* and *Aeromonas*) to native birds and to humans[2, 7].

ERADICATION AND CONTROL PROGRAMMES
Mahé

Between 1977 and 1994, the Division of Environment carried out a successful eradication on Mahé using a combination of shooting and poisoning[9]. Two out of five birds which colonised in 1977 were shot but a remaining pair succeeded in breeding. Between 1977 and 1986, another 9 birds were shot; and between 1986 and 1994, more birds were shot (for example one was shot and one injured on a small roost of three birds at Glacis). A poisoning attempt resulted in two crows poisoned. Poisoning was conducted by first attracting the crows to a rock with small chicken chicks. Then bait treated with alphachloralose (a chemical that slowly puts birds into a comatose state) was provided (while dogs and non-target species were constantly chased away).

The crows were considered to have been eradicated in 1994 when the last two remaining birds (from a breeding population) were shot at Victoria. However, new birds arrived after that, most of which were shot, including: in 1998, one new arrival near Victoria; between 2002 and 2005, three more birds; in 2009, another two[9]; in 2013, two more[6]. In addition, a few other House crows disappeared by themselves.

This long term programme has been driven by the Conservation section of the Department of Environment, with shooting conducted by marksmen from the police or the army (Special Support Unit now called Public Security Support Wing). Since 1992, a bounty of SR500 (€30) has been offered for each dead bird, and repeated appeals through newspaper articles and television for crow sightings have encouraged public participation and contributed to the success of the programme. Incursions are likely to become more frequent as the species is now well established in the Gulf of Aden, where most ships navigating between Seychelles and Europe transit[9].

Small granitic islands
Crows were occasionally observed on some of the small granitic islands, but they generally disappeared after a short period of time without having to be trapped or shot. On Aride (in 1988), it was repelled by nesting terns and was only seen for three days.

Details on control and eradication programmes conducted elsewhere can be found on the GISD and ISC databases[2]. Starlicide has been widely used with effectiveness to control the species, and Alphachloralose is considered a second choice. Destruction of crow nests has also been widely implemented and traps have been used successfully in Malaysia, Tanzania and Kenya, in conjunction with other methods including shooting and a bounty system. Control of garbage, so that it is unavailable to crows, appears to be the most effective limiting factor. The only two known cases of successful eradication are in Seychelles and Yemen (Socotra Island, in 2009). Australia has used public awareness and an early response shooting programme to prevent colonisation despite repeated invasions.

MANAGEMENT RECOMMENDATIONS
Control and eradication protocol (small islands)
- If a small breeding population succeeds in establishing itself, a campaign to shoot the adults and destroy the nests needs to be organised immediately. On large inhabited islands, this is best done by ED.
- Crows that turn up on small inner islands managed for nature conservation can be eliminated using a .22 air rifle. However, shooting in areas with intense humans are activity on Mahé needs to be done with the help of police marksmen. Feeding stations baited with fish can be used to attract the birds to a place where the marksmen can shoot them safely.
- Shooting is effective only to eliminate isolated individuals. Shooting at groups of birds is strongly discouraged as this will create shyness and compromise eradication.

- Poisoning (e.g. with alphachloralose) is not recommended because of the very smalls numbers of crows, and the likelihood of impacting non-target species.
- Trapping may be conducted to capture crows at their feeding sites. Larson or other traps have been used with success in other countries.
- If poisoning at feeding sites is necessary, permanent surveillance must be conducted to avoid casualties of non-target species (see p. 265).
- Human food waste or garbage at dumpsites should be properly managed so that it is not available to crows. In order to limit their breeding success.

Euthanasia
- Captured birds should be euthanized in a humane way through shooting or the use of carbon dioxide, or physically by concussion.
See p. 127 and selected references[10-12].

PREVENTING REINFESTATION (INNER ISLANDS)
- Given the high risk that crows will reinvade from passing ships, permanent vigilance needs to be maintained.
- Keep the public informed through press articles and television about the risks of reinvasion, how to recognise Indian house crows and the need to report any sightings to DoE or local NGOs.
- Offering a public bounty of SR500 for a dead bird helps to keep the public interested in preventing the re-establishment of a breeding population (even if the bounty itself is not effective in controlling the crows).
- Regularly survey Victoria harbour and adjacent areas looking for crows.
- Eliminate invaders immediately to prevent establishment of a breeding population. On Mahé and other large inhabited islands, arrangements can be made for specialised marksmen to shoot them.
- Keep records of (re)invasion occurrences and inform the Seychelles Bird Records Committee. Report the numbers of individuals shot, trapped or poisoned.
- Keep a police-licensed rifle in good working condition (.22 air rifle) on small islands managed for nature conservation, to be used by licensed staff or police marksmen.

References / Further reading
1 Skerrett et al. 2001; 2 GISD www.issg.org/database/welcome and ISC www.cabi.org/isc/ (accessed on 10.03.14); 3 Ryall 1986; 4 Skerrett et al. 2007; 5 SBRC database and DoE 2009; 6 G. Rocamora, pers. obs. & ED 2013; 7 Michèle Martin & Andrew Jean-Louis, pers. obs.; 8 Nevill 2009; 9 Beaver & Mougal 2009; 10 Reilly 2001; 11 Sharp & Saunders 2005; 12 Close et al. 1997; 13 A. Skerrett/ SBRC, pers. comm.; 14 Linders & Langrand, 2014.

CASE STUDY The eradication of introduced Red whiskered bulbuls and Madagascar fodies from Assomption and Aldabra

Contributed by Nancy Bunbury, Jessica Moumou, Nick Page, Chris Feare, Janske van de Crommenacker, Wilna Accouche & Frauke Fleischer-Dogley (Seychelles Islands Foundation)

A Red whiskered bulbul from Assomption.
Denis Hansen/SIF

Until recently, Aldabra was one of the largest atolls in the world with no introduced avian species. A potential threat had been recognised from neighbouring Assomption Island, 27 km to the south-east, where Red-whiskered bulbuls *Pycnonotus jocosus* and Madagascar fodies *Foudia madagascariensis* had been introduced in the mid-1970s[1, 2]. The Invasive Species Specialist Group of the IUCN Species Survival Commission reported in the 1990s that it was only a matter of time before one of Assomption's introduced bird species arrives on the atoll, and highlighted the importance of an eradication programme for these species on Assomption before they reach the atoll[3]. More recently, the presence of exotic birds on Assomption was considered to be possibly the biggest threat to Aldabra's avifauna[4].

In March 2012, small numbers of both species were discovered at Takamaka, one of the least visited parts of Aldabra Atoll, in the location closest to Assomption. It appeared that only one Red-whiskered bulbul was present, but some of the Madagascar fodies were discovered to be breeding and so had presumably been on the atoll but had remained undiscovered for some time. While steps were immediately taken to eradicate these birds, their continued presence on Assomption clearly represented a threat of repeated invasion. This emphasised the urgent need for the eradication of both species from Assomption.

This Red-whiskered bulbul is native to southern Asia. Omnivorous, the species is considered an agricultural pest for damaging fruit in countries where it has been introduced, such as La Réunion, where control programmes have been tested successfully[5]; and also Mayotte (Dzaoudzi) where a small colonising population of c. 50 birds was eradicated at the end of the 1990s (the only such case recorded for this species)[6]. The Red-whiskered bulbul population on Assomption had increased from a mere handful of 6 individuals in 1977, when a few birds were released[1]. By 1986, there were c. 400[1], by 1996 1,000-1,500 birds were estimated[7], and point counts conducted in 2011 indicated a population size of over 3,000 birds[14]. This population increase suggests that the species would also do well in the similar terrain of Aldabra, and would likely compete with the endemic avifauna, in particular the endemic subspecies of Madagascar bulbul. The Madagascar fodies on Assomption grew from 20-30 birds released in 1977[1] and by 2011 they were estimated at more than 1,200 birds[14]. Introduction of the Madagascar fody to Aldabra also seriously threaten the endemic Aldabra fody *Foudia aldabrana*[8] through competition or hybridization; as fodies are known to have hybridised elsewhere (e.g. the Seychelles fody population on D'Arros with Madagascar fodies[9, 10]) (see Hybridisation p. 49). Any avian species introduction may also transmit novel avian diseases to Aldabra's endemic landbirds, which could have catastrophic impacts on their populations. The risk had been considered sufficiently serious to warrant eradication

Trapping with mist-nets was the most effective method to capture Red-whiskered bulbuls and Madagascar fodies.
SIF

of these exotic species from Assomption but no consolidated action had previously been taken.

Since October 2011, the Seychelles Islands Foundation (SIF), custodians of Aldabra, have been implementing with funding from the European Union, and in partnership with the Islands Development Company and the Island Conservation Society, an eradication programme for both species on Assomption. The programme aimed to eliminate this threat to Aldabra's native avifauna. The project has consisted of three phases including a feasibility study, an intensive eradication stage, and a follow-up and monitoring phase.

During Phase 1, feasibility studies included surveys and testing multiple capture methods, including use of traps, avicides, and trials to ensure minimal impact on native landbirds, essentially the only surviving indigenous bird of Assomption, the Souimanga sunbird *Nectarinia sovimanga abbottii*). Initial results suggested that the mist-netting of birds is an efficient control method, whereas trapping trials proved largely ineffective[14]. Eradication strategies were developed based on the results of this phase.

Eradication methods used during Phase 2 employed techniques that proved to be most successful in Phase 1, including mist-netting, shooting by professional hunters at night roosts or of isolated birds, and destroying nests. This combination of different methods maximised catching efficiency and the chances of successful eradication. All birds caught throughout the programme were measured, sexed, and screened for pathogens.

It took several months to eliminate the last few bulbuls, and the last one was killed on 18th December 2014 on Asssomption. The final phase of monitoring lasted 6 weeks, during which the SIF team intensively surveyed the 11 km² of this island to confirm the absence of any survivors, and in January 2015 the eradication of the Red-whiskered bulbul from Asssomption was announced[11]. Over more than three long years of efforts with a permanent team of 3 to 10, combining both local and foreign staff on the island, more than 5,200 bulbuls were eliminated, mainly by mist-netting. This appears to be the world's largest eradication of birds from any island to date[12]. On Aldabra, the single Red-whiskered bulbul that had been detected in 2012 was finally caught and eliminated in July 2013.

Madagascar fodies have been similarly eliminated combining mist-netting and shooting of individual birds. By November 2014, there were less than 30 birds remaining on Assomption, and ongoing breeding of the species made further reduction of their numbers challenging[13]. During the first quarter of 2015, the number of fodies was down to a handful of birds, and by mid-march 2015, the last known male was shot. Monitoring of this species needs to continue for a few more months and it is hoped that the success of the eradication can be declared in early 2016.

THE ERADICATION OF THE MADAGASCAR FODY ON ALDABRA

The population of Madagascar fodies found at Takamaka, Aldabra, was estimated to be less than 200 birds. A permanent camp holding 2 to 6 people was established and Madagascar fodies were eliminated mainly through mist-netting, and some shooting. Despite a difference of size between the Madagascar fody and the Aldabra fody, distinguishing between the two species was difficult. In addition, some individuals showing characters from both species indicated that hybridisation was taking place between the two species, which was confirmed by genetic analysis (it was also confirmed that birds originated from Assomption and that their colonisation was recent)[15]. During the second season of the eradication (November 2013-April 2014), intensive surveys took place and more shooting (using .22 air rifle) was conducted. In total, over 200 Madagascar fodies and hybrid specimens have been culled. The monitoring phase will continue for more seasons and it is hoped that the success of the eradication can be confirmed before the end of 2016.

A male Madagascar fody in a mist-net. SIF

The successful eradication of these two invasive birds from Aldabra and Assomption will mark the first time that established populations of either species have been eradicated from sizeable tropical islands. Moreover, the eradication of Red-whiskered bulbuls (which occurred nowhere else in Seychelles) has completely eliminated this invasive species from the country. This is a very significant first step that opens the door to the ecological rehabilitation of Assomption. Future activities could include the eradication of other invasive species such as rats, cats, vegetation management, and the reintroduction to Assomption of rare endemic species of birds and plants that are currently restricted to Aldabra. New biosecurity policies and quarantine protocols that are currently being established in Seychelles will need to be respected to prevent both re-invasion of the eradicated species and introduction of other alien invasive species that could pose similar risks to the environment.

A male Aldabra fody next to its nest. Adrian Skerrett

The results of this successful project are yielding important lessons for the management and control of invasive birds, with substantial scope for replicability and applications for other tropical islands facing problems with similar species. An integral component of the project has been the capacity building of local staff: intensive training in all procedures has allowed them to transfer their knowledge and skills to other islands within the Seychelles that have similar problems with alien invasive birds.

References / Further reading

1 Roberts, 1988; 2 Rocamora & Skerrett, 2001; 3 GISD 2011; 4 BirdLife International 2011; 5 Amiot et al., 2007; 6Louette, 1999; 7 based on a line transect count along a band of vegetation of 25m all along the southern border of the airstrip (G. Rocamora, unpublished); 8 Safford & Hawkins, 2013; 9 Rocamora, 2003a, 2003b; 10 Rocamora & Richardson, 2004; 11 SIF, 2015; 12 DIISE, 2014; 13 SIF, 2014; 14 SIF, unpublished data; 15 van de Crommenacker et al., 2015.

CASE STUDY Addressing the threat of the Ring-necked parakeet to the Seychelles black parrot

Contributed by the Seychelles Islands Foundation and the Environment Department (MEECC).

The Ring-necked parakeet (or Rose-ringed parakeet, *Kato ver kolye roz* in Creole, *Psittacula krameri*) is the most widely introduced and successful parrot species in the world with breeding populations in nearly 40 countries outside its native range in southern Asia and Africa. It is highly adaptable and poses threats to native biodiversity in both temperate and tropical regions. Even within its native range it is considered a major crop pest of grain and fruit. Ring-necked parakeets have been shown to have a detrimental effect on native cavity nesting birds and are vectors of deadly avian diseases such as Newcastle disease and Psittacine Beak and Feather Disease (PBFD)[1].

The species was introduced to the Seychelles fairly recently, probably between 30 and 40 years ago, although the exact date is unknown[2]. By 2011, he population had shown over the last 15 years clear signs of increase in both numbers and range on Mahé, causing problems for fruit farmers[3], whereas its impacts on Seychelles' native wildlife were unknown. The parakeet has no natural predators to limit population growth in Seychelles so, if left unchecked, the population was likely to spread rapidly to other islands. At least one individual reached Silhouette in 1995 and remained present there for several years[2], and one individual was seen and quickly eliminated on Praslin in 2011. One individual was confirmed on Silhouette in June 2014 and was killed the following month[4]. If this parakeet establishes on Praslin, it could have catastrophic effects on the endemic Seychelles black parrot *Coracopsis barklyi*[5-7]. The Black parrot is one of the rarest birds on the Seychelles archipelago and breeds only on Praslin, where the mature palm forest around the Vallée de Mai World Heritage Site, forms its main breeding area. The Black parrot's limited range and small total population size (520-900 birds)[8, 9] make it highly vulnerable and among the most concerning threats is the presence of the Ring-necked parakeet on Mahé (approx. 40 km from Praslin). Thus far, the species do not overlap in range but parakeets are strong flyers and, without action, it is likely to be only a matter of time before the introduced parakeets reach and establish a population on Praslin and directly threaten the endemic black parrots.

The main threats to the Seychelles black parrot posed by a Ring-necked parakeet range expansion to Praslin would be twofold; firstly the parakeets could aggressively compete for food and nest sites as the species have similar diets and both are cavity nesters; and secondly, the parakeet could introduce pathogens to which the endemic parrots are likely to be highly susceptible[5-7]. Introduced Ring-necked parakeets are considered responsible for introducing and spreading fatal PBFD virus among endemic parrots in the region (e.g. echo parakeets *Psittacula eques* in Mauritius, and Cape parrots *Poicephalus robustus* in South Africa), causing population declines and major setbacks for conservation[1]. Human intervention and control was considered essential to curb the species' spread and eliminate it from the country, which needed to be done before the species spreads to other islands and the population becomes too large to manage or eliminate.

As a result of these concerns and its impacts in other countries, eradication of the Ring-necked parakeet has been a priority in Seychelles since the late 1990s and the population was controlled by the Environment Department with some success. A government- supported study in 2008 indicated a population size of 150-200 birds on Mahé and identified main roosting and feeding areas[3]. The relatively low total population size, combined with its restricted distribution to a single island, made an eradication attempt of this highly invasive species a feasible and cost-effective solution, particularly when compared to other countries where, despite heavy impacts on native biodiversity and agriculture, the species cannot be eliminated. An increase in population size or range would make eradication more expensive and less likely to succeed so action was

An adult Ringed-necked parakeet.
Jacques de Spéville

considered an urgent priority. Eradication of the Ring-necked parakeet from Mahé would also mean national elimination of this Invasive Alien Species.

An eradication programme, which started in 2013, is being implemented by the Seychelles Islands Foundation (SIF), the Environment Department and the Seychelles People's Defence Force, as part of a major European Union funded project awarded to SIF[8]. Since the eradication is being carried out entirely on Mahé, public assistance and sensitisation are essential for success and the programme includes publicity campaigns, with the aim of ensuring everybody on the island is aware of the project and its needs. An intensive eradication phase was launched after an initial sensitisation period. A combination of control methods were trialled to determine the most effective one to avoid 'educated' birds and optimise chances of success. These included mist-netting, trapping using cages with and without decoys, and shooting at small groups of foraging, perching or flying birds. By September 2015, more than 95% of Ring-necked parakeets (over 530 birds) have been eliminated on Mahé since the start of the programme and less than 10 birds may remain in the wild[10]. Intensive training in all procedures and skills was provided to new staff, and local staff have led the eradication for the majority of its duration. All targetted birds are measured, sexed and have blood or tissue collected for DNA analysis and pathogen screening. An extended monitoring and follow-up phase is envisaged after the intensive eradication. Public support and engagement will remain vital throughout the activity for continued reports of sightings, which can then be verified and followed up by project staff. In the later stages of the project, a bounty system may also be assessed, applied if suitable, and carefully monitored to increase the chances of success. The use of recently passed biosecurity legislation will also be crucial to keep this species out of the country (including in captivity) once it has been eradicated.

The international interest in and replicability potential of this activity are high, due to the global distribution and considerable impacts of Ring-necked parakeets on both native biodiversity and communities. Dissemination of information and lessons learned through multiple appropriate channels will therefore be essential.

Eradications take dedication and commitment over a long period of time. It is therefore essential that the momentum of this project is continued until success is confirmed. The project has faced major challenges, mainly because this is the first large-scale attempt in the world to eradicate this species and lessons are being learned along the way. While the status of the Ring-necked parakeet in many other countries makes effective action in these countries difficult, in Seychelles the parakeet's limited range, small population size and success with control efforts so far, makes successful eradication increasingly likely. Should the population spread to other islands, the chance to eradicate this pest and threat to Seychelles' native biodiversity could still be lost.

References / Further reading

1 GISD, 2011; 2 Skerrett et al., 2001; 3 Beaver & Mougal, 2009; 4 ICS Silhouette, pers. comm.; 5 Rocamora & Laboudallon, 2009; 6 Rocamora & Laboudallon,2013; 7 Reuleaux et al. 2014; 8 Reuleaux, 2011; 9 Reuleaux et al. 2013; 10 SIF Newsletter, Sept. 2015.

The Crested tree lizard is a voracious predator posing a threat to island native wildlife. Dane Marx

Crested tree lizard

Agame à gorge rouge

Lezar lakret (proposed name in Creole)

Calotes versicolor

IDENTIFICATION AND BIOLOGY

Named after the crest running from the head almost to the tail in both males and females, the Crested tree lizard resembles a small iguana, 40-60 cm in length, with a very long tail. Adults have large heads, massive shoulders and laterally flattened bodies. Coloration is generally light brownish olive with irregular dark brown speckles or bands, but breeding males have pale yellow bodies and develop a bright red head and shoulders, and a black patch on each side of the throat. Although the Crested tree lizard is predominantly insectivorous, the species is regarded as an opportunistic omnivore that can feed on everything from plants to invertebrates (large insects such as dragonflies, butterflies, grasshoppers; also spiders and earthworms) and small vertebrates including other lizards, young birds as well as eggs robbed from nests. Able to breed at about 1 year old, it lays approx. 10-20 eggs buried in moist soil (c. 7 cm deep)[1,2]. A semi-arboreal, sun-loving lizard that spends a lot of time on tree-trunks and rocks, it is commonly found in open habitats, wasteland, gardens and 'man-made habitats' between sea level and an altitude of 1000 m[3]. The species appears not to favour dense forest with closed canopies. Active early morning and late afternoon, it is also seen sun-bathing during mid-day hours. Breeding in Seychelles apparently occurs during the wetter North-West monsoon (November to April) when the species is more often recorded.

ORIGIN

Native to South and Southeast Asia. Within the genus *Calotes,* the Crested tree lizard is the species with the widest geographic range, extending from the Middle East to South and Southeast Asia[3]. It has been introduced to several parts of the world including the Comores, Diego Garcia in the Chagos[4], Maldives, Mauritius, Réunion and the Seychelles.

DISTRIBUTION AND ABUNDANCE
IN SEYCHELLES

It was first reported at Barbarons (Mahé) in 1982[3, 5], when an individual was captured and killed. Two more individuals were reported three years later on Mahé. However, the species appears to have subsequently died out on Mahé.

There seems to have been two separate introductions in the Seychelles; one in the early 1980s to Mahé, and a second in the early 2000s to Ste Anne. In 2003, a lizard was reported at Anse Aux Pins[2]; but the Conservation Section of the Division of Environment determined that the individual was initially captured on Ste Anne Island, where a tourism development was being built, and was transported to Mahé.

Workers on Ste Anne Island reported that they first saw the Crested tree lizard in 2001, apparently confined to the beachfront area where construction materials were being offloaded. The species is believed to have been accidently introduced amongst uninspected cargo from Mauritius during the construction of the Ste Anne hotel. The lizard population apparently increased rapidly: from 20-25 adults around the jetty in 2003 to 40-45 individuals in 2004. In March 2006, after a two-year control programme, at least 4 adults and possibly some juveniles still remained.

Since then, the species appeared restricted to Ste Anne Island, mainly around the hotel, where small

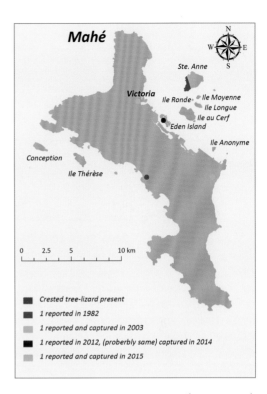

numbers were still present until 2014[6] (see Figure 25). However, one was seen in April 2012 next to the hotel property, near the first old buildings of Ste Anne 2 (NYS ruins). In 2014, two individuals were observed around the hotel, one in January and one in May 2014[8].

A single adult was also photographed on Eden Island on 17 June 2012[7], and one (probably the same animal) captured in October 2014 in the same area[9]. One adult was also captured in Victoria in March 2015. These records probably represent new introductions from outside Seychelles through the importation of construction materials and goods

DAMAGE AND THREATS

The Crested tree lizard is a voracious predator, which poses a threat to island native wildlife because of its unlimited appetite and capacity to feed on endemic lizards, geckos and snakes, insects including dragonflies, small birds, and eggs. Because it is larger and more aggressive, it can outcompete native lizards. The species also harbours parasites that can be transmitted to native reptiles, thus endangering their survival.

Establishment of viable populations of this lizard on other islands could have very negative ecological impacts on native wildlife. In Singapore and Mauritius, the species is considered to be an invasive alien that

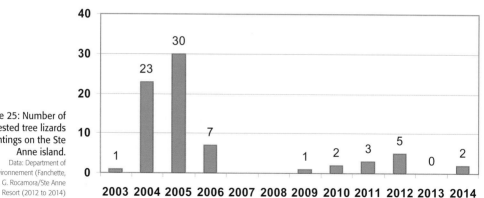

Figure 25: Number of Crested tree lizards sightings on the Ste Anne island.
Data: Department of Environnement (Fanchette, 2012) & G. Rocamora/Ste Anne Resort (2012 to 2014)

competes with and feeds on native biota. In Singapore it has displaced the native Green crested lizard, *Bronchocela cristatella;* and in Mauritius it is thought to have a significant impact on native biodiversity where it competes with native geckos and consumes native invertebrates. It may also have caused population declines of phasmids in both Mauritius and Réunion.

Because of its affinity for open habitats, it may easily become established in coastal areas and perhaps even on some of the high-altitude 'inselbergs' of the granitic islands, as well as in the coralline islands. Its dispersal to other islands would represent a serious new threat to native biodiversity considering the wide range of animal groups that the species preys upon. The smaller, rat-free islands with extensive seabird colonies and populations of endemic terrestrial birds, invertebrates, amphibians and reptiles would be particularly vulnerable. There is a significant risk for the species to recolonise Mahé through the transport of waste material from Ste Anne.

ELIMINATION AND CONTROL PROGRAMMES
An eradication campaign led by the Conservation Section of the Division of Environment was launched in 2003 on Ste Anne Island, where a small population of 15-25 individuals had established (around the jetty/hotel area)[6]. In January 2004, the Government announced a SR50 bounty on any Crested tree lizard caught. This greatly encouraged the participation of the island staff. The method most frequently used to successfully capture the lizards was the 'flush and chase' in which once a lizard was spotted, it was flushed out and pursued until caught. Unfortunately, this method caused the lizards to become wary and secretive, and more difficult to detect. Other capture

methods used were nets, rat glue and bait stations but these were not as effective. A trap was also designed, to be baited with live insects and having several entry points; but it was never actually built.

During the early stages of the eradication campaign, trapping and capture were done four times daily within a week, alternating with a full week of no lizard capture. Between 2003 and 2006, 80 lizards including 48 adults and 32 juveniles and hatchlings (indicating the presence of a breeding population) were caught. The population was thought to be reduced to a minimum of four individuals; but these were elusive and extremely difficult to capture. No sightings were reported in 2007 and 2008 from Ste Anne, but in 2009 a few individuals were seen and one was captured. Since 2010, a dozen sightings have been recorded, almost exclusively around Ste Anne Resort (SAR) and a minimum of three animals were captured and killed by its staff. Individuals captured at Eden Island or in Victoria were eliminated by the Environment Department.

To reduce the risk of re-introducing the Crested tree lizard to Mahé, inspection of equipment and cargo leaving Ste Anne took place in 2003-04, but this was later discontinued.

MANAGEMENT RECOMMENDATIONS
Control and elimination protocol
(on Ste Anne Island)
- Re-establish and advertise an increased bounty (e.g. SR100 or more per individual captured) to encourage public participation, particularly of Ste Anne Resort staff.
- Continue long-term monitoring and occasionnal captures to reduce the population size and distribu-

tion during periods when low numbers are present.
- If numbers increase, undertake an effective control of the population through a more active programme of capture, using the 'flush and chase' method.

The above methods have been used successfully on Ste Anne Island to reduce the lizard population, but eradication has not yet been achieved. To attempt complete eradication, the following efforts are recommended:
- Research and trial new capture methods that might be effective at low population densities. These include pit-fall traps baited either with food or living captive lizard specimens (especially during the breeding season). These methods should be used in the areas where sightings have been made, and aim at catching systematically remaining individuals.

PREVENT REINFESTATION
(TO MAHÉ AND OTHER ISLANDS)
- Improve inspection of cargo coming from Ste Anne or overseas (in particular Mauritius) to detect possible invaders, and wherever feasible conduct

fumigation with Aluminium Phosphide in imported containers.
- Where there is suspicion of infestation, implement early detection efforts to locate aliens before they establish and spread.
- Maintain awareness amongst Ste Anne island staff to prevent the (re)introduction (either accidental or voluntary) from Ste Anne to Mahé or to other islands.
- Increase vigilance on Ste Anne when transporting waste or any cargo from Ste Anne to Mahé or elsewhere, and on Mahé or other islands when importing materials from Ste Anne.
- The Division of Environment needs to enhance information campaigns informing the public about legal restrictions and precautions against introduction of animals, and why they are important for maintaining ecosystem health.
- Encourage members of the public to actively report sightings of unusual animals.

The Crested-tree lizard, is a semi-arboreal, sun-loving lizard that spends a lot of time on tree-trunks and rocks.
Vikash Tatayah

References / Further reading
1 Enge *et al.*, 2004; 2 Tan, 2001; 3 Matyot, 2004; 4 J. Mortimer *pers. comm*; 5 Beaver & Mougal, 2009; 6 Fanchette, 2012; 7 Dane Marx, *pers. obs.* (Eden Island); 8 G. Rocamora & Ste Anne Resort, *pers. comm.* (Ste Anne); 9 Ronley Fanchette & F. Sophola, *pers. comm.*

The Red-eared slider's omnivorous diet and adaptability to a variety of habitats give it great potential to impact indigenous habitats.
Maxime Briola/Biotope

ONE OF WORLD'S WORST **100** INVASIVE SPECIES

Red-eared slider

Trachémyde à tempes rouges / Tortue à tempes rouges / Tortue de Floride

Torti latanp rouz

Trachemys scripta elegans

IDENTIFICATION AND BIOLOGY

The Red-eared slider is a freshwater turtle named for the distinctive red stripe that runs from behind its eye. 'Slider' refers to its ability to quickly slide off rocks and logs into the water. It lives almost entirely in water, but comes out to bask in the sun and lay eggs in nests constructed on land. Male turtles are usually smaller (20–25 cm) than females (25–33 cm) with longer and thicker tails. The carapace is oval and flattened dark green/olive brown with yellow markings. The plastron (bottom shell) is yellow with dark, paired and irregular markings in the centre of most scutes. The head, legs, and tail are green with fine, yellow and irregular lines[1].

Red-eared sliders are omnivorous.They eat a variety of animal and plant materials: fish, crustaceans, insects, worms and numerous aquatic plant species; although younger turtles tend to be more carnivorous with a ferocious appetite. In the USA, reproduction occurs between March and July. A female may lay from two to 30 eggs, and up to five clutches in a single year. It prefers quiet waters but is highly adaptable and can tolerate a wide variety of habitats including rivers, ditches, swamps, lakes and ponds, freshwater or brackish water, natural or manmade. It can live at least 20-40 years in the wild[1].

ORIGIN

Native to the Mississippi Valley of the United States. But, it has become established in other parts of the world because of its popularity in the international pet trade[1].

DISTRIBUTION AND ABUNDANCE IN SEYCHELLES

It was first reported in 1995 when an individual was seen at Beau Vallon. All sightings and captures (10 between 2004 and 2006) were made in the north of

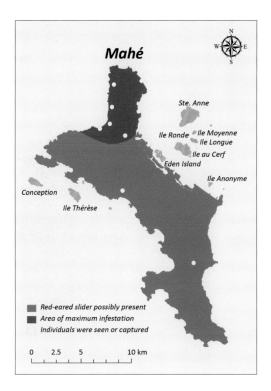

Map legend:
- Red-eared slider possibly present
- Area of maximum infestation
- Individuals were seen or captured

Map labels: Mahé, Ste. Anne, Ile Moyenne, Ile Ronde, Ile Longue, Ile au Cerf, Eden Island, Ile Anonyme, Conception, Ile Thérèse

Scale: 0 2.5 5 10 km

aggressiveness. In Washington (USA), the species is a potential threat to the Pacific pond turtle (*Clemmys marmorata*), a declining species endemic to the Pacific states[1]. In Europe, an experimental study showed that it successfully out-competed the European pond turtle (*Emys orbicularis*)[6]. The study recommended applying the precautionary principle and prohibiting the introduction of slider turtles into European wetlands.

In Seychelles, the Red-eared slider may pose a threat to the two native subspecies of terrapins locally called 'Torti soupap': Black mud terrapin *Pelusios subniger parietalis* (which indigenous status is being questioned)[7, 8] and Yellow-bellied terrapin *Pelusios castanoides intergularis*. Releasing exotic pet turtles into natural ecosystems increases the risk of parasite transmission to native species. The Red-eared slider is known to carry nematodes and can be a reservoir for *Salmonella*, which can cause human salmonellosis[1]. Despite being classified one of the 100 world's worst invasive species[4], little is known of its impact on native ecosystems, and research and education on the dangers of releasing pet turtles into the wild are needed.

ELIMINATION AND CONTROL PROGRAMMES
In 2005, the Conservation Section of the Division of Environment conducted surveys of most major rivers and marshes in the northern regions of Mahé, to determine the status of this introduced species. Traditional Seychellois fish traps baited either with fish or coconut were used to capture the turtles, but the only species caught during the surveys were the two native ones.

There has not been any control or eradication attempt of the Red-eared slider in the Seychelles, but the individuals captured by the public since 1995 were euthanised by the Veterinary Services and kept as specimens at the Division of Environment and the Natural History Museum in Victoria. The Division of Environment made several public appeals against the illegal introduction of animals and asking people to use the Greenline to report sightings of any such unusual creatures.

MANAGEMENT RECOMMENDATIONS
The following protocols are based on what has been tried in other parts of the world.

Mahé (Bel Ombre, Beau Vallon, Victoria, NE Point, Pointe Conan, Glacis), with two exceptions where one individual was captured at Mont Plaisir, southwest Mahé (2005), and one at Grande Anse (March 2012). The species is restricted to Mahé and has not been reported since then[3]. Because of its secretive behaviour (no reports in 2007-2011), some individuals may still be present undetected.

DAMAGE AND THREATS
It has been popular in the pet trade because of its aesthetic appearance and ease of maintenance especially when still a hatchling. The introductions to the wild are probably unwanted pets that have been released or escaped into fresh water bodies[1]. The species has become invasive in many parts of the world[4]. Its omnivorous diet and adaptability to a variety of habitats, give it great potential to impact indigenous habitats. In Europe, it preys on aquatic organisms, including insects such as dragonflies and their larvae, crayfishs, shrimps, worms, snails, amphibians and small fishs, as well as young ground birds which they pull under water and drown[1, 5]. The species may compete for food, egg-laying sites, or basking places with other native turtle species which it can out-compete by means of its greater

Control and elimination protocol

Physical control

- Hand capture any Red-eared sliders encountered opportunistically, and arrange with the Division of Environment or the Veterinary services to euthanize them.

When conducting a programme aimed at eradicating Red-eared sliders, you can:

- Follow females to their nesting areas and destroy their eggs.
- Trap them using a variety of nets and traps – e.g. hand nets, fyke nets, fish traps, and hoop traps.
- Place baited trap cages on top of the floating boards they use as basking sites.

PREVENTING (RE)INFESTATION

- Respect the Seychelles legislation (Animals and Diseases Act) that prohibits the importation of animals except for authorised farming purposes.
- Enact and enforce stricter regulations for the aquarium trade.
- Enact legislation prohibiting the release of pets into the wild.
- Do not transfer pond turtles (or any other species) between islands.
- Enhance public awareness campaigns to sensitise the local population about the dangers of illegal introduction of exotic animals and their potential impacts on our ecosystems.

The Red-eared slider is popular in the pet trade because of its aesthetic appearance, but the species has become invasive in many parts of the world. Quim Soler

References / Further reading

1 GISD, 2011; 2 Nevill, 2009; 3 Fanchette, 2012; R. Fanchette/Environment Department, 2014, *pers. comm.*; 4 Lowe *et al.*, 2000; 5 Soubeyran *et al.*, 2011; 6 Cadi *et al.*, 2004; 7 Silva *et al.*, 2010; 8 Fritz *et al.*, 2013.

ONE OF
WORLD'S WORST
100
INVASIVE
SPECIES

Big-headed ants living in symbiotic relationship with meally bugs. Rene Gaigher

Big-headed ant, African big-headed ant

Fourmi à grosse tête

Fourmi grolatet

Pheidole megacephala

IDENTIFICATION AND BIOLOGY

The common name of Big-headed ant derives from the soldier's disproportionately large head[1,2]. Individuals are 2-4 mm long, yellowish-brown or reddish-brown to nearly black. The body has sparse, long hairs. Antennae are curved and have club-like tips. Head and gaster are commonly darker than the rest of the body. The waist is two-segmented with the node immediately behind conspicuously swollen[1]. The species forms large colonies with multiple nests and multiple fertile queens[2]. Each queen lays up to 300 eggs per month. It constructs debris-covered foraging tunnels with numerous entrances (similar to termite tubes) on the surface of the ground. It is a generalist feeder consuming fish, reptiles and other insects, and harvesting honeydew[2].

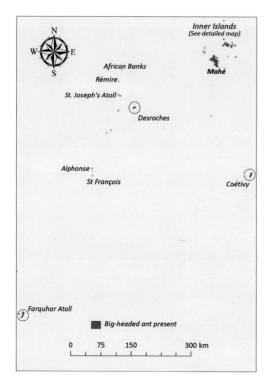

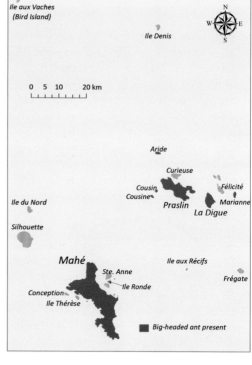

ORIGIN

Big-headed ant is possibly native to southern Africa from where it has spread throughout the temperate and tropical zones of the world. The species prefers warm tropical and subtropical environments. It generally inhabits disturbed habitats, particularly agricultural and urban areas, but it is capable of invading mature rainforests[1].

DISTRIBUTION AND ABUNDANCE
IN SEYCHELLES

It is found on Aride, Cousine, Cousin, La Digue, Mahé, and Praslin. There are records from Île Ronde (near Mahé), Marianne, Desroches, Coëtivy and Farquhar but the exact distribution is unclear because of a history of misidentification and difficulty of distinguishing from similar species. It may reach very high densities (see following case study).

DAMAGE AND THREATS

The Big-headed ant is highly invasive and a serious threat to biodiversity in agricultural and natural ecosystems[3, 4, 5]. In 2008-2010, very high densities were observed on certain parts of Cousine Island, corresponding with hemipteran mutualisms, resulting in foliage damage, tree dieback and significant negative impact on the *Pisonia* forests, and displacement of

certain native invertebrates[4, 6]. *Pisonia* forest makes up an important habitat type on Cousine. The loss of such forest would result in the decline of many valuable endemic and keystone species[4].

In early 2010, these ants were noticed on the plateau on Aride Island and a preliminary survey was carried out[7, 8]. In May 2012, a more thorough survey was carried out which indicated low to moderate densities of infestations mainly on the plateau and extending up to the surrounding hills on the northern and eastern side of the island. The problems are the same as Cousine Island with hemipteran mutualisms and negative impacts on the *Pisonia* forest and certain invertebrates[5, 8].

The Big-headed ant may also pose a threat to seabird populations (especially vulnerable hatchlings) which nest and roost on the ground and in the forest canopy. High levels of these ants also directly threaten other invertebrate fauna in the system, as demonstrated by the decrease in ant species diversity[4].

This ant is a pest of agriculture as it harvests seeds and harbours phytophagous insects that reduce crop productivity[9, 10]. Moreover, being a major harvester of honeydew from Hemiptera, it is considered to have a negative economic impact on agriculture. Like the Crazy ant, Big-headed ants live in symbiotic relationship with coccids, harvesting their sucrose-rich honeydew. The effect of this association on host plants is often negative due to the direct withdrawal of plant nutrients and the growth of sooty mould and other fungi on accumulated honeydew which may lead to leaf death. These ants may also contribute to the transmission of viral plant pathogens and other diseases. For example, Big-headed ants increase the abundance of the corn delphacid, *Peregrinus maidis*, which is a vector of maize viral diseases[11]. The ants also tend pest insects: in Zimbabwe, *Hilda patruelis* on peanuts[12]; in Australia, the sugarcane pest *Saccharicoccus sacchari*[13]; and in Hawaii, mealy bugs associated with pineapple wilt disease[9, 10].

Big-headed ants can also negatively impact agriculture by displacing beneficial ants. In Tanzania, *Oecophylla longinoda* protects coconut trees from the coconut bug *Pseudotheraptus wayi*. But, infestation by Big-headed ants can reduce the abundance of *Oecophylla*, enabling damage from coconut bugs and resulting in reduced revenues[14].

Another negative economic impact is direct damage to agricultural equipment. For example, the worker ants damage the drip irrigation tubes in sugarcane in Hawaii, and are also known to chew on telephone cabling and electrical wires[10].

Bait stations attached to strings along the path in the treatment area on Aride Island. Melinda Curran / ICS

ERADICATION AND CONTROL PROGRAMMES
Cousine: A chemical control, using 2% hydramethylnon-based granular corn bait, 'Siege®' was used to control Big-headed ants in 2010 over some 8 ha (see following case study). This chemical spreads throughout the colony during communal feeding, killing the queen, and thus effectively destroying the entire colony. The active constituent rapidly breaks down into harmless metabolites after exposure to light and the product is reported to have low potential for bioaccumulation in the environment. Such a specific and short-lived product is ideal for use in conservation areas. However, the active ingredient is highly soluble and toxic to aquatic invertebrates[6, 15, 16].

Aride: An eradication attempt started in 2014 (see following case study).

Kakadu, Australia: Big-headed ants were treated with hydramethylnon applied as the commercially available formicide 'Amdro®' (same formulation as 'Siege®')[17, 18]. The effects of this treatment were relatively quick. Most target ants were killed within 24 hours, but there was enough time for the toxicant

Bait stations baited with 'Siege®' were effective in reducing ant densities to low levels. Melinda Curran

to also be passed to the queens, thereby effectively killing the colony. 'Amdro®' is well known for its ability to eradicate the species and its effects are very rapid. The 'Amdro®' was spread by a team of people aligned in a row walking from one edge of the infested area to the other in parallel paths. A 5 m buffer zone was also treated to ensure complete coverage. The 'Amdro®' granules were spread evenly across open landscapes at the recommended rate of 2.5 kg/ha, but was applied more generously at the bases of trees, logs, rocks and thick vegetation where ant populations were greater. The spread of 'Amdro®' over open landscapes was done by hand in the early stages of the project, but was later conducted using hand-held fertilizer spreaders. The spreaders achieved the same eradication outcome as the hand dispersal method, but with the use of much less product and probably with a more even spread. All infested areas were treated prior to the start of the rainy season (since the active constituent is highly soluble). Intensive surveys using attractant (non-toxic) were conducted in the first 12 months post-treatment. Attractant baits were placed in grids approx. 5 m apart over entire treated areas and inspected after 15 minutes. Assessments were conducted at 3-monthly intervals. After two years of post-eradication monitoring without finding any Big-headed ants, the eradication was considered successful[17].

Hawaii: The formicide hydramethylnon was effectively used to eradicate the Big-headed ant on offshore islets of Hawaii. The eradication was followed by changes in the ant community, including the colonization by another invasive ant *Anoplolepis gracilipes*[19, 20].

Cuba: The uses of traps have also been trialled in Cuba. The traps were baited with banana leaves, banana pseudostems and sweet potato. Traps baited with banana leaf were most effective in capturing the largest numbers of Big-headed ants[21].

MANAGEMENT RECOMMENDATIONS
The following methods have been used on Cousine Island and other countries such as Australia.

Control and elimination protocol
Chemical control
- Determine the spatial distribution of the Big headed ant along transects in the infested area.
- Place bait stations in a grid system and place the bait 'Siege®' in the stations at a calculated rate of 2.5 kg/ha (and up to 5 kg/ha in case of very high densities). The distance between the bait stations and the amount of bait in grammes that is placed in the bait station depends on the size of the infested area, to achieve the recommended broadcasted rate (for example, space stations every 4.5 m to achieve a density of c. 500 stations/ha, and a rate of of 2.5 kg/ha with 5 g per station).
- Alternatively, 'Siege®' can be broadcast with hand spreaders, but may then be ingested by non-target taxa. That is why bait stations may be more appropriate, plus the bait is protected from moisture and UV light, prolonging the life of the bait and making application more cost-effective.
- An alternative formicide is 'Amdro®' which is well known for its ability to eradicate the Big-headed ant.
- Conduct 1-2 years of post-treatment monitoring to evaluate the success of the control programme.
- Post-treatment monitoring can be done using a non-toxic attractant placed in grids over entire treated areas and inspected after 15 minutes. Assessments can be conducted at 3-monthly intervals.

PREVENTING REINFESTATION
Some protocols can be implemented to minimise the risk of (re)infestation.
- Public knowledge of the existence and threats of the Big-headed ants and public awareness campaigns can be done on a regular basis.
- Control Big-headed ant populations in infested areas to prevent spread to key conservation areas.
- Protocols for prevention, early detection and intervention (inspection of trails, landing areas etc.) should be adopted by all islands currently free of Big headed ants.
- Inter-island transfer of soil and plant materials should be prohibited, or subject to strong insecticide treatment followed by thorough inspection.
- Visitors to the island must pay close attention to any products, goods or personal possessions they carry with them.

References / Further reading
1 Global Invasive Species Database, 2011; 2 CAB international, 2011; 3 Hoffmann *et al.*, 1999; 4 Lawrence, 2011; 5 Wetterer, 2007; 6 Gaigher & Samways, 2012; 7 Rene Gaigher *pers.comm*; 8 Johnson, 2010; 9 Beardsley *et al.*, 1982; 10 Chang, 1985; 11 Dejean *et al.*, 2000; 12 Weaving, 1980; 13 de Barro, 1990; 14 Oswald, 1991; 15 Gaigher *et al.*, 2011; 16 Gaigher *et al.*, 2012; 17 Hoffmann & O'Connor, 2004; 18 Stanley, 2004; 19 Plentovich *et al.*, 2009; 20 Plentovich *et al.*, 2010; 21 Castineiras *et al.*, 1985.

CASE STUDY Cousine Island Big-headed ant management programme

Contributed by Rene Gaigher
(Faculty of AgriSciences, Stellenbosch University, RSA)

COUSINE ISLAND
SEYCHELLES

The Big-headed ant has been present on Cousine Island since the 1980s, but a considerable increase in its population levels was noticed in 2008. In highly infested areas Big-headed ant abundance was 16 times higher than the rest of the ant community combined in uninvaded areas, and tree foraging activity was as high as 500 ants per minute per tree (Gaigher & Samways unpublished data). This highly invasive species is notorious for its negative influence on native ecosystems[1]. On Cousine it has displaced native invertebrates (Gaigher & Samways unpublished data) and has caused serious indirect damage to native trees via a mutualism with exotic scale insects[2]. A Big-headed ant management programme was developed to reduce population levels. As Cousine is an island of conservation significance, the challenge was to develop a control method that would be effective, yet targeted and with minimal impact on the environment. The programme consisted of pre- and post-treatment ant surveys, a strategic, low-impact toxic baiting phase and continuous monitoring of ecosystem response to the ant management.

Figure 26: The 8 ha on Cousine Island that were delineated to reduce high population levels of Big-headed ants. R. Gaigher / Image©2015 DigitalGlobe

PRE-TREATMENT SURVEYS

An island-wide survey of Big-headed ant densities was done in May-June 2010, with ant activity recorded on 494 trees across the island and mapped with a GPS. Activity levels were categorised as absent, low, medium or high. Only areas of medium and high Big-headed ant densities would be targeted for treatment, as ant damage was most obvious at these densities. The area delineated for treatment was a continuous 8 ha area which included the northern plateau and the north hill (Fig. 26).

TREATMENT

The demarcated area was treated in June and July 2010 with toxic ant bait, 'Siege®' (also known as 'Amdro®'), which consists of maize granules, soybean oil and the active ingredient, hydramethylnon, a metabolic inhibitor. 'Siege®' is known to be highly effective for controlling Big-headed ants, but with low toxicity to non-target organisms and minimal risk of environmental contamination[3]. The conventional method for applying the bait is broadcasting with hand spreaders. But, despite the bait's high specificity, there have been reports of non-target effects of toxic bait broadcasting in other studies[4, 5]. To minimise exposure of the bait to non-target organisms, it was deployed in plastic bait stations that allowed ant access to the bait, but excluded larger organisms. These bait stations were placed throughout the treatment area at the bases of trees where ant nest holes were concentrated. Station density was adapted to ant density in the field and bait was applied at an average rate of 4.9 kg per ha, higher than the recommended rate of 2.5 kg/ha because of the exceptionally high ant densities. Spent bait stations were collected after one week.

POST-TREATMENT SURVEY

The island was resurveyed in October 2010 and in May 2011 to assess the treatment efficacy using the same ant activity survey method as in the pre-treatment survey. Isolated localities where high or medium Big-headed ant densities still persisted received spot-treatments.

ADDITIONAL DATA COLLECTION

Throughout the management programme, additional data were collected in 40 permanent monitoring plots positioned throughout the island. Various methods were used to collect data on the response of the rest of the ecosystem to the ant treatment programme, which would allow a more holistic assessment of treatment efficacy. Data collected included ant and other soil-surface arthropod abundance, mutualist scale insect densities, canopy arthropod abundances and tree condition.

OUTCOMES

The Big-headed ant population was successfully reduced to low and harmless levels throughout the 8 ha treated area and required only isolated follow-up treatments where nests persisted. The suppression of the target species resulted in a rapid decline of exotic mutualist scale insects, dramatic improvement in *Pisonia* tree condition (Fig. 27) and recovery of various soil-surface arthropods and canopy insects[6, 7]. No non-target effects were detected. As the Big-headed ant still occurs on the island at low densities, continued monitoring will be essential to prevent future resurgence. Follow-up treatment may be necessary and, considering the efficacy of the baiting method, an eradication attempt may be a viable option for future management of these ants on Cousine.

Figure 27: *Pisonia* tree condition (A) before baiting in May 2010 indicating high levels of sooty mold and leaf loss indirectly caused by BHA and (B) one year after baiting in May 2011 indicating tree recovery

References / Further reading
1 Wetterer, 2007; 2 Gaigher et al., 2011; 3 Stanley, 2004; 4 Plentovich et al., 2010a; 5 Plentovich et al., 2010b; 6 Gaigher et al., 2012; 7 Gaigher & Samways, 2012.

CASE STUDY Aride Island Big-headed ant eradication attempt

Contributed by Melinda Curran, Gérard Rocamora & Pierre-André Adam, Island Conservation Society

Aride Island is very similar to Cousine in respect to its native species, with Big-headed ants (*Pheidole mega-cephala*) having caused serious indirect damage to native trees via a mutualism with exotic scale insects. An eradication programme was developed in an effort to eliminate all Big-headed ants from the island[1]. Because Aride is an island of very high conservation significance, the challenge was to develop a control method that would be effective, yet targeted and with minimal impact on the environment, following the Cousine example. The programme consists of: pre- and post-treatment ant surveys; a strategic, low-impact toxic baiting phase; and continuous monitoring.

PRE-TREATMENT SURVEYS

The Big-headed ant population was first assessed in early 2010, and the species was found to be present only on the plateau. In May 2012, tuna baited traps were left out for 2 hours every 5 m in transects starting from the heart of the infestation on the plateau and moving outwards up onto the surrounding hill. This was combined with visual surveys all around the hill area bordering the plateau. Big-headed ants were found on a large part of the plateau and were spreading up the north and west hills with 1-350 ants found at each bait station. Ants were found in an area totaling about 6.1 ha, with the heaviest infestation near human habitation. In July 2013, transects and visual surveys were repeated and the results were very similar to 2012 without a noticeable spread. But, in 2014, a visual survey was carried out all around the perimeter and the ants were found to have spread into certain rocky areas where they had not previously occurred.

TREATMENT

In July 2014, a hotspot around a large Banyan (*Ficus bengalensis*) tree at the edge of the perimeter was treated with 'Siege®' (also known as 'Amdro®'; see composition in species account and Cousine case study). To avoid negative effects and to minimize exposure of the bait to non-target organisms, plastic bait stations similar to the ones used on Cousine (which allowed ants access to the bait but excluded larger organisms) were deployed. These bait stations were placed throughout the treatment area, preferably at the bases of trees where ant nest holes were concentrated.

To ensure bait stations were not lost in the field they were attached to the strings by elastic bands at 3 m intervals. This also made it easier to move in the stations an orderly manner. Several parallel lines of strings arranged 3 m apart were deployed (hence a density of 1,000 stations/ha). Each bait station contained 5 g of 'Siege®'. The principle of this eradication protocol was to deploy 6 lines at a time around the entire perimeter of the infestation and, after checking for Big-headed ant absence, to move them 3 lines at a time, and then redeploy them towards the inside of the infested perimeter until the ants become restricted to a small area on the plateau where they can be more easily and completely eradicated. During August 2014, 4,000 bait stations were deployed on the hills all around the infested perimeter. A large section of that perimeter, running through the East hill and cliffs and characterised by difficult terrain was treated first while the weather was still dry. The entire operation is expected to be interrupted during the wet season, as humidity may affect the bait.

Acknowledgements: ICS wishes to thank the Cadbury family and the 'US Embassy Self Help Program' for covering the costs of this ant eradication programme.

Reference / Further reading
1 Adam *et al.*, 2012.

Crazy ant

Fourmi jaune

Fourmi maldiv

Anoplolepis gracilipes

ONE OF
WORLD'S WORST
100
INVASIVE
SPECIES

IDENTIFICATION AND BIOLOGY

Crazy ants have a yellow-brown body of 1-5 mm, remarkably long legs and antennae, and are characterised by erratic movement (very rapid, often with many changes of direction). The Crazy ant is a generalist feeder with a broad diet. It is a scavenger, and also preys on a variety of invertebrates, plant nectar, honey dew, and even larger animals like crabs, reptiles and chicks. Colonies have many queens (polygynous). It also forms super-colonies with huge numbers of workers. Nesting requirements are generalised and nests may occur under leaf litter, in cracks and crevices in the soil, or in land crab burrows and canopy tree hollows[1].

ORIGIN

Native range is uncertain; but it may have originated from Africa or Asia. It has been introduced into parts of Africa, Asia, South America, Australia, some Caribbean islands, some Indian Ocean islands and some Pacific islands. It was accidentally introduced to Mahé around 1962[2, 3, 4].

DISTRIBUTION AND ABUNDANCE
IN SEYCHELLES

Present on at least 9 of the inner islands: Anonyme, Longue, Mahe and Ste Anne, Bird and Denis; Cousin, Félicité, La Digue, Marianne, Petite Soeur and Praslin (introduced in 1975)[4, 5]. A study in 1978 indicated densities of Crazy ants in the invaded woodland on Bird as high as 0.6 million ants per hectare with a range of 0.08-0.79 million (as indicated by leaf litter sampling with Tullgren funnels). This was comparable to densities estimated for parts of Mahé during that same year (Union Vale 0.25-0.5 million and Mamelles

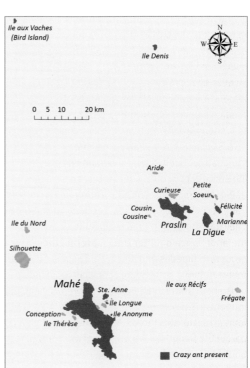

0.9 million), but lower than the introduction site at Maldive (1.3 million per hectare)[3, 6]. In 2000, the species occurred on Bird at densities more than 80 times higher than in the granitic islands (41-53 ants/m² equivalent to 0.4-0.5 million/ha on leaves, and mean pitfall trap catch of 3,800-5,500 ants/m² equivalent to

With its long legs, the Yellow crazy ant is one of the largest invasive ants.
Gérard Rocamora

38-55 million/ha ants on the ground over 3 nights). Densities were very low in some islands (e.g. Cousin)[7].

DAMAGE AND THREATS

The Crazy ant is a major environmental and secondary agricultural pest, as well as a human nuisance. Where densities are high, direct impacts can occur on invertebrate communities (e.g. in Samoa)[16] including native keystone species[1] and on species of special conservation value (including endemic reptiles, birds, and mammals). The Crazy ant invasion of Christmas Island (Australia) perhaps provides the best known example of the cascading ecosystem effects that this invasive species can cause[8-10]. Crazy ants have been implicated in the population crash of the Christmas Island Red land crab *Gecarcoidea natalis,* a keystone species on Christmas Island, which consequently altered the ecosystem in terms of habitat structure, species composition and ecosystem process. The invader reduced native bird, reptile and mammal life, and frigatebird numbers on Christmas Island are predicted to decline by 80% in the next 30 years due to predation of their young by Crazy ants[8].

In Seychelles, Crazy ants reportedly have caused the death of Sooty tern chicks and reduced breeding success of Sooty tern colonies, and the death of native skinks, blind snakes, turtle hatchlings and Fairy tern chicks that had fallen from their nests. They have been implicated in significant declines in the abundance of other ants, several other insect groups, millipedes and spiders[11]. The ant threatens many endemic and endangered species, especially on islands[5], and undermines many potential or actual tourism investments. For example, on Bird Island the tourism sector was threatened by the ants displacing part of the Sooty tern colony, one of the main attractions of the island[11].

Crazy ants can also impact vegetation. The ants tend honeydew-producing coccoid bugs; and this leads to high infestation by these bugs. This results in an increased level of honeydew residues on tree surfaces and an increase in the growth of sooty moulds on trees. That, in turn, can lead to difficulties in photosynthesis and in extreme cases death and forest dieback[6]. In agriculture, this may lead to a reduction in harvest[5]. On Bird Island, there was reported die-back of *Pisonia grandis* and *Guettarda speciosa* trees, and the ant significantly impacted invertebrate communities,

Yellow crazy ant feasting on a dead Wolf snake in the Valée de Mai, Praslin. SIF

both foliar and on the ground[6]. Invasion by this ant may facilitate secondary invasions; for example, on Christmas Island the Giant African land snail and some woody weeds increased in invaded areas.

Following the first observations of Crazy ants in the Vallée de Mai in 2009, research was initiated by the Seychelles Islands Foundation to assess distribution and potential impact of *A. gracilipes* on arboreal species within the palm forest[12]. The species invaded and remained confined to the north-east of the site, with isolated outbreaks. By 2012, the Crazy ants had spread to the east and southeast including the northeastern sides, but then contracted back to its original area of occupation. Overall, abundance and species richness of arboreal endemics (slugs, snails and five species of gecko) were significantly lower in the *A. gracilipes* infested area than in the non-infested area. Molluscs were most affected, with the White slug *Vaginula seychellensis* being absent from infested areas. The Dwarf bronze gecko *Ailuronyx tachyscopeaus* was also significantly less abundant in the invaded area. Continuous monitoring of Crazy ants is now underway at the site because further incursion into the Vallée de Mai and potential consequences for its unique palm forest ecosystem is a serious concern.

ELIMINATION AND CONTROL PROGRAMMES

In 1969 the Department of Agriculture undertook trials to control the Crazy ants by using 'Dieldrin' (the active ingredients of various commercial products like 'Dieldrite®', Dieldrex® and Octalox®) baited

Yellow crazy and tending to scale insects from which they harvest honeydew. C. Kaiser-Bunbury

with fish. Results were inconclusive[2]. More trials were conducted in 1976 using 1% Aldrin baited with coconut coir waste, yeast extract and animal fat. The bait was scattered evenly over the area (at a rate of 10 kg per hectare). Some 90% of ants were killed within the first few days (more effective during dry weather conditions and in relatively open habitats). Although more effective compared to other poisons tested, 'Aldrin (in the form of commercial products like 'Aldrec®', 'Drinox®', and 'Seedrin®') was banned in Seychelles in 1994 due to its high toxicity. Since 2009, local pest control companies have been using 'Dursban®' (active ingredient Chlorpyrifos) which is now known to have harmful effects on aquatic life (fish), birds and mammals. Other methods used include physically destroying the ant nests, and better hygiene (e.g. removal of dried grass, piles of coconut husks)[3, 5].

On Bird Island, the following methods have been trialed: Cypermethrin-based insecticidal sprays combined with burning to clear ants from the area used by the Sooty tern colony. This has successfully prevented the ant from occupying the area of the tern colony since 2000[9].

MANAGEMENT RECOMMENDATIONS
Control and elimination protocol
Chemical control

Control and management of Crazy ants has primarily involved the broadcast of toxic chemical baits. The following treatments have been used with success:

- Fish meal bait laced with an invertebrate toxicant, e.g. 'Fipronil®'[13, 14]. Concentration of the toxicant is low so that foragers will live long enough to return the bait to the colony and disperse the toxicant among other workers, larvae, and queens[15].
- 'Fipronil' along with Hydramethylnon 1 and Pyriprox-ifen are apparently now being used with success to control Crazy ants in other parts of the world. 'Fipronil' is very effective for the ants and the baiting programme has proved successful on Christmas Island[13, 15].
- An aerial baiting campaign was executed in Australia in 2002, where a helicopter was used to distribute 11 tonnes of fish-based 'Fipronil' bait over 25 km^2 of infested rainforest. The campaign resulted in a reduced ant activity of 99% within two weeks, with very few identifiable non-target impacts. Baiting achieved immediate control of all known Crazy ant supercolonies[13, 14].

Physical control
- Destroy the ant nests.
- Remove dried grass, leaves, piles of coconut husks and other dried debris.

PREVENTING REINFESTATION
- Protocols for prevention, early detection and intervention should be adopted by all islands currently free of Crazy ants.
- Inter-island transfer of soil and plant materials should be prohibited, or subject to strong insecticide treatment followed by thorough inspection.
- Visitors to the island must pay close attention to any products, goods or personal possessions they bring with them, making sure that they are ant free.
- Enact and implement laws involving agricultural quarantine. The Seychelles Department of Agriculture is mandated to regulate importation of all plants and animals into the country.
- Improve border protection and establish a protocol and infrastructure for rapid response. These measures are critical to preventing additional losses to our native biodiversity and national economy likely to result from such unwanted newcomers.

References / Further reading
1 GISD, 2009; 2 Beaver & Mougal, 2009; 3 Haines, 1978a; 4 Nevill, 2009; 5 Dunlop, 2005; 6 Haines, 1978b; 7 Hill, Holm et al., 2003; 8 O'Dowd et al., 2003; 9 Lowe et al., 2000; 10 Abbott, 2006; 11 Gerlach, 2004; 12 Kaiser-Bunburry et al., 2014; 13 Boland et al., 2011; 14 Green et al., 2004; 15 Slip et al., 2003; 16 Hoffmann et al., 2014.

A Pink-lipped (left) and a Pale-lipped (right) Giant African snail.
Gérard Rocamora

Giant African snail (Pale-lipped / Pink-lipped)

Escargot géant d'Afrique (à lèvre pâle / à lèvre rose)

Kourpa zean afriken (lalevpal / lalevroz)

Lissachatina fulica & *Achatina immaculata*

Two different species of Giant African snail: Pale-lipped *Lissachatina fulica* and Pink-lipped *Achatina immaculata*.

The easiest feature to distinguish between the two species is the colour of the "inner lip" (columellar lip). The inner lip of *Achatina immaculata* (= *A. panthera*) is pink; whereas the one of *Lissachatina fulica* (= *A. fulica*) is white or bluish-white, with no trace of pink (see arrows on above picture). The shell of *Achatina immaculata* also tends to be paler, thicker and heavier (and tougher).

Rosy wolf snail

Euglandine

Kourpa moustas

Euglandina rosea

The Rosy wolf snail introduced in the Seychelles as a biocontrol of the Giant African snail. Olivier Gargominy/MNHN

IDENTIFICATION AND BIOLOGY

The Giant African snail can reach a height of c. 7 cm, and a total length of 20 cm. Shell colouration is highly variable, but it typically has a brown conical-shaped shell with white, grey, and light brown vertical stripes[1]. Two species are present: *Lissachatina fulica*, with a dark reddish-brown body. The shell is generally reddish-brown with white, yellowish, grey and brown vertical stripes and a white or bluish inner lip (referred to as Pale-lipped Giant African snail) and *Achatina immaculata*, which body is pale yellowish-brown and the shell has white, grey, and brown stripes. Columella and inner lip pink (referred to as Pink-lipped Giant African snail)[1-4]. *Achatina immaculata* also tends to be thicker and heavier. Herbivorous, they eat a wide range of plant materials, fruits and vegetables; and they pose a threat to agricultural crops. Their life cycle facilitates rapid invasion, as the species are hermaphroditic (each individual with both male and female sexual organs), lay 5-6 clutches of 200 eggs each per year and are fast growing[1].

The Rosy wolf snail grows up to 8 cm in height and 3 cm in diameter. It has a brownish-pink thick shell with prominent growth lines. Carnivorous, it eats other snails and slugs; and poses a threat to native species. It is also hermaphroditic and lays 25-35 eggs per clutch[1,2].

ORIGINS

The Giant African snail is native to Kenya and Tanzania. The species has been widely introduced (both intentionally and accidentally) to Asia, the Pacific, the United States, the Caribbean islands and the Indian Ocean islands[1].

The Rosy wolf snail is native to the United States. It was introduced to Indian and Pacific Ocean Islands and the Caribbean beginning in the 1950s as a biological control agent for the Giant African snail[1,5]. In the Seychelles, the Rosy wolf snail was imported from Mauritius in the late 1950s[6].

DISTRIBUTION AND ABUNDANCE IN SEYCHELLES

The Giant African snail is extremely abundant and found on several granitic islands[7].

The pale-lipped species *Lissachatina fulica* is found on: Mahé, Anonyme, Conception, Ile au Cerf, Ile aux Vaches Marines, Ronde, Longue, L'Islette, Ste Anne, Thérèse; Praslin, Curieuse, Félicité, Grande Soeur, La Digue; Silhouette, Frégate and Bird Island[7-9]. Also known from the outer islands D'Arros and

Desroches[6, 10, 11]. It is apparently extinct on Cousin and Cousine[9] where many empty shells have been found[12].

The pink-lipped species *Achatina immaculata* has been reported from Mahé, Ile au Cerf, Ronde, Ste Anne, Thérèse, Praslin, Curieuse, Félicité, Grande Soeur and La Digue[8]. Only empty shells have been found on Ile aux Récifs[8, 13]. The species is apparently not present on Silhouette[2].

Both snail species are possibly present on some other islands but their presence needs to be confirmed. Giant African snails have also been observed on Ile du Nord, Moyenne and Marianne (although there is uncertainty on which of the two species are on these islands). They are known to be absent on Bancs Africains, Rémire, Farquhar-Providence and the Aldabra group.

The Rosy wolf snail was once restricted to the area near Victoria where it occured in lower numbers[2, 14, 15]. It has now spread to the north of Mahé and has been seen at Glacis, La Misère, Fairview and St. Louis[8, 16].

DAMAGE AND THREATS

The Giant African snail is a major agricultural pest feeding on seedlings and delicate plants[1, 6, 17, 18]. There are no reports of that species feeding on other snails in Seychelles); but it is also a vector for several

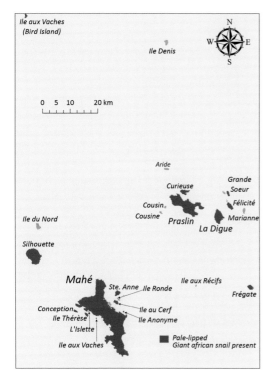

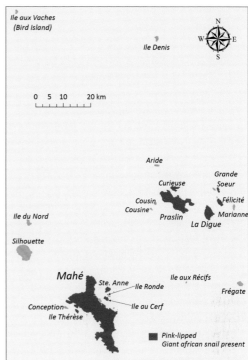

pathogens and parasites[1]. In Mauritius, where it can occur at phenomenal densities in certain areas, it is reportedly both a plant pest and the cause of mortality in birds. Where it is most abundant, it enters nest boxes and climbs over the chicks of the critically endangered Echo parakeet (*Psittacula eques*), leaving a trail of slime that may cover the beaks and nostrils of the chicks, suffocating them[19].

Rosy wolf snails which prey on native mollusc species[5] might be expected to put them at risk of extirpation or extinction. But, in the Seychelles, it has had no apparent impact on biodiversity. This is probably because the endemic snails have evolved with many predators and so have defences and high reproductive rates; and they are rare in the area colonised by the alien snails[14, 15]. There are several introduced populations of this species in Madagascar and there is concern that it may provoke local extinctions from native snails[20].

CONTROL PROGRAMMES
Physical control
Control efforts have involved: collection and destruction of adult snails and their eggs; the establishment of physical barriers, like sprinkling ash or sawdust and scattering pieces of broken egg shells around plants; and construction of slit-trap trenches around vegetable plots or copper barriers around susceptible plants[7, 17]. In Seychelles, snails were purchased in the 1990s from individual collectors to produce food for prawn farms; but this practice disappeared along with prawn farms in the mid 2000s. These snails may also be used as food components for other animals. In Mauritius, a practical method of excluding Giant African snails from trees was developed by using a copper strip around trees. The small electrical charge from the metal surface prevents the snails from crossing over[19]. In the 1950s, children were paid to collect snails, a method that apparently was effective in other places like Hawaii and Japan.

Chemical control
Snail pellets have been used in Seychelles since before the 1950s to control the pest[7]. These pellets, containing the molecule 'methaldehyde' are still promoted by the Seychelles Agricultural Agency as a control measure[17]. In the 1990s, however, freshwater clams at Cascade, Mahé were reported to have one of the highest concentrations of toxic methaldehyde anywhere in the world, probably due to excessive use of slug pellets[21]. Balasubramaniam (1990), warned that high applications of chemicals such as metaldehyde could lead to persistence in coastal environments[22].

Biological control

The Department of Agriculture introduced two species of predatory snails in the 1950s in an effort to control the Giant African snail[7]: *Gonaxis quadrilateris*, (known to prey on eggs and juveniles of Giant African snail[23]) imported from Kenya and released on Cerf Island; and the Rosy wolf snail. Initially, *G. quadrilateris* was reported to be controlling the Giant African snail on Cerf, and so it was also released on Mahé and Praslin, and unofficially released on Ste Anne, Ile du Nord and Frégate. In fact, neither of the two introduced predatory snails have been effective in controlling the Giant African snail[7]; and the Rosy wolf snail is now considered to be an invasive species.

MANAGEMENT RECOMMENDATIONS
Control and elimination protocol
Physical control

- Handpicking and crushing. Look for snails daily, paying careful attention to potential hiding places. Collect and destroy them by crushing. Handpicking can be very effective if done thoroughly on a regular basis. This method could be used to eradicate the Rosy wolf snail which has a restricted distribution.
- Barrier trunk treatments using copper strips. Prune tree skirts 60 cm above the ground and apply a band of copper foil or strip wrapped around the trunk. Staple a nail at one end of the strip to the tree, wrap the sheet around and fastened with a paper-clip. The copper foil should be affixed to the trunk with 20 cm overlap for growth expansion of the trunk. Copper strips can repel snails for several years.

Chemical control

- Barrier trunk treatments using slurry of basic copper sulphate. Paint or spray it on the tree trunks in about a 10-15 cm wide band. Not all copper compounds are approved for use in organic production so be sure to check individual products. This kind of treatment may last for a year.
- Heap ash in a band of 3 cm high and 8 cm wide around susceptible plants. This barrier method can be effective, but loses its effectiveness after becoming damp, making the barriers difficult to maintain.
- Sprinkle 'Epsom salt' (magnesium sulphate) around vulnerable plants to control the snails. Note that this method may alter soil pH.
- Use snail pellets carefully to control the population, and never in areas where native snails are present. Use of snail chemical pellets is not recommended on a permanent ongoing basis, as these can be dangerous to other animals.

PREVENTING REINFESTATION

- Implement strict quarantine to prevent introduction and further spread of the Rosy wolf snail on islands other than Mahé, and of the Giant African snail on outer islands where it is absent.
- Prohibit inter-island transfer of soil and plant materials, or subject such materials to strong insecticide treatment followed by thorough inspection.
- Properly inspect merchandise before island transfers to prevent spread of the Rosy wolf snail or the Giant African snail to other islands.

Chemicals can be used to eliminate invasive snails in areas where native snails are absent (e.g. in vegetable gardens).
Gérard Rocamora

References / Further reading
1 GISD, 2009; 2 Gerlach, 2006; 3 http://www.petsnails.co.uk; 4 http://home.global.co.za/~peabrain/achatina.htm; 5 Lowe *et al.*, 2000; 6 Nevill, 2009; 7 Beaver & Mougal, 2009; 8 P. Matyot *pers.comm.*; 9 S. Hill *pers,comm.*; 10 P. Nogués *pers. comm.*; 11 P-A. Adam *pers.comm.*; 12 Hill, 2002; 13 D.Dine *pers. obs.*; 14 Gerlach, 1994a; 15 Gerlach, 1994b; 16 L. Chong Seng *pers. obs.*; 17 Dogley, 2004; 18 Integrated Pest Management, 2011; 19 Tatayah *et al.*, 2007a; 20 Gerlach, 2014; 21 Gerlach *pers comm.*; 22 Balasubramaniam, 1990; 23 Barker, 2002.

Spiralling whitefly adults resemble tiny moths of 2 mm long, are white and coated with a fine dust-like waxy secretion. Elvina Henriette

Spiralling whitefly

Mouche blanche à spirale

Bigay blan / Mous blan

Aleurodicus dispersus

Coconut whitefly

Aleurode du cocotier

Bigay blan koko / Mous blan koko

Aleurotrachelus atratus

The Coconut whitefly attacks primarily palm species and poses a threat to native palms. Gérard Rocamora

IDENTIFICATION AND BIOLOGY

The common name for Spiralling whitefly arises from the habit of adult females laying eggs in a spiralling pattern of waxy material. Adults resemble tiny moths of 2 mm long, are white and coated with a fine dust-like waxy secretion. In fact, the Spiralling whitefly is not a fly but a Hemipteran bug. It is a phytophagous species, meaning it feeds on plant sap of plant species; and it is considered highly polyphagous, meaning that it can infest a variety of plant species. The polyphagy of the Spiralling whitefly is the major reason it is able to spread so quickly from one crop to another across the tropics[1].

The Coconut whitefly is similar to the Spiralling whitefly except that the females lay their eggs in a non-spiralling pattern. Coconut whiteflies feed on palm sap[1].

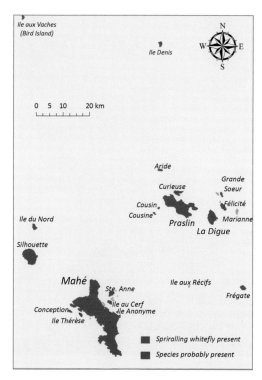

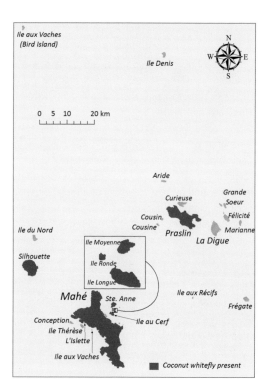

The common name for Spiralling whitefly arises from the habit of adult females laying eggs in a spiralling pattern of waxy material.
Elvina Henriette

ORIGIN

The Spiralling whitefly is native to Central America and the Caribbean region. Recently it has spread to various parts of the world including India, Africa, Indonesia and northern Australia. The species is known to thrive in warm, dry weather conditions[1, 9].

The Coconut whitefly may originate from Brazil and has spread across the tropics and subtropics, to places that include the Caribbean, Hawaii, Mayotte, Mauritius, Madagascar and Seychelles[1].

Whiteflies (of undetermined species) were first reported in the Seychelles in the early 1980s[2, 3]. It was only in 2003 that the occurrence of the Spiralling whitefly was confirmed[2, 4, 5]. Both species could have been introduced through ornamental plants or agricultural products imported into the country.

DISTRIBUTION AND ABUNDANCE IN SEYCHELLES

The Spiralling whitefly spread rapidly after its introduction, and it is now present on Mahé, Anonyme, Conception, Ile au Cerf, Ste Anne, probably Thérèse; Praslin, Aride, Cousin, Cousine, Curieuse, Grande Soeur, Félicité, La Digue; Silhouette, Ile du Nord, Frégate and the inner coralline islands (Denis and Bird)[3, 13]. The Spiralling whitefly is also present on some outer islands including Alphonse and Desroches[18]. Leaf densities can be extremely high especially during the drier South-East monsoon and may fluctuate with seasons.

The Coconut whitefly was initially confined to Mahé in 2007[10, 2] and later spread to Praslin (where it was reported in the vicinity of Baie Ste Anne jetty)[5, 9], to Silhouette and all the islands of the Ste Anne group[13]. At Silhouette, Spiralling whitefly and Coconut whitefly were first recorded in 2005 and 2007 respectively[6]. Their spread was subsequently monitored, and natural predators (mainly Neuroptera) were observed to increase; and by 2010, severity of infestations had declined (with no treatments).

Whiteflies have been reported on Frégate although it is unclear which species is present.

DAMAGE AND THREATS

The Spiralling whitefly is a major pest responsible for huge financial losses in yields of agricultural crops across the tropics[1, 17], including vegetables, fruits, ornamental plants as well as native plants where numbers of individual whiteflies on a single plant may reach several thousands. Spiralling whitefly damages the host by feeding on plant sap from leaves. This causes leaves to turn yellow, appear dry, or fall off plants prematurely (defoliation), and reduces plant vigour and yields[1, 7, 8]. Spiralling whitefly may also act as a vector for disease[3].

Indirect damage is due to excreted honeydew that encourages the development of black sooty moulds, which hinder photosynthesis and reduce yields. Cosmetic damage is due both to sooty moulds and to the white wax secreted by immature stages of the pest which reduces the market-value of crops[1, 7, 8]. The Spiralling whitefly poses a threat to native plant species and is of management concern in protected areas and island nature reserves such as Aride and Cousin[16].

Like the Spiralling whitefly, the Coconut whitefly is also a pest but one that specifically targets palms[9]. It secretes honeydew, on which sooty mould fungi develop, impeding the palm's ability to photosynthesise[1]. In Seychelles, the Coconut whitefly attacks several endemic palms posing serious threats to biodiversity[4, 9, 5, 10]. Cosmetic damage to the palms also reduces the aesthetic values of palm trees – the coconut is no longer a major economic crop, but it remains a vital tourism emblem. The Coconut whitefly can lead to reduced yield of coconuts and in Comoros, the loss of revenues from Coconut whitefly damage has been estimated at €3-5 million[9-11].

On Silhouette in 2010, both whiteflies were considered as threats to agriculture but not as significant threats to native species[6].

CONTROL PROGRAMMES

Attempts have been made to control the pest using several methods. The Seychelles Biodiversity Centre effectively controls whiteflies by using petroleum oil 'Caltex®' (active ingredient paraffinic oil 815 g/L) twice weekly on infested plants[2]. Private households spray plants with dilute aqueous solutions of soaps and detergents (one teaspoon of liquid soap in 5 litres of water). This may be an effective control on a small scale. Gardeners also practice plant hygiene by regular pruning of plants and spraying a jet of water onto the underside of infested leaves[7]. A combination of methods may help to reduce Spiralling whitefly numbers. However, a suitable method for the effective control of the Spiralling whitefly still needs to be found in Seychelles.

In other parts of the world, biological control has proved successful. A successful biological control programme was implemented in Hawaii after the introduction of the Spiralling whitefly in 1978[1]. The coccinellid beetle *Nephaspis oculatus* and the parasitoid *Encarsia haitiensis* were introduced in Hawaii and they successfully controlled the Spiralling whitefly. Controls have also been successful in the Pacific where *Encarsia haitiensis* was introduced.

Similar programmes have also been implemented in Africa. In Benin, the parasitoids *E. haitiensis* and *E. guadeloupae*, helped control the whiteflies[1]. In the Maldives, the Ministry of Fisheries and Agriculture introduced two parasitoid *Encarsia* wasps from the Pacific, to control the whiteflies after they arrived in 1990. The *Encarsia* wasps proved effective in suppressing the population of the Spiralling whitefly. In Mauritius, the coccinellid predator *Nephaspis bicolor* introduced from Trinidad in 2003 has been used to successfully control the Spiralling whitefly[12].

Although few native natural enemies of the whiteflies appear to be present in Seychelles. *Encarsia basicincta*, a parasitoid wasp, was found to be naturally present (and abundant) on Mahé, Praslin and La Digue[15]. On Ile du Nord (North Island), the introduced Seychelles white-eye feeds on the Spiralling whitefly. Other birds like the Seychelles sunbird and Madagascar fody may also prey on them.

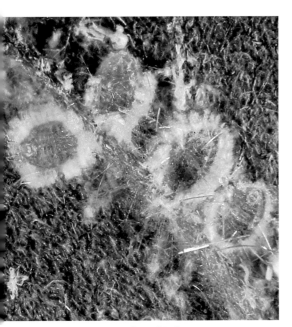

Tiny larvae of the Spiralling whitefly. Vikash Tatayah

As a way of controlling the spread of Coconut whitefly between islands in the Seychelles, the Department of Environment (DoE) imposed a ban on the movement of all palm species in mid 2007. DoE also advised against the removal and burning of heavily infested coconut leaves so as not to generate the movement of adult whiteflies to other host plants.

A biological control programme of the Coconut whitefly was started in 2008 in the Seychelles[15]. In Comoros, the introduction of a natural enemy of the Coconut whitefly, *Eretmocerus cocois,* a parasitoid wasp, has achieved encouraging preliminary results[10]. Its introduction was envisaged in Seychelles before the presence of *Encarsia basicincta* was discovered[15]. Even if biological control of the whiteflies looks promising, we must first evaluate the risk that introducing the biological control agents themselves may cause further problems (see Barn owl, p. 248). One principle of biocontrol is that the biocontrol agent should be host-specific and it is better to introduce agents that have been vigorously tested and proven in other countries.

In general, an integrated pest management approach relying on multiple tactics may prove most effective in controlling whiteflies.

MANAGEMENT RECOMMENDATIONS

A series of methods are proposed below based on experiences from the Seychelles and abroad[2, 7, 10].

Control and elimination protocol
Physical Control
- Improve plant hygiene by regular pruning of plants, removal of infected leaves, and power spraying with water or detergent (e.g. liquid soap) onto the underside of infested leaves.
- Open plant canopy to allow the penetration of sunlight, wind and rain which disturbs the whiteflies. Canopy opening also eliminates shelter for the whiteflies.
- Remove plants that repeatedly host high populations of whiteflies.
- Use a small, hand-held, battery-operated vacuum cleaner to vacuum adults off leaves. Vacuum in the early morning or other times when it is cool and whiteflies are sluggish. Kill vacuumed insects by placing the vacuum bag in a plastic bag and freezing it overnight.
- Light traps coated with Vaseline and yellow sticky traps can be placed around the garden to trap adults. Such traps will not eliminate damaging populations but may reduce them. Whiteflies do not fly very far so many traps may be needed over a wide area: one trap for every two large plants (for sticky traps, place the yellow part of the trap at the level of the whitefly infestation). Place traps so the sticky side faces the plants but is out of direct sunlight. However, these traps are not selective and may trap all sorts of invertebrates, including those that are beneficial (bees, whitefly predators, etc.).
- Use aluminium foil or reflective mulches to repel whiteflies from vegetable gardens, and use sticky traps to monitor or, to reduce high levels of whitefly numbers.
- Add mulch (dead plant materials) at the base of plants to counter moisture loss in cases where plants are infested and are showing signs of wilting.
- Fertilise to improve plant vigour and to enable them to better resist infestations.

Chemical Control
Because the whitefly has a wide host-plant range, and insecticides also impact natural enemies, chemical control is not practical and economic in the long-term. Moreover, insecticides may have only a limited effect

on whiteflies, which may quickly build up resistance to them. Nonetheless, some measures have been recommended:

- Spray with dilute aqueous solutions of soaps and detergents (one teaspoon of liquid soap in 5 litres of water).
- Spray with tobacco or Sisal extract, neem oil, and petroleum oil. Because these products only kill whitefly nymphs that are directly sprayed, plants must be thoroughly covered with the spray solution. Make sure to cover undersides of all infested leaves.
- Spraying with contact and systemic insecticides like Malathion (in the form of 'Celthion®', 'Maltox®' or 'Dielathion®', 'Rogor®' (active ingredient: Dimethoate 400 g/L), 'Decis®' (active ingredient: Deltamethrin 27.5 g/L), 'Ultracide®' (active ingredients: pyrethrin, permethrin and pyriproxyen), 'Vertimec®' (active ingredient: abamectin) and 'Confidor®' (active ingredient: imidacloprid) has also been recommended by the Natural Resource Department, Seychelles.

Biological Control
- Introduce natural enemies such as ladybirds and parasitic wasps following comprehensive risk assessment and management of the introduction of biocontrol agents. *E. guadeloupae*, naturally present in Réunion, could be used against the Spiralling whitefly[14].

Symptoms of Coconut whitefly on a coconut palm leaf.
Gérard Rocamora

PREVENTING INFESTATION AND REINFESTATION OF SMALL ISLANDS

- Thoroughly examine legitimate consignments of living plant material (fruit, vegetables, cuttings for propagation, flowers and ornamental plants).
- Implement strict procedures to deal with airline and ferry passengers who illegally carry plant material in their luggage.
- Avoid transferring plant materials (including palms) between islands, especially during replanting activities.

Encarsia basicincta, a tiny parasitoid wasp naturally present and relatively abundant in Seychelles, could be used for biological control on the Coconut whitefly. Nicolas Borowiec/INRA

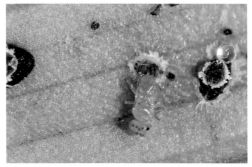

Biological control with *Eretmocerus cocois*, a parasitoid wasp and a natural enemy of the Coconut whitefly, has achieved encouraging preliminary results in Comores. Nicolas Borowiec/INRA

References / Further reading

1 CAB international, 2011; 2 Beaver & Mougal, 2009; 3 Nevill, 2009; 4 Matyot, 2004; 5 Matyot, 2007; 6 Gerlach, 2011; 7 Dogley, 2004; 8 Integrated Pest Management, 2011; 9 Borowiec *et al.*, 2010; 10 Borowiec, 2007; 11 Streito *et al.*, 2004; 12 Indranee Buldawoo pers. comm. 13 P. Matyot & G. Rocamora, *pers. obs.*; 13 P. Matyot & G. Rocamora, *pers. obs.*; 14 Borowiec, *pers. comm.*; 15 Borowiec *et al.*, 2009; 16 Hazell, 2005; 17 Hazell *et al.*, 2008; 18 Pep Nogués & Sam Balderson/ICS, *pers. comm.*).

Tiger mosquito

Moustique tigre

Moustik tig

Stegomyia albopicta; synonym: *Aedes albopictus*

IDENTIFICATION AND BIOLOGY

The Tiger mosquito is characterised by its black and white striped legs, small black and white striped body and dorsal band of silvery scales on the thorax. Its striped appearance is similar to a tiger – hence the name. Like any mosquito, it has four distinct life stages comprising egg, larva, pupa and adult. The first three stages develop in water. The adult is the free-flying insect. The average body size of adult mosquitoes does not exceed 10 mm. Both males and females feed on nectar and other sweet plant saps. Only the females feed on blood which is a necessity to develop their eggs[1].

ORIGIN

Native to Southeast Asia. However, it has invaded many countries throughout the world during the transport of goods (especially tires in which females lay eggs that resist desiccation), and increasing international travel. This includes East Africa and all western Indian Ocean countries[1]. It was first reported in Seychelles in 1905 on Desroches, then on Mahé, Silhouette, Praslin and Denis in 1908-1909[2].

DISTRIBUTION AND ABUNDANCE
IN SEYCHELLES

The Tiger mosquito is present in almost all granitic islands. It is widespread in the four larger granitic islands: Mahé, Praslin, La Digue and Silhouette and some of their satellites including Anonyme, Conception, Thérèse, Ile au Cerf, Ronde, Longue, Moyenne, Ste Anne, Ile du Nord; and Félicité[2, 3]. It has also been reported on Bird, Denis, Plate, Desroches and

is probably also present on D'Arros. It has not been reported on Cousin, Cousine, Curieuse, Frégate, Grande Soeur, Marianne, Petite Soeur, Ile au Récif, and Alphonse where surveys will be required to confirm its presence of absence. It seems to be absent from Aride and most outer islands, particularly those of the Aldabra group (confirmed absent on Aldabra and Assomption in December 2008)[2]. Overall, mosquito vector densities on Mahé, Praslin and La Digue are high, and this could lead to serious consequences in the event of an epidemic disease outbreak[4]. This species usually reaches highest abundances in urbanised areas with high density of human residents[9].

DAMAGE AND THREATS

This species is a significant pest, an epidemiologically important vector for the transmission of many viral pathogens and viruses, including Chikungunya, Yellow fever, encephalitis and dengue fever[1, 5]. Vector-borne disease such as Chikungunya can significantly impact the economy. The 2005-2008 outbreak cost the Seychelles an estimated total of US$1.9 million or an equivalent SR10.4 million in lost revenues in terms of the Gross Domestic Product, medicine and disease control[5]. The high densities of Tiger mosquitoes are a concern as they relate to future disease outbreaks which could lead to epidemics, particularly in the context of climatic change.

Tiger mosquito is a vector for the transmission of many viral pathogens and viruses, including Chikungunya.
Centers for Disease Control and Prevention (CDCP)

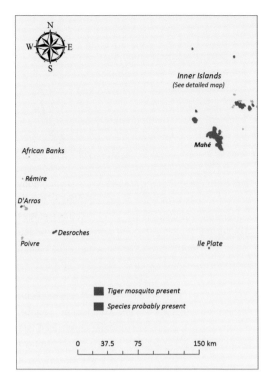

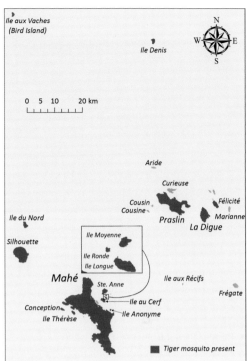

- Tiger mosquito present
- Species probably present

0   37.5   75   150 km

- Tiger mosquito present

ELIMINATION[a] / CONTROL PROGRAMMES

With the opening of the airport in the early 1970s the Ministry of Health dedicated significant effort to preventing the introduction of new species of mosquito into the country by spraying all aircraft coming from foreign destinations[3]. During the 2005-2008 Chikungunya outbreak the Ministry of Health initiated a series of management measures that included: reducing mosquito breeding sites associated with human infrastructure at the community level; educating the public on matters relating to mosquitoes and diseases; training District Environmental Health Officers to be more efficient and motivated in regard to vector control; and controlling mosquito populations through the use of organophosphate or pyrethroid pesticides[3, 4]. The campaign was considered a success. In 1963, Tilapia fishes had been introduced to biologically control the mosquito on Praslin and La Digue (although Seychelles native fishes also feed upon mosquitoes and help, see next section)[6, 7].

Nowadays, pest control companies routinely apply insecticides over vegetation and buildings for all kinds of mosquitoes (using compression pumps, thermal foggers or mist blowers), and larvicides on water

bodies and water-collecting vegetation.

MANAGEMENT RECOMMENDATIONS

The Tiger mosquito has proven difficult to control due to its remarkable ability to adapt to a variety of environments, its affinity to humans, and its reproductive biology. Nonetheless some control and prevention measures are suggested below[3, 4, 7, 8].

Control and elimination protocol

Physical control

- Destroy breeding sites: puddles or pools of stagnant water, roof gutters, old tires, litter, containers of standing water, flower pots and vases, empty tins, and coconut shells.
- Crevices that can collect water should be filled with sand or fine gravel to prevent mosquitoes laying their eggs in them.
- Because litter can hold rain water it should be removed and deposited into public bins.
- Use mosquito nets over windows and beds to limit the access of mosquitoes to you and your house, particularly during disease outbreaks.
- Plant mosquito repellent vegetation – e.g. Lemon grass - 'Citronelle' or 'Sitronel' - *Cymbopogon citratus* - around the outside of your house and below your windows. To enhance the repellent efficacy

a   in medical entomology, the term 'elimination' is used for local eradication, as 'eradication' means worldwide elimination.

Adult female Tiger mosquito having a blood meal.
A. Franck, L. Bagny/Cirad

of these plants, one needs to rub on or crush the plant to release its scent.

Chemical control

- Treat ditches and other standing water bodies with insecticides such as pyrethroids that are relatively more environmentally friendly.
- Use pesticides under strict control to limit environmental impact, as recommended by the Ministry of Health.
- Use mosquito repellent sprays or lotions with cutaneous application or to impregnate clothes.

Biological control

- Propagate dragonfly populations. Aquatic dragonfly larvae eat mosquito larvae in the water, and adults snatch adult mosquitoes as they fly.
- The endemic freshwater fish, the Golden panchax Pachypanchax playfairii ('Gourzon' in Creole) will feed on mosquito larvae[8]. Tiger mosquito larvae, however, tend to develop in temporary breeding sites that lack fishes (e.g. containers of non-permanent water). Nevertheless, laboratory experiments conducted in Seychelles in 1912 indicated the value of Golden panchax as a predator of mosquitoes. Though never used in the Seychelles in a practical way, Golden panchax specimens were sent to East Africa (Zanzibar) in c. 1912, where good results were apparently obtained in wells and tanks. Some of these cyprinodont fish were also

released to nearby swamps, but results are unclear. Interestingly, Golden panchax is reported to have been tested against mosquitoes in Azerbaidjan and Tadjikistan as well[7].

- Larvicides Bacillus thuringiensis israeliensis (e.g. Vectobac®), or methoprene, a growth retardant, may be used in stagnant waters or water-collecting plants but long-term effects on the environment are unknown.

Awareness and Education

- Implement an early warning system for vector disease outbreaks to alert the population and relevant authorities in time so they can take appropriate action against mosquitoes.
- Develop specific intervention plans including public education and awareness campaigns to address outbreaks of disease and to maintain and evaluate the system and its components.

PREVENTING REINFESTATION

- Implement efficient monitoring and surveillance to prevent the spread and establishment of the species, particularly at islands where it does not yet occur.
- Encourage airline companies to develop a program to certify that their aircraft have undergone residual spraying in accord with disease control programmes.

Water that accumulates on vegetation is also a breeding ground for mosquitoes. Vincent Robert/IRD

References / Further reading
1 GISD 2011; 2 Le Goff et al., 2012; 3 S. Julienne/MoH pers.comm; 4 Henriette-Payet & Jullienne, 2009; 5 Lowe et al., 2000; 6 Greathead, 1971; 7 Vitlin & Artem'ev, 1987; 8 Keith et al., 2006; 9 Bagny-Beille et al. 2012.

Discarded bottles and wastes act like good breeding grounds for mosquitoes. Vincent Robert/ARS Mayotte

Taking samples of larvae allows to identify the species of mosquito present and to adapt control and mitigating measures to their biology. Vincent Robert/ARS Mayotte

Broadleaf trees

The rapid growth of Albizia allows it to out-compete slow-growing native trees, thus impacting on native biodiversity. Thomas Le Bourgeois/Cirad

Albizia

Albizia

Albizya

Falcataria moluccana

IDENTIFICATION AND BIOLOGY

Albizia is a fast growing tree that reaches heights up to 40 m. It is characterized by: smooth whitish-grey bark, high broad crown; and alternate leaves bipinnately compound with many small oblong leaflets c.1 cm long (15-25 pairs per pinna). Clusters of small flowers are creamy-white. Fruits are in the form of pods containing small flat seeds dispersed by wind. Seedlings are shade tolerant, grow rapidly and can colonise forest gaps[11]. Albizia is a nitrogen fixing plant able to grow in acidic and nutrient-poor soils. Its ability to alter ecosystem function makes it a problematic invader. It occurs in forests, riparian zones and disturbed areas from sea-level to 1,500 m[1].

ORIGIN

Native to Moluccas (of Indonesia), New Guinea, New Britain (of Papua New Guinea) and the Solomon Islands[1, 2]. It was introduced throughout the humid tropics as an ornamental[2] and for reforestation purposes; but it has become a problematic invader in its introduced range[3].

DISTRIBUTION & ABUNDANCE IN SEYCHELLES

Albizia occurs on Mahé, Anonyme, Conception, Ile au Cerf, Longue, Ste Anne and Thérèse[4, 16]; Praslin, Curieuse, La Digue and Félicité; Silhouette and Ile du Nord; and Frégate[4, 13-16]. It was apparently introduced in 1911 for its timber to rehabilitate degraded lands, and to improve the poor soil conditions through its nitrogen-fixing abilities[5].

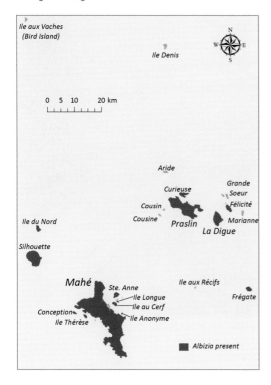

IMPACTS AND THREATS

Being a fast-growing coloniser, Albizia poses a threat to native forest ecosystems. Its rapid growth allows it to out-compete slow-growing native trees, reducing light-levels and shading out other plants[3]. In Seychelles, it has invaded and is especially abundant in mid-altitude forest, particularly in valleys and along streams. It is prone to wind-fall when mature, posing a risk to infrastructure particularly during storms[6, 7 8] but also creating gaps in the canopy, hence promoting the spread of other invasive plant species[5].

In Palau, Albizia has caused a reduction in native biodiversity where it has invaded extensive portions of wetland-mangrove forest, and poses a critical threat to this forest type[1]. In Hawaii, it has been invading the few remaining stands of native-dominated wet lowland forest on early-successional lava flows. It is known to dramatically alter forest structure and litter inputs in forests it invades. Enhanced leaf litter quality and quantity of Albizia compared to native species causes increases in soil nutrient levels, decomposition rates, microorganism community composition and soil invertebrates[1, 9]. In Hawaii, its leaf litter was found to contain 400% more non-native fragmenters (Amphipoda and Isopoda) and 200% more non-native predaceous ants compared to litter under native vegetation. This was attributed to greater nitrogen and phosphorous concentrations in Albizia litter. Its Nitrogen-rich leaves also have the potential to alter ecosystem processes in aquatic environments. Decomposition of leaves may lead to increases in nitrogen concentration of streams which may stimulate algal production[9].

ERADICATION OR CONTROL PROGRAMMES

Various control campaigns have been undertaken in the past, mainly ring-barking (see Management recommendations). Approximately 2,000 large trees were ring-barked by the Forestry section in the 1990s as part of a control and elimination programme of Albizia from important water catchments on Mahé: the species is thought to consume large quantities of water thus lowering the water table[5]. Ring-barking, however, causes the tree to seed before dying and with its large seed bank, propagules can rapidly colonise forest gaps. Other mechanical methods used have been clear cutting or felling of trees using a chainsaw. But, this method causes significant damage to the surrounding vegetation and creates huge forest gaps which increase light levels and favour the establishment of other invasive plants.

On Silhouette, Albizia were effectively controlled by ring barking conducted by NPTS. Ring barking seems to be effective in controlling Albizia but caution is necessary in public areas where falling dead trees or branches can endanger infrastructure and human lives. Poisoning with herbicide has not been tested[10] to determine its efficacy relative to other control measures.

In other island territories like Tahiti of the French Polynesia, Albizia has been controlled by handpulling seedlings and young plants, and chemical treatment through a herbicidal injection (triclopyr and/or dicamba) into the trunk or by spraying the debarked trunk[1, 11]. In Hawaii, Albizia was highly susceptible to 10% 'Milestone®' (active ingredient: aminopyralid). The herbicide was applied through the Incision Point Application (IPA) and the control was successful. Ring barking the tree followed by spraying with glyphosate or 20% dilution of triclopyr in biodiesel ('Garlon®') was also effective[12].

MANAGEMENT RECOMMENDATIONS
Refer to page 326 for management recommendations.

Albizia has caused reduction in native biodiversity where it has invaded natural forests. Gérard Rocamora

References/Further reading
1 PIER, 2012; 2 CAB International, 2011; 3 GISD, 2011; 4 Hill, 2002; 5 Beaver & Mougal, 2009; 6 http://khon2.com/2014/08/17/albizia-trees-part-of-problem-of-puna-recovery/; 7 http://hawaiitribune-herald.com/news/local-news/albizia-enemy-no-1-troublesome-trees-made-iselle-s-impact-much-worse; 8 http://westhawaiitoday.com/news/local-news/iselle-paves-way-albizia-clearing; 9 Tuttle *et al.*, 2009; 10 Gerlach, 2011; 11 Meyer, 2008; 12 Leary *et al.*, 2012; 13 Hansen & Laboudallon, 2013; 14 Senterre *et al.*, 2013; 15 Nevill, 2009; 16 G. Rocamora & P. Matyot, *pers. obs.*

Cinnamon is the most widespread exotic tree species in the granitic Seychelles where it has invaded all habitat types from coastal to mountain forest. E. Henriette

Cinnamon

Cannelier de Ceylan

Kannel

Cinnamomum verum

IDENTIFICATION AND BIOLOGY

Cinnamon is a 10 to 25 m tall evergreen tree, with opposite green ovate-oblong coriaceous leaves which are red and soft when young. Creamy-yellowish flowers have a distinct odour. The purple ovoid fruit contains a single seed. The bark is light brown to brown-black, and the central bark has a cinnamic aldehyde aroma used as food flavouring. It is found in woodland and mature secondary forests, in rocky outcrops, and forest gaps; and it is dispersed mainly by frugivorous birds[1].

ORIGIN

Native to Sri Lanka[1].

DISTRIBUTION AND ABUNDANCE IN SEYCHELLES

Cinnamon was introduced in 1772 as part of the spice garden 'Jardin du Roi' at Anse Royale, Mahé[2, 3]. By the late 19th century, having been spread through cultivation, it came to dominate much of the secondary forests of Mahé. It is also present on Anonyme, Conception, Ile au Cerf, Longue, Moyenne, Ste Anne, and Thérèse

in the Mahé group[4, 13]; Praslin, Curieuse, Félicité, Grande Soeur, La Digue, Marianne, Petite Soeur; Silhouette and Ile du Nord; and Frégate islands[4, 10-13]. In the early to mid 20th century it was a crop of major economic importance; and its leaves were used to produce oil, while the bark was exported for use as food flavouring[2].

DAMAGE AND THREATS

A rapid coloniser forming a dense canopy which partially shades out other plants, Cinnamon creates

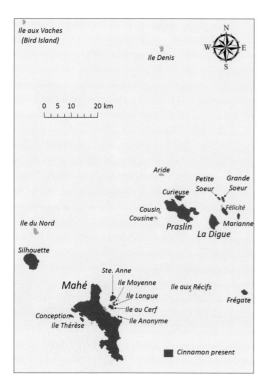

Dense monostands prevent the growth of native species (top). The prolific production of fruits results in dense carpets of Cinnamon seedlings on the forest floor (bottom). Thomas Le Bourgeois/Cirad

species-poor stands covering large areas[1]. Its rapid colonisation is facilitated by its prolific production of fruits that are preferentially dispersed by native frugivorous birds and fruit-bats. Once established, it becomes the dominant tree, but according to Kueffer *et al.* (2012) Cinnamon is not a particularly aggressive invader of undisturbed native vegetation[5]. However, even these moderate levels of invasion can be a major threat to the remaining patches of native vegetation. Adult Cinnamon trees form a dense surface root mat that may prevent regeneration of seedlings of other species and efficiently compete for scarce nutrients[3]. Cinnamon is considered a fire risk because its oily sap and dead leaf matter can act as an accelerant. It is the most widespread exotic tree species in the granitic Seychelles where it has invaded all habitat types from coastal to mountain forest[2,5].

ELIMINATION AND CONTROL PROGRAMMES

Frégate Island: A small scale programme was undertaken in 2003 to control Cinnamon that had invaded patches of forest after the death of stands of mature Sandragon (*Pterocarpus indicus*) due to the Sandragon wilt disease. Cinnamon were cut and uprooted using a 'backhoe loader' (an excavating equipment) and

sometimes by hand. It was successfully controlled in the invaded stands, and from some stands it was completely eradicated[6].

Silhouette: Some localised control of Cinnamon was done by NPTS and Labriz Hotel during the period 1998-2010, initially by tree felling, followed by repeated removal of shoots. This was successful but took 2-3 years of frequent shoot removal. Ring-barking was used at a later stage, and found to require much less shoot removal than was needed using the tree felling method. Old trees were very effectively killed using ring barking, but younger trees vigorously re-sprouted from the base (re-growth). Glyphosate was also applied to the areas ring barked and to the cut stumps using the recommended dosage (1.5-3%).

Although stumps still showed re-growth, ring bark-ing and application of Glyphosate resulted in a high proportion of dead trees; and those that did not die had much less vigorous re-growth. There were no problems of healing of the exposed wood after ring barking which sometimes occurred if trees were not properly ring barked[7].

Mahé: In 1995-1996, there was an experiment funded by the Indian Ocean Commission to investigate control methods for Cinnamon (and Strawberry guava) at high altitudes of Morne Seychellois National Park (Congo Rouge). Three types of physical control were trialed: uprooting by hand; cutting using machete, axe and chainsaw; and ring barking using machete and axe. The uprooting of Cinnamon seedlings within a 100 m^2 plot caused severe land degradation whilst cutting of trees and young plants led to vigorous re-sprouting of the cut stump. Felling of large trees also damaged native plants in the understory and the opening of the canopy negatively impacted shade tolerant species like mosses and ferns. Ring barking of 10 cm diameter trees was ineffective as new bark was formed after a year[3].

The same experiment also tested chemical treat-ment with Roundup® – a glyphosate herbicide applied using five dosages (5, 10, 15, 20 and 25 ml per litre of water, hence 0.5% to 2.5%). Roundup® was applied to the following parts of the plant: leaves, stumps, and through injection. Painting Cinnamon leaves with Roundup® was not effective. The stump treatment with a high dose of herbicide prevented re-sprouting, but was ineffective in stopping the growth of root shoots. The injection method using a tree borer to drill into the tree and then inject a 15 ml concentrate or more into the tree was more effective. Most of the trees treated died but some non-targeted species were also affected, possibly through root to root contact[3].

In 1996, several projects were started by the Forestry Section of the Division of Environment to restore and manage habitats by removing invasive woody plants such as Cinnamon, Cocoplum and Strawberry guava. Some 5 ha were cleared without follow-up replanting at L'Exile, a site where invasive species had overgrown glacis vegetation under harsh climatic and poor soil conditions. A positive impact was observed relatively quickly because the remaining native vegetation was substantial. However, due to a lack of funding the control was discontinued.

In another attempt to rehabilitate upland forests dominated by alien species, the Division of Environ-ment trialed the control of Cinnamon in the Mare aux Cochons area. The trial was maintained over 10 years (beginning in 1998) during which Cinnamon was cleared with chainsaws and the fallen trees were then removed and piled up at the edge of the forest gap. Replanting with native plants was then undertaken[8].

Sainte Anne: A control programme that combined cutting and two paintbrush applications of glyphosate to stumps in December 2011 gave encouraging results, although some trees needed follow up treatment. Pure commercial product was mixed with diesel in equal parts. One month later for trees that resisted the first treatment, pure product with 10% water was applied (i.e. 900 ml pure product plus 100 ml water). Although, diesel was used in this particular control, it is not recommended to use diesel due to the risk it poses to people and the environment. Instead, biodiesel should be used as the dilutant or surfactant.

Important note: Complete removal of Cinnamon where it dominates may pave the way for invasion by other more serious invaders like the Strawberry guava (Psidium cattleianum) and Koster's curse (Cli-demia hirta). So, the complete removal of Cinnamon is unlikely to be more beneficial than simply mitigating against its negative impacts. Moreover, one needs to take into account the role of well established Cinna-mon forest in ecosystem functioning before deciding to eliminate Cinnamon-dominated mature forests, referred as 'novel ecosystems' by some authors[5, 9]. Refer to management recommendations for further information.

MANAGEMENT RECOMMENDATIONS
Refer to page 326 for management recommendations.

References/Further reading
1 PIER, 2011; 2 Kueffer & Vos, 2004; 3 Beaver & Mougal, 2009; 4 Hill, 2002; 5 Kueffer et al., 2012; 6 S. Hill pers. comm.; 7 Gerlach, 2011; 8 Simara et al., 2008; 9 Kueffer et al., 2010b; 10 Hansen & Laboudallon, 2013; 11 Senterre et al., 2013; 12 Nevill, 2009; 13 G. Rocamora pers. obs.

Dense thickets of Cocoplum preventing the growth of other plants and natural vegetation regeneration. E. Henriette

Cocoplum

Prune de France

Prin de frans

Chrysobalanus icaco

IDENTIFICATION AND BIOLOGY

Cocoplum is an evergreen shrub generally 1 m tall, but which can grow into a tree 4 to 6 m high. It is characterised by broad-oval to nearly round leathery leaves, green to light red in colour. The bark is greyish or reddish brown with white specks. Flowers are small, white, and occur in clusters. Fruits are round (globose), pale-yellow with rose blush or pink-red and in clusters, edible and with a single large seed inside. The shrub is propagated by its seeds. It is found from coastal habitats to mid-altitude elevations and is tolerant of nutrient poor soils. It grows in dense monospecific thickets that impede growth of native plants[1].

ORIGIN

Native to Tropical America[1, 2].

DISTRIBUTION AND ABUNDANCE IN SEYCHELLES

Cocoplum occurs on Mahé, Anonyme, Conception, Moyenne, Ile au Cerf, Longue, Ronde, Ste Anne and Thérèse[3, 13, 14]; Praslin, Curieuse, Félicité, Grande Soeur, La Digue and Petite Soeur; Silhouette and Ile du Nord; and Frégate[3, 11-14]. It was introduced during the 20th century to promote stabilisation and conservation of soil and to control erosion[3, 4].

DAMAGE AND THREATS

Cocoplum forms dense thickets that prevent the growth of other species of plants and the regeneration of natural vegetation[1]. It is considered a fire hazard since the thickets contain and retain a lot of dry materials. In Seychelles, it has invaded coastal, inselberg and mid-altitude habitats, especially on

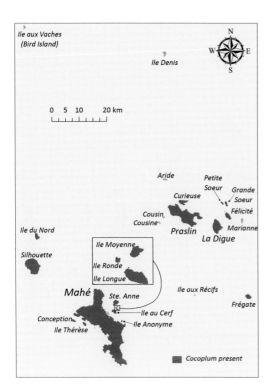

In Seychelles, Cocoplum has invaded coastal, inselberg and mid-altitude habitats, especially on dry soils.
Thomas Le Bourgeois / Cirad - Pl@ntInvasion

dry soils. It is widespread in the granitic islands and presumed to be spreading[4]. In French Polynesia, the species has been declared a threat to biodiversity[1].

ELIMINATION AND CONTROL PROGRAMMES

Mahé: At Sans Soucis, Cocoplum was repeatedly cut as part of a programme to replant endemic plants between 1997 and 2002. Frequent re-cutting was undertaken to control Cocoplum until the native plants were big enough to compete with the Cocoplum. As for Cinnamon, a trial was done in 1996 at l'Exile (see section on Cinnamon for details).

Frégate: Some areas dominated by Cocoplum have been restored successfully since 1997 and turned into native woodland as part of the reha-bilitation programme to replant native species on the island. Cocoplum was cut using machete and uprooted[5].

Ile du Nord: In 2007, the North Island landscaping team, as advised by PCA, initiated manual removal of small patches of Cocoplum in inselbergs, which they replanted with suitable native seedlings (mainly broadleaf species), as part of the ICS FFEM project 'Rehabilitation of Island Ecosystems'. Regrowth was substantial but the rehabilitated areas were success-fully maintained, and will be extended, probably using herbicide as well as manual cutting[6].

Anonyme: About 0.5 ha dominated by Cocoplum was hand removed by workers from DoE and ICS in 2005-2006 as part of the same (see above) FFEM project. No herbicide was applied and re-growth was observed in much of the area in subsequent years[7].

Moyenne: Cocoplum was removed in some parts of Moyenne Island by the late Brendan Grimshaw, but no details are available.

Praslin and Curieuse: Cocoplum dominates large areas on these two islands, mainly on the dry laterite soils where removal of large patches of Cocoplum would present an imminent risk of erosion, especially in the absence of replanting. Hence, no targeted removal attempts have been made. However, the control of Cocoplum was done on Curieuse as part of a Coco-de-Mer replanting project funded by the Dutch Trust Fund in 1997-1998. Cocoplum was cut back and cleared to make way for the planting of the Coco-de-Mer. One thousand nuts were targeted for replanting on approximately 10 ha at Baie Laraie, but due to poaching of the valuable Coco-de-Mer nuts the target was not achieved and both the replanting and clearing of cocoplum stopped prematurely[8].

Sainte Anne: By the end of November 2011, small patches of Cocoplum along the botanical trail were cleared by hand. Glyphosate was then applied on two occasions one month apart, to the cut stumps using a paintbrush. During the first application, pure

Fruits of the Cocoplum are round, pale-yellow with rose blush or pink-red and in clusters, edible and with a single large seed inside. E. Henriette

product was mixed with diesel in equal parts; and in the second application, (for plants that resisted the first treatment) pure product was mixed with 10% water and applied to re-cut stumps. Diesel increases the toxicity of the herbicide, and in addition

the diesel/kerosene acts as a carrier that takes the herbicide deep into the system faster than would otherwise occur by the normal translocation process. A large majority of Cocoplum plants died but some required follow-up treatment (rainy weather during the period of the first treatment might have affected its efficacy)[9].

The side effects and non-target impacts of composite applications with diesel outweigh the benefits, especially when it has to be spread on large areas. Diesel/kerosene is toxic to many plants and animals (it has for example been reported to substantially reduce egg fertility in birds). Diesel/kerosene can damage the rubber/washer parts of the applicator equipment (sprayer) sometimes rendering it useless even after the first application. In Hawaii, biodiesel is used instead of the toxic diesel to control woody invasives[10].

MANAGEMENT RECOMMENDATIONS
Refer to page 326 for management recommendations.

A large patch of Cocoplum in the South of La Digue.
Bruno Senterre

References/Further reading
1 PIER, 2011; 2 GISD, 2011; 3 Hill, 2002; 4 Kueffer & Vos, 2004; 5 S. Hill *pers. comm.*; 6 PCA/North Island, 2009; 7 Rocamora & Jean-Louis, 2009; 8 M.Vielle *pers. comm.*; 9 R. Nolin, *pers. comm.*; 10 J. Leary *pers. comm.*; 11 Hansen & Laboudallon, 2013; 12 Senterre *et al.*, 2013; 13 Nevill, 2009; 14 G. Rocamora, *pers. obs.*

Devil's tree ability for rapid spread is mainly through its prolific production of wind dispersed seeds. E. Henriette

Devil tree

Bois jaune

Bwa zonn

Alstonia macrophylla

IDENTIFICATION AND BIOLOGY

Devil tree is a tall fast growing evergreen tree reaching up to 30 m high with smooth, grey-whitish trunk. It has simple leaves that are shiny above, paler beneath, with numerous veins (15-20 pairs of lateral nerves). It has small funnel-shaped white flowers in clusters. It produces and is propagated by large amounts of wind-dispersed seeds throughout the year, and can regenerate from the seed bank, vegetative growth or re-sprouting. Found from the lowlands up to mid-altitude forests, it can rapidly colonise dry-ridge forests. Known to grow on degraded or infertile soils, it is drought tolerant, and is considered a commercial timber tree[1].

ORIGIN

Native to tropical and subtropical Africa, Central America, Southeast Asia, Polynesia and Australia[1].

DISTRIBUTION AND ABUNDANCE IN SEYCHELLES

Devil tree was introduced to Seychelles in the 1950s or 1960s as a timber tree. It occurs on Mahé, Anonyme, Conception, Ile au Cerf, Longue, Ste Anne and Thérèse[11, 15]; Praslin, Curieuse, Félicité, Grande Sœur, La Digue, Petite Sœur (only few small specimens[14]), Silhouette and Ile du Nord; and Frégate[2, 10-12]. On Félicité, there are no large trees, but some seedlings presumably introduced in soil or compost were spotted and removed[3]. Devil tree had also been recorded on one outer island: Farquhar, but has not been reported recently[13, 16].

DAMAGE AND THREATS

Devil tree poses a potentially high risk to biodiversity and is considered to be one of the most abundant

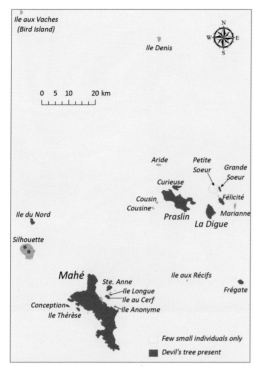

Devil tree poses a potentially high risk to biodiversity and is considered to be one of the most abundant invasive woody plant species in the Seychelles. E. Henriette

invasive woody plant species in the Seychelles[4, 5]. It was listed as a priority species for control at an IAS prioritisation workshop (UNDP-GEF-GOS Biosecurity Programme; November 2010, Seychelles)[6]. It has rapidly invaded inselbergs, mid-altitude, palm and dry-ridge forests, and has the potential for becoming an aggressive invader where it is present. It can invade habitats with harsh conditions that are usually resistant to invasion[4]. It exhibited a rapid natural increase in dry, exposed environments on scrub-covered hillsides on Ile du Nord and on Frégate island inselbergs. Previously considered to be primarily a species of disturbed, secondary or open habitats, it has also invaded closed-canopy forests. Its capacity for rapid spread is mainly through its prolific production of wind dispersed seeds.

ELIMINATION AND CONTROL PROGRAMMES
Mechanical control has been the main mode of control for Devil tree.

Silhouette: A control trial by NPTS showed that Devil tree could be killed by felling and also by ring barking, with no re-sprouting. The problem though, was the vast numbers of seedlings which kept germinating from the seed bank for at least 4 years[7].

Mahé: The Division of Environment controlled the species through repeated cutting until the plant died[8, 9].

Frégate: A few hundred Devil trees were cut down and burnt between 1997 and 2007 in an effort to control them. But the species is so numerous on the island, and even dominant in some places, that it will require a more integrated approach. In addition, endangered species like the Seychelles white-eye utilise Devil tree for both feeding and nesting and its control would have to be managed carefully[3].

Félicité: There are no large trees, but some seedlings, which may have been introduced in soil or compost. These have been hand pulled[3].

MANAGEMENT RECOMMENDATIONS
Refer to page 326 for management recommendations.

The species remains on the Protected Species list because of its value for timber, hence permission must be sought to the Environment Department before undertaking its removal.

References/Further reading
1 PIER, 2011; 2 Nevill, 2009; 3 S. Hill *pers. comm.*; 4 Kueffer & Vos, 2004; 5 Kueffer *et al.*, 2004; 6 Government of Seychelles, 2010; 7 Gerlach, 2011; 8 W. Andre *pers. comm.*; 9 M. Vielle, *pers. comm.*; 10 Hansen & Laboudallon, 2013; 11 Hill, 2002; 12 Senterre *et al.*, 2013; 13 Robertson, 1989; 14 Roland Nolin, *pers. comm.*; 15 G. Rocamora, pers. obs; 16 ICS/IDC, *pers. comm.*

Red sandalwood invades intact, undisturbed forests as well as disturbed sites and can quickly form large colonies and open thickets. Gérard Rocamora

Red sandalwood/coralwood

Bois de condori, Bois noir de Bourbon, Agati

Lagati

Adenanthera pavonina

IDENTIFICATION AND BIOLOGY

Red sandalwood is an open branching tree reaching up to 20 m high. Roots have numerous large nodules and nitrogen-fixing abilities. Leaves are compound, up to 30 cm long with 20-50 leaflets in pairs, which can be opposite to alternate on the same leaf. Flowers are white, yellowish, rose-pink or red. Pods are linear and twisting as they mature, each containing 15-50 bright red seeds. Red sandalwood grows fast, and is tolerant of a wide range of soils including alkaline, poorly drained, saline, and infertile soil; but it prefers heavy clay soils. It is well adapted to hot, humid environments. It is found in coastal habitats, natural forests and disturbed sites[1], but also in places where it has been planted for timber.

ORIGIN

Native to Tropical Asia (Southeast Asia, India and Malaysia). Its current distribution is widespread and pan-tropical[1]. Although some scientists consider it to be native to Seychelles, this status is doubtful.

DISTRIBUTION AND ABUNDANCE IN SEYCHELLES

Red sandalwood occurs on Mahé, Anonyme, Conception, Ile au Cerf, Moyenne, Ronde, Ste Anne and Thérèse[3, 10]; Praslin, Cousin, Cousine, Curieuse, Félicité, Grande Soeur, La Digue, Marianne and Petite Soeur; Silhouette and Ile du Nord, Frégate and Denis[2, 3, 8-10]. In the outer islands, it is found on Alphonse, Coëtivy, D'Arros, Poivre, but not in the Aldabra group[8, 11, contra 2]. It may have been introduced for ornamental and wood production purposes, but some believe that this species could also be indigenous to Seychelles.

DAMAGE AND THREATS

Its rapid early growth and erect habit usually enable it to access sunlight by overtopping neighbouring

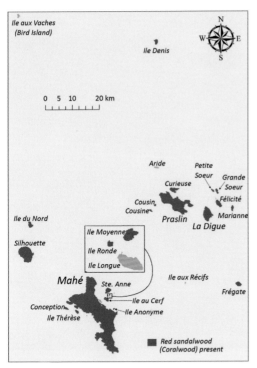

Red sandalwood
(Coralwood) present

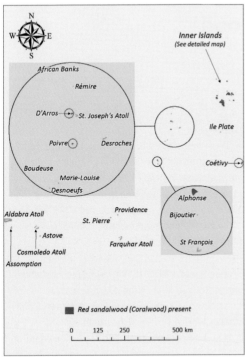

Red sandalwood (Coralwood) present

0   125   250   500 km

Red sandalwood produces linear and twisting pods each containing 15-50 bright red seeds. E. Henriette

plants. It invades intact, undisturbed forests as well as disturbed sites and can quickly form large colonies and open thickets[1]. In Seychelles, it has rapidly extended its distribution in the palm, lowland and intermediate altitude forests of Mahé and Silhouette and inselbergs[2,4] where there are deeper soil pockets. However, its impact appears more limited than that of other invasives, as it does not form dense stands and dominates only in former plantations[4].

ELIMINATION AND CONTROL PROGRAMMES
No control or elimination programme has been conducted by the Division of Environment for this species[5,6]. On Cousin Island, a control programme was started in 2004 to eliminate a small number of trees in the plateau woodland[7]. This mainly involved mechanical hand pulling and chemical treatment with 'Vigilant®' which contains 5% Picloram as the active ingredient. 'Vigilant®' was applied directly to the freshly cut stem with promising results. By 2005, no mature trees existed but regeneration of seedlings from the seed bank was apparent and they had to be continuously removed. Red sandalwood is still present on Cousin.

MANAGEMENT RECOMMENDATIONS
Refer to page 326 for management recommendations.

References/Further reading
1 GISD, 2011; 2 Nevill, 2009; 3 Hill, 2002; 4 Kueffer & Vos, 2004; 5 W. Andre pers. comm; 6 M. Vielle pers. comm.; 7 Dunlop, 2005; 8 Hansen & Laboudallon, 2013; 9 Senterre et al., 2013; 10 G. Rocamora & Pat Matyot, pers. obs ; 11 Mike Betts & Catherina Onezia/ SIF pers. comm.

ONE OF
WORLD'S WORST
100
INVASIVE
SPECIES

Strawberry guava forms dense thickets posing a major threat to endemic flora by competing for light and soil nutrients. Thomas Le Bourgeois / Cirad - Pl@ntinvasion

Strawberry guava/Chinese guava

Goyavier

Gouyav desin

Psidium cattleianum

IDENTIFICATION AND BIOLOGY

Strawberry guava is an evergreen slender tree growing up to 8 m high. It has smooth reddish-brown peeling bark. Leaves are opposite, simple, shiny dark green, elliptic to oblong, and up to 8 cm long. Flowers are numerous and white with yellow stamens. Fruit is a globose berry, 3-6 cm long, purple-red, with whitish flesh, usually sweet-tasting when ripe. It propagates both by seeds and vegetatively. It engages in expansive vegetative reproduction through root sprouts whilst its prolific seed production is widely dispersed by birds and mammals[1].

Strawberry guava occurs in a range of habitats such as agricultural land, coastland, natural forests, planted forests, grasslands, riparian zones, disturbed areas, shrublands, urban areas, and wetlands. It is found from near sea level to high altitude forests. It is shade tolerant and known to grow on degraded or infertile soils[1].

ORIGIN

Native to Brazil. It has been introduced to various parts of the tropical world because of its edible fruit[1], which is greatly appreciated in western Indian Ocean islands.

DISTRIBUTION AND ABUNDANCE IN SEYCHELLES

Strawberry guava occurs on Mahé, Anonyme, Conception, Ile au Cerf, Moyenne, Ste Anne and Thérèse[12, 13]; Praslin, Aride, Curieuse, La Digue, Félicité, Grande Soeur and Marianne; Silhouette and Ile du Nord; and Frégate[2, 10-12]. It appears to have been introduced in 1870.

DAMAGE AND THREATS

Strawberry guava is a serious invasive species[3] that alters habitats and poses a major threat to endemic flora by competing for light and soil nutrients[1]. It

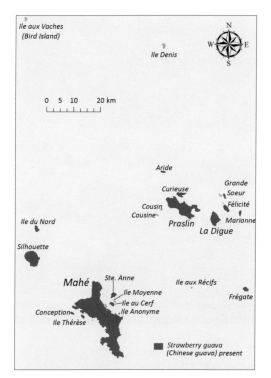

forms dense thickets that shade out native vegetation; meanwhile its very extensive surface root mat inhibits the growth of other species nearby. Thus, it can out-compete native plants and prevent natural regeneration. It is able to invade areas with low disturbance. In Seychelles, the Strawberry guava is classified as one of the main woody invasive species in semi-natural to natural habitats. It has invaded montane forest, inselberg, mid-altitude forest and palm forest, and is considered a threat to native biodiversity[4]. It was listed as a priority species for control at the 'IAS prioritisation workshop' (November 2010)[5].

In Mauritius it has had a devastating effect on native habitats, and in Hawaii it has invaded a variety of natural areas including the dry-mesic forest and wet forest, at a variety of altitudes, and is considered to be the worst plant pest[1, 6].

ELIMINATION AND CONTROL PROGRAMMES
No large scale programme to control or eliminate the species has ever been undertaken because it is so widespread. In 1995-1996, there was an experiment

The Strawberry guava fruit is a globose berry, 3-6 cm long, purple-red, with whitish flesh, usually sweet-tasting when ripe, and highly appreciated in La Réunion and Mauritius. E. Henriette

funded by the Indian Ocean Commission to investigate control methods for Strawberry guava on Mahé in the Morne Seychellois National Park (Congo Rouge)[7]. Refer to text on Cinnamon and Cocoplum for details on physical and chemical treatment.

On Silhouette, NPTS did some cutting and shoot removal but very little control was done overall[8].

In other parts of the world including Hawaii, effective chemical control was achieved using Triclopyr in the following concentrations: 4% foliar spray, effective on saplings of less than 2 m high; 20%, for basal bark and cut-stump; 100%, for undiluted injection into a stem cut. The commercial product Milestone®, comprising 10% Aminopyralld (a broadleaf-selective herbicide), was also injected[9].

MANAGEMENT RECOMMENDATIONS

Refer to page 326 for management recommendations.

A monostand of Strawberry guava preventing regeneration of native plants.
Thomas Le Bourgeois / Cirad - Pl@ntInvasion

References/Further reading
1 GISD, 2011; 2 Nevill, 2009; 3 Lowe *et al.*, 2000; 4 Kueffer & Vos, 2004; 5 Government of Seychelles, 2010; 6 J. Beachy *pers. comm.*; 7 Beaver & Mougal, 2009; 8 J. Gerlach, *pers.comm.*; 9 Leary & Hardman, 2012; 10 Hansen & Laboudallon, 2013; 11 Senterre *et al.*, 2013; 12 Hill, 2002; 13 G. Rocamora & Pat Matyot, *pers. obs.*

Left: A dense stand of White cedar. Right: Details of flowers (top) and fruits (bottom).
E. Henriette

White cedar

Calice du pape

Kalis dipap

Tabebuia pallida

IDENTIFICATION AND BIOLOGY

White cedar is a deciduous tree reaching up to about 25 m high. It has a grooved bark and a narrow crown. Leaves are opposite, palmately compound with 3-5 leaflets, 6-16 cm long, and rounded at their extremity (unlike the cultivated Kalis dipap roz, *T. heterophylla/rosea*). It has large white to light purple flowers that are borne in terminal and lateral clusters, or occasionally as individuals. The flowers produce a lot of nectar and are highly attractive to endemic sunbirds and other native and introduced pollinators. Fruits are dark brown cigar-like pods (when ripe), about 8 to 20 cm long, and contain many winged seeds each about 2 cm long. White cedar is widespread in natural tropical forests, secondary forests and abandoned pastures in its native range. It grows on steep slopes and ridges but also occurs on flat land adjacent to rivers from sea level to higher elevations. It can tolerate any soil type including poor or degraded soils but seems to prefer deep clays. It is fast growing and forms pure monotypic stands[1].

ORIGIN

Native to South America[1].

DISTRIBUTION AND ABUNDANCE IN SEYCHELLES

White cedar occurs on Mahé, Anonyme, Conception, Ile aux Cerf, Longue, Moyenne, Ronde, Ste Anne and Thérèse[3,9]; Praslin, Curieuse, Félicité, Grande Soeur and Marianne; Silhouette and Ile du Nord; Frégate, Bird and Denis islands[2,3,8]. It is very abundant on Praslin and along all coastal roads on Mahé. It also occurs on some outer islands: Coëtivy, Desroches,

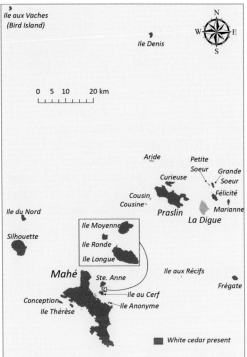

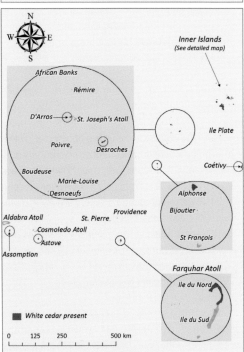

D'Arros, Alphonse, Farquhar (Ile du Nord) and Astove[11] but not Aldabra[2, 8, 10 contra 4]. White cedar was introduced during the 20th century to stabilise and conserve soil, and is also used as timber[5].

DAMAGE AND THREATS

Proliferous production of wind dispersed seeds, high propagule pressure and its fast growing habit enable it to easily outcompete native plants[1, 5]. It bears leaves and branches almost to the base and casts a deep shade under which virtually no other species can grow. Its thick leaf litter may also prevent the growth of native seedlings. In the seedling and sapling stages, it is an aggressive pioneer, and it can maintain viable populations in both dry and moist forest habitats. In Seychelles, it has invaded coastal, mid-altitude and palm forests[4, 5].

ELIMINATION AND CONTROL PROGRAMMES

There has been no formal programme to control or eliminate the species, but exploitation of the tree for economic timber is a viable means to control populations of such invasive timber tree species.

On Silhouette, control of White cedar was done by NPTS through ring barking which was effective with no re-sprouting, but very difficult to do properly due to the shape of the trunk, and its ability to recover when the ring barking was not done properly Trials with Glyphosate were ineffective. On Ile du Nord, ring barking has also been tried and proven to be effective, provided that the bark and the cambium were properly removed including from the grooves of the trunk[6].

In the Vallée de Mai, Praslin, SIF is experimenting to control White cedar and other woody invasive species using both physical control (ring barking) and chemical control (using Picloram). But it is still too early to draw any conclusions about the effectiveness of these methods[7].

It is noteworthy that White cedar is a fire resistant species that has previously been used to rehabilitate burnt areas on Praslin.

MANAGEMENT RECOMMENDATIONS

Refer to page 326 for management recommendations.

References/Further reading

1 GISD, 2011; 2 Hansen & Laboudallon, 2013; 3 Hill, 2002; 4 Nevill, 2009; 5 Kueffer & Vos 2004; 6 Gerlach, 2011; 7 Seychelles Island Foundation *pers. comm.*; 8; Fosberg & Renvoize 1980; 9 G. Rocamora & P. Matyot, *pers. obs.*; 10 Mike Betts & Catherina Onezia/SIF *pers. comm.*; 11 V. Laboudallon, *pers. obs.*

Management recommendations for broadleaf exotic species

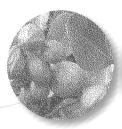

For widespread species such as Cinnamon and Strawberry guava, effective control may not be possible. Nonetheless, the targeted management of key core biodiversity areas can be achieved. Unfortunately, the very limited information available from control trials in Seychelles does not allow us yet to reach firm conclusions in regard to best recommended practice (chemical, dosage, type of application, etc.). Hence, only broad general recommendations can be provided, based on the outcomes of the trials conducted so far, which are either described in the 'Elimination and Control' sections of species accounts, or below when they concern a variety of plant species.

CONTROL AND ELIMINATION PROTOCOL

CHEMICAL CONTROL

As a general rule, chemicals should not be seen as an ideal one-off solution as they often require repeated treatments and follow-up and may be costly. That is why effective dosage and application protocols need to be established. Repeated use of herbicides, especially at high concentrations, generates pollution and may be dangerous to the environment (certain plants and animals are particularly sensitive) and to humans. It follows that use of herbicides should be minimised, precautions of use strictly respected, and long term exposure avoided (especially around residential areas with children or pregnant women).

Some herbicides are strongly suspected to be toxic to a number of insect species including bees (and to indirectly affect pollination) and are carcinogenic. Older toxic herbicides like 2,4-D or 2,4,5-T (both in commercial product 'Agent Orange') were developed as defoliants by the US military, and are still widely used around the world.

Herbicides used in the control of broadleaf weeds are Glyphosate (Roundup®), Picloram (Tordon®, Grazon®), Dicamba (Vanbel®, Oracle®, Vanquish®) and Triclopyr (Garlon®, Release®)[1-6]. Glyphosate is a widespread and highly effective herbicide used in agriculture. Some authors consider Glyphosate as one of the least dangerous herbicides as it does not bioaccumulate and degrades rapidly[7].

Cut stump treatment: Cut the whole tree as close to the ground as possible and apply the recommended dose of herbicide to the cut stumps using a paintbrush or a sprayer. Apply the herbicide to the cut stump as quickly as possible for more effective treatment. A tracer dye can be incorporated into the herbicide solution to ensure treatment of the entire cut stump and of all individual stumps[3, 4].

Drill and plug - for large trees: Using a drill, place holes every 15 cm around the trunk at a slightly declining angle (15-30%) and inject the recommended herbicide and dosages into the sapwood with a syringe. Close the holes with a wood plug to prevent rainwater from diluting the herbicide. This method has been applied in Mauritius, La Réunion and Rodrigues islands.

Foliar spray: Foliar spray introduces the herbicides directly onto the leaves of a plant. Use spray equipment like a backpack sprayer with the recommended herbicide and dosage to spray and cover the entire foliage. Spraying from multiple angles provides good coverage. Thoroughly wet all leaves, but not to the point of runoff. Incomplete coverage of all the foliage may result in ineffective control. It may be necessary to repeat the foliar spraying following consecutive intervals before complete control is accomplished. To avoid non-targeted species spray drift needs to be controlled by adjusting the nozzle to control the squirting of herbicides. Otherwise this method will have high non-target impacts; and it is tricky to use when desirable plants are located in the immediate vicinity of untargetted plants. A dye can be included into the herbicide to trace its application and ensure that plants are not missed and that there is complete coverage of the foliage[3, 4]. A 'sticker' can be added to improve adherence and reduce rain wash.

Incision Point Application (IPA) method– for plants less than 8 cm in diameter: Using a machete, cut several small incisions at a 45° angle penetrating just beyond the cambium layer (about 5 cm deep) to create a clean notch. Widen the notch, if necessary, by wiggling the machete. The incisions should be around the base of the tree

RESULTS OF ONGOING TRIALS COMBINING PHYSICAL CONTROL AND HERBICIDE APPLICATION

Mahé: Seychelles National Parks Authority, in collaboration with Dr Christopher Kaiser-Bunbury, has conducted extensive replicated herbicide application trials since March and April 2011. During the trials, several removal techniques were applied to 10 alien invasive tree species (including Cinnamon, Coco-plum, Devil tree, *Albizia*, Strawberry guava, and White cedar) at three locations on Mahé (Salazie, Tea Plantation and Trois Frères glacis). Two different sets of trials were conducted to test the most effective removal technique for a set of invasive alien plant species, and to assess the potential threat of herbicide application to non-target species. Treatments included several concentrations of herbicide (5, 10 and 15%) and different techniques including cut, cut and spray the stump, and foliar spray. The herbicide used was Picloram ('Tordon®'). Results suggest that this herbicide is effective when applied to cut stumps (mortality increased from 42 to 98%), but foliar spray resulted in high mortality of native saplings (53%) and adults (18%), and both treatments reduced seedling abundance and native adult growth[8,15].

Sainte Anne: Experiments, conducted in late November 2011 by Roland Nolin and Ste Anne Resort, combined cutting and application of the herbicide glyphosate to stumps using a paintbrush. The concentration used was: pure commercial Roundup® product (Glyphosate acid eq. 360 g/L/Glyphosate iso-propylamine salt 480 g/L) mixed with kerosene in equal parts. One month later, those plants surviving the first treatment were given another treatment. This time, pure product with 10% water was applied. The results have been encouraging for Indian laurel *Litsea glutinosa* (Bwa zozo), White leadtree *Leucaena leucocephala* (Kasi), Cocoplum and Cinnamon. The majority of small trees with trunks of more than 2-3 cm of diameter died, whereas smaller ones survived (particularly Leadtree) and had to be physically uprooted, and some large ones also required follow-up treatment. A similar operation was conducted in April 2012 on small leadtrees. The trunk sections were sliced to facilitate product absorbance before applying a solution of 75% diesel + 25% of pure commercial Roundup®. More than 80% of small trees were dead within a month[9].

D'Arros: White leadtree - and also Papaya *Carica papaya* - were cut off at the base using a machete or chainsaw and the cuts of the stump painted with a mixture of Roundup® (25%) and diesel (75%) to prevent re-sprouting. This was generally effective (particularly on Papaya) but several leadtree stumps resprouted some months later. These trees had their stem cortex exposed and were successfully retreated with undiluted Roundup®. The drill and plug method has also been applied successfully to eliminate Coconut trees, through the injection at the base of each trunk of 7.5 ml or 10 ml of 'Roundup®' into a single hole of 1-2 cm of diameter and up to 25 cm deep[10]. Coconuts that were injected with 2.5 or 5.0 ml did not die.

Foliar spraying can be effective in controlling woody invasive species.

Christopher Kaiser-Bunbury

20-50 cm above ground at equidistant points (with 10- 30 cm between each incision). Use a syringe to deliver 0.5 ml of undiluted herbicide to the centre of each cut. This method has been effectively trialed on invasive woody species in Hawaii using herbicides that include triclopyr, Glyphosate and aminopyralid[5].

Recommended concentrations of herbicide vary depending on the brand of commercial product used and type of application; generally between 1% to 10% for foliar spray or paintbrush application on stumps, and up to 50% for very big trees. It is very important to apply the herbicide immediately after the cut is made, and that treatment is done in dry conditions, outside rainy periods. Under wet conditions, concentrations will need to be higher to reach the optimal effect, but higher concentrations are likely to negatively impact the surrounding environment, especially if sprayed and it is not recommended to apply herbicides in wet conditions. For very resistant species or trees, herbicide (such as Glyphosate or Picloram) may have to be applied at higher concentrations on stumps and multiple times before the root system can be killed, especially when the absorption area (i.e. the cut or debarked area) is small. However, increasing concentration will make the operation less cost effective and not necessarily more efficient. Increasing the absorption surface by cutting the trunk or stump at an angle, or by slicing with a machete appears to enhance the efficacy of treatment to control White leadtree *Leucaena leucocephala* (*Kasi*).

PHYSICAL CONTROL

Ring barking - for trees more than 8 cm in diameter: Using a machete or an axe strip away a portion of the bark and the cambium (30 cm wide) all around the trunk at a height of 1 m above the ground[11]. In contrast to tree felling, ring barking allows the tree to die gradually within six to seven months, at the same time providing a longer shading period for shade dependent native species to re-establish themselves within

the area. Ringed-barked trees are often left to die and fall by themselves, but we recommend instead to cut them down, or pull them down using ropes – after they have died and dried, so as to avoid accidents and injuries caused by their fall (one death was recorded in South Mahé in the late 1990s). Ring-barking should therefore **not** be used in public areas such as along roadsides, footpaths or near habitations for safety reasons. Sometimes, a herbicide is applied to the exposed ringed area (using a combination of physical and chemical methods). Ring barking is not effective on some invasives, including Cinnamon and White cedar, which can grow back their bark, especially if the cambium is not properly removed.

Ring barking can be efficient when done properly by cutting away the cambium around the trunk.

Elvina Henriette (left);
Lucia Latorre/SIF (right)

Diminish the seed bank: Continued removal of invasive species can diminish the seed bank enough to encourage native regeneration. An additional strategy is to conduct supplemental native species planting or seed spreading (sowing). Follow-up maintenance by regular removal of surviving invasive plants and any new seedlings is essential for success. However, treating and sustaining follow-up maintenance is very labour intensive, and may not be feasible on a large scale.

Tree felling - of large trees: Use a chainsaw to cut down trees as close to the ground as feasible. Sometimes the tree may resprout, and in such cases, a herbicide may be applied across the entire cut surface.

Hand-pulling: Hand-pulling is effective for seedlings, small herbaceous shrubs and weeds. But sometimes, the soil disturbance caused may promote invasion by other alien plant species.

INTEGRATED MANAGEMENT AND ECOSYSTEM APPROACH

Combining tree removal and herbicide: Trees can be removed by hand or using chainsaws and stumps treated with a herbicide to prevent re-sprouting (see the Cut stump

PERSONAL PROTECTION FOR PESTICIDE USE

- Protective clothing, properly functioning equipment, and careful application methods all help minimise exposure to pesticides during all phases of handling, including storage, mixing, transportation, application, and cleanup.
- Any time you handle pesticides, wear at least a long-sleeved shirt and long-legged pants made of sturdy material. Fasten the shirt collar completely to protect the lower part of your neck. A hat is also recommended and coveralls may be useful. Also, bring along an extra change of clothing to avoid contaminating car seats or chairs.
- Wear chemical resistant and easy to clean footwear, such as rubber boots, that come at least halfway to the knee.
- Wear face mask or breathing mask, goggles or safety glasses when spraying chemical solutions and when mixing or pouring herbicides. They should be rinsed after each use, dried, and stored in a clean place.
- Wear chemical-resistant gloves at all times. To reduce exposure further, sleeves should be tucked into gloves that should reach up the forearm.
- Make sure gloves are clean, in good condition, and worn properly; replace gloves often. Wash gloves thoroughly before taking them off, and wash your hands thoroughly and dry them before you put the gloves on again.
- Wash hands thoroughly before eating, drinking, smoking, or going to the bathroom.

Modified from Hillmer, Steward & Liedtke 2003.

Herbicide sprayer: Foliar spraying using a back-pack sprayer; Application of herbicide to a cut stump.

C. Kaiser-Bunbury

treatment above). Foliage and branches can be left as mulch on the ground around native species saplings to facilitate their growth and prevent rapid re-establishment of unwanted invasive species. Using herbicides at lower concentrations may still reduce resprouting considerably (probably over 60%) and may represent a compromise solution which balances efficiency, treatment cost, pollution and the need to follow-up in an acceptable manner. Follow up maintenance should therefore be seen as indispensable after removal of broadleaf exotic trees, even when herbicides have been used.

Small scale management: Another way to efficiently and cost effectively manage broadleaf invasive species, is to either create small gaps through the felling of a few of the invasive species trees and then planting the gaps with native plants, or by weeding stands of mostly native trees. Small scale management may be more beneficial than the removal of whole patches of invasive species which can lead to a rapid reinvasion by other non-native species and increased erosion. At the same time, the creation of native vegetation patches may act as sources of native seeds, which may increase the proportion of native biodiversity in the invasive-dominated forest. Moreover, some invasive species have long been established in the Seychelles and may play a positive role in ecosystem functioning[12]. For instance the dominance of most forests by Cinnamon since the 19th century may have kept out other more serious invaders; and the prolific production of fruits has been a source of food for several native birds and the Seychelles fruit bat.

Tree injector for herbicide treatment.
Thomas Le Bourgeois/CIRAD

PREVENTING REINFESTATION

Educate the public about the problem of invasive species and their environmental and economic impact. Commitment of the general public is prerequisite to successful control measures.

Train Custom services at border control (international airport and Victoria port) and personnel from other organizations concerned with biosecurity about the threat of invasive plant species.

Improve the capacity of the Plant Protection Unit (in terms of human and financial resources) to fulfill its mission and to provide effective border control to prevent the introduction of exotic plant species.

Update the legislative framework to prevent new introductions of potentially invasive woody plant species (e.g. Breadfruit and Other Trees Act, Plant Conservation Act).

Actively restore habitat for the protection of the native flora and prevent reinfestation by exotic species.

References/Further reading
1 GISD, 2011; 2 CAB International, 2011; 3 Ferrell *et al.*, 2006; 4 Leary *et al.*, 2012, 5 Leary *et al.*, 2013; 6 Motooka *et al.*, 2003; 7 Mensink & Janssen, 1994; 8 C. Kaiser-Bunbury (University of Technology, Darmstadt, Germany) & J. Mougal (SNPA) *pers. comm.* (Mahé trials); 9 R. Nolin & G. Rocamora *pers. comm.* (Ste Anne trials); 10 von Brandis *et al.*, 2012; 11 Beaver & Mougal, 2009; 12 Kueffer *et al.*, 2012; 13 Hillmer *et al.*, 2003; 14 Kaiser-Bunbury *et al.* 2015.

Creepers

Devil's ivy covers forest floor and climbs up tree trunks into the forest canopy, smothering and eliminating native plants. E. Henriette

Devil's ivy

Pothos, Liane du Diable

Filodendron (not specific)

Epipremnum pinnatum syn. aureum

IDENTIFICATION AND BIOLOGY

Devil's ivy is a vine that climbs and clings to tree trunks by means of aerial roots. It grows to 20 m tall, with stems up to 4 cm in diameter. The leaves are evergreen, alternate, and heart-shaped. They are entire on juvenile plants, but irregularly pinnatifid (i.e feather-like) on mature plants. It is found in disturbed habitats and along roadsides. Although drought and shade tolerant, it also develops well in moist places and in partial sun making it able to colonise a variety of habitats. Reproduction is mainly vegetative. Berries contain seeds that are eaten and dispersed by various animals[1, 2].

ORIGIN
Temperate Asia to Australia[1].

DISTRIBUTION AND ABUNDANCE IN SEYCHELLES
Devil's ivy has been recorded on Mahé and Ile au Cerf; Praslin, La Digue and Félicité; Silhouette and Ile du Nord; Frégate[3] and Denis islands[4, 6]. The species dominates open and closed secondary forests near roads and abandoned settlements, but its spread in closed forest seems very slow[5]. On Denis Island, it is restricted to only a small area in the north which suggests its introduction may be recent[4].

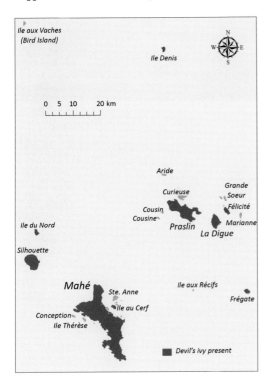

DAMAGE AND THREATS
Devil's ivy covers forest floor and climbs up tree trunks into the forest canopy. It smothers and eliminates native plants[1, 2].

ELIMINATION AND CONTROL PROGRAMMES
In the Seychelles, the following management methods have been applied.

Physical control
Long-term removal in the Vallée de Mai (Praslin) started in 1980. Creepers were cut at 1 m above ground, and then removed from the ground and from the trees by pulling and hooking. Regular follow-up work had to be undertaken to remove re-sprouts. To date the creepers have almost been eliminated from the Vallée de Mai, although there are small pockets of re-sprouting individuals[2].

Chemical treatment
The Forestry Section of the Division of Environment trialed the use of salt water and the herbicide Roundup® (Glyphosate)[2]. These were not effective against the creeper. However, trials with 'Vigilant®' (active ingredient Picloram, as potassium salt in gel form) produced good preliminary results. Unfortunately, the trial could not be extended due to insufficient herbicide supply and hence the long-term effectiveness of Vigilant® could not be ascertained. Nonetheless, the following experimental protocol was proposed for future trials:
- Cut stems 10 cm above the ground.
- Treat both parts of the cut stem by applying 3-5 ml of 'Vigilant®' gel. The product should kill both the root system and the aerial parts, preventing the shooting of aerial roots.
- Send tree loppers to apply Vigilant® to the main stem remaining high up in the trees.
- If aerial shoots are close to the ground, treat them as well.
- In order to avoid leakage of the product to the ground, treated stems should be covered with plastic bags.
- Avoid doing the treatment during rainy periods.

MANAGEMENT RECOMMENDATIONS
Refer to page 340 for management recommendations of alien invasive creepers.

References/Further reading
1 GISD, 2011; 2 PIER, 2011; 3 Beaver & Mougal, 2009; 4 E. Henriette *pers. comm.*; 5 Senterre, 2009; 6 Hansen & Laboudallon, 2013.

Japanese climbing fern can invade all types of habitats from wetlands, forests to agricultural lands. E. Henriette

Japanese climbing fern

Fougère grimpante, Fougère volubile

Fouzer Zaponnen

Lygodium japonicum

IDENTIFICATION AND BIOLOGY

Japanese climbing fern is a hardy, fast-growing twinning fern with fronds (leaves of ferns) extending to 30 m in length, and lobed leaflets. It has slender, dark brown rhizomes. When it climbs onto trees and forms dense mats, it smothers the trees. It grows in several types of habitats including: shady as well as sun-lit areas, gardens, roadsides, natural forests, plantations, wetlands, lakes and agricultural lands. It is a problematic invader of both forest and glacis habitats. Because it reproduces by spores that are wind dispersed, it can colonise vast areas. Spores can also be dispersed to other areas by attaching to animals, sticking onto clothes or being transported on equipment used during mechanical removal. Japanese climbing fern persists and colonises by rhizomes[1, 2, 3, 4].

ORIGIN

Eastern Asia, Australia and the East Indies[1, 4].

DISTRIBUTION AND ABUNDANCE IN SEYCHELLES

Restricted to Mahé, Japanese climbing fern seems to be a recent new introduction in the Seychelles and is spreading fast. It was found to be abundant on exposed trails above Cascade, and on the Montagne Posée road. It was also found in a garden along the road at Saint Louis, and in Victoria. After this (late 2010), the Terrestrial Restoration Society of Seychelles (TRASS) discovered the species along Sans Soucis Road. It has also been recorded along Beau Vallon and Bel-Ombre roads[5, 6].

DAMAGE AND THREATS

Japanese climbing fern is one of the worst invasive creepers and is a serious threat to biodiversity. It rapidly climbs to the tree canopy where it forms dense mats, shading out the host trees and any other supporting vegetation. The thick mats make it difficult for trees to grow, weakening or even killing smothered trees. It can invade all types of habitats: marshes, mangroves, streams, forests, and agricultural lands. It has the potential to impact native species and is able to dominate all other species and all ecological niches, from understory to canopy[1-4]. It is a fire hazard and can alter fire behaviour by facilitating the rapid spread of fires, which is a problem for fire-prone areas such as Praslin Island, Seychelles. In Florida, Japanese climbing fern causes fires when its thick mats enclose trees and serve as ladders to carry fire into tree canopies[3, 4]. It was listed as a priority species for control in the Seychelles (IAS prioritisation workshop, UNDP-GEF-GOS Biosecurity Programme, November 2010)[7].

ELIMINATION AND CONTROL PROGRAMMES

No elimination or control programmes exist for this species in the Seychelles. However, TRASS in association with the Plant Conservation Action group (PCA) and with funding from UNDP, conducted trials on the management of invasive creepers including the Japanese climbing fern[8].

The trial tested both physical treatments (cutting and uprooting) and chemical application (using Glyphosate in the form of Roundup®) on 2 m² plots

The Japanese climbing fern, one of the worst invasive creepers in the Seychelles is posing a serious threat to biodiversity.
E. Henriette

at Cascade, Mahé. Physical treatment consisted of cutting with the machete along the perimeter of the plot and uprooting all stems from within the plot. The chemical treatment consisted of foliar spraying of the plots. Leaves and stems were thoroughly sprayed using the recommended dosage 1.5% Glyphosate. Post-treatment monitoring after 1 and then 3 months indicated that physical removal was ineffective for the Japanese climbing fern which was vigorously resprouting in all plots, and that the dosage used for the chemical treatment was too low to kill the creepers. Chemical treatment with a physical method cutting stems to kill fronds above the cut location; then applying higher doses of the herbicide (e.g. 2%, 3% and 5%) by targeted spraying to the rooted portion of the plant should be trialed in future.

A trial conducted in Florida in 2010 tested three herbicides (glyphosate, imazapyr, and metsulfuron methyl) through foliar spraying at different concentrations, alone and in combination, to evaluate their efficacy for fern control[9, 10]. All herbicide treatments had reduced Japanese climbing fern cover by 77-98% within one year following treatment; but there was considerable re-growth on imazapyr and metsulfuron-methyl treated plots after two years. Glyphosate applied alone as a 2 or 4% solution provided 91-98% fern control after 2 years. There was no significant improvement when glyphosate was used in combination with other herbicides. In summary, the study recommended direct spray applications using 2% glyphosate solutions for efficient control of Japanese climbing ferns.

Other studies in the USA suggested that the most common control method has been application of glyphosate and metsulfuron herbicides. These can be used either individually or in combination, and either as foliar-spray application of 2% 'Roundup®' (glyphosate) or as a cut-stump treatment[11].

Trials in Florida showed that physical control of the fern through cutting resulted in death of fronds above the cut location but fronds re-grew from material below the cut and after hand pulling. In addition, cut fronds may harbour viable spores easily dispersed by disturbance. Dead fronds need to be removed to reduce fire-risk and to enhance recovery of native plants.

Biocontrol of the Japanese climbing fern has also been considered such as the rust fungus, *Puccinia lygodii* which severely infect the foliage causing wilting and death[1]. Surveys in search of natural predators of *Lygodium* were conducted in West Africa, the Dominican Republic, and Argentina where only one natural enemy was discovered-the *Tenuapalpis* mite[12, 13]. The mite damages the fern, but it is not a specialist; moths, sawflies and beetles have also been identified as natural enemies but primarily for another species of fern (*L. microphyllum*)[14]. Although the use of insects as biocontrol agents seems to be a promising approach, it is nonetheless important to consider only species that have restricted hosts or that are themselves host specific in order to prevent damages to other non-targeted ferns and other plants.

MANAGEMENT RECOMMENDATIONS
Refer to page 340 for management recommendations of alien invasive creepers.

References/Further reading
1 GISD, 2011; 2 CAB International, 2011; 3 Ferriter, 2001; 4 PIER, 1999; 5 Mougal & Henriette, 2012; 6 Senterre, 2009; 7 Government of Seychelles, 2010; 8 Henriette *et al.*, 2012; 9 Minogue, 2010; 10 Bohn *et al.*, 2011; 11 Langeland *et al.*, 2009; 12 Hutchinson *et al.*, 2006; 13 Pemberton *et al.*, 2002; 14 Boughton *et al.*, 2011.

Merremia crawls up and covers trees, forming thickets that smothers and strangles other vegetation. G. Rocamora

Merremia

Liane d'argent

Lalyann darzan

Merremia peltata

Note: the origin of *Merremia peltata* in the Seychelles is being debated. For a long time the species was believed to be introduced, but some authors now believe that it may be native[4, 12]. There are some cases in which a species considered to be native has become invasive, although the reasons behind this sudden change are not clearly understood. Over the last two decades *Merremia* has heavily invaded disturbed areas mainly along roadsides, forest edges, abandoned agricultural lands and human habitation, but invasion is not apparent in secondary and undisturbed forests where it seems to occupy a historical niche in natural forest succession processes. *Merremia* may have become invasive by adapting to the considerable extensions of open landscapes created by the human-induced alteration of natural habitats (see also p. 51 and p. 183).

IDENTIFICATION AND BIOLOGY

Merremia is a climbing vine with underground tubers that make it difficult to control unless tubers are uprooted. It has smooth stems reaching up to 20 m high. Leaves are heart shaped and alternate with purple veins beneath. White funnel shaped flowers are borne in clusters on stalks 15-30 cm long. The species name 'peltata' comes from the leaves that are peltately attached, i.e. attached to the stalk near the center. It disperses either vegetatively or by seeds, and can resprout from stem fragments after cutting. Merremia may provide rapid ground cover following land disturbance, thereby reducing erosion and nutrient loss. It can be found along roadsides and forest fringes. It is known to colonise forest gaps where it acts like a gap specialist, i.e. invading newly opened gaps, and then playing an important function in forest dynamics. It is a lowland species but may invade submontane forests if disturbed[1, 2].

ORIGIN

Native to Pemba Island in Tanzania, Madagascar, Mauritius, La Réunion and possibly the Seychelles; also native to Indonesia, Malaysia, the Philippines, northern Queensland of Australia and French Polynesia. It has been introduced and subsequently became invasive on some Pacific islands, such as Vanuatu. Merremia is sometimes promoted (especially through agriculture) as a means of providing rapid ground cover (thereby reducing erosion and nutrient losses following disturbance of the land), and also as a source of food for cattle[1, 2].

DISTRIBUTION AND ABUNDANCE
IN SEYCHELLES

Merremia *is* known from Mahé, Silhouette[3, 4] and Denis[5]. On Mahé, the creeper is considered a major problem because it has invaded many disturbed areas (including agricultural land, residential and peri-urban habitats)[4] but on Silhouette it is currently only a problem along the roads and footpaths. On Denis, Merremia is found in small numbers in the Badamier forest *Terminalia catappa* near the Airstrip[5].

Merremia, a native species, has only recently become invasive particularly in human-modified landscapes. E. Henriette

DAMAGE AND THREATS

The invasiveness of the species is most noticeable in disturbed habitats where land and forests have been cleared[4]. An important threat is to agricultural production where it can have an economic impact. The environmental threat is huge. As the vine crawls up and covers trees, forming thickets either on the ground or over the canopy, it smoothers and strangles other vegetation[1]. In Seychelles, it invades disturbed areas of lowland and mid-altitude woodland and forest that are dominated by secondary non-native plants[4]. It has spread to forest edges of important biodiversity sites like Copolia, Mare Aux Cochons, Congo Rouge and La Reserve where it poses a threat to native biodiversity. It seems to have established a real foothold at La Reserve and its rate of invasion should be of real concern[6, 7]. As ground cover, Merremia suppresses non-native weeds that would likely be present as ground cover in its absence.

In Samoa, Merremia is considered native but it is an aggressive vine that is covering stands of native lowland rainforest[1]. In the Comoros archipelago, the invasive Merremia is one of the major threats to biodiversity[2, 7]. In Vanuatu, it is invading previously un-infested forests and causing the death of large numbers of canopy trees, covering entire trees and forming 'myriads of stems lying tangled like a mass of electric wiring over the forest floor'[8]; it threatens the Vatthe Conservation Forest, home to several globally endangered species of birds, 92% (2300 ha) of which

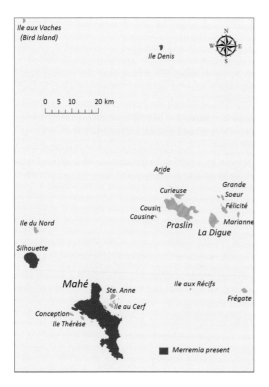

Ile aux Vaches
(Bird Island)

Ile Denis

0 5 10   20 km

Aride

Curieuse *Grande*
            Soeur
Cousin *Félicité*
Cousine
Ile du Nord *Praslin* *Marianne*
                La Digue
Silhouette

Mahé *Ile aux Récifs*
Ste. Anne
            Frégate
Ile au Cerf
Conception
Ile Thérèse

■ Merremia present

had been invaded by 2010[8, 9]. This included 1,300 ha that had become so infested that they needed to be entirely cleared and replanted. Merremia suppresses species diversity and aids the spread of other creepers when it forms a ground cover.

ELIMINATION AND CONTROL PROGRAMMES

Physical and chemical controls were trialled on Mahé (at Le Niole, Sans Soucis, Port Glaud, Port Launay and Intendance) between 1999 and 2003 by the Forestry Section of the Division of Environment[6]. The vine was cut on the lower part and the remnants were uprooted. All plant materials removed were burnt in a container on site or left to dry on rocks on site. At each location, physical removal was repeated after a three months interval. Post-treatment replanting was partially done at only one site at Sans Soucis with no follow-up.

At Intendance all creepers were successfully suppressed, but reinvasion later occurred with land clearance for a new hotel development. Unfortunately, Merremia reinvaded the trial sites after the control and continued maintenance were stopped. It is noteworthy that the control of Merremia requires constant effort combined with replanting of fast growing species that can close the forest gaps and prevent sunlight from reaching the Merremia. Physical control alone is not an effective method[6, 10].

For the chemical control the stem was cut near the ground or the root and both ends were dipped in 3-5% 'RoundUp®'. When this dosage did not kill the creeper, it was increased. Unfortunately, non-targeted species nearby were affected and the chemical treatment was stopped[6].

In places like Vanuatu, Merremia is readily grazed by cattle, which can be used to control the creeper. Grazing by cattle is an economical way to prepare the ground for permanent cover crops and tree crops[11]. In the Vatthe Conservation Area on the Island of Espiritu Santo, in 2009, trial treatments were conducted that combined cutting the vine and injecting herbicides into the main stem of the vine. The results suggest that eradication may be possible for small newly established infestations, and also that control to manageable levels is possible over larger areas by injecting Glysophate into the large stems and cutting all the small stems[8, 9].

Herbicide treatment with 'Weedmaster Duo®' (active ingredient 360 g/L glyphosate) applied by injecting 10 ml (diluted 50/50 in water) into the main stem of large vines proved to be the most effective and suitable for large areas. The control resulted in a positive response from the ecosystem in which previously unhealthy trees were able to regenerate their crowns and the forest floor was covered with tree seedlings. Trials of direct injection of 2.5 ml to 5 ml of undiluted 'Weedmaster Duo®' in 2010, reduced the presence of Merremia at Vatthe by more than 70%. The herbicide was injected into the ground running stems using a specially invented walking stick injector, and into the thicker vines growing up into the canopy using an animal injector (an injector comprised of a syringe and a needle, developed for use mainly in veterinary practice). However, the herbicide takes 9 or more months to completely kill a vine[8, 9].

Treatments with 'Victory Gold®' (active ingredient: 50 g/L picloram plus 100 g/L triclopyr) and 'Ultimate®' (active ingredients: 25 g/L metsulfuron-methyl and 75 g/L triclopyr) were less effective. Large stems were scraped (20 cm long strips along both sides, with each strip overlapping by 10 cm, i.e. not a full ring bark) and 'Vigilant®' gel was applied. This did kill the plant but was too time consuming. Special hole punches (using a slide hammer to puncture the larger vine stems), injection kits and a walking stick injector were developed for applications to large areas. This allowed people to walk through the forest easily and quickly inject the ground running stems on the forest floor. Large scale trials (over 100 ha) of vine cutting at ground level (without herbicide application) were very effective in making it possible for the forest canopy to recover, but it is expected that the vines will have recovered within a few years[8, 9].

MANAGEMENT RECOMMENDATIONS

Refer to page 340 for management recommendations of alien invasive creepers. Note that some authors recommend not to eliminate it in natural and semi-natural forests, particularly in sloppy areas to prevent erosion, and in areas where its presence may protect the forest by acting as a firebreak.

References/Further reading

1 GISD, 2011; 2 CAB International, 2011; 3 Nevill, 2009; 4 Senterre, 2009.5 E. Henriette *pers. obs.*; 6 Beaver & Mougal, 2009; 7 Katulic *et al.*, 2005; 8 Maturin, 2010; 9 Maturin, 2013; 10 M.Vielle, *pers.comm.*; 11 FAO, 2011; 12 Gerlach, 2006.

Management recommendations for alien invasive creepers

Management of alien invasive creepers requires a variety of control techniques in order to be effective. Elements of effective management for these species include: early detection, and the use of a combination of herbicidal, mechanical and physical controls. For more effective results, control programmes should be conducted during the dry season. Management recommendations are proposed based on trials that have been done in the Seychelles and abroad[1-6].

CONTROL AND ELIMINATION PROTOCOL

CHEMICAL CONTROL

The active ingredients of herbicides such as 2,4-D, Dicamba, Triclopyr, Picloram and especially Glyphosate are effective in controlling creepers[1,2,5,6]. The chemicals can be foliar-sprayed (e.g. using 3% commercial 'Roundup®') or more effectively applied to the rooted part of the cut stems (cut stump application) or through injection to kill the tubers (in the case of Merremia) or rhizomes (in the case of Japanese climbing fern). Trials have shown Glyphosate to be an effective herbicide for use against Merremia spp. in the Solomon and Vanuatu Islands[5] (see specific text for dosages used with success), and against Japanese climbing fern in Florida[2,4,7]. Although creepers can be controlled by herbicides, multiple treatments may be required for successful elimination unless the dosage is strong enough to kill the creepers after the first application.

The following protocol can be followed for chemical control of creepers:
For cut stump application:
- Cut the stems at 10 cm above the ground.
- Treat both parts of the cut stem by applying the recommended dosage of the herbicide and leave the plant to die.
- Cover the treated stems with plastic bags to avoid leakage of product onto the ground.
For foliar-spray application:
- Spray the whole plant with the recommended dosage of the herbicide.
- Follow-up by repeating the control programme 2-3 times at 3 monthly intervals until all creepers have been eliminated.
- Avoid doing the treatment during rainy periods.

PHYSICAL CONTROL

Physical control alone is not effective in controlling and eliminating invasive creepers like Merremia, Devil's ivy and Japanese climbing fern. This is because any roots, stems or pieces left behind will sprout. Nonetheless, the following protocol can be followed for physical control of creepers:
- In the dry season, cut the lower part of the creeper and remove it from the trunk as high up as possible by hooking and pulling the parts in the trees.

A group of agricultural and environmental officers testing methods to control invasive creepers.
E. Henriette

- Uproot the remaining parts ensuring that the tubers and rhizomes are removed from underground.
- Place all plant materials removed on a rock to dry or in gunny bags to rot and use it as compost.
- Contain all plant materials removed to prevent re-sprouting.
- Conduct post-treatment planting in gaps to prevent re-colonisation of invasive creepers and soil erosion. The decision whether or not to replant will depend on the ecology of the creeper. For climbing creepers with small footprints there may be no need to replant, but in the case of creepers that grow and cover the ground, replanting may be necessary.

- Repeat the control programme on successive occasions and monitor at least once a year until the creeper is eliminated.

PREVENTING REINFESTATION

Early detection of alien invasive creepers is important because their eradication is often very difficult in a later stage of spread[8].

- Make the general public and especially people who encounter creepers in their work (e.g. roadside maintenance personnel) aware and get them involved.
- Conduct frequent monitoring and immediate removal of newly established creeper populations. This may be the best strategy for controlling their spread.
- Enforce stricter border control at customs to prevent the introduction of new alien species that usually come in as ornamentals.
- Educate the general public about the environmental impact of and control measures for creepers.
- Minimise disturbance inside forests. Disturbance enhances the growth of creepers, and encourages their spread and invasion.
- Conduct a rehabilitation programme with native plants immediately after the creepers are cleared in order to prevent reinvasion and soil erosion. As rehabilitation may be labour intensive and costly, explore more efficient ways to propagate native plants. For example, scattering of seeds, rather than establishing seedlings in nurseries for replanting, may be a more efficient and self-maintaining strategy. Replanting can also shade out sun-loving creepers like Merremia and prevent their spread.
- Manage the habitat of the invaded ecosystem. Conduct gap restoration by eliminating open canopies, notably by planting native palms. This will close the forest canopy which is probably the most effective way to control the spread and establishment of shade intolerant creepers like Merremia.

References/Further reading
1 GISD, 2011; 2 Ferriter, 2001; 3 Kline & Duquesnel, 1996; 4 Langeland et al., 2009; 5 Maturin, 2010; 6 PIER, 1999; 7 Minogue et al., 2010; 8 Katulic et al., 2005.

A group of agricultural
and environmental
officers monitoring
the effectiveness of
creeper control.
E. Henriette

Injection of herbicide can be effective to control creepers.
Emily Tasale

Creeper control on Praslin. E. Henriette

Other plants

ONE OF WORLD'S WORST **100** INVASIVE SPECIES

Koster's curse invading a forest gap after the felling of mature Cinnamon trees. E. Henriette

Koster's curse

Clidémie hérissée

Fo watouk

Clidemia hirta

IDENTIFICATION AND BIOLOGY

Koster's curse is a coarse perennial shrub up to 2 m tall, but sometimes 5 m depending on habitats. Stems are covered with red bristles the colour of which lightens with age. Leaves are opposite, hairy, with ovate-to-oblong blades and have crenate (rounded) margins. Flowers are small and white. Fruit is hairy (facilitating its adherence to animals and clothing), ovoid, bluish-black with each fruit containing over 100 seeds. A mature plant produces more than 500 fruits each season, further enhancing its invasiveness through dispersal. Seeds can remain dormant for up to four years in the soil. Flowering and fruiting occur throughout the year. Fruit dispersal is mainly done by frugivorous birds, but also by humans, e.g. seeds attached to shoes and clothes.

Koster's curse does not occur in forests in its native range but is a vigorous invader of tropical forests in its introduced range. It grows mainly in open disturbed areas, along paths and gaps in the understory of forest. It has a high tolerance for sunlit areas and hence easily invades forest gaps and clearings[1-3].

ORIGIN

Originates in humid tropical Central and South America, extending from southern Mexico to Argentina, including Venezuela, and the Caribbean Islands. It is known to have invaded wet and dry regions of the tropics and subtropics including Hawaii, Fiji, Singapore and the Seychelles[1, 3].

DISTRIBUTION AND ABUNDANCE IN SEYCHELLES

Koster's curse was first reported on Silhouette in 1987, where it occurs in all forest habitats[4-6]. On Mahé, a single plant was first recorded and uprooted in 1993 at Le Niole. In 1999, a small patch was discovered along the Mt Sébert trail at Cascade. The species is believed to be rapidly spreading on Mahé and some of the largest patches are found at Barbarons, Dans Bernard, Mare aux Cochons, Mt Sébert, Salazie, along Sans Soucis road and Vingt Cinq Sous[5]. Koster's curse is also found in and around the Morne Seychellois National Park where single plants or patches have been found in remote and undisturbed places such as Congo Rouge, Glacis Deros, and Trois Frères[6]. It also occurs on Ile du Nord[6, 7]. It is unclear whether the species was

The probably native Indian Rhododendron *Melastoma malabathricum sechellarum* (Watouk). Do not confuse it with Korster's curse! E. Henriette

introduced to Seychelles for ornamental purposes or accidentally. In 2011, two plants were found on Praslin (along the road near the waterfall) and were removed[8].

DAMAGE AND THREATS

Koster's curse was listed as a priority species for control at an IAS prioritisation workshop (November 2010)[9]. It is a proficient invader[10] of major concern to nature conservation in the Seychelles because it is apparently able to invade sensitive areas of high biodiversity.

It invades clearings where light penetrates the canopy, forming dense thickets that smother native vegetation. Where endemic species formerly predominated, they become threatened with extinction. The affinity of Koster's curse for open degraded areas makes the degraded forests of Mahé, Praslin, La Digue and Curieuse particularly susceptible to its invasion and spread. Although probably not resistant to fire, it rapidly colonises burned areas and could be a serious threat if introduced to fire-prone Praslin Island. On Silhouette, Koster's curse may be a threat to the largest remaining wild population of *Impatiens gordonii* ('Belzamin maron')[12].

Koster's curse has negatively affected native ecosystems and proven difficult to control in the Hawaiian archipelago[3]. There is concern that it will have a similar effect in the Seychelles. The species is problematic as an agricultural weed in the Comoros archipelago, and in sugar cane fields in La Réunion[1].

ELIMINATION AND CONTROL PROGRAMMES

A control and monitoring programme for Koster's curse was undertaken by the Forestry and National Park Section in 1999, within the Morne Seychellois

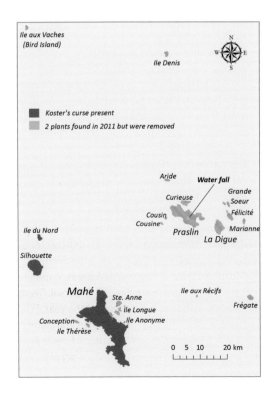

Ile aux Vaches
(Bird Island)

Ile Denis

N
W   E
S

■ Koster's curse present

■ 2 plants found in 2011 but were removed

Aride

Water fall

Curieuse   Grande Soeur

Cousin   Félicité
Cousine

Ile du Nord   Praslin   Marianne
La Digue

Silhouette

Mahé   Ile aux Récifs

Ste. Anne   Frégate
Ile Longue
Conception   Ile Anonyme
Ile Thérèse

0   5   10   20 km

National Park on Mahé[6, 13]. Plants were uprooted and left hanging over other vegetation to dry in the sun. In 2001, the Forestry Section expanded the programme of control, and in early 2003 widened it to involve the participation of local communities. A database and early detection system were established with the aim of containing the further spread of the species, especially to areas of high biodiversity and to new islands, but this practice seems to have been discontinued. A brochure was also produced to raise public awareness about the ecology and management of the species.

Appeals by the Ministry of Environment to eliminate Koster's curse resulted in several organisations and environmental clubs conducting activities to eliminate the species. For instance, Forestry Section staff together with 30 students from the Plaisance and Pointe Larue secondary Wildlife Clubs, joined forces in January 2003 in an exercise to eliminate the invasive plant at Mont Sébert (Cascade, Mahé). The plants were uprooted in a large invaded area which was immediately replanted with endemic palms. Club members of the Natural History Museum physically removed the plants from some sites on Silhouette; and school children and staff from private companies participated in an eradication programme around the historic site of Mission Lodge on Mahé[6]. On Silhouette, uprooting was conducted in selected areas by NPTS, and was found to be effective[12].

The awareness campaign was deemed successful; but due to a lack of follow-up activities the species quickly re-established itself within a couple of months.

MANAGEMENT RECOMMENDATIONS
Control and elimination protocol
Chemical control
- Foliar spraying with Glyphosate is an effective control method particularly for small populations.
- Herbicide application to the cut stump is also effective. Cut stem close to the ground, and apply a herbicide such as 'Garlon 4®' (20% dilution in biodiesel) to the cut surface.

Physical control
- Remove plants manually, which may be effective only for small populations. This is best done before fruiting occurs.
- Uproot the plant and hang it over the branch of a shrub or tree so that it dries and dies.
- Re-visit treated areas every 4-6 months to remove new plants. Growth rates of Koster's curse are very fast and time to reach maturity is very short. So follow-up is critically important.

Preventing reinfestation
- Enact effective border control to prevent new introductions. Most alien populations of this plant are probably the result of deliberate ornamental introductions.
- Conduct an awareness campaign to teach hikers and field workers how to remove and dispose of any Koster's curse fruits that stick to their clothes and shoes after passing through or working in infested areas. They should carefully remove any fruits that have become stuck to their clothes and shoes and wash them especially before hiking in new areas. Shoes should be cleaned with any disinfectant, and a brush used to remove spores or seeds.
- Minimise forest clearance and slash-and-burn in agricultural areas. Such disturbance is a key element in the establishment and invasion of Koster's curse.
- Properly dispose of all invasive plant material to prevent their regeneration (e.g. destroy them into an incinerator or burn in a kiln; bagging by placing them in gunny bags or black plastic bags and allow to rot; chipping using either a machete or a chipper; air-drying by leaving plants on rocks, branches or any raised surfaces to dry).
- Ensure that cut Koster's curse plant material is not further distributed, e.g. when maintaining a roadside, by ensuring that equipment are properly cleaned before moving onto another site.

Government and NGOs need to conduct public awareness campaigns to educate the general public about the environmental impact and control measures for Koster's curse.

References/Further reading
1 GISD, 2011; 2 CAB International, 2011; 3 PIER, 1999; 4 Kueffer & Vos, 2004; 5 Kueffer & Zemp, 2004; 6 Beaver & Mougal, 2009; 7 Nevill, 2009; 8 G. Jessie & V. Laboudallon, *pers. comm.*;9 Government of Seychelles, 2010; 10 Lowe *et al.*, 2000; 11 Kueffer *et al.*, 2004; 12 Gerlach, 2011; 13 Simara *et al.*, 2008.

Sisal stand on Grand Polyte (Cosmoledo atoll)
C. Rocamora

Green aloe invading forests in La Réunion.
T. Le Bourgeois/CIRAD (detail by E. Henriette)

Sisal

Sisal/Agave

Lalwa or Lalwes

Agave sisalana

Green aloe or Yucca

Choka vert/Cadère

Yuka or Pti Lalwa

Furcraea foetida

IDENTIFICATION AND BIOLOGY

Sisal is a large succulent plant, 1.5 to 2 m high, with thick green sword-shaped leaves in a rosette. Leaves are up to 150 cm long, 10 cm wide and have dark brown terminal spines of 2-2.5 cm. The margins can be either smooth or with numerous prickles especially in young plants. Inflorescences form 5-6 m tall panicles (shoots), with 10-20 lateral branches bearing greenish yellow flowers at the ends. The fruit is a capsule of c. 6 cm length and has a beak at the end. Sisal propagates vegetatively by rhizomes/stolons (suckers) and plantlets (bulbils at the end of lateral branches which then fall to the ground); this makes it a prolific invader[1-4].

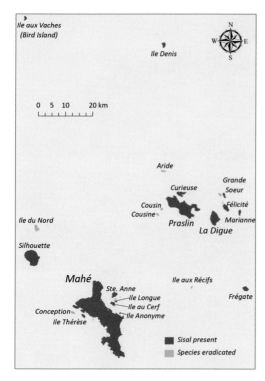

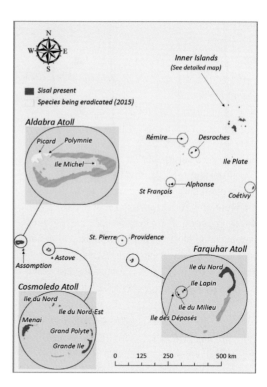

It is found on coastal land, and in deserts, forests, grasslands, shrublands and rock outcrops; and it prefers sandy soils[2]. Sisal is used both commercially as a source of fiber[2, 5] and as an ornamental plant. Parts of the plant are also used medicinally[2].

Green aloe is a smaller plant with a small stem 20-30 cm long. Its has green to yellow-green sword-shaped leaves that are 1-1.8 m long, up to 10 cm wide, and narrowing to 6-7 cm at the leaf base. Leaves are fleshy with a few widely spaced marginal spines 4-10 mm long (especially near the base); they lack a terminal spine. The inflorescences (panicles) are terminal (over 5 m high) containing many pendulous, fragrant, white, greenish-white, yellowish-green, or pale blue green, 4 cm long flowers. Green aloe produces numerous bulbils after flowering. It is found in agricultural areas, natural and disturbed forests and shrublands. It can invade many habitat types, given that it grows in all types of soil from well-drained to poor and eroded soils, as well as on rocky terrain. Green aloe has been introduced to many parts of the world both for its fiber and as an ornamental plant[2-4].

ORIGIN

Sisal is native to Mexico. It has been introduced to Africa (naturalised in Madagascar), United States (Hawaii and Florida), the Caribbean (cultivated in Bahamas and Bermuda), Australia, Spain and some Pacific islands[1-4]. Green aloe is native to the Caribbean and the northern part of Latin America[2-4].

DISTRIBUTION AND ABUNDANCE IN SEYCHELLES

Sisal was introduced to almost all inhabited islands where it was used by people to make strings and fishing lines, or to attach hooks to other types of fishing lines. Except where they have been eradicated, sisal plants still survive on most of these islands, and in many cases they have become invasive. In the inner islands, Sisal still occurs on Mahé, Anonyme, Ile au Cerf, Longue, Ste Anne and Thérèse; Praslin, Curieuse, Félicité, Grand Soeur, La Digue and Marianne; Silhouette, Frégate, Bird and Denis[7, 8, 12, 25]; and in the outer islands at Coëtivy; Desroches and Rémire; Alphonse; Farquhar and St. Pierre; Assomption, Astove and Cosmoledo, and Aldabra where it is being eradicated (see map)[9, 11, 12, 24-26]. It is reported absent from D'Arros[6, 28]. It was eradicated from Aride (where the last plant was removed in 2010)[26], and probably also from Ile du Nord[9]. In the inner islands, Green aloe has been reported on Mahé, Anonyme, Conception, Sainte Anne and Thérèse; Praslin, Cousin (now extinct), Curieuse, Félicité, Grande Soeur, La Digue and Marianne; Silhouette and Ile du Nord, Frégate, and Denis[7-9, 12, 25]; and in the outer islands at Coëtivy; Desroches, D'Arros, Poivre, Rémire (unconfirmed), Alphonse, Marie-Louise and Farquhar (Ile du Nord)[6, 23, 25, 26, 28] but not in the Aldabra group[23, 26, 27] although it used to occur at Aldabra[11].

DAMAGE AND THREATS

Sisal poses a serious threat to native plant species. It has the potential to exclude and out-compete native plant species due to its large size and rapid propagation[2-4]. Both Sisal and Green aloe can form dense impenetrable stands which can prevent the regeneration of native plants thus excluding native vegetation. Sisal is one of the most problematic non-woody invasive plants in the Seychelles[13], particularly on the islands of the Aldabra group[10]; and is also considered a threat in Mauritius and Rodrigues[14], as well as in Réunion and Europa where it was cultivated[15, 22]. In South Africa it has been declared as a weedy invader[16] and as one of the worst invasive species in the USA[2]. The Green aloe has become one the most invasive species on Réunion Island[15]. Some countries like Tahiti and French Polynesia have banned new imports, propagation, planting and transfer of Green aloe from one island to another[4].

ERADICATION AND CONTROL PROGRAMMES
Cosmoledo atoll

The eradication of Sisal was envisaged on Grand Polyte (Cosmoledo atoll) by the Island Conservation Society in November 2007 as part of the FFEM Rehabilitation of Islands Ecosystems programme[9]. A mechanical trial was implemented in which Sisal was cut on c. 40% of the single patch occupied by the species (c. 0.5 ha at the time), just before the rainy season. This was not the best time to do the control and by November 2008 most plants had regrown from root shoots and seeds, but not from the main large stems (also called 'heart' or 'trunk') that had been cut (leaves were first cut with machetes, then stems with a chainsaw). Regardless of whether control had been done during the dry season, regrowth from seedlings and root shoots would probably have occurred, although perhaps less vigorously, and would have required repeated control, probably for several years. On the larger island, Grand Ile, where 4 large patches of Sisal covered several hectares, eradication by hand was considered not to be feasible. A proposal for mechanical control using machinery

Detail of one Sisal plant from La Réunion. E. Henriette

such as a small caterpillar was submitted to IDC but could not be implemented. An alternative chemical method using a tree injector with an extended arm of 2 m (see below) and metsulfuron herbicide which gave very good results in Australia was identified as the best next step. Herbicides and equipment to eradicate Sisal on Grand Polyte were sourced. Unfortunately, mechanical control on Grand Polyte could not be continued and alternative methods could not be implemented due to transportation problems in 2008 complicated by the threat of piracy.

Grande Soeur

Sisal distribution is limited to patches along part of the eastern beach (Grande Anse). Mechanical control, uprooting, and repeated control of seedlings and offshoots are ongoing in an effort to progressively reduce the extent of patches, and eventually to eradicate the species.

Aldabra atoll

Substantial manual removal (presumably by uprooting and cutting) of Sisal was done in the 1970s by the British Royal Society. Most of it was removed, but unfortunately it was not eradicated. Recently, as part of an EU funded project on IAS management, a feasibility and mapping study was undertaken by SIF. Three large patches of Sisal were identified on three islands: Picard, Polymnie and Ile Michel (by far the largest, with hundreds of plants).

The Picard patch was monitored monthly for flowering, beginning in 2012, to determine the best timing for intervention and to find out how long it takes for the inflorescence and bulbils to develop. Manual uprooting of the Polymnie patch in 2013 was not totally effective as the roots were very difficult to extricate from the limestone. Follow-up monitoring was done to confirm no new growth. A herbicide trial was done on the Picard patch between November 2013 to August 2014 using Tordon 101® diluted in 50% or 90% of water. Results suggest that low concentrations are not effective and neither is foliar spraying. Herbicide must be applied to the central stem after it has been cut using a machete or a chainsaw; spraying alone is not effective. Moreover, targetted herbicide application to the cut central stem has no non-target effects. No sign of impacts on the surrounding vegetation was observed. SIF is currently removing Sisal from Ile Michel and Polymnie using the methods trialed on Picard. Monitoring and follow-up treatment of remaining plants will be necessary for all patches[10]. Complete eradication from last remaining patch (on Ile Michel) is expected in 2015[29].

ELIMINATION AND CONTROL PROGRAMMES IN OTHER REGIONS

Australia

On Peel Island, Morton Bay, Queensland, a chemical treatment using 'Brush Off®' herbicide (active constituent metsulfuron) was effectively used on Sisal[17-19]. The herbicide was applied using two treatments: one with sea water and one with freshwater at the rate of 2 g per litre; and in each case 5 ml of mixture were injected into the stem. The two treatments, using sea water and freshwater mixtures, were conducted on adjacent sites with 3 replications. The mixture was applied using a 'Sidewinder Injection Apparatus' which has a 5 m long flexible chemical hose and which can be used by one person (or two in difficult locations)[18]. The hose is part of a backpack assembly wherein a fluid pump and chemical container is mounted. The pump is hand lever operated and puts out discreet doses for each operation of the lever. The other end of the hose is connected to a 2 m long aluminium pole, the front end fitted with a sharp pointed hollow stainless steel spike which is thrust into the lower part of the basal stem of the plant to be treated. The output of the injector pump is 5 ml per swing of handle but this may be reset to 15 ml per swing. The advantages of the Sidewinder system over

Controlling Sisal on Cosmoledo Atoll, Seychelles. G. Rocamora

other herbicide applicators are: improved operator safety; reduced quantity of herbicide required; reduced impact on non-targeted species; and ability to be used in all weather conditions including during rain. The dosage of the mixture injected depended on the height of the plant. Some 5 ml of mixture or 0.025 ml of 'Brush-Off®' was injected into 1 m high plants; 10 ml of mixture or 0.05 ml of 'Brush-Off®' into 2 m plants; and 15 ml of mixture or 0.075 ml of 'Brush-Off' into >2 m high plants. Further treatment was not necessary since, after the first treatment, all plants treated had atrophied (wasted away) in-situ. During the 1980s, the following treatments had also proved relatively successful: application of pure herbicide fluroxypyr to the crown after removal of the growth tip; application of dowel (a solid cylindrical rod) soaked in herbicide and hammered into the stem[20]; and spraying of triclopyr and diesel (which appears to break down the thick leaf cuticle) on slashed sisal hemp plants.

New Zealand

In 2008, the Auckland Council recommended the following control measures for Green aloe[1]:
- Dig out small plants.
- Inject stem or leaf with 4-10 syringes of 5 ml undiluted glyphosate (depending on size of plant). Leave plants to rot on site.
- Slash 4-10 lengthwise cuts (depending on size of plant) and apply 5 ml undiluted glyphosate to each cut. Leave plants to rot on site.
- If disposal of the leaves is possible, cut at ground level and stump paint with 200 ml glyphosate/L.

Hawaii

Sisal was controlled in the Hawaii Volcanoes National Park by the foliar application of herbicide ('Garlon 4®'). The treatment successfully decreased the reproduction and population size of the Sisal[2].

Florida

About 12 ha of Sisal were destroyed during a pre-scribed fire treatment. Other management techniques used in Florida included spraying with herbicide (20% Garlon 4) subsequent to breaking the 'heart' of the plant (presumably by cutting with a machete or a chainsaw) and hand removal of small plants[2, 5].

La Réunion

Green aloe is subject to ongoing control in efforts to contain the invasion. The flowering stem is cut to help prevent new colonisation by the numerous bulbils. For this management method to be effective in reducing or controlling the invasion, it needs to be perfectly synchronised with the flowering season - i.e. the cut must be performed just after blooming and before the plant sheds its bulbils[15].

Mayotte

Green aloe is currently being mechanically eradicated from the Réserve Naturelle de l'Ilot Mbouzi. For some plants, the stems are cut with a machete, while others are uprooted. The plants are left to dry in the sun, so cutting needs to be done during the dry season. Green aloe is smaller and less difficult to control than Sisal[21]

MANAGEMENT RECOMMENDATIONS
Control and elimination protocol
Physical control

- Eliminate both Sisal and Green aloe by first cutting the leaves, then the heart (stem or 'trunk'). On the dry sandy soils of the southern islands of Seychelles, this may be enough to kill the mother plants, but uprooting is normally required.
- Destroy root shoots and seedlings, as these will regrow.

- Do not leave removed tips standing upright nor lying in depressions where water collects, as they can remain alive for many months.
- Uprooted plants can be left drying in the sun and may be burned, but heaps need to be turned as necessary to ensure that the roots do not regenerate from the stems and grow into the soil.
- Continue to periodically remove regenerating bulbils and root shoots even after the adult plant has been destroyed.
- Ensure that you also extricate all roots and rhizomes when small plants and small patches are either removed by hand or dug out.
- Use caution when removing Sisal, as its sap contains saponin and oxalic acid that can cause skin irritation and eye damage.

Chemical control

- Effective herbicides against both species are glyphosate, or triclopyr plus picloram, or metsulfuron applied to cut plants (sprayed or with paintbrush) or injected into the basal stem (see details above in Australia section).
- Good results can be obtained by breaking the 'heart' from the main plant and spraying it with a suitable herbicide.
- Herbicide application appears to be most effective after removal of the growth tip.
- Persistent follow up work may be required, especially when using physical control methods; as large apparently 'dead' Sisal plants can produce sprouting juveniles at their base for many months.

Preventing reinfestation

- Discourage inter-island transfer, and be particularly careful about seeds, in order to avoid accidental introduction.
- Conduct public awareness programmes to educate communities about the dangers of these invasive plants.

References/Further reading

1 BioNET-EAFRINET, 2011; 2 GISD, 2011; 3 HEAR, 2011; 4 PIER, 2011; 5 Brown, 2002; 6 Fosberg, 1979; 7 Fosberg, 1983; 8 Hill, 2002; 9 Rocamora & Jean-louis 2008, 2009; 10 SIF unpublished data; 11 A.Constance (Mazarin), *pers. comm*; 12 Pat Matyot, *pers. comm.*; 13 Kueffer & Vos, 2004; 14 Strahm, 1983; 15 Hivert, J. 2003; 16 Working for wetlands, 2011; 17 Harris *et al.*, 2011; 18 Sidewinder, 2011; 19 Foley & Bolton, 1990; 20 Bickerton, 2006; 21 S. Said, *pers. comm.* (Mayotte); 22 Le Corre & Jouventin, 1997; 23 Hansen & Laboudallon, 2013; 24 van Dinther *et al.* 2015. ; 25 Robertson 1989; 26 ICS unpublished data ; 27 Mike Betts & Catherina Onezia/SIF *pers. comm.*; 28 R. von Brandis, *pers. comm.*; 29 SIF newsletter, Issue 30, April 2015.

Extensive field of water lettuce after invading a river. E. Henriette

Water lettuce

Laitue d'eau

Leti lanmar

Pistia stratiotes

IDENTIFICATION AND BIOLOGY

Water lettuce is a freshwater invasive plant resembling a floating open head of lettuce. Thick, soft, green leaves form a rosette (a circular arrangement of leaves). Leaves are up to 14 cm long with parallel veins, wavy margins, and covered in short hairs. Water lettuce floats on the water surface with its long, feathery roots hanging submersed beneath the leaves. Flowers are few, inconspicuous and small (up to 1.5 cm). The plant produces small green berries and also reproduces vegetatively. Its rapid vegetative reproduction enables it to form dense mats on the water surface. It is usually found in lakes and rivers, but can also survive in muddy water[1, 2].

ORIGIN

Native distribution is uncertain. It probably originated in South America, but is now found throughout the tropics and subtropics[1, 2].

DISTRIBUTION AND ABUNDANCE IN SEYCHELLES

Water lettuce occurs on Mahé, Praslin and La Digue islands[3, 4]. On Frégate it is currently found only in arti-ficial ponds[5]. It has become over-abundant at all sites where it occurs, covering much of the water surface. The species was introduced as an ornamental plant for ponds and aquarium, and may have spread to natural habitats by the dumping of unwanted plants.

DAMAGE AND THREATS

Water lettuce can have a severe impact on both the environment and the economy of infested areas. The

Water lettuce rapid vegetative reproduction means that it is capable of forming dense mats on water surfaces.
Thomas Le Bourgeois/Cirad

extremely dense mats can block waterways, hindering flood control efforts and making fishing impossible. In some countries it interferes with navigation. Mats of water lettuce can also severely impact the environment by blocking the air-water interface which can drastically reduce the concentration of oxygen in the water and sediments. This can lead to the death of fish and other marsh creatures including invertebrates. Extremely thick mats of water lettuce can even prevent sunlight from reaching underlying water, killing native submerged plants, and altering immersed plant communities by crushing them. The cumulative effect of these negative outcomes is a loss of biodiversity in invaded habitats. Water lettuce mats can also serve as a breeding place for mosquitoes[1, 2]. Water lettuce was listed as a priority species at the IAS prioritisation workshop (November 2010) in the Seychelles[6].

ELIMINATION AND CONTROL PROGRAMMES

During 2000-2004, Ministry of Environment conducted a programme to control and eliminate Water lettuce from important wetlands. Manual removal of plants was carried out on Mahé at North-East Point and Anse

Royale[3], on La Digue, and on Praslin at Jalousie (Anse Kerlan) and Côte d'Or[7]. The programme managed to keep the plants under control but did not eliminate them due to insufficient financial resources. On Praslin, at Jalousie (Anse Kerlan), water lettuce was progressively removed from the marsh. After a section had been cleared, a second visit the following day allowed for more complete removal of any small plants or remaining plant parts. Unfortunately, the impressive seed bank at the bottom of the marsh prevented complete elimination. Despite multiple visits to clear seedlings that popped up at the surface - first weekly during 6 months, then bi-monthly for approximately 18 months - seeds continued to germinate for more than 2 years after the removal. During this time, fish, terrapins and other marsh creatures were observed to repopulate the marsh, and the functioning of the marsh became normal again. However, within three months after the visits stopped, the water lettuce had recovered over the entire marsh to how it was before the start of the operation._

In 2007-08, the Marsh Unit of the DoE conducted some relatively more limited operations to remove water lettuce: on Mahé, at North-East Point marsh; and

A bed of Water hyacinth in a lake where it has become a problematic invader. Thomas Le Bourgeois/Cirad

at La Digue marsh[3]. In 2009, at the Banyan Tree Hotel and Resort, Takamaka (Mahé), an active management programme was conducted to manually eliminate the species. Increasingly, public authorities have attempted to hand over the control and elimination of water lettuce to private contractors.

MANAGEMENT RECOMMENDATIONS
Control and elimination protocol
Chemical control
The following herbicides have been used to treat Water lettuce:

- In the United States, the contact herbicide endothall® (dipotassium salts) in the form of Aquathol (Aquathol K and Aquathol Super K). Contact herbicides act quickly and kill all plant cells that they contact. These herbicides are known to be toxic to some fish, but have low toxicity to crustaceans, and a medium toxicity to aquatic insects.
- 'Reward®' is a Diquat formulation liquid that also has been effective on Water lettuce in the United States. Diquat is a non-selective contact herbicide and crop desiccant. Because of its rapid degradation

in water and strong adsorption onto sediments, Diquat has rarely been found in drinking-water.
- Glyphosate in the form of Rodeo, Aquamater and Aquaneat also have been effective on water lettuce in the United States. These are broad spectrum, systemic herbicides that are absorbed and move within the entire plant and kill it.

One danger with any chemical control method is the chance of oxygen depletion caused by the decomposition of the dead plant material after the treatment. Oxygen depletion can kill fish. In order to minimize oxygen depletion in cases of heavy infestations, it may be possible to treat the water body in sections and let each section decompose for about two weeks before treating a new one.

Physical control
Physical control has proven to be temporarily effective for Water lettuce when conducted as follows:

- Remove the Water lettuce from the surface of the water by raking or seining it (using a large fishing net).
- Take the plants from the waterways to the shore and

cut them up into small pieces for disposal. These can apparently be safely disposed of as slurry on the surface of the water.

The Praslin DoE experiment showed that the existence of a huge seed bank on the floor of the marsh will prevent complete eradication, despite regular and continuous clearing of seedlings for over two years. Unless follow-up can be maintained until the seed bank is exhausted, such control will only protect the marsh throughout the duration of the operations.

Preventing reinfestation
- Wash all the equipment used during a control operation at one site before attempting to use it again for control at a new site.
- Conduct public awareness campaigns about the dangers of Water lettuce.
- Properly dispose of unwanted Water lettuce that has been kept in aquarium or fish ponds. Water lettuce can be chopped into small pieces and disposed off before they produce seeds.

WATER HYACINTH

Jacinthe d'eau
Lisdo anvaisan
Eichhornia crassipes

Another invasive aquatic plant, the Water hyacinth ('Jacinthe d'eau' in French; 'Lisdo anvaisan' in Creole), also originating from South America, is present in Seychelles on Mahé, Praslin and La Digue. Like the Water lettuce, it has an extremely high productivity of biomass, and it may be used for various agricultural/biomass production purposes (including production of animal food, or as an ornamental plant). Fortunately, Water hyacinth is much less widespread in Seychelles; and its mats are less dense and thinner. This allows more light and oxygen to enter the water, enabling better survival of marsh creatures, and some functioning of the marsh ecosystem. Water hyacinth is also more easily controlled. However, some marshes on Mahé (e.g. Anse gouvernement marsh) are entirely covered by this species that needs to be controlled and if possible eradicated.

In East Africa, Water hyacinth is problematic in Lake Victoria, but there it has been successfully reduced by biocontrol weevils *Neochetina bruchi* and *N. eichhorniae*. Mechanical and chemical methods were not as successful as the biocontrol.

References/Further reading
1 GISD, 2011; 2 CAB International, 2011; 3 Beaver & Mougal, 2009; 4 Nevill, 2009; 5 E. Henriette *pers. obs.;* 6 Government of Seychelles, 2010; 7 V. Laboudallon *pers. comm.*

Fungal disease

Symptom of a Takamaka tree affected by the wilt disease. E. Henriette

Takamaka wilt disease

Verticilliose du Takamaka

Maladi Takamaka

Leptographium calophylli

IDENTIFICATION AND BIOLOGY

The Takamaka wilt disease is a fungus that attacks Takamaka (*Calophyllum inophyllum*) trees causing its host to wilt[1,2]. This disease is characterised by the wilting of part of the crown of trees (or the whole of it); normally starting by the extremities and progressively affecting entire branches. The leaves of affected trees lose their shine, curl inwards and dry out, whilst brownish streaks form in the tracheids (xylem cells). The fungus resides in the xylem – the vascular tissue through which the sap circulates upwards - of its host. Its growth blocks the xylem tissues. In response to injury or infection the plant produces tyloses (ingrowth that protrudes into the xylem) which further block the xylem. Blockage of the xylem leads to wilting of the host tree. Symptoms of Takamaka wilt disease are: wilting and dieback of shoots; abnormal colours, forms (rolled or folded) and wilting of leaves; internal discolouration and wilting of stems; as well as wilting of the entire plant[1]. In the Seychelles, the endemic Bark beetle *Cryphalus trypanus* is believed to be the likely vector of the wilt fungus *L. calophylli*[2,3]. High population densities of the Bark beetle previously recorded only on the endemic tree

Northea hornei (Kapisen) more than 300 m above sea level, occur in trees affected by *L. calophylli*[3]. The fungus was previously designated as *Verticillium* or *Haplographium*[4].

ORIGIN

Takamaka wilt disease was first recorded in Mauritius in the 1930s, and similar diseases have been recorded on *Calophyllum* species in El Salvador, India and Indonesia[2, 4].

DISTRIBUTION AND ABUNDANCE
IN SEYCHELLES

After the symptoms of the disease were first recorded in the Seychelles in 1994[4], a small sample of the timber was collected and sent to the Forestry Commission in Oxford UK, as well as to the Seychelles Agricultural Department and the Seychelles Bureau of Standards to identify the pathogen[5]. The origin of the disease is unknown, but it is speculated to have been introduced through a shipment of untreated timber from Senegal on its way to Sri Lanka late in 1993. The ship had to be quarantined and eventually left the port with its cargo[5]. The fungus was first

reported on the following islands: Mahé, Ste Anne, Ile au Cerf, Longue and Thérèse; Praslin, Curieuse and La Digue; Silhouette and Ile du Nord islands[4, 6]. After 2001, the disease reached Grande Soeur, Félicité, Denis and possibly Marianne and Anonyme. There are no confirmed records of the disease at Conception[6] and Petite Soeur, where it may be absent.

DAMAGE AND THREATS

Takamaka wilt disease causes wilting and dieback of Takamaka trees, and may lead to complete defoliation and death of trees within months[1, 2, 4]. Takamaka is a native plant of Seychelles, present throughout much of the Indian Ocean coastlands (East Africa, Southern India, South-East Asia and Australia). Its timber is highly priced and used as a construction material. In the Seychelles, it is protected under the 'Breadfruit and other Trees Act'. Takamaka plays a vital role in island ecosystems and is an important component of the few remaining lowland forests. Being among the most widespread of the native woody plant species, it has an important function in forest dynamics, contributing to fallen timber and leaf litter. On beach crests, it controls erosion and provides much appreciated shade.

Rapid loss of mature Takamaka trees threatens the existence of lowland forest in addition to facilitating the establishment of invasive alien species in areas formerly dominated by Takamaka. The death of Takamaka trees leads to habitat loss affecting other species that are dependent on such habitats including the Critically Endangered Seychelles paradise flycatcher, restricted to La Digue and Denis islands. Hence, the Takamaka wilt disease is a major threat to the survival of the flycatcher and the loss of Takamaka to wilt disease appears likely to have a range of significant negative effects on conservation.

ELIMINATION AND CONTROL PROGRAMMES
Legal measures

As early as 1995, legal measures were implemented to abolish the transport and trade of Takamaka timber and seedlings within and between islands[4]. Unfortunately, not enough attention may have been paid to the need to disinfect cutting equipment before moving it between islands to avoid spreading the disease.

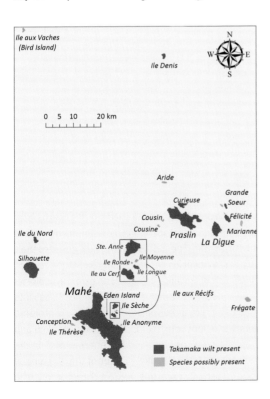

Pre-control trials

In 1994, prior to the implementation of the control programme, experimental treatment sites were established to test the tolerance of the trees to different doses of chemical treatments[5]. Sites were established on Mahé, Praslin and La Digue and more than 3,000 trees were selected, numbered and mapped for injection. The selection of trees was based on their age (young and old), aesthetic value, and what role they played in the ecosystem (e.g. provision of micro-habitats for other species, roosting habitats etc).

Two fungicides were tested: 'Tilt', a preventative fungicide used before symptoms become apparent, and 'Storite', a curative fungicide used after the manifestation of symptoms. The tests concluded that the choice of fungicides should depend on the progression of the disease. The dosage should also be based on the diameter of the tree trunk, as an overdose could kill the tree. For trees greater than 30 cm in diameter, it was determined that 10 ml/L be used; and trees smaller than 20 cm in diameter would not be treated, out of concern that the treatment could be fatal. 'Storite' would be administered at a dosage of 10 ml/L of water; while the dosage for 'Tilt', a relatively weaker fungicide, was lower[5].

The experiment also showed that the treatment was less effective on old trees and those where the infection was advanced. This was apparently because the production of tyloses as a reaction to the disease reduced the capacity of the tree to absorb the fungicide[5].

Another experiment done in 1996 showed that when Bark beetles were provided cut branches of Takamaka, they showed a preference for entry at the cut sites[5]; hence the need to seal off all cut ends during treatments.

Control programmes

In the mid 1990s, infected trees were initially cut and burnt on site through controlled burning in kilns that had been fabricated for that purpose. This was believed to be a measure that might halt the spread of the fungus. Unfortunately, the disease seemed to spread more rapidly instead, possibly because felling diseased trees may have prompted the Bark beetles carrying the fungus to take flight and land on nearby healthy trees[5].

So, from 1997 onwards, affected trees were only pruned but neither did pruning stop the spread of the disease, probably because sufficient care was not taken to protect the cut surfaces. Infected trees remained a source of inocculum (i.e. further infection), as their unprotected cut surfaces and damaged tissue were more easily available for colonisation by the Bark beetle. When the outbreak started, cuts were systematically protected by bitumen (coal tar) to seal the wounds, but this practice was later discontinued and not done in a sufficiently systematic way[5].

Not until 1996 were affected trees injected with systemic fungicides ('Storite' and 'Tilt') by the Forestry Section. To administer the fungicide, a T-piece (the syringe) was used to inject either 'Tilt' or 'Storite' into the tree. The T-piece was made of fibre-glass, synthetic plastic and bronze metal, with a tapered end. The use of another T-piece made essentially of plastic was discontinued due to breakage. A 4-5 cm hole was hand-drilled into the tree trunk. Next, the T-piece was hammered in, and interconnected to 7 other T-pieces giving a total of 8 T-pieces per tree (5 on small trees). The T-pieces were then connected to a 20 litre heavy-duty aluminium cylinder containing the fungicide. The cylinder was hand pumped to 8 bars of pressure and the valve opened to inject the fungicide. The treatment lasted for 2 hours and was monitored every half hour. This type of control was deemed effective, and most Takamaka trees treated across the country have survived and are still alive 19 years later. In some cases, however, an overdose of chemical resulted in the death of the tree[5].

By 2003, control of the disease in the Seychelles by national authorities was largely limited to the removal of dead trees especially in places where there was a public danger of the dead trees falling. Alternative measures were also taken to mitigate against negative effects of dead Takamaka trees on the functioning ecosystem. To offset the removal of dead Takamaka trees, the Division of Environment implemented an extensive replanting programme to replace the missing Takamaka with other species of coastal native trees (such as Badamier *Terminalia catappa* and Bois blanc *Hernandia nymphaeifolia*) on beach crest and lowland plateaus where Takamaka mortality had been high. On La Digue, replanting of affected areas of Takamaka forest was slow and complicated by multiple private landowners[5].

Since 2010, the Seychelles National Park Authority (SNPA) became responsible for the management of

Wilting and death of Takamaka trees on North Island as a result of the wilt disease. E. Henriette

the Takamaka wilt disease[7]. SNPA has been injecting infected trees with 'Bumper®' (active ingredient 41.8% Propiconazole). Sanitation of the trees through removal and burning of dead branches is also being done. Follow-up visits have indicated the survival of most treated trees, and the disease is considered to be under control (although its eradication is uncertain).

Chemical control of the Takamaka wilt disease depends on a number of factors including: the availability and ability to adequately stock the relevant fungicide; regular and timely visits to inject diseased trees; and public vigilance. Most cases of Takamaka wilt disease now treated by SNPA are in response to requests made by members of the public.

In Cuba, chemicals were used to stop the spread of a wilt disease on a *Callophyllum* tree but with limited success[1].

MANAGEMENT RECOMMENDATIONS
Control and elimination protocol
Chemical control
Chemical treatment is expensive and its viability in the long term has been questioned. However, trials done in the Seychelles have showed that treatment

with 'Storite' and 'Tilt' is effective in controlling the disease.

The following treatment is recommended:
• In the initial stage of disease, i.e. before the tree shows any marked symptons (i.e. for trees nearby infected ones that may potentially develop the disease), use 'Tilt' at a dose of less than 10 ml/L of water.
• If the tree is already showing clear symptons of illness, then use 10 ml 'Storite' per every litre of water.
• Use 8 interconnected T-pieces (the syringe) for injecting either 'Tilt' or 'Storite' into the tree.
• The T-piece should be robust, and preferably made of fibre-glass or synthetic plastic (easier to insert and remove, and cheaper) and bronze metal (but not made entirely of plastic as it will break).
• Drill a hole (diameter equal to the tip of the T-piece) 3.5-4 cm deep into the tree trunk through the bark until the wood is reached.
• Next, hammer in the T-piece so that it becomes inserted into the cambium (soft wood below the bark where all the active sap vessels are concentrated).
• Connect the T-piece to a 20 litre heavy-duty

aluminium cylinder containing the fungicide (a 20 L cylinder will typically contain 200 ml of fungicide for 20 L of water).

- Hand-pump the cylinder to 8 bars of pressure and open the valve to inject the fungicide.
- Allow the injection to last for 2 hours and monitor it every half hour.

Physical control and habitat management

The following are recommended when cutting infected trees:

- Strictly implement phytosanitary measures, in particular the use of disinfected cutting equipment (using flame or alcohol to destroy fungus spores).
- Seal cut surfaces to avoid adverse new infections and further spread the disease. Covering cuts with a protective paste (that may have some contact fungicide in it) has helped to contain other bark fungal tree diseases.
- Remove dead and dying trees by cutting at ground level so as not to leave any bark on.
- Replace felled Takamaka with other native high canopy trees such as *Terminalia catappa*, *Barringtonia asiatica*, *Hernandia nymphaeifolia*, *Mimusops sechellarum* and *Ficus* spp., in order to ensure continuity of native forest cover on plateaus and the functionality of lowland forest (e.g. erosion control, vegetation biomass).

Biological control

No biological control methods have yet been developed for this disease. Other bark fungus diseases, however, have been successfully contained by developing an hypovirulent variety of the fungus that the tree can overcome and live with. This was done in the 1980s with *Cryphonectria* (*Endothia*) *parasitica* (an alien invasive fungus imported from Asia) on the Chesnut tree *Castanea sativa* in southern Europe[8, 9].

PREVENTING INFESTATION

Strict phytosanitary measures are needed to prevent establishment and further spread of the disease. These include the following:

- Regulate the importation of live plants, soil and timber and especially, unprocessed timber.
- Strict sanitation of sites (e.g. removal and burning of affected branches, sealing off cut ends with bitumen) after the first report of any disease is vital in preventing establishment and further spread of the disease.
- **Systematically disinfect all forestry cutting equipment such as machetes and chainsaws before transporting them to another site, and especially to another island.** Department of Environment and ICS have strictly implemented this policy since 1998 and successfully prevented the disease from reaching Conception Island.
- Ensure that staff and personnel treating infected trees are well trained on the management of the disease and the biology of the fungus. Inadequate understanding will facilitate its spread. Technical mistakes made in the past by the Forestry Section occurred because the only senior staff member trained at the Forestry Commission in UK was not always able to adequately supervise the staff he trained to apply the techniques in the field.
- Rapid response measures need to be in place to control the establishment and spread of such plant diseases as soon as possible after they arrive.

References/Further reading

1 Plantwise, 2011; 2 Wainhouse *et al.*, 1998; 3 Vielle, 1999; 4 Hill *et al.*, 2003; 5 M. Vielle, *pers.comm.*; 6 Rocamora & Labiche, 2009b, c; 7 E. Sophola, *pers.comm.*; 8 Heiniger & Rigling, 1994; 9 Milgroom & Cortersi, 2004.

Following the eradication of Brown rats on Conception in Sept. 2007, a spectacular forest of 8-10m high dominated by seabird-dispersed native tree Pisonia grandis has established spontaneously over c. 1.5 ha (May 2014). E. Henriette

Insert: The same lower plateau in Nov. 2008 dominated byTabebuia pallida woodland, with planted native sapplings and small Pisonia trees. Roland Nolin

Bibliography

Abbott, K.L. (2006). Spatial dynamics of supercolonies of the invasive yellow crazy ant, *Anoplolepis gracilipes*, on Christmas Island, Indian Ocean. Diversity and Distributions. 12: 101-110.

Abdulla, A., Floerl, O., Richmond, M., Johnston, O., Bertzky, M., Birch, S. and Walsh, A. (2007). Enhanced Detection and Management of Marine Introduced Species in the Seychelles. Final Project Report. IUCN Global Marine Program. www.iucn.org. 125 pp.

Abrol, D.P. and Shankar, U., Eds. (2012). Integrated Pest Management: Principles and Practice. CAB International, Cambridge, Massachussetts.

ACVM (2002). Controlled pesticides. DRC 1339 for bird control. Agricultural Compounds and Veterinary Medicines Group for ERMA New Zealand. New Zealand Food Safety Authority, Wellington. 4 pp.

Adam, P.A. and Gendron, G. (2014). Marine Biodiversity. Survey at Providence Industrial Estate. Unpublished Impact Assessment Report . 9th March 2014.

Adam, P.A., Skerrett, A. and Rocamora, G. (2009). First confirmed breeding record of Black-naped Tern Sterna sumatrana from St François Atoll and a new population estimate for Seychelles and the Afrotropical region. Bull. ABC. 16 (1): 78-82.

Aguirre-Muñoz, A., Samaniego-Herrera, A., Luna-Mendoza, L., Ortiz-Alcaraz, A., Méndez-Sánchez, F. and Hernández-Montoya, J. C. (2015, in press). La restauración ambiental exitosa de las islas de México: una reflexión sobre los avances a la fecha y los retos por venir. In: Ceccon, E. and Martínez-Garza, C. C. Experiencias mexicanas en restauración de ecosistemas. UNAM.

Angel, A., Wanless, R.M. and Cooper, J. (2009). Review of impacts of the introduced house mouse on islands in the Southern Ocean: are mice equivalent to rats? Biological Invasions. 11: 1743-1754.

Anonyme (2010). Stratégie de lutte contre les espèces invasives de la Réunion. DIREN, Parc National et Conseil Régional, La Réunion. 67 pp. + annexes pp.

Anonymous (2015a). Authorities coordinate effort to tackle stinging caterpillar infestation. Seychelles Nation, 20th February. vol. 37 (no. 35): pp. 3.

Anonymous (2015b). Proper study of pest needed before effective plan of action. Seychelles Nation, 30th March. vol. 37 (no. 45): pp. 3-4.

Anonymous (2015c). Hairy caterpillar infestation – students sent home yesterday. Today in Seychelles, 3rd June: pp. 2.

Anonymous. (undated). Indian Myna Handbook. Information about the Myna. The Indian Myna Control Project. NSW mid North Coast, Australia. Retrieved 2011 www.sres.anu.edu.au/associated/myna/.

Asconit and ICS. (2010a). France – Seychelles. Rehabilitating Island Ecosystems. Eradication of invasive exotic species and Re-introduction of threatened endemic species. Asconit Consultants, ICS and FFEM, Paris and Seychelles. 12 pp.

Asconit and ICS. (2010b). France - Seychelles. Rehabilitation des Ecosystemes Insulaires. Eradication des espèces exotiques envahissantes et réintroduction d'espèces endémiques menacées. Asconit Consultants and FFEM, Paris and Seychelles. 12 pp.

Atkinson, I.A.E. (1985). The spread of commensal species of Rattus to oceanic islands and their effects on island avifaunas. pp 35-81. In P. J. Moors. Conservation of island birds: case studies for the management of threatened island species. International Council for Bird Preservation, Cambridge. No. 3.

Balasubramaniam, A. (1990). Review and improvements proposed to pesticide management procedures in the Republic of Seychelles (unpublished report). FAO Fertiliser Programme, FAO, Rome. 51 pp.

Balbaa, M. and Zakaria, M. (1983). Trapping as an effective method for mouse control in poultry farms. Proceedings of the First Symposium on Recent Advances in Rodent Control, Ministry of Public Health, State of Kuwait. 19-26 pp.

Barker, G.M. (2002). Molluscs as crop pests. CABI, ISBN: 978-0851993201 DOI:10.1079/9780851 993201.0000.

Beardsley, J.W., Su, T.H., McEwen, F.L. and Gerling, D. (1982). Field investigations on the interrelationships of the big-headed ant, the gray pineapple mealybug, and pineapple mealybug wilt disease in Hawaii. Proceedings of the Hawaiian Entomological Society. Vol. 24 (No. 1): 51-67.

Beaver, K. and Mougal, J. (2009). Review and evaluation of Invasive Alien Species (IAS) control and eradication activities in Seychelles. Mainstreaming Prevention and Control Measures for Invasive Alien Species into Trade, Transport and Travel across the Production Landscape. Consultancy Report. Plant Conservation Action group (PCA), GoS-UNDP-GEF Project, Victoria, Mahé. 88 pp.

Beaver, K., Vanherck, L. and Wepener, G. (2007). North Island Vegetation Management Plan. 2007-2011. North Island Ltd & Plant Conservation Action Group, Seychelles

Bednarczuk, E., Feare, C.J., Lovibond, S., Tatayah, V. and Jones, C.J. (2010). Attempted eradication of house sparrows Passer domesticus from Round Island (Mauritius), Indian Ocean Conservation Evidence. 7: 75-86

Behle, R.W., McGuire, M.R. and Shasha, B.S. (1997). Effects of sunlight and simulated rain on residual activity of Bacillus thuringiensis formulations. Journal of Economic Entomology. 90 (6): 1560-1566.

Bell, B.D. (2002). The eradication of alien mammals from five offshore islands, Mauritius, Indian Ocean. pp 40-45. In C. R. Veitch and M. N. Clout. Turning the tide: the eradication of invasive species. IUCN SSC Invasive Species Specialist Group, Gland, Switzerland and Cambridge, UK.

Bennett, D. (1999). Expedition Field Techniques: Reptiles and Amphibians. Expedition Advisory Centre, Royal Geographical Society, London

Betts, M. (1997). Aride Island Nature Reserve, Seychelles. Annual Report 1997. Unpublished report. Royal Society for Nature Conservation / Island Conservation Society, Seychelles

Betts, M. (2009). Alphonse Conservation Management Plan. Unpublished report. Island Conservation Society, Seychelles.

Bibby, C., Martin, J. and Marsden, S. (1998). Expedition Field Techniques: Bird Surveys. Expedition Advisory Center. Royal Geographical Society, London

Bibby, C.J., Burgess, N.D., Hill, D.A. and Mustoe, S.H. (2000). Bird Census Techniques, 2nd ed. Academic Press, London

Bickerton, D.C. (2006). Using herbicide to control century plant (*Agave americana*): implications for management. Proceedings of the Fifteenth Australian Weeds Conference, (Crop Science Society of South Australia, Adelaide). 219-222 pp.

BioNET-EAFRINET. (2011). Agave sisalana. Retrieved 12 December, 2011. http://keys.lucidcentral.org/keys/v3/eafrinet/weeds/key/weeds/Media/Html/Agave_sisalana_(Sisal).htm.

BirdLife International. (2004). Persistent organic pollutants stay in the environment, affecting humans and wildlife. Presented as part of the BirdLife State of the world's birds website. Retrieved 2nd May, 2015. http://www.birdlife.org/datazone/sowb/casestudy/178.

BirdLife International (2008). Threatened birds of the world 2008 CD-ROM.

BirdLife International. (2010). The BirdLife checklist of the birds of the world, with conservation status and taxonomic sources. Version 3. http://www.birdlife.info/docs/SpcChecklist/Checklist_v3_June10.zip [.xls zipped 1.6 MB].

BirdLife International. (2011). Important Bird Areas factsheet: Aldabra atoll. Retrieved 15 December, 2011. http://www.birdlife.org.

Blondel, J., Hoffman, B. and Courchamp, F. (2013). The end of Invasion Biology: intellectual debate does not equate to nonsensical science. Biol. Invasions. 16: 977-979. Published online on 26 September 2013. DOI 10.1007/s10530-013-0560-6.

Bohn, K.K., Minogue, P.J. and Pieterson, E.C. (2011). Control of invasive Japanese climbing fern (*Lygodium japonicum*) and response of native ground cover during restoration of a disturbed longleaf pine ecosystem. Ecological Restoration Vol. 29 (4): 346-356.

Boland, C.R.J., Smith, M.J., Maple, D., Tiernan, B., Barr, R., Reeves, R. and Napier, F. (2011). Helibaiting using low concentration fipronil to control invasive yellow crazy ant supercolonies on Christmas Island, Indian Ocean. Island invasives: eradication and management. IUCN, Gland, Switzerland: 152-156.

Bonnaud, E., Medina, F.M., Vidal, E., Nogales, M., Tershy, B., Zavaleta, E., Donlan, C.J., Keitt, B., Le Corre, M. and Horwath, S.V. (2011). The diet of feral cats on islands: A review and a call for more studies. Biological Invasions. 13 (3): 581-603.

Bonnaud, E., Zarzoso-Lacoste, D., Bourgeois, K., Ruffino, L., Legrand, J. and Vidal, E. (2011). Top predator control on islands boosts endemic prey but not mesopredator. Animal Conservation. 13: 556-567.

Booth, L.H., Eason, C.T. and Spurr, E.B. (2001). Literature review of the acute toxicity & persistence of brodifacoum to invertebrates. pp 1-9. In Department of Conservation 2001. Literature review of the acute toxicity & persistence of brodifacoum to invertebrates & studies of residue risks to wildlife & people. Science for Conservation 177, vi + 23p. Department of Conservation, Wellington, New Zealand.

Borowiec, N. (2007). Présence d'Aleurotrachelus atratus Hempel (Hemiptera: Aleyrodidae) aux Seychelles: Evaluation des risques économiques et écologiques sur les iles de Mahé, Praslin et La Digue. Compte-rendu de la mission entomologique effectuée du 1er au 8 juillet 2007. CIRAD, La Réunion.

Borowiec, N., Jorcin, A., Brutus, S., Petrousse, B., Moustache, M., Quilici, S. and Reynaud, B. (2009). Project de lutte biologique contre l'Aleurode du Cocotier, Aleurotrachelus atratus Hempel (Hemiptera: Aleyrodidae) aux Seychelles. Rapport technique de fin d'étape 3 (15/04/09 – 15/10/09) et préconisations pour un contrôle durable d'A. atratus aux Seychelles. CIRAD, La Réunion.

Borowiec, N., Quilici, S., Martin, J., Issimaila, M.A., Chadhouliati, A.C., Youssoufa, M.A., Beaudoin-Ollivier, L., Delvare, G. and Reynaud, B. (2010). Increasing distribution and damage to palms by the Neotropical whitefly, Aleurotrachelus atratus (Hemiptera: Aleyrodidae). Journal of Applied Entomology. 134 (6): 498-510.

Boudjelas, S. (2011). Rodent and cat eradication resource kit: a new support tool for invasive species management. Island Net Newsletter. No.7, August 2011: pp 19-21.

Boudjelas, S., Ritchie, J., Hughes, B. and Broome, K. (2011). The Pacific Invasives Initiative Resource Kit for planning rodent and cat eradication projects. pp 523. In C. R. Veitch, M. N. Clout and D. R. Towns. Island invasives: eradication and management. IUCN, Gland, Switzerland.

Boughton, A., Kula, R. and Center, T. (2011). Biological control of Old World climbing fern by Neomusotima conspurcatalis in Florida: post-release impact assessment and agent monitoring XIII International Symposium on Biological Control of Weeds.

Bourn, D., Gibson, C., Augeri, D., Wilson, C.J., Church, J. and Hay, S.I. (1999). The rise and fall of the Aldabran giant tortoise population. Proceedings of the Royal Society of London. Series B-Biological Sciences 266 (1424): 1091-1100.

Bovet, P., Gédéon, J., Louange, M., Durasnel, P., Aubry, P. and Gaüzère, B.A. (2013). Health situation and issues in the Seychelles in 2012. Med Sante Trop. 23 (3): 256-266. DOI: 10.1684/mst.2013.0222.

Bowler, J. (2006). Wildlife of Seychelles. Wild Guides, Hampshire

Bristol, R. (2003). Recent seabird censuses on Recif Island. Seychelles Seabird Group Newsletter, Issue 2, p.6. Nature Seychelles, Victoria.

Bristol, R. and Millett, J. (2002). Census of Wedge-tailed shearwaters Puffinus pacificus on Fouquet

Island, St Joseph Atoll, November 2002. Nature Seychelles. Unpublished report.

Bristol, U. (2006). North Island Myna Eradication. Zwazo. 15: 34-35.

Brondeau, A. and Triolo, J. (2007). Etablir des strategies de lutte opérationnelles contre les plantes exotiques invasives: examples à l'île de La Réunion. 13ème forum des gestionnaires. Espèces Exotiques Envahissantes: une menace majeure pour la Biodiversité. MNHN, Paris

Brooks, T.M., Mittermeier, R.A., Mittermeier, C.G., Da Fonseca, G.A., Rylands, A.B., Konstant, W.R., Flick, P., Pilgrim, J., Oldfield, S., Magin, G. and Hilton-Taylor, C. (2002). Habitat loss and extinction in the hotspots of biodiversity. Conservation biology. 16 (909-923).

Broome, K. (2007). Island bio-security as a pest management tactic in New Zealand. pp 104-108. In G. Witmer, W. Pitt and K. Fagerstone. Managing vertebrate invasive species: an international symposium. USDA National Wildlife Research Center, Fort Collins, CO, U.S.A.

Broome, K., Cox, A., Golding, C., Cromarty, P., Bell, P. and McClelland, P. (2014). Rat Eradication Using Aerial Baiting: Current Agreed Best Practice Used in New Zealand (Version 3.0). New Zealand Department of Conservation internal document. Department of Conservation, Wellington, New Zealand

Brouwer, L., Tinbergen, J.M., Both, C., Bristol, R., Richardson, D.S. and Komdeur, J. (2009). Experimental evidence for density-dependent reproduction in a cooperatively breeding passerine. Ecology. 90 (3): 729-741. <Go to ISI>://000263776800014.

Brown, K. (2002). Agave sisalana Perrine. Wildland Weeds, Summer: 18-21.

Buckle, A.P. and Smith, R.H., Eds. (1994). Rodent Pests and Their Control. CAB International, Wallingford, Oxon, UK.

Bunbury, N., Von Brandis, R., Currie, J., Jean-Baptiste, M., Accouche, W., Souyave, J., Haupt, P. and Fleischer-Dogley, F. (2013). Goats eradicated from Aldabra Atoll. Aliens 33. 33: 18-22.

Burbidge, A.A. (2011). 2001 to 2010 and beyond: Trends and future directions in the eradication of invasive species on islands. pp 515-519. In C. R. Veitch, M. N. Clout and D. R. Towns. Island invasives: eradication and management. IUCN, Gland, Switzerland.

Burger, A.E. and Lawrence, A.D. (2003). Seabird Monitoring Handbook for Seychelles. Second Edition. Nature Seychelles.

Burns, B., Innes, J. and Day, T. (2012). The use and potential of pest-proof fencing for ecosystem restoration and fauna conservation in New Zealand. pp 65-90. In M. W. Hayward and M. J. Somers. Fenc-

ing for conservation. Springer, New York.

Burt, A., Gane, J., Olivier, I., Calabrese, L., de Groene, A., Liebrick, T., Marx, D. and Shah, N. (in prep). Current population status of and trends for the Endangered Seychelles Magpie Robin Copsychus sechellarum.

Butaud, J.F. and Rahariveloma-nana, A. (2007). Rat impacts on the recruitment of the endangered endemic sandalwood (Santalum insulare) in French Polynesia. Université de la Polynesie Francaise, Tahiti

Buxton, R.T., Jones, C., Moller, H. and Towns, D.R. (2014). Drivers of seabird population recovery on New Zealand islands after predator eradication. Conservation biology 28: 333-344.

CAB International. (2011). Invasive Species Compendium / Glossary. http://www.cabi.org/isc/glossary.

CAB International. (2014). Invasive Species Compendium / Glossary. Retrieved 12th November, 2014. http://www.cabi.org/isc/glossary.

Cadi, A. and Joly, P. (2004). Impact of the introduction of the red-eared slider (Trachemys scripta elegans) on survival rates of the European pond turtle (Emys orbicularis). Biodiversity and Conservation. 13 (13): 2511-2518. DOI: 10.1023/B:BIOC.0 000048451.07820.9c.

Calabrese, L. and Maggs, G. (2012). Aride annual report 2011. Aride Island Nature Reserve. Island Conservation Society, Seychelles

Campbell, K.J., Beek, J., Eason, C.T., Glen, A.S., Godwin, J., Gould, F., Holmes, N.D., Howald, G.R., Madden, F.M., Ponder, J.B., Threadgill, D.W., Wegmann, A.S. and Baxter, G.S. (2015). The next generation of rodent eradications: innovative technologies and tools to improve species specificity and increase the feasibility on islands. Biol. Conserv. 185: 47-58.

Campbell, K.J., Harper, G., Algar, D., Hanson, C.C., Keitt, B.S. and Robinson, S. (2011). Review of feral cat eradications on islands. pp 37-46. In C. R. Veitch, M. N. Clout and D. R. Towns. Island invasives: eradication and management. IUCN, Gland, Switzerland.

Canning, G. (2011a). Population assessment of the Fregate Island giant tenebrionid beetle Polposipus herculeanus. Phelsuma. 19: 69-78.

Canning, G. (2011b). Eradication of the invasive common myna, Arcidotheres tristis, from Fregate Island, Seychelles. Phelsuma. 19: 43-53.

Capizzi, D., Baccetti, N. and Sposimo, P. (2010). Prioritizing rat eradication on islands by cost and effectiveness to protect nesting seabirds. Biological conservation. 143: 1716-1727.

Carlström, A. (1996). Endemic and Threatened Plant Species on the Granitic Seychelles. Consultancy Report. Ministry of Foreign Affairs, Division of Environment, Victoria

Castineiras, A., Obregon, O. and Borges, A. (1985). Evaluation of traps for transplanting colonies of Pheidole megacephala (Hymenoptera: Formicidae). Ciencia y Tecnica en la Agricultura, Proteccion de Plantas. 8 (2): 99-106.

Castle, G.E. and Mileto, R. (1994). Flora of Aride Island, Seychelles. ECO TECH, Shrewsbury. 74 pp.

Caut, S., Casanovas, J.G., Virgós, E.S., Lozano, J., Witmer, G.W. and Courchamp, F. (2007). Rats dying for mice: modelling the competitor release effect. Austral Ecology. 32: 858-868.

CEPF (2014). Ecosystem Profile. Madagascar and Indian Ocean Islands. Final version. December 2014. CEPF, Washington DC

Chang, V.C.S. (1985). Colony revival, and notes on rearing and life history of the big-headed ant. Proceedings of the Hawaiian Entomological Society. 25: 53-58.

Chapuis, J.-L., Barnaud, G., Bioret, F., Lebouvier, M. and Pascal, M. (1995). L'éradication des espèces introduites, un préalable à la restauration des milieux insulaires. Cas des îles françaises. Natures Sciences et Sociétés,. Hors série 3: 51-65.

Chapuis, J., Boussès, P. and Barnaud, G. (1994). Alien mammals, impact and management in the French Subantartic Islands. Biological Conservation. 67: 97-104.

Cheke, A. (2010). The timing of arrival of humans and their commensal animals on Western Indian Ocean oceanic islands. Phelsuma. 18: 38-69.

Cheke, A. (2013). Extinct birds of the Mascarenes and Seychelles - a review of the causes of extinction in the light of an important new publication on extinct birds. Phelsuma. 21: 4-19.

Cheke, A.S. and Rocamora, G. (2014). Duchemin's 'Linnet': Was there a second species of native fody Foudia sp. in the Granitic Seychelles? – with additional evidence for the mid-19th century introduction of F. madagascariensis. Malagasy Nature. 8: 73-79.

Christianson, E. (1944). Outlying Islands: Report to the Governor from Edward Christianson, Temporary Visiting Magistrate for the Outlying Islands 4th December 1944. Typed manuscript. Seychelles National Archives, Victoria

Clavero, M. and García-Berthou, E. (2005). Invasive species are a leading cause of animal extinctions. TRENDS in Ecology and Evolution 20 (3): 110.

Climo, G. (2004a). The Eradication of Cats and Rats from D'Arros Island, Seychelles and Proposed Strategies to Reduce the Risks of Re-introductions. Unpublished internal report for D'Arros Research Center.

Climo, G. (2004b). The Eradication of Cats and Rats from North Island, Seychelles and Proposed Strategies to Reduce the Risks of

Re-introductions. Unpublished internal report for North Island Co. Ltd.

Climo, G., Duncan, J., Rocamora, G. and Vanherck, L. (2006). Proposed Indian Myna Eradication Plan for North Island, Seychelles. January 2006. Island Conservation Society & North Island, Seychelles

Climo, G. and Rocamora, G. (2006a). The successful eradication of Black rats from North Island (Seychelles) in September 2005 and proposed strategies to reduce the risks of reinvasions. Réhabilitation des Ecosystèmes Insulaires (FFEM) Project. Island Conservation Society / North Island, Seychelles

Climo, G. and Rocamora, G. (2006b). Report on a first mission to eradicate rats and cats from Grande Ile and Grand Polyte, Cosmoledo atoll, October 2005. Seychelles. Project FFEM Rehabilitation of Islands Ecosystems. Island Conservation Society, Seychelles

Close, B., Banister, K., Baumans, V., Bernoth, E.M., Bromage, N., Bunyan, J., Erhardt, W., Flecknell, P., Gregory, N., Hackbarth, H., Morton, D. and Warwick, C. (1997). Recommendations for euthanasia of experimental animals: Part 2. Laboratory Animals. 31: 1-32.

Clout, M.N. and Hay, J.R. (1989). The importance of birds as browsers, pollinators and seed dispersers in New Zealand forests. New Zealand Journal of Ecology. 12: 27–33.

Clout, M.N. and Williams, P.A., Eds. (2009). Invasive species management: a handbook of principles and techniques. Oxford University Press, Oxford, UK. 308 pp.

Coblentz, B.E. and Van Vuren, D. (1987). Effects of feral goats (Capra hircus) on Aldabra Atoll. Atoll Research Bulletin. 306: 1-6.

Coblentz, B.E., van Vuren, D. and Main, M.B. (1990). Control of feral goats on Aldabra Atoll. Atoll Research Bulletin. 337 (1-13).

Coffey, E.D.D., Froyd, C.A. and Willis, K.J. (2011). When is an invasive not an invasive? Macrofossil evidence of doubtful native plant species in the Galápagos Islands. Ecology. 92 (4): 805-812.

Cole, R.G. and Singleton, R.J. (1996). Monitoring of reef fish populations on Kapiti Island during aerial poisoning for rats, 1996. National Institute of Water and Atmospheric Research, PO Box 893, Nelson. (an unpublished report prepared for Wellington Conservancy). Department of Conservation, Wellington, New Zealand

Comité français de l'UICN. (2012). Atelier de Travail sur les Espèces Exotiques Envahissantes dans l'Océan Indien. Mamoudzou, Mayotte 23-26/01/12. 2012. http://www.especes-envahissantes-outremer.fr/pdf/atelier_ocean_Indien_2012.

Conservacion de Islas (2015). Pushing the boundaries of rodent eradications on tropical islands: Ship rat eradication on Cayo Centro, Banco Chinchorro, Mexico. Press release June 2015. Conservacion de Islas, Baja California. Mexico.

Conservation International (Madagascar) (2014). Profil d'écosystème. Hotspot de Madagascar et des Iles de l'Océan Indien. Critical Ecosystem Partnership Fund, Washington & Antananarivo

Cook, C.J., Eason, C.T., Wickstrom, M. and Devine, C.D. (2001). Development of antidotes for sodium monofluoroacetate (1080). Biomarkers. 6 (1): 72-76. DOI: 10.1080/135475001452814.

Cory, J.S., Hirst, M.L., Sterling, P.H. and Speight, M.R. (2000). Narrow host range nucleopolyhedrovirus for control of the browntail moth (Lepidoptera: Lymantriidae). Environmental entomology. 29 (3): 661-667.

Courchamp, F., Langlais, M. and Sugihara, G. (1999). Cats protecting birds: modelling the mesopredator release effect. Journal of Animal Ecology. 68: 282-292.

Courchamp, F. and Sugihara, G. (1999). Biological control of alien predator populations to protect native island prey species from extinction. . Ecological Applications 9:112-123.

Courchamp, F., Chapuis, J.L. and Pascal, M. (2003). Mammal invaders on islands: impact, control and control impact. Biological Review. 78: 347-383.

Courchamp, F., Caut, S., Bonnaud, E., Bourgeois, K., Angulo, E. and Watari, Y. (2011). Eradication of alien invasive species: surprise effects and conservation successes. pp 285-289. In C. R. Veitch, M. N. Clout and D. R. Towns. Island Invasives: Eradicationand Management. Proceedings of the International Conference on Island Invasives. IUCN, Gland, Switzerland and Auckland, New Zealand.

Courchamp, F., Hoffmann, B.D., Russell, J.C., Leclerc, C. and Bellard, C. (2014). Climate change, sea-level rise, and conservation: keeping island biodiversity afloat. Trends in Ecology & Evolution. 29: 127-130.

Cowan, P. and Warburton, B. (2011). Ethical issues in island pest eradication. pp 418-421. In C. R. Veitch, M. N. Clout and D. R. Towns. Island invasives: eradication and management. IUCN, Gland, Switzerland.

Cresswell, W., Irwin, M., Jensen, M., Mee, A., Mellanby, R., McKean, M. and Milne, L. (1997). Population estimates and distribution changes of landbirds on Silhouette Island, Seychelles. Ostrich. 68 (2-4): 50-57.

Crook, J.H. (1960). The present status of certain rare land birds of the Seychelles Islands. Seychelles Government Bulletin.

Csóka, G. (1996). Aszályos évek - fokozódó rovarkárok erdeinkben [Drought years - increasing damage by insect pests in Hunga-rian forests]. Növényvédelem. 32 (11): 545-551.

Culliney, T.W. (2005). Benefits of Classical Biological Control for Managing Invasive Plants Critical Reviews in Plant Sciences. 24: 131-150.

Cunningham, D.M. and Moors, P.J. (1993). Guide to the identification and collection of New Zealand rodents, 2nd Edition. New Zealand Department of Conservation.

Cuthbert, R. and Hilton, G. (2004). Introduced house mice Mus musculus: a signifi cant predator of threatened and endemic birds on Gough Island, South Atlantic Ocean? Biological Conservation. 117: 483-489.

Daltry, J.C., Hillman, J.C. and Meier, G.G. (2007). Chagos Ecological Restoration Project: Follow up visit to Eagle Island, March 2007. Report to the Foreign & Commonwealth Office and Fauna & Flora International. UK. 20 pp.

Davis, M.A., Chew, M.K., Hobbs, R.J., Lugo, A.E., Ewel, J.J., Vermeij, G.J., Brown, J.H., Rosenzweig, M.L., Gardener, M.R., Carroll, S.P., Thompson, K., Pickett, S.T., Stromberg, J.C., Del Tredici, P., Suding, K.N., Ehrenfeld, J.G., Grime, J.P., Mascaro, J. and Briggs, J.C. (2011). Don't judge species by their origin. Nature. 474: 153- 154.

Dawson, P. (2003). The history of the D'Arros fodies. Birdwatch. 49 (1): 5-7.

De-Long, S. (1981). Mulberry tussock moth dermatitis: A study of an epidemic of unknown origin. Journal of Epidemiology and Community Health. 35 (1): 1-4.

De Jong, C.J.M., Hoedemaeker, P.J., Jongebloed, W.L. and Nater, J.P. (1976). Investigative studies of the dermatitis caused by the larva of the brown-tail moth (Euproctis chrysorrhoea Linn.). Archives of Dermatological Research. 255 (2): 177-191.

Dejean, A., Orivel, J., Durand, J.L., Ngnegueu, P.R., Bourgoin, T. and Gibernau, M. (2000). Interference between ant species distribution in different habitats and the density of a major pest. Sociobiology. 35 (1): 175-189.

Dhillon, M.K., Singh, R., Naresh, J.S. and Sharma, H.C. (2005). The Melon Fruit Fly, Bactrocera cucurbitae: a review of its biology and management. Journal of Insect Science. 5: 40.

Diamond, A.W. and Feare, C.J. (1980). Past and present biogeography of central Seychelles birds. Proc. IV Pan-Afr. Orn. Congr: 89-97.

Diaz, J.H. (2005). The evolving global epidemiology, syndromic classification, management, and prevention of caterpillar envenoming. The American Journal of Tropical Medicine and Hygiene. 72 (3): 347-357.

DIISE. (2014). The Database of Island Invasive Species Eradications, developed by Island Conservation, Coastal Conservation Action Laboratory UCSC, IUCN SSC Invasive Species Specialist Group. University of Auckland and Landcare Research New Zealand. Retrieved 14th September 2014. http://diise.islandconservation.org.

Division of Environment (2009). Indian house crows and their eradication from Mahé. Seychelles Nation. 12-October-2009. Victoria (Seychelles).

Dobson, A.P. (1988). Restoring Island Ecosystems. The Potential of parasites to Control Introduced Mammals. Conservation Biology. 2 (1): 31-38. DOI:10.1111/j.1523-1739.1988.tb00333.x.

DoC (2003, reviewed 2010)). Best Practice. Island Biosecurity Best Practice Manual. Department of Conservation, New Zealand, Wellington

Dogley, W. (2004). Plant Pest and Disease Management: A manual for farmers and gardeners. Department of Natural Resources. Ministry of Environment and Natural Resources, Victoria, Seychelles. 108 pp.

Dogley, W. (2007). Melon Fruit Fly eradication Programme. Project No. 8ACP SEY 009. Final Report. Department of Natural Ressouces, MENRT, Victoria

Dogley, W. (2009). Evaluation of the threats of introduction and spread of IAS through production sector activities in Seychelles. Report. Mainstreaming Prevention and Control Measures for Invasive Alien Species into Trade, Transport and Travel across the Production Landscape. UNDP/GEF/GoS Biosecurity Project, Victoria. 60 pp.

Dolbeer, R.R., Fiedler, L.R. and Rasheed, H. (1988). Management of fruit bat and rat populations in the Maldive Islands, Indian Ocean. pp 112-118. In A. C. Crabb and R. E. Marsh. Proceedings of the Thirteenth Vertebrate Pest Conference. University of California, Davis. http://digitalcommons.unl.edu/vpcthirteen/24

Donlan, C.J., Howald, G.R., Tershy, B.R. and Croll, D.A. (2003). Evaluating alternative rodenticides for island conservation: roof rat eradication from the San Jorge Islands, Mexico. Biological Conservation. 14: 29-34.

Drake, D.R. (2007). Impacts of alien rats on plant recruitment: positive, negative, direct, and indirect. University of Hawaii at Manoa, USA

Drake, C.M., Lott, D.A., Alexander, K.N.A. and Webb, J. (2007). Surveying terrestrial and freshwater invertebrates for conservation evaluation. Natural England Research report NERR005. 139 pp.

Driesche, R.V., B., B., Hoddle, M., Lyon, S. and Reardon, R. (2002). Biological control of invasive plants in the Eastern United States 139 pp.

Dubois, N.R., Rajamohan, F., Cotrill, J.F. and Dean, D.H. (1997). On understanding the mechanism of Bacillus thuringiensis insecticidal activity. pp 35-36. In S. L. C. Fosbroke and K. W. Gottschalk. Proceedings, U.S. Department of Agriculture Interagency gypsy moth research forum. U.S. Department of Agriculture, Forestry Service, , General Technical Report, 240.

Dubois, N.R., McManus, M.L., Huntley, P.J. and Newman, D. (2001). Implementation of a program to optimize the use of Bacillus thuringiensis against the Browntail Moth (Euproctis chrysorrhoea). pp 37-44. In A. M. Liebhold, M. L. McManus, I. S. Otvos and S. L. C. Fosbroke. Proceedings: Integrated management and dynamics of forest defoliating insects, August 15-19,1999. Department of Agriculture, Forest Service, Northeastern Research Station, Victoria, British Columbia. U.S.

Dunlop, E., Hardcastle, J. and Shah, N. (2005). Cousin and Cousine islands Status and Management of Alien Invasive Species. World Bank/ GEF funded project: –Improving management of NGO and Privately-owned Nature Reserves and High Biodiversity islands in Seychelles. Project Report September 2005.

Duplantier, J.-M., Orth, A., Catalan, J. and Bonhomme, F. (2002). Evidence for a mitochondrial lineage originating from the Arabian peninsula in the Madagascar house mouse (Mus musculus). Heredity. 89: 154-158.

DWCT / MWF (2010). Invasive Species Management 2010, Mauritius (8th – 13th November 2010). Workshop Syllabus & Course Pack. Durrell Wildlife Conservation Trust, Jersey.

Elton, C.S. (1958). The ecology of invasions by animals and plants. Methuen, London.

Eltringham, H. (1914). On the urticating properties of Porthesia similis, Fuess. Transactions of the Royal Entomological Society of London. 61 (3): 423-427.

Emerton, L. (1997). Seychelles Biodiversity: Economy Assessment. IUCN, Nairobi

Emerton, L. and Howard, G. (2008). A Toolkit for the Economic Analysis of Invasive Species. Global Invasive Species Programme, Nairobi

Enge, K.M. and Krysko, K.L. (2004). A new exotic species in Florida, the Bloodsucker Lizard, Calotes versicolor (Daudin 1802) (Sauria: Agamidae). Biological Sciences (No. 3).

Environment Department. (2013). Eradication of 'Korbo' still a challenge. Seychelles Nation. 21-January-2013. Victoria (Seychelles).

Erickson, W. and Urban, D. (2002). Potential risks of nine rodenticides to birds and non-target mammals : a comparative approach. Unpubl.

report. United States Environmental Protection Agency, Washington, DC.

Escoda, R. (1996). Memories d'un poble: La Vilella d'Amunt - Priorat. Institut d'Estudis Tarraconenses Ramon Berenger IV. Diputació de Tarragona, Catalonia.

Fanchette, R. (2003). Successful eradication of an IAS from Mahe island, Seychelles, House Sparrow (Passer domesticus) 2002-2003. Unpublished report, Government of Seychelles.

Fanchette, R. (2006). Indian Mynah Bird Control Programme. Sainte Anne Resort & Spa. Status report. January 2006. Internal report. Conservation section, Division of Environment, Ministry of Environment and Natural Resources, Seychelles

Fanchette, R. (2006). Status report. Crested tree lizard Calotes versicolor Ste Anne Island. Unpublished report. Division of Environment, Ministry of Environment and Natural Resources. Government of Seychelles, Seychelles

Fanchette, R. and Kerridge, F. (2009). Effect of forest type on morphometrics and diet of Rattus species. Morne Seychellois National Park. Master Thesis School of Health and Social Studies, Biology (Ecology and Conservation). Department of Psychology and Life Sciences. University of Bolton, Bolton.

Fanchette, R. (2012). Invasive Alien Species: threats to humans and biodiversity. Seychelles Strategy. PowerPoint Presentation. Atelier de Travail sur les Espèces Exotiques Envahissantes dans l'Océan Indien. Mamoudzou, Mayotte 23-26/01/12. Comité français de l'UICN. Retrieved 2012 http://www.especes-envahissantes-outremer.fr/pdf/atelier_ocean_Indien_2012/Seychelles.pdf.

FAO. (2011). Common weeds in Vanuatu. Retrieved on 19 December, 2011. http://www.fao.org/ag/AGP/AGPC/doc/Publicat/FAOBUL2/B201.htm.

Fauvel, A.A. (1909). Unpublished documents on the History of Seychelles anterior to 1810. Govt. Printer, Victoria, Mahé,Seychelles

Feare, C.J. (1979). Ecology of Bird Island, Seychelles. Atoll Res. Bull. 226: 1-29.

Feare, C.J. (1999). Ants take over from rats on Bird Island, Seychelles. Bird Conservation International. 9: 95-96.

Feare , C.J. and Gill, E.L. (1995). The turtle doves of Bird Island, Seychelles. Bull Brit. Orn. Club. . 115: 206-210.

Feare, C.J., Jaquemet, S. and Le Corre, M. (2007). An inventory of Sooty Terns (Sterna fuscata) in the western Indian Ocean with special reference to threats and trends. Ostrich. 78 (2): 423-434.

Feare, C.J. (2010). The use of Starlicide® in preliminary trials to control invasive common myna

Acridotheres tristis populations on St Helena and Ascension islands, Atlantic Ocean. Conservation Evidence. 7: 52-61.

Feare, C.J., French, G.C.A., Nevill, J.E.G., Pattison-Willits, V.S., Wheeler, V., Yates, T.L., Hoareau, C. and Prescott, C. (2015). Attempted re-establishment of a sooty tern Onychoprion fuscatus breeding colony on Denis Island, Seychelles. Conservation Evidence: 19-24.

Ferrell, J., Langeland, K. and Sellers, B. (2006). Herbicide application techniques for woody plant control. IFAS Extension Bulletin SS-AGR-260. University of Florida. http://edis.ifas.ufl.edu/.

Ferriter, A. (2001). Lygodium Management Plan for Florida. A Report from the Florida Exotic Pest Plant Council's Lygodium Task Force. First Edition

FFEM and ICS (2010). France – Seychelles. Rehabilitating Island Ecosystems. Eradication of invasive exotic species and Re-introduction of threatened endemic species. Asconit Consultants and FFEM, Paris.

Fiedler, L.R. (1984). Recommendations for reducing coconut losses due to rodents in the Maldive Archipelago. FAO Unpubl. Report, Rome. 19 pp + 2 appendices.

Fleischmann, K. (1997). Invasion of alien woody plants on the islands of Mahé and Silhouette, Seychelles. Journal of Vegetation Science 8:5-12.

Fleischmann, K., Kollmann, J., Ramseier, D. and Edwards, P.J. (2005). Stand structure, species diversity and regeneration of an endemic palm forest on the Seychelles. African Journal of Ecology. 43: 291-301.

Foley, M.J. and Bolton, M.P. (1990). Control of sisal hemp (Agave spp.) in native bushland. Proceedings of the 9th Australian Weeds Conference, (Crop Science Society of South Australia, Adelaide). 145 pp.

Fosberg, F.R. (1979). Plants of D'Arros island. pp 43-48. In D. R. Stoddart, M. J. Coe and F. R. Fosberg. D'Arros and St Joseph, Amirante islands. Atoll Research Bulletin. 223.

Fosberg, F.R. (1983). Natural history of Cousin island. pp 7-37. In M. H. Sachet, D. R. Stoddart and F. R. Fosberg. Floristics and ecology of western Indian Ocean islands. Atoll Research Bulletin. 273.

Fosberg, F.R. and Renvoize, S.A. (1980). The Flora of Aldabra and Neighbouring Islands. Kew Bulletin Additional Series VII, HMSO

Foster, S., King, C., Patty, B. and Miller, S. (2011). Tree-climbing capabilities of Norway and ship rats. New Zealand Journal of Ecology. 38 (4): 285-296.

Foxcroft, L.C., Pyšek, P., Richardson, D.M. and Genovesi, P. (2013). Plant invasions in protected areas. Patterns, problems and challenges. Dordrecht, Springer.

Franklin, K. (2013). Informational report on the use of Goodnature®A24 rat traps in Hawaii. Pacific Cooperative Studies Unit. Oahu Army Natural Resources Program, University of Hawaii. 22 pp.

Gaigher, R., Samways, M.J., Henwood, J. and Jolliffe, K. (2011). Impact of a mutualism between an invasive ant and honeydew-producing insects on a functionally important tree on a tropical island. Biol Invasions. Publiched online: 19 January 2011.

Gaigher, R. and Samways, M.J. (2012). Strategic Management of an Invasive Ant-scale Mutualism Enables Recovery of a Threatened Tropical Tree Species. Biotropica. 45 (1): 128-134. DOI: 10.1111/j.1744-7429.2012.00898.x http://dx.doi.org/10.1111/j.1744-7429.2012.00898.x.

Gaigher, R., Samways, M., Jolliffe, K. and Jolliffe, S. (2012). Precision control of an invasive ant on an ecologically sensitive tropical island: a principle with wide applicability. Ecological Applications. 22 (5): 1405-12.

Galman, G. (2011). Suivi spatio-temporel des communautés d'Arthropodes, effets de l'éradication des rats et tentative de réintroduction d'un insecte rare dans des îles en cours de réhabilitation des Seychelles. Muséum National d'Histoire Naturelle, Paris.

Galman, G., Rocamora, G., Labiche, A. and Russell, J. (in prep.). Invertebrate's response to the eradication of rats Rattus rattus on Ile du Nord, Seychelles.

Galván, J.P., Tershy, B., Howald, G., Samaniego, A., Keitt, B., Browne, M., Russell, J., Pascal, M. and Parkes, J. (2005). A Review of Commensal Rodent Eradication on Islands. Pacific Seabird Group Conference. Portland, U.S.A.

Gamberale, G. and Tullberg, B. (1998). Aposematism and gregariousness: the combined effect of group size and coloration on signal repellence Proceedings of the Royal Society of London. B, 265 (1399): 889-894.

Gargominy, O., Ed. (2003). Biodiversité et conservation dans les collectivités françaises d'outre-mer. Comité français pour l'UICN, Paris, http://www.uicn.fr/Biodiversite-outre-mer-2003.html.

GEIR. (2014). Groupe Espèces Invasives de La Réunion. Retrieved 21st May, 2015. www.especesinvasives.re. DIREN, St Denis de La Réunion.

Genovesi, P. (2011). Are we turning the tide? Eradications in times of crisis: how the global community is responding to biological invasions. pp 5-8. In C. R. Veitch, M. N. Clout and D. R. Towns. Island invasives: eradication and management. IUCN, Gland, Switzerland.

Gerlach, J. (1994a). The ecology of the carnivorous snail Euglandina rosea. Unpublished PhD thesis, University of Oxford.

Gerlach, J. (1994b). The Distribution and status of Seychelles land snails. Papustyla. 8 (3): 12-14.

Gerlach, J. (2004). Impact of the invasive crazy ant *Anoplolepis gracilipes* on Bird Island, Seychelles. Journal of Insect Conservation. 8: 15-25.

Gerlach, J. (2005a). The impact of rodent eradication on the larger invertebrates of Fregate Island, Seychelles Phelsuma. 13: 43-54.

Gerlach, J. (2005b). The status of invertebrates on Fregate island, Seychelles following rat eradication. Phelsuma. 13.

Gerlach, J. (2006). Native or introduced plant species? Phelsuma. 4: 70-74.

Gerlach, J. (2006). Terrestrial and freshwater Mollusca of Seychelles islands. Backhuys Publishers, Leiden

Gerlach, J. (2011). Attempting to balance development and conservation on Silhouette Island. Phelsuma. 19: 79-114.

Gerlach, J. (2014). Snailing round the South Seas - the Partula story. Retrieved September, 2014. www.islandbiodiversity.com/pnews.htm

Gerlach, J. and Florens, V. (2000). Toxicity of 'specific' rodenticides and the risk to non-target taxa. Phelsuma. 8: 75-76.

Gerlach, J. and Gerlach, R. (2011). Re-introduction and supplementation of terrapins on Silhouette Island, Seychelles. pp 102-105. In P. Soorae. Global Re-introduction Perspectives: More case studies from around the globe. IUCN/SSC Re-introduction Specialist Group and Abu-Dhabi, UAE: Environment Agency- Abu-Dhabi, UAE, Gland, Switzerland.

Gerlach, J. and Gerlach, R. (2011). Translocation of giant tortoises in the Seychelles Islands. pp 98-101. In P. Soorae. Global Re-introduction Perspectives: More case studies from around the globe. IUCN/SSC Re-introduction Specialist Group and Abu-Dhabi, UAE: Environment Agency- Abu-Dhabi, UAE, Gland, Switzerland.

Gerlach, J., Lawrence, J. and Canning, L. (2005). Mortality, population changes and exceptional behaviour in a giant millipede. Phelsuma. 13: 85-94.

Gerlach, J., Rocamora, G., Gane, J., Jolliffe, K. and Vanherck, L. (2013). Giant tortoise distribution and abundance in the Seychelles islands: past, present and future. Chelonian Conservation & Biology. 12 (1): 70-83.

Géry, R. (1991). Étude expérimentale de la predation de la cochenille des Seychelles Icerya seychellarum (Westwood) par deux Coccinellidae: Rodolia cardinalis (Mulsant) et Rodolia chermesina Mulsant. Application à la lutte biologique sur l'atoll d'Aldabra. Thèse de doctorat de l'Université de Rennes I 91 + annexes. l'Université de Rennes I.

Gibson, C.W.D. and Hamilton, J. (1983). «Feeding Ecology and Seasonal Movements of Giant Tortoises on Aldabra Atoll» Oecologia. 56: 84-92.

GISD. (2009). Global Invasive Species Database. Retrieved 2011 http://www.issg.org/.

GISD. (2011). Global Invasive Species Database. Retrieved 20 February, 2011. http://www.issg.org/.

GISD. (2014). Global Invasive Species Database. Retrieved 20th August, 2014. http://www.issg.org/database/welcome/.

Glen, A.S., Atkinson, R., Campbell, K.J., Hagen, E., Holmes, N.D., Keitt, B.S., Parkes, J.P., Saunders, A., Sawyer, J. and Torres, H. (2013). Eradicating multiple invasive species on inhabited islands: the next big step in island restoration? Biological Invasions. 15: 2589-2603.

Global Invasive Species Programme. (2011). Retrieved 10th November 2011. http://www.gisp.org/ecology/invasives.asp.

González-Cano, J.M. (1981). Predación de procesionaria del pino por vertebrados en la zona de Mora de Rubielos (Teruel). Boletin de la Estacion Central de Ecología. 10: 53-77.

GoS/UNDP/GEF (undated). Integrated Ecosystem Management Programme: Mainstreaming Prevention and Control Measures for Invasive Alien Species into Trade, Transport and Travel across the Production Landscape. Project Document. PIMS No: 3820, Proposal ID: 00045017, Project ID: 00053109.

Gould, M.S. and Swingland, I.R. (1980). The tortoise and the goat: interactions on Aldabra Island. Biol . Conserv. . 17: 267-279.

Government of Mauritius (2010). The National Invasive Alien Species Strategy and Action Plan for the Republic of Mauritius: 2010-2019.

Government of Seychelles (2011a). The National Invasive Alien Species (Biosecurity) Strategy for Seychelles 2011-2015. Ministry of Home Affairs, Environment and Transport & Ministry of Industry, Natural Resources and Industry, Victoria, Seychelles

Government of Seychelles (2011b). Fourth National Report to the United Nations Convention on Biological Diversity. Department of Environment, P.O. Box 445, Botanical Gardens, Mont Fleuri, Victoria, Republic of Seychelles

Government of Seychelles (2014). Republic of Seychelles. General Emergency Response Plan for Pests, Diseases and Alien Invasive Species. Government of Seychelles/UNDP/GEF, Victoria

Grant, P.R. (1972). Interspecific competition among rodents. Annual Review of Ecol. And Syst. 3: 79-106.

Grant, P.R. and Grant, B.R. (2002). Unpredictable evolution in a 30-year study of Darwin's finches. Science. 296: 707-711.

Greathead, D. (1971). Review of biological control in the Ethiopian region. Technical communication - Commonwealth Institute of Biological Control. no. 5: 162.

Greaves, J.H. (1994). Resistance to anticoagulant rodenticides. pp 197-217. In A. P. Buckle and R. H. Smith. Rodent pests and their control. Cab. International, Oxon G.B.

Green, P.T., Comport, S. and Slip, D. (2004). Shock and awe: rapid response to alien ant invasion on an isolated oceanic island.

Gregory, R.D., Gibbons, D.W. and Donald, P.F. (2004). Bird census and survey techniques. pp 17-53. Chapter 2. In W. J. Sutherland, I. Newton and G. R. Bird Ecology and Conservation: A Handbook of Techniques. Oxford University Press, Oxford, UK.

Gregory, S.D., Henderson, W., Smee, E. and Cassey, P. (2014). Eradications of vertebrate pests in Australia: A review and guidelines for future best practice. PestSmart Toolkit publication, Invasive Animals Cooperative Research Centre, Canberra, Australia http://www.feral.org.au/wp-content/uploads/2014/02/AusEradications_2014.pdf.

Griffiths, C.J. (2014). Rewilding in the Western Indian Ocean. pp In J. Gerlach. Tortoises of the Western Indian Ocean. Siri Scientific Press, Manchester.

Griffiths, C.J., Jones, C.G., Hansen, D.M., Puttoo, M., Tatayah, R.V., Müller, C.B. and Harris, S. (2010). The Use of Extant Non-Indigenous Tortoises as a Restoration Tool to Replace Extinct Ecosystem Engineers. Restoration Ecology. 18: 1-7.

Griffiths, R. (2011). Targeting multiple species – a more efficient approach to pest Eradication. pp 172-176. in C. R. Veitch, M. N. Clout and D. R. Towns. Island invasives: eradication and management. IUCN, Gland, Switzerland.

Griffiths, R., Miller, A. and Climo, G. (2011). Addressing the impact of land crabs on rodent eradications on islands. Pacific Conservation Biol. . 17: 347.

Grobler, J.H. (1957). Some aspects of the biology, ecology and control of the pine brown tail moth, Euproctis terminalis, Walk. Department of Agriculture, Pretoria. ii+186 pp.

Gruner, D.S. (2004). Attenuation of top-down and bottom-up forces in a complex terrestrial community. Ecology. 85: 3010-3022.

Hansen, D.M., Donlan, C.J., Griffiths, C.J. and Campbell, K.J. (2010). Ecological history and latent conservation potential: large and giant tortoises as a model for taxon substitutions. Ecography. 33: 272-284.

Harper, G. (2013). Population ecology and impacts of black rats and feral cats on Aldabra Atoll, Seychelles. Lead Consultant's Report on initial research for Seychelles Islands Foundation, January-April 2013 (Project objective 1d). SIF EU funded project. Seychelles Islands Foundation, Seychelles

Harper, G.A. (2014). Aldabra Atoll Biosecurity Plan. Seychelles Islands Foundation, Republic of Seychelles.

Harper, G. and Bunbury, N. (2015). Invasive rats on tropical islands: their population biology and impacts on native species. Global Ecology and Conservation. 3: 607-627.

Harper, G.A., Van Dinther, M., Russell, J.C. and Bunbury, N. (2015). The response of black rats (Rattus rattus) to evergreen and seasonally arid habitats: informing eradication planning on a tropical island. Biol. Conserv. 185: 66-74.

Harris, L., Anderson, T. and Bennett, D. (2011). Eradication trials on Peel Island, Australia.

Hazell, S. P. (2005). The spiralling whitefly Aleurodicus dispersus in Seychelles. Nature Seychelles, Seychelles.

Hazell, S. P., Vel, T. and Fellowes, M. D. E. (2007). The role of exotic plants in the invasion of Seychelles by the polyphagous insect Aleurodicus dispersus: a phylogenetically controlled analysis. Biol Invasions 10 (2): 169-175. DOI 10.1007/s10530-007-9120-2

HEAR. (2012). Global Compendium of Weeds. Retrieved 20 January, 2012. http://www.hear.org/.

Heiniger, U. and Rigling, D. (1994). Biological control of chestnut blight in Europe. Annual Review of Phytopathology. 32 (1): 581-599.

Henri, K., Milne, G.R. and Shah, N.J. (2004). Costs of ecosystem restoration on islands in the Seychelles. Ocean and Coastal Management. 47: 409-428.

Henriette, E. (2011). Conservation introductions as a tool for the recovery of endangered island species: territoriality and demography of the Seychelles white-eye Zosterops modestus on Frégate Island. Thesis Doctor of Philosophy. Muséum National d'Histoire Naturelle, Paris/ICS/MERN.

Henriette, E. (2013). Rat Control and Seychelles White-Eye monitoring in collaboration with the local community in SWE territories on Mahé. Final report: March 2013. US Embassy Self Help Fund Project. Island Conservation Society, Seychelles

Henriette, E. and Dine, D. (2014). Population assessment of the Seychelles white-eye Zosterops modestus on North Island, Seychelles. Report for the period 20th September to 14th November 2014. Unpublished report. North Island and GIF, Seychelles

Henriette, E. and Julienne, S. (2009). Impact of climate change on the health sector. Enabling activities for the preparation of the Seychelles second national communication to the United Nations Framework Convention on Climate

Change. UNEP-GEF project, Victoria. 37 pp.

Henriette, E., Senterre, B., Kaiser-Bunbury, C., Athanase, M., Mougal, J., Mirabeau, M. and Laboudallon, V. (2012). Management of Invasive Alien Creepers in Selected Sites in Seychelles. Technical report on the management of Invasive Alien Creepers at Casse Dent, Salazie and Cascade, Mahé, Seychelles.

Hiebert, R. and Stubbendieck, J. (1993). Handbook for Ranking Exotic Plants for Management and Control. U.S. Natural Resources Report NPS/NRMWRO/NRR-93/08. Department of the Interior, National Park Service, University of Nebraska, Lincoln

Hill, M. (2002). Biodiversity Surveys and Conservation Potential of Inner Seychelles Islands. Atoll Research Bulletin, No. 495.

Hill, M., Currie, D. and Shah, N. (2003a). The impacts of vascular wilt disease of the takamaka tree Calophyllum inophyllum on conservation value of islands in the granite Seychelles. Biodiversity and Conservation. 12: 555-566.

Hill, M., Holm, K., Vel, T., Shah, N. and Matyot, P. (2003b). Impact of the introduced yellow crazy ant Anoplolepis gracilipes on Bird Island, Seychelles. Biodiversity and Conservation 12 (9): 1969-1984.

Hill, M.J., Vel, T. and Shah, N.J. (2003c). The morphology, distribution and conservation implications of introduced rats, Rattus spp. in the granitic Seychelles. African Journal of Ecology. 41 (2): 179-186.

Hill, M.J., Shah, N.J. and Parr, S. (2004). Handbook for Biodiversity Assessment of Small Tropical Islands, Methodologies Developed to Assist in Ecological Restoration in Seychelles. Nature Seychelles, Victoria, Seychelles

Hill, M. and Currie, D. (2007). Wildlife of Seychelles. Collins, UK

Hillmer, J., Steward, N.O.L. and Liedtke, D. (2003). Safe herbicide handling in natural areas. The Nature Conservancy. Northeast Ohio Field Office.

Hivert, J. (2003). Plantes exotiques envahissantes: état des méthodes de lutte mises en oeuvre par l'Office National des Forêts à La Réunion. Rapport de l'Office National des Forêts (ONF). l'Office National des Forêts (ONF), La Réunion 2842072847.

Hoare, J.M. and Hare, K.M. (2006). The impact of brodifacoum on non-target wildlife: gaps in knowledge. New Zealand Journal of Ecology. 30: 157-167.

Hobbs, R.J., Higgs, E.S. and Hall, C.M., Eds. (2013). Novel Ecosystems - Intervening in the New Ecological World Order. Wiley-Blackwell.

Hoffmann, B.D., Andersen, A.N. and Hill, G.J.E. (1999). Impact of an introduced ant on native rain forest invertebrates: Pheidole mega-

cephala in monsoonal Australia. Oecologia 120: 595-604.

Hoffmann, B.D., Auina, S. and Stanley, M.C. (2014). Targeted Research to Improve Invasive Species Management: Yellow Crazy Ant Anoplolepis gracilipes in Samoa. PLoS ONE. 9 (4). DOI: e95301. doi:10.1371/journal.pone.0095301.

Hoffmann, B.D. and O'Connor, S. (2004). Eradication of two exotic ants from Kakadu National Park. Ecological management & restoration. Vol 5 (No 2).

Holmes, N.D., Griffiths, R., Pott, M., Alifano, A., Will, D., Wegmann, A.S. and Russell, J.C. (2015). Factors associated with rodent eradication failure. Biol. Conserv. 185: 8-16.

Holzmueller, E.J. and Jose, S. (2009). Invasive plant conundrum: What makes the aliens so successful? Journal of Tropical Agriculture. 47 (1-2): 18-29.

Hostetter, D.L. and Bell, M.R. (1985). Natural dispersal of baculoviruses in the environment. pp 249-284. In K. Maramorosch. Viral insecticides for biological control. Academic Press, USA.

Howald, G., Donlan, C.J., Galvan, J.P., Russell, J., Parkes, J., Samaniego, A., Wang, Y., Veitch, D., Genovesi, P., Pascal, M., Saunders, A. and Tershy, B. (2007). Invasive rodent eradication on islands. Conservation Biology. 21: 1258-1268.

http//:www.mynamagnet.com. (2014). Myna Magnet. www.mynamagnet.com.

http://home.global. co.za/~peabrain/achatina.htm. (2014). Land snails. Retrieved 4 September, 2014.

http://home.global. co.za/~peabrain/achatina.htm

Hutchinson, J., Ferriter, A., Serbesoff-King, K., Langeland, K. and Rodgers, L. (2006). Old world climbing fern (Lygodium microphyllum) management plan for Florida. Florida Exotic Pest Council: 115.

ICS (2013). Annual Report 2012. Island Conservation Society, Seychelles

ICS (2014). Annual Report 2013. Island Conservation Society, Seychelles

ICS Alphonse (2008). Monthly report. May 2008. www.islandconservationseychelles.com.

Ikin, R. and Dogley, W. (2009). Institutional Review of Quarantine and Control Functions for Invasive Alien Species in the Seychelles. Report. Mainstreaming Prevention and Control Measures for Invasive Alien Species into Trade, Transport and Travel across the Production Landscape. UNDP/GEF/GoS Biosecurity Project, Victoria

Innes, J. and Saunders, A. (2011). Eradicating multiple pests: an overview. pp 177-181. In C. R. Veitch, M. N. Clout and D. R. Towns. Island invasives: eradication and management. IUCN, Gland, Switzerland.

Invasive Species Compendium. Retrieved 10th September, 2014. www.cabi.org/isc/.

IPPC/FAO (1996). Code of Conduct for the Import and Release of Exotic Biological Control Agents. ISPM no. 3. IPPC Secretariat, FAO, Rome

IPPC / FAO (2004). ISPM 11: Pest risk analysis for quarantine pests including analysis of environmental risks and living modified organisms. FAO, Rome, Italy

IPPC / FAO (2005). Guidelines for the export, shipment, import and release of biological control agents and other beneficial organisms. ISPM 3 Secretariat of the International Plant Protection Convention, FAO, Paris. 14 pp.

IPPC / FAO (2005). Identification of risks and management of invasive alien species using the IPPC framework. Proceedings of the workshop in Braunschweig, Germany, 22-26 September 2003. FAO, Rome, Italy

IPPC / FAO (2007). ISPM 2: Framework for pest risk analysis, Pp 27-42 In International Standards for Phytosanitary Measures. ISPM compilation (1-29), 2007 edition. FAO, Rome, Italy

Island Conservation. (2012). Retrieved 15 June, 2012. www.island-conservation.org

IUCN. (2009). Seychelles Coral Reef Invasives. Seychelles introduced species. http://www.iucn.org/about/work/programmes/marine/marine_our_work/marine_invasives/seychelles/seychelles_introduced_species/

IUCN. (2012). SSC Invasive Species Specialist Group. Retrieved February, 2012. http://www.issg.org/.

IUCN. (2014). Island Ecosystems. Retrieved 15th August, 2014. http://www.iucn.org/about/union/commissions/cem/cem_work/tg_islands/.

IUCN. (2014). IUCN Red List of Threatened Species. Retrieved 20th November, 2014. http://www.iucnredlist.org.

IUCN. (undated). Biological invasions: a growing threat to biodiversity, human health and food security. IUCN's Policy Brief on Invasive and Alien Species, Biodiversity, Huyman Health and Food security. Retrieved 15th October, 2014. http://cmsdata.iucn.org/downloads/policy_brief_on_invasive_and_alien_species.pdf.

IUCN/GoS (2006). Marine Introduced Species: towards better detection and management in the Seychelles. Draft Report IUCN & Government of Seychelles, August 2006.

IUCN/SSC (2013). Guidelines for Reintroductions and Other Conservation Translocations. Version 1.0. IUCN Species Survival Commission, Gland, Switzerland. viiii + 57 pp.

Johnson, K.S., Scriber, J.M., Nitao, J.K. and Smitley, D.R. (1995). Toxicity of Bacillus thuringiensis

var. kurstaki to three nontarget Lepidoptera in field studies. Environmental Entomology. 24 (2): 288-297.

Johnson, R. (2010). African bigheaded ants (Pheidole megacephala) on Aride Island, Seychelles. Internal Report. Island Conservation Society, Seychelles

Johnson, S. and Threadgold, R. (1999). Report on the monitoring & status of the coccid Icerya seychellarum on Aldabra from 1980 to 1999. March 1999. Aldabra Research Station. Seychelles Islands Foundation, Seychelles

Jolliffe, K., Jolliffe, S. and Henwood, J. (2008). Third post-release report of the Seychelles White-eye on Cousine Island. Report Period: August 2007 up to March 2008. 60 pp.

Jones, A.G., Chown, S.L. and Gaston, K.J. (2003). Introduced mice as a conservation concern on Gough Island. Biodiversity and Conservation. 12: 2107-2119.

Jones, A.G., Chown, S.L., Ryan, P.G., Gremmen, N.J.M. and Gaston, K.J. (2003). A review of conservation threats on Gough Island: a case study for terrestrial conservation in the Southern Oceans. Biological Conservation. 113: 75-87.

Jones, H.P. (2010). Seabird islands take mere decades to recover following rat eradication. Ecol. Appl. 20 (8): 2075-2080.

Jones, M.G.W. and Ryan, P.G. (2010). Evidence of mouse attacks on albatross chicks on sub-Antarctic Marion Island. Antarctic Science. 22: 39-42.

Jones, H.P. and Kress, S.W. (2012). A review of the world's active seabird restoration projects. Journal of Wildlife Management. 76: 2-9.

Jones, C. and Merton, D. (2012). A tale of two islands: The rescue and recovery of endemic birds in New Zealand and Mauritius. pp 33-72. In J. Ewen, D. Armstrong, K. Parker and P. Seddon. Reintroduction biology: Integrating science and management. Blackwell Publishing Ltd, Oxford.

Jones, H.P., Tershy, B.R., Zavaleta, E.S., Croll, D.A., Keitt, B.S., Finkelstein, M.E. and Howald, G.R. (2008). Severity of the effects of invasive rats on seabirds: a global review. Conservation Biology. 22: 16-26. DOI:10.1111/j.1523-1739.2007.00859.x.

Jupiter, D. (2011). Desroches Island Conservation Center. Annual Report 2010. Island Conservation Society, Seychelles

Kaiser-Bunbury, C.N. and Mougal, J. (2014). How do pollinators and seed dispersers respond to restoration of inselberg (glacis) vegetation? Kapisen, Journal of the Plant Conservation Action group, Seychelles. 16: 6-7.

Kaiser-Bunbury, C.N. and Senterre, B. (2014). Manual on the Key Biodiversity Area CyberTracker application. Report, Government of Seychelles, United Nations Development Programme. Victoria, Seychelles. 30 pp.

Kaiser-Bunbury, C.N., Traveset, A. and Hansen, D.M. (2010). Conservation and restoration of plant–animal mutualisms on oceanic islands. Persp Plant Ecol Evol Syst. 12: 131-143.

Kaiser-Bunbury, C.N., Valentin, T., Mougal, J., Matatiken, D. and Ghazoul, J. (2011). The tolerance of island plant–pollinator networks to alien plants. Journal of Ecology 99: 202-213.

Kaiser-Bunbury, C.N., Mougal, J. and Matatiken, D. (2015). Restoration of Seychelles upland forest biodiversity. WIOMSA 7: http://www.wiomsa.org/wiomsa-magazine/.

Kaiser-Bunbury, C.N., Mougal, J., Valentin, T., Gabriel, R. and Blüthgen, N. (2015). Herbicide application as a habitat restoration tool: impact on native island plant communities. Applied Vegetation Science. 8 (4): 650-660. DOI: 10.1111/ avsc.12183.

Kappes, M., Coustaut, K. and Le Corre, M. (2013). Census of Wedge-tailed shearwaters Puffinus pacificus breeding at D'Arros and Saint-Joseph atoll, Seychelles. Marine Ornithology. 41: 29-34.

Kappes, P. and Jones, H.P. (2014). Integrating seabird restoration and mammal eradication programs on islands to maximize conservation gains. Biodiversity Conservation. 23: 503-509.

Katulic, S., Valentin, T. and Fleischmann, K. (2005). Invasion of creepers on Mahé, Seychelles. Kapisen (3): 10-13.

Keith, P., Marquet, G., Valade, P., Bosc, P. and Vigneux, E. (2006). Atlas des poissons et crustacés d'eau douce des Comores, Mascareignes et Seychelles. Muséum national d'Histoire naturelle, Paris

Keitt, B., Campbell, K., Saunders, A., Clout, M., Wang, Y., Heinz, R., Newton, K. and B., T. (2011). The Global Islands Invasive Vertebrate Eradication Database: A tool to improve and facilitate restoration of island ecosystems. pp 74-77. In C. R. Veitch, M. N. Clout and D. R. Towns. Island invasives: eradication and management. IUCN, Gland, Switzerland.

Keitt, B., Griffiths, R., Boudjelas, S., Broome, K., Cranwell, S., Millett, J., Pitt, W. and Samaniego-Herrera, A. (2015). Best practice guidelines for rat eradication on tropical islands. Biol. Conserv. 185: 17-26.

Kier, G., Kreft, H., Lee, T.M., Jetz, W., Ibisch, P.L., Nowicki, C., Mutke, J. and Barthlott, W. (2009). A global assessment of endemism and species richness across island and mainland regions. Proceedings of the National Academy of Sciences 106: 9322-9327. DOI: 106: 9322-9327.

Kirk, D. and Bathe, G. (1994). Population size and home range of black-naped hares Lepus nigricollis

nigricollis on Cousin Island (Seychelles, Indian Ocean) Mammalia. 58: 557-562.

Kirk, D. and Racey, P. (1992). Effects of the introduced black-naped hare Lepus nigricollis nigricollis on the vegetation of Cousin Island, Seychelles and possible implications for avifauna. . Biological Conservation, . 61: 171-179.

Kline, W.N. and Duquesnel, J.G. (1996). Management of invasive exotic plants with herbicides in Florida. Down To Earth. 51 (2).

Knight, P. (2008). Economic Feasibility Study: Feasibility of Integrating the Sterile Insect Technique to the ongoing Area-wide Melon Fly Eradication Programme. Mission Report (21st-25th April 2008), Project No. SEY5003. Department of Technical Cooperation (TC), IAEA

Komdeur, J. (1994). Conserving the Seychelles warbler Acrocephalus sechellensis by translocation from Cousin Island to the islands of Aride and Cousine. Biological Conservation. 67 (2): 143-152. <Go to ISI>://A1994MT61400008

Komdeur, J. (1996). Breeding of the Seychelles magpie robin Copsychus sechellarum and implications for its conservation. Ibis. 138 (3): 485-498.

Komdeur, J. and Pels, M.D. (2005). Rescue of the Seychelles warbler on Cousin Island, Seychelles: The role of habitat restoration. Biological Conservation. 124 (1): 15-26. <Go to ISI>://000228426700002

Kueffer, C. and Vos, P. (2004). Case Studies on the Status of invasive Woody Plant Species in the Western Indian Ocean: 5. Seychelles. Forest Health & Biosecurity Working Papers FBS/4-5E. Forestry Department, Food and Agriculture Organization of the United Nations, Rome, Italy

Kueffer, C., Vos, P., Lavergne, C. and Mauremootoo, J. (2004). Case Studies on the Status of Invasive Woody Plant Species in the Western Indian Ocean: 1. Synthesis. Forest Health and Biosecurity Working Papers FBS/4-1E. Forestry Department, Food and Agriculture Organization of the United Nations, Rome, Italy

Kueffer, C. and Zemp, S. (2004). Clidemia hirta (Fo Watouk): A factsheet. Kapisen (1).

Kueffer, C. and Schumacher, E. (2005). Consultancy on invasive species management in Seychelles - Final report. Geobotanical Institute (ETH), Zurich, Switzerland. 50 pp.

Kueffer, C., Schumacher, E., Fleischmann, K., Edwards, P.J. and Dietz, H. (2007). Strong belowground competition shapes tree regeneration in invasive Cinnamomum verum forests. Journal of Ecology. 95: 273-282.

Kueffer, C., Daehler, C.C., Torres-Santana, C.W., Lavergne, C., Meyer, J.-Y., Otto, R. and Silva, L. (2010a). A global comparison of plant invasions on oceanic islands.

Perspectives in Plant Ecology, Evolution and Systematics. 12 (2): 145-161.

Kueffer, C., Schumacher, E., Dietz, H., Fleischmann, K. and Edwards, P.J. (2010b). Managing successional trajectories in alien-dominated, novel ecosystems by facilitating seedling regeneration: a case study. Biological Conservation. 143 (7): 1792-1802.

Kueffer, C. (2011). Preventing and managing plant invasions on oceanic islands. Journal of Botanic Gardens Conservation International. 8 (2): 14-17. http://www.geobot.umnw.ethz.ch/publications/PDF_publications/1105.pdf.

Kueffer, C., Beaver, K. and Mougal, J. (2013). Management of novel ecosystems in the Seychelles. pp 228-238. In R. J. Hobbs, E. S. Higgs and C. Hall. Novel ecosystems: intervening in the new ecological world order. Wiley-Blackwell.

Kueffer, C. and Kaiser-Bunbury, C.N. (2014). Reconciling conflicting perspectives for biodiversity conservation in the Anthropocene. Frontiers in Ecology and the Environment. 14: 131-137.

Kull, C.A. and Tassin, J. (2008). Invasive Australian acacias on western Indian Ocean islands: a historical and ecological perspective. African Journal of Ecology. 46 (4): 684-689.

Kull, C.A. and Shackleton, C. (2011). Adoption, use, and perception of Australian acacias around the world. Diversity and Distributions. 17: 822-836.

Labiche, A. and Rocamora, G. (2009). Rat control using poisoning and trapping techniques at the properties of the United Arab Emirates President (ex-Tracking Station & Haut Barbarons) from May 2008 to May 2009, and overall synthesis of the results obtained since October 2006.Pp. 8-23. In Rocamora G, Jean-Louis A. 2009. Final report to the FFEM secretariat : 4th year of operation (1st May 08 to 30th June 09), synthesis for the four years and perspectives. FFEM Project Rehabilitation of Island Ecosystems. Island Conservation Society, Seychelles

Labisko, J., Maddock, S.T., Taylor, M.L., Chong Seng, L., Gower, D.J., Wynne, F., Wombwell, E., Morel, C., French, G.C.A., Bunbury, N. and Bradfield, K.S. (2015). Chytrid fungus (Batrachochytrium dendrobatidis) undetected in the two orders of Seychelles amphibians Herpetological Review. 64: 41-45.

Lablache, J. and Mériton-Jean, S. (2015). Biological control planned to eradicate hairy caterpillar pest in Seychelles. Seychelles News Agency article of 30th March 2015.

Laboudallon, V. (1987). Cat eradication on Cousine Island. Unpublished report. ICBP, (BirdLife International)

Lambertini, M., Leape, J., Marton-Lefevre, J., Mittermeier, R.A., Rose, M., Robinson, J.G., Stuart, S.N., Waldman, B. and Genovesi,

P. (2011). Invasives: a major conservation threat. Science 333: 404-405.

Langeland, K.A., Ferrell, J.A., Sellers, B., Macdonald, G.E. and Stocker, R.K. (2009). Control of Nonnative Plants in Natural Areas of Florida.

Lavers, J.L., Wilcox, C. and Donlan, C.J. (2010). Bird demographic responses to predator removal programs. Biological Invasions 12: 3839-3859.

Lawrence, J., Samways, M., Henwood, J. and Kelly, J. (2011). Effect of an invasive ant and its chemical control on a threatened endemic Seychelles millipede. Ecotoxicology. 20 (4): 731-738. DOI: 10.1007/s10646-011-0614-4 http://dx.doi.org/10.1007/s10646-011-0614-4.

Le Corre, M., Danckwerts, D.K., Ringler, D., Bastien, M., Orlowski, S., Morey Rubio, C., Pinaud, D. and Micol, T. (2015). Seabird recovery and vegetation dynamics after Norway rat eradication on Tromelin Island, western Indian Ocean. Biol. Conserv. 185: 85-94.

Le Corre, M. and Jouventin, P. (1997). Ecological significance and conservation priorities of Europa Island (western Indian Ocean), with special reference to seabirds. Revue d' ecologie. 52 (3): 205-220.

Le Goff, P., Boussès, P., Julienne, S., Brengues, C., Rahola, N., Rocamora, G. and Robert, V. (2012). The mosquitoes (Diptera: Culidae) of Seychelles: taxonomy, ecology, vectorial importance and keys of determination. Parasites & Vectors. 5: 207. http://www.parasitesandvectors.com/content/5/1/207.

Le Roux, V., Chapuis, J.L., Frenot, Y. and Vernon, P. (2002). Diet of the house mouse (Mus musculus) at Guillou Island, Kerguelen archipelago, Subantartic. Polar Biology. 25: 49-57.

Leary, J. and Hardman, A. (2012). Practitioner's guide for effective non-restricted herbicide techniques to control and suppress invasive woody species in Hawaii. Weed Control. 10.

Leary, J., Beachy, J.R., Hardman, A. and Lee, J.G. (2013). A Practitioner's guide for testing herbicide efficacy with the Incision Point Application (IPA) technique on invasive woody plant species. Weed Control. 11.

Legros, J.P. and Argelès, L. (1986). La Gaillarde à Montpellier. Association des anciens élèves de l'ENSAM. Ecole Nationale Supérieure Agronomique de Montpellier, France. 343 pp.

Leonhardt, B.A., Mastro, V.C., Schwarz, M., Tang, J.D., Charlton, R.E., Pellegrini-Toole, A., Warthen Jr., J.D., Schwalbe, C.P. and Cardé, R.T. (1991). Identification of sex pheromone of browntail moth, Euproctis chrysorrhoea (l.) (Lepidoptera: Lymantriidae). Journal of Chemical Ecology. 17 (5): 897-910.

Linders, E.W. and Langrand, O. (2014). First record of House Crow Corvus splendens for Madagascar—potential impacts and suggested management of an invasive bird species. Bull. African Bird Club. 21 (2): 216-219.

Lionnet, G. (1966). La lutte contre les rats. Agrichelles. Vol.2, Issue 7, pp.97-101. Seychelles Archives, Government of Seychelles

Lionnet, G. (1984). Observations d'histoire naturelle faites aux Seychelles en 1768 par l'expedition Marion-Dufresne. Mauritius Inst. Bulletin. 10: 15-73.

Littin, K.E. and Mellor, D.J. (2005). Strategic animal welfare issues: ethical and animal welfare issues arising from the killing of wildlife for disease control and environmental reasons. Rev. sci. tech. Off. int. Epiz 24 (2): 767-782.

Lorvelec, O. and Pascal, M. (2005). French attempts to eradicate non-indigenous mammals and their consequences for native biota. Biological Invasions. 7: 135-140.

Louda, S.M., Pemberton, R.W., Johnson, M.T. and Follet, P.A. (2003). Nontarget effects – the Achilles' heel of biological control? Retrospective analyses to reduce risk associated with biocontrol introductions. Annual Review of Entomology. 48 (1): 365-396.

Louette, M., Ed. (1999). La Faune terrestre de Mayotte. Annales Musée Royal de l'Afrique Centrale. Tervuren, Belgique.

Lucking, R. (2013). Seychelles Magpie-robin Copsychus sechellarum. pp 690-692. In R. J. Safford and A. F. A. Hawkins. The Birds of Africa. Volume VIII: The Malagasy Region. Christopher Helm, London.

Lucking, R.S. (1997). Hybridization between Madagascan Red Fody Foudia madagascariensis and Seychelles Fody Foudia sechellarum on Aride Island, Seychelles. Bird Conservation International. 7: 1-6.

Luken, J.O. and Thieret, J.W., Eds. (1997). Assessment and management of plant invasions. Springer-Verlag Environmental Management Series, New York, NY.

Lundrigan, B. and Foote, S. (2003). "Lepus nigricollis" Animal Diversity Web. Retrieved 22nd March 2015. http://animaldiversity.ummz. umich.edu/site/accounts/information/Lepus_nigricollis.html.

Mac Gregor, L. (2007). Veterinary report on the health screening of SWEs during island transfers from Conception Island to Cousine and North Islands between 5th and 21th of July 2007. Island Conservation Society (Seychelles) / North Island / Cousine Island, Seychelles

MacArthur, R.H. and Wilson, E.O. (1967). The theory of island biogeography. Princeton University Press, N.J. 215 pp. 0691088365.

Macdonald, D.W. and Barrett, P. (1993). Mammals of Europe. Princeton University Press, New Jersey ISBN 0-691-09160-9.

Macdonald, I.A.W., Reaser, J.K., Bright, C., Neville, L.E., Howard, G.W., Murphy, S.J. and Preston, G., Eds. (2003). Invasive alien species in southern Africa: national reports and directory of resources. Global Invasive Species Programme, Cape Town, South Africa.

MacKay, J.W.B., Russell, J.C. and Murphy, E.C. (2007). Eradicating house mice from islands: successes, failures and the way forward. pp In K. A. Fagerstone and G. W. Witmer. Managing vertebrate invasive species: an international symposium. USDA, National Wildlife Research Center, Fort Collins, Colorado, USA.

MacKay, J.W.B., Murphy, E.C., Anderson, S.H., Russell, J.C., Hauber, M.E., Wilson, D.J. and Clout, M.N. (2011). A successful mouse eradication explained by site-specific population data. pp 198-203. In C. R. Veitch, M. N. Clout and D. R. Towns. Island invasives: eradication and management. IUCN, Gland, Switzerland.

Maggs, G., Nicoll, M., N., Z., White, P.J.C., Winfield, E., Poongavanan, S., Tatayah, V., Jones, C.G. and Norris, K. (2015). Rattus management is essential for population persistence in a critically endangered passerine: Combining small-scale field experiments and population modelling. Biological conservation. 191: 274-281.

Martin, L. and Pinchart, J. (2015). 'ICS Cosmoledo Expedition 8th November - 6th December, 2014. Final Technical Report. ICS & UNDP/GEF, Victoria

Maturin, S. (2010). Break through in controlling an Invasive Vine in the Pacific. Retrieved on 9 December, 2011. www.forestandbird.org.nz.

Maturin, S. (2010). Vatthe Vine Control Project – Vanuatu – Summary of activities and results 2009-2010. Retrieved 9 November, 2011. www.forestandbird.org.nz.

Maturin, S. (2013). Merremia peltata (Big leaf) control at Vatthe Conservation Area - Vanuatu. Technocal report for GEF/SGP.

Matyot, P. (1999). The arrowhead vine, Syngonium podophyllum Schott (Family Araceae), a potential invader in Seychelles. Phelsuma. 7: 70-72.

Matyot, P. (2004a). The establishment of the crested tree lizard, Calotes versicolor (DAUDIN, 1802) (Squamata: Agamidae), in Seychelles. Retrieved 23 November, 2011. Phelsumania. http://www. phelsumania.com/.

Matyot, P. (2004b). Good and bad news from Praslin and La Digue. Seychelles Nation. Monday 9th August (vol 27, no. 147), p.4.

Matyot, P. (2007). The menace of the coconut whitefly. Seychelles Nation. Monday 16th July (vol. 30, no. 132), p. 4.

Mauremootoo, J.R. (2003). Proceedings of the regional workshop on invasive alien species and terrestrial ecosystem rehabilitation for western Indian Ocean island states - sharing experience, identifying priorities and defining joint action. Indian Ocean Commission Regional Environment Programme, 13-17 October 2003, Seychelles. 207 pp.

McGavin, G.C. (1997). Expedition Field Techniques. Insects and other terrestrial arthropods. Geography Outdoors; Royal Geographical Society. 96 pp.

McNeely, J., Ed. (2001). The great reshuffling: human dimensions of invasive alien species. IUCN, Gland, Switzerland.

McNeely, J.A., Mooney, H.A., Neville, L.E., Schei, P.J. and Waage, J.K., Eds. (2001). Global strategy on invasive alien species. IUCN, Cambridge, U.K., in collaboration with the Global Invasive Species Programme. X + 50 pp.

Medina, F.M., Bonnaud, E., Vidal, E., Tershy, B.R., Zavaleta, E.S., Donlan, C.J., Keitt, B.S., Le Corre, M., Horwath, S.V. and Nogales, M. (2011). A global review of the impacts of invasive cats on island endangered vertebrates. Global Change Biology. 17: 3503-3510.

Medina, F.M., Bonnaud, E., Vidal, E. and Nogales, M. (2014). Underlying impacts of invasive cats on islands: not only a question of predation. Biodiversity and Conservation. 23 (2): 327-342.

Meise, W. and Schifter, H. (1972). The cuckoos and their relatives. pp In B. Grzimek. Grzimek's Animal Life Encyclopedia, vol. 8. Van Nostrand Reinhold, New York.

Mensink, H. and Janssen, P. (1994). International Programme on Chemical Safety. Environmental Health Criteria 159, Glyphosate. World Health Organisation, Geneva. ISBN 9241571594

Merton, D. (1987). Eradication of rabbits from Round Island, Mauritius: a conservation success story. Dodo, Journal of the Jersey Wildlife Preservation Trust 24: 19-44.

Merton, D. (1999). Ecological restoration in the Seychelles - a proposal to eradicate rats from Denis Curieuse, Frégate, Thérèse and Conception Islands and cats from Denis, Curieuse and Thérèse Islands. Unpublished report to Director of Conservation, MET, Republic of Seychelles

Merton, D. (2001). Report on eradication of rats from Curieuse, Denis and Fregate Islands and cats from Curieuse and Denis islands, May – August 2000. Unpublished report, for Ministry of Environment, Seychelles

Merton, D.V., Atkinson, I.A.E., Strahm, W., Jones, C., Empson, R.A., Mungroo, Y., Dulloo, E. and Lewis, R. (1989). A management plan for the restoration of Round Island, Mauritius. Jersey Wildlife Preservation Trust,

Merton, D., Climo, G., Laboudallon, V., Robert, R. and Mander, C. (2002). Alien mammal eradication and quarantine on inhabited islands in the Seychelles. pp 182-198 In C. R. Veitch and M. N. Clout. Turning the tide: the eradication of invasive species. IUCN. 182-198.

Meyer, A.N. (1994). Control of rats, rodent-borne disease and rat damage in Seychelles. Unpublished Report. World Health Organisation, UNDP

Meyer, J.-Y. and Butaud, J.F. (2007). Rats as transformers of native forests in the islands of French Polynesia (South Pacific). Université de la Polynesie Francaise, Tahiti

Meyer, J.-Y. and Fourdrigniez, M. (2011). Conservation benefits of biological control: the recovery of a threatened plant subsequent to the introduction of a pathogen to contain an invasive tree species. Biological conservation. 144 (1): 106-113.

Meyer, J.-Y., Fourdrigniez, M. and Taputuarai, R. (2012). Restoring habitat for native and endemic plants through the introduction of a fungal pathogen to control the alien invasive tree Miconia calvescens in the island of Tahiti. BioControl. 57 (2): 191-198. Published online 23 August 2011. DOI 10.1007/s10526-011-9402-6.

Micol, T. and Jouventin, P. (1995). Restoration of Amsterdam Island, South Indian Ocean, following control of feral cattle. Biological Conservation. 73: 199-206.

Micol, T. and Bernard, F. (2010). Seychelles. Réhabilitation des écosystèmes insulaires. Eradication des espèces exotiques envahissantes et réintroduction d'espèces endémiques menacées. Rapport d'évaluation finale. Secrétariat du FFEM. Asconit consultants. 118 pp.

Milgroom, M.G. and Cortesi, P. (2004). Biological control of chestnut blight with hypovirulence: a critical analysis. Annu. Rev. Phytopathol. 42: 311-338.

Millennium Ecosystem Assessment (2005). Ecosystems and Human Well-being: Synthesis. Island Press, Washington, DC

Miller, J.H. (2003). Nonnative Invasive Plants of Southern Forests. A Field Guide for Identification and Control.

Millett, J.E., Hill, M.J., Parr, S.J., Nevill, J., Merton, D.V. and Shah, N.J. (2001). Eradication of mammalian predators in the Seychelles in 2000. CBD Technical Series (UNEP). No.1: 69-70.

Millett, J., Climo, G. and Shah, N.J. (2005). Eradication of common mynah Acridotheres tristis populations in the granitic Seychelles: successes, failures and lessons learned. Advances in Vertebrate Pest Management. 3: 163-183.

Mills, K.L., Pyle, P., Sydeman, W.J., Buffa, J. and Rauzon, M.J. (2002). Direct and indirect effects of house mice on declining populations of a small seabird, the ashy storm-petrel (Oceanodroma homochroa), on Southeast Farallon Island, California, USA. pp 406-414. In C. R. Veitch and M. Clout. Turning the tide: the eradication of invasive species. IUCN SSC Invasive Species Specialist Group, Gland, Switzerland and Cambridge, UK.

Ministry of Health (1994). Eradication of rats in Seychelles ? Newsletter, N°16, p.4. Ministry of Health, Seychelles

Ministry of Health (Disease Surveillance & Response Unit) Republic of Seychelles (2015). Caterpillar skin rash. Communique issued on 12th February 2015.

Minogue, P.J., Bohn, K.K., Osiecka, A. and Lauer, D.K. (2010). Japanese Climbing Fern (*Lygodium japonicum*) Management in Florida's Apalachicola Bottomland Hardwood Forests. Invasive Plant Science and Management. 3 (3): 246-252. http://dx.doi.org/10.1614/IPSM-D-09-00023.1.

Mittermeier, R.A., Robles Gil, P., Hoffmann, M., Pilgrim, J.D., Brooks, T.M., Mittermeier, C.G. and Fonseca, G.A.B.d. (2004). Hotspots Revisited: Earth's Biologically Richest and Most Endangered Ecoregions. CEMEX, Mexico City

Mooney, H.A., Mack, R.N., McNeely, J.A., Neville, L.E., Schei, P.J. and Waage, J.K., Eds. (2005). Invasive Alien Species: A New Synthesis. Island Press, Washington, DC. 368 pp.

Morris, K.D. (2002). The eradication of the black rat (Rattus rattus) on Barrow and adjacent islands off the north-west coast of Western Australia. pp 219-225 *In* C. R. Veitch, M. N. Clout and D. R. Towns. Island invasives: eradication and management. IUCN, Gland, Switzerland.

Morton, J.F. (1995). Plants poisonous to people in Florida and other warm areas. Hallmark Press, Miami

Motooka, P. (2003). Hiptage benghalensis. Retrieved 19 February, 2011. http://www.ctahr.hawaii.edu/invweed/WeedsHI/W_Hiptage_benghalensis.pdf.

Mougal, J. and Henriette, E. (2012). Management of Invasive Alien Creepers. Recommended practices for the management of a selection of priority Invasive Alien Creepers: A Guidebook

Mulder, C., Anderson, W., Towns, D. and Bellingham, P. (2011). Seabird Islands. Ecology, Invasion, and Restoration. Oxford University Press, New York

Mulder, C., Grant-Hoffman, N., Towns, D., Bellingham, P., Wardle, D., Durrett, M., Fukami, T. and Bonner, K. (2009). Direct and indirect effects of rats: does rat eradication restore ecosystem functioning of New Zealand seabird islands? . Biological Invasions. 11 (7): 1671-1688.

Mullen, G.R. (2009). Moths and butterflies (Lepidoptera). pp 345-362. *In* G. R. Mullen and L. A. Durden. Medical and veterinary entomology (2nd edition). Academic Press, Elsevier, USA.

Muñoz-Fuentes, V., Green, A.J. and Negro, J.J. (2013). Genetic studies facilitated management decisions on the invasion of the ruddy duck in Europe. Biological Invasions. 15: 723-728. DOI: 10.1007/s10530-012-0331-9.

Muñoz-Fuentes, V., Vilà, C., Green, A.J., Negro, J.J. and Sorenson, M.D. (2007). Hybridization between white-headed ducks and introduced ruddy ducks in Spain. Molecular Ecology. 16: 629-638.

Murphy, E.C., Shapiro, L., Hix, S., MacMorran, D. and Eason, C.T. (2011). Control and eradication of feral cats: field trials of a new toxin. pp 213-216. *In* C. R. Veitch, M. N. Clout and D. R. Towns. Island Invasives: Eradication and Management. Proceedings of the International Conference on Island Invasives, Gland, Switzerland: IUCN and Auckland, New Zealand.

Murray, M. and Henri, K. (2005). An economic assessment of Seychelles biodiversity. Consultancy report for the UNDP-GEF PDF-B Mainstreaming Biodiversity.

Mwebaze, P., MacLeod, A. and Barois, H. (2009). Economic Valuation of the Influence of Invasive Alien Species on the National Economy. Final Report for the Government of Seychelles (GOS)-UNDP-GEF "Mainstreaming Prevention and Control Measures for Invasive Alien Species into Trade, Transport and Travel across the Production Landscape". Food and Environment Research Agency, San Hutton, York, UK.

Mwebaze, P., MacLeod, A., Tomlinson, D., Barois, H. and Rijpma, J. (2010). Economic Valuation of the Influence of Invasive Alien Species on the National Economy of the Seychelles. Ecological Economics. 69 (12): 2614-2623.

Nahaboo (2011). ICS Alphonse Annual Report 2010, Island Conservation Society, Seychelles.

Nevill, J. (2001). Case study 5.34. Ecotourism as a source of funding to control invasive species. pp 202. *In* R. Wittenberg and M. J. W. Cock. Invasive Alien Species: A Toolkit of Best Prevention and Management Practices. CAB International, Wallingford, Oxon, UK.

Nevill, J. (2001). Case study 5.41. Invasive Species Mitigation to Save the Seychelles Black Parrot. pp xii - 228. *In* R. Wittenberg and M. J. W. Cock. Invasive Alien Species: A Toolkit of Best Prevention and Management Practices. CAB International, Wallingford, Oxon, UK.

Nevill, J. (2004). Ecotourism as a Source of Funding to Control Invasive Species: the case of Seychelles. Insula. International Journal of Island Affairs: 99-102.

Nevill, J. (2009). National IAS Baseline Report. GOS - UNDP - GEF. Mainstreaming Prevention and Control Measures for Invasive Alien Species into Trade, Transport and Travel across the Production

Landscape. Government of Seychelles/UNDP/GEF.164 pp.

Nevill, J. (2011). Case Study 6. Ecosystem Rehabilitation. The case of small islands in Seychelles – A government, NGO and private sector partnership. pp 84-87. *In* Goverment of Seychelles. Fourth National Report to the United Nations Convention on Biological Diversity. Department of Environment, P.O. Box 445, Botanical Gardens, Mont Fleuri, Victoria, Republic of Seychelles.

Newman, D.G. (1994). Effects of a mouse, Mus musculus, eradication programme and habitat change on lizard populations on Mana Island, New Zealand with special reference to McGregor's skink, Cyclodina macgregori. New Zealand Journal of Zoology. 21: 443-456.

Nicoll, M. (1984). Tenrecs. pp 744-747. *In* D. Macdonald. The Encyclopaedia of Mammals. Volume 2. Unwins, London.

Nicoll, M.E. and Racey, P.A. (1985). Folicular development, ovulation, fertilization and fetal development in tenrecs (*Tenrec ecaudatus*). Journal of Reproduction and Fertility. 74: 47-55.

Nishijima, S., Takimoto, G. and Miyashita, T. (2014). Roles of Alternative Prey for Mesopredators on Trophic Cascades in Intraguild Predation Systems: A Theoretical Perspective. The American Naturalist. Online publication date: 3-Mar-2014. DOI:10. 1086/675691.

Noble, T., Bunbury, N., Kaiser-Bunbury, C.N. and Bell, D.J. (2011). Ecology and co-existence of two endemic day gecko (Phelsuma) species in Seychelles native palm forest. Journal of Zoology. 283: 73-80.

Nogales, M., Martin, A., Tershy, B.R., Donlan, C.J., Veitch, D., Puerta, N., Wood, B. and Alonso, J. (2004). A review of feral cat eradication on islands. Conservation Biology. 18: 310-319.

Nogales, M., Vidal, E., Medina, F.M., Bonnaud, E., Tershy, B.R., Campbell, K.J. and Zavaleta, E.S. (2013). Feral cats and biodiversity conservation: the urgent prioritization of island management. BioScience. 63 (10): 804-810.

O'Dowd, D.J., Green, P.T. and S., L. (2003). Invasional meltdown on an oceanic island. Ecology Letters. 6 (6): 812-817.

Oerke, E.C. (2006). Crop losses to pests. Journal of Agricultural Science. 144: 31-43.

Orueta, G.F. (2003). Manual práctico para el manejo de vertebrados invasores en islas de España y Portugal. Proyecto LIFE2002NAT/CP/E/000014. Govern Balear / Govierno de Canarias / UE.

Orueta, G.F. (2007). Vertebrados invasores. Ministerio de Medio Ambiente de España. Madrid.

Orueta, G.F. and Ramos, Y.A. (2001). Methods to control and eradicate non-native terrestrial vertebrate

species. Nature & Environment. 118. Council of Europe.

Pacific Island Initiative. (2011). Guidelines on Cat Eradication and Monitoring Techniques Version 1.2.2. Resource Kit for Rodent and Cat Eradication. www.pacificinvasivesinitiative.org/rk.

Pacific Island Initiative. (2011). Resource Kit for Rodent and Cat Eradication. Retrieved 7th August, 2014. http://rce.pacificinvasivesinitiative.org.

Pacific Island Initiative. (2012). Resource Kit for Invasive Plant Management. Retrieved 7th September, 2014. http://ipm.pacificinvasivesinitiative.org.

Pain, D.J., Brooke, M.d.L., Finnie, J.K. and Jackson, A. (2000). Effects of Brodifacoum on the land crab of Ascension Island. J.Wildlife Management. 64 (2): 380-387.

Parkes, R. and Forrester, G. (2011). Diagnosing the cause of failure to eradicate introduced rodents on islands: brodifacoum versus diphacinone and method of bait delivery. Conservation Evidence 8:100-106.

Parks, R. and Townsend, M. (2011). Tree inspection and control of infestations of Oak Processionary Moth Thaumetopoea processionea (Linnaeus) (Lepidoptera: Thaumetopoeidae) (OPM) in the UK in 2010. Plant Health Service, Forestry Commission, U.K. 32+xxxiii pp. http://www.forestry.gov.uk/pdf/OPMsurveyandcontrolreport2010.pdf.

Parr, S.J., Hill, M.J., Nevill, J., Merton, D.V. and Shah, N.J. (2000). Alien species case-study: eradication of introduced mammals in Seychelles in 2000. Unpublished report to CBD Secretariat, IUCN, Gland, Switzerland. 22 pp.

Payet, R.A. (2006). Sustainability in the context of coastal and marine tourism in Seychelles. PhD Thesis. Department of Biology and Environmental Science, University of Kalmar, Sweden

PCA / North Island (2009). Final Report. FFEM – ICS Project: "Rehabilitation of Island Ecosystems". Plant Conservation Action Group & North Island, Seychelles.

Peck, D.R., Faulquier, L., Pinet, P., Jaquemet, S. and Le Corre, M. (2008). Feral cat diet and impact on sooty terns at Juan de Nova Island, Mozambique Channel. Animal Conservation. 11: 65-74.

Pemberton, R.W. (2002). Old World Climbing Fern. Biological Control of Invasive Plants in the Eastern United States, USDA-FS, FHTET-2002-04, Morgantown, West Virginia, US: 139-147.

PetSnails.co.uk. (2014). Snail and Slug. Retrieved 4 September, 2014. http://www.petsnails.co.uk.

Pham, T. and van Son, J. (2009). Indian Myna control project handbook. Managing the invasion of Indian Mynas in Northern NSW. NSW Environmental Trust, Australia. 18 pp. http://www.indianmynaproject.com.au/sites/default/files/media/docs/Indian%20Myna%20Handbook.pdf.

Phisalix, M. (1922). Animaux venimeux et venins, vol. 2. Masson & Cie, Paris

PIER. (2011). Invasive species. Retrieved 20 December, 2011. http://www.hear.org/pier/species/ Last updated 16th January 2011.

Pigott, C.J. (1961). A soil survey of Seychelles. Directorate of Overseas Surveys, Land Resources Division, Tech. Bull. 2: 1-89.

Pimentel, D., McNair, S., Janecka, J., Wightman, J., Simmonds, C., O'Connell, C., Wong, E., Russel, L., Zern, J., Aquino, T. and Tsomondo, T. (2001). Economic and environmental threats of alien plant, animal, and microbe invasions. Agr. Ecosyst. Environ. 84: 1-20.

Pitt, W.C., Driscoll, L.C. and Sugihara, R.T. (2011). Efficacy of Rodenticide Baits for the Control of Three Invasive Rodent Species in Hawaii. Arch Environ Contam Toxicol. 60: 533-542. DOI 10.1007/s00244-010-9554-x.

Plantwise. (2011). Plantwise technical factsheet: Takamaka disease (*Leptographium calophylli*). Retrieved 21 December, 2011. http://www.plantwise.org.

Plentovich, S., Hebshi, A. and Conant, S. (2009). Detrimental effects of two widespread invasive ant species on weight and survival of colonial nesting seabirds in the Hawaiian Islands. Biological invasions. 11 (2): 289-298.

Plentovich, S., Eijzenga, J., Eijzenga, H. and Smith, D. (2010). Indirect effects of ant eradication efforts on offshore islets in the Hawaiian Archipelago. Biological invasions. 13 (3): 545-557.

Plentovich, S., Swenson, C., Reimer, N., Richardson, M. and Garon, N. (2010). The effects of hydramethylnon on the tropical fire ant, Solenopsis geminata (Hymenoptera: Formicidae), and non-target arthropods on Spit Island, Midway Atoll, Hawaii. Journal of Insect Conservation. 14 (5): 459-465. 10.1007/s10841-010-9274-6 http://dx.doi.org/10.1007/s10841-010-9274-6.

Pott, M., Wegmann, A.S., Griffiths, R., Samaniego-Herrera, A., Cuthbert, R.J., Brooke, M.D.L., Pitt, W.C., Berentsen, A.R., Holmes, N.D., Howald, G.R., Ramos-Rendón, K. and Russell, J.C. (2015). Improving the odds: assessing bait availability before rodent eradications to aid in selecting bait application rates. Biol. Conserv. 185: 27-35.

Prescott, J., Shah, N.J. and Jeremie, M. (2013). Seychelles National Biodiversity Strategy and Action Plan to 2020. Government of Seychelles-UNDP-GEF, Victoria.

Racey, P.A. and Nicoll, M.E. (1984). Mammals of the Seychelles. pp 607-626. In D. R. Stoddart. Biogeography and ecology of the Seychelles Islands. W. Junk Publishers, The Hague.

Rainbolt, R.E. and Coblentz, B.E. (1999). Restoration of insular ecosystems: control of feral goats on Aldabra Atoll, Republic of Seychelles. Biological Invasions. 1: 363-375.

Raposo de Resende, T. and Canning, G. (2012). Population assessment and distribution of the endemic Enid snail Pachnodus fregatensis on Frégate Island, Seychelles. Phelsuma. 20: 1-8.

Rayner, M.J., Hauber, M.E., Imber, M.J., Stamp, R.K. and Clout, M.N. (2007). Spatial heterogeneity of mesopredator release within an oceanic island system. Proceedings of the National Academy of Sciences, USA. 104: 20862 - 20865.

Reaser, J.K., Meyerson, L.A., Cronk, Q., de Poorter, M., Eldrege, L.G., Green, E., Kairo, M., Latasi, P., Mack, R.N., Mauremootoo, J., O'Dowd, D., Orapa, W., Sastrutomo, S., Saunders, A., Shine, C., Thrainsson, S. and Vaiutu, L. (2007). Ecological and socioeconomic impacts of invasive alien species in island ecosystems. Environmental Conservation. 34 (2): 1-14.

Recher, H.F. and Majer, J.D. (2006). Effects of bird predation on canopy arthropods in wandoo Eucalyptus wandoo woodland. Austral Ecology. 31: 349-360.

Reilly, J.S. (2001). Euthanasia of animals used for scientific purposes. ANZCCART. pp 136.

Reuleaux, A. (2011). Population, feeding and breeding ecology of the Seychelles Black Parrot (*Coracopsis nigra barkyli*). Georg-August Universitaet Goettingen, Germany

Reuleaux, A., Bunbury, N., Villard, P. and Waltert, M. (2013). Status, distribution and recommendations for monitoring of the Seychelles black parrot Coracopsis (nigra) barklyi. Oryx. 47: 561-568.

Reuleaux, A., Richards, H., Payet, T., Villard, P., Waltert, M. and Bunbury, N. (2014). Insights into the feeding ecology of the Seychelles Black Parrot Coracopsis barklyi using two monitoring approaches. Ostrich - Journal of African Ornithology. 53 (3): 245-253.

Reuleaux, A., Richards, H., Payet, T., Villard, P., Waltert, M. and N., B. (2014). Breeding ecology of the Seychelles black parrot Coracopsis barklyi. Ostrich - Journal of African Ornithology. 85: 255-265.

Rew, L.J., Maxwell, B.D., Aspinall, R.J. and Dougher, F.L. (2006). Searching for a needle in a haystack: evaluating survey methods for sessile species. Biological Invasions. 8: 523-539.

Richardson, D. and Rocamora, G. (2004). Genetic study of differences between the four different populations of Seychelles Fody, Foudia sechellarum. Unpublished Research report April 2004 – Department of Animal and Plant Sciences. University of Sheffield.

Richardson, D.S., Bristol, R. and Shah, N.J. (2006). Translocation of the Seychelles warbler Acrocephalus sechellensis to establish a new population on Denis Island, Seychelles. Conservation Evidence. 3: 54-57.

Ringler, D., Russell, J., Jaeger, A., Pinet, P., Bastien, M. and Le Corre, M. (2014). Invasive rat space use on tropical islands: implications for bait broadcast. Basic and Applied Ecology. 15 (2): 179-186.

Ringler, D., Russell, J.C. and Le Corre, M. (2015). Trophic roles of black rats and seabird impacts on tropical islands: mesopredator release or hyperpredation? Biol. Conserv. . 185: 75-84.

Roberts, P. (1988). Assumption Island – a threat to Aldabra. Oryx. 22: 15-17.

Robertson, S.A. (1989). Flowering plants of Seychelles (An annotated checklist of angiosperms and gymnosperms with line drawings). Royal Botanic Gardens, Kew, UK

Rocamora, G. (1997a). Rare and Threatened Species, Sites and Habitats Monitoring Programme in Seychelles. Project G1 EMPS. Final Report. Vol. 1. Monitoring methodologies and recommended priority actions; Vol. 2. The database: results and applications; Vol. 3. Programme achievements, training sessions, public awareness & conservation projects. Ministry of Environment/Birdlife International/European Union. 400 pp.

Rocamora, G. (1997b). The Seychelles Grey White-eye. Red Data Book. World Birdwatch. 19 (2): 20-21.

Rocamora, G. (comp. 1997). A conservation strategy for Mediterranean forest, schrubland and rocky habitats. pp 239-266. In G. Tucker and M. Evans. Habitats for Birds in Europe : a conservation strategy for the wider environment. BirdLife Conservation Series (6). BirdLife International, Cambridge.

Rocamora, G. and François, J. (2000). Seychelles White-eye Recovery Programme. Phase 1. Saving the Seychelles grey white-eye. Final report. Ministry of Environment and Transport /Dutch Trust Fund/ IUCN. 182 pp.

Rocamora, G. and Solé, A. (2000). Sounds of Seychelles. Fauna of the granitic islands / Faune des îles granitiques. Audio CD and colour illustrated booklet. Island Nature (Pty) Ltd. 32 pp.

Rocamora, G. (2001). Conservation introduction of Zosterops modestus from Conception Island to Frégate Island (Seychelles). Seychelles White-eye Recovery Programme Phase 2. Report SWERP Phase 2 Oct 2001. Ministry of Environment/DTF. 19 pp. + annexes.

Rocamora, G., François, J., Henriette, E. and Youpa, R. (2001). Habitat suitability and carrying capacity of Frégate Island and Curieuse Island for the Seychelles white-eye Zosterops modestus. Ministry of Environment and the Dutch Trust Fund. 31 pp.

Rocamora, G. and Skerrett, A. (2001). Seychelles. pp 751-768. In L. Fishpool. Important Bird Areas in Africa and associated islands. BirdLife Conservation Series, BirdLife International, Cambridge.

Rocamora, G. and Payet, R. (2002). Sustainable ecotourism as a financing mechanism for the conservation of Cosmoledo Atoll. Proceedings of the Conference on Sustainable Development of Ecotourism in Small Islands Developing States (SIDS) and Other Small Islands. Preparatory Conference for the International Year of Ecotourism, Mahé (Seychelles), 8-10 December 2001. World Tourism Organisation & Ministry of Tourism, Seychelles.

Rocamora, G. and Matyot, P. (2002). Short report on a first visit to D'Arros island and considerations on its biological value and conservation potential (15.06.02 to 18.06.02). Internal Report. Island Conservation Society, Seychelles. 9 pp.

Rocamora, G. (2003a). The spectacular colorations of D'Arros Fodies. Birdwatch, NPTS (Seychelles). 46.

Rocamora, G. (2003b). Seychelles Fodies with spectacular colorations discovered on D'Arros. Bull. ABC. 1: 9.

Rocamora, G. (2005). Risk assessment to the Seychelles White-eye population of Conception from a proposed aerial spread of Pestoff cereal green dyed coconut flavoured pelleted baits at 20ppm Brodifacoum to eradicate the Norway rat. Project FFEM Rehabilitation of Islands Ecosystems – Island Conservation Society.

Rocamora, G., Labiche, A. and Matyot, P. (2005). Habitat suitability and carrying capacity of Anonyme Island for the Seychelles White-eye Zosterops modestus. Projet 'Réhabilitation des Ecosystèmes Insulaires'. Island Conservation Society / FFEM.

Rocamora, G. and Said, S. (2005). Eradication complète des rats sur les trois îlots d'Hajangoua (Mayotte). Rapport Direction de l'Agriculture et de la Forêt / Collectivité de Mayotte.

Rocamora, G., Labiche, A. and Vanherck, L. (2006). Conservation introduction of the (critically) endangered Seychelles White-eye Zosterops modestus to North Island (Seychelles). Project proposal. 50 pp.

Rocamora, G. (2007a). Eradication plan of rats and cats for Grande Ile and Grand Polyte, Cosmoledo atoll (Seychelles) in 2007. Report Project FFEM Rehabilitation of Islands Ecosystems – Island Conservation Society, Seychelles.

Rocamora, G. (2007b). Preliminary report on the eradication of Rattus rattus conducted in November 2007 on Cosmolédo atoll (Grande Ile, Grand Polyte and Petit Polyte). Annex 9. pp 51-58. *In* G. Rocamora and A. Jean-Louis. Annexes. Final report to the FFEM secretariat: 4th year of operation (1st May 08 to 30th June 09), synthesis for the four years and perspectives. FFEM Project Rehabilitation of Island Ecosystems, Island Conservation Society, Seychelles.

Rocamora, G. (2007c). The impact of rats in Seychelles: historical distribution, threats to ecosystems and humans, island eradications and control measures. Abstract for the Conference Rats, humans and their impacts on islands: integrating historical and contemporary ecology. 27-31 March 2007. University of Hawai'i.

Rocamora, G. (2007d). A method for three dimentional measurements of vegetation composition and structure to determine bird habitat preferences. Report, ICS, FFEM Project 'Rehabilitation of Island Ecosystems'. Victoria. 13 pp.

Rocamora, G. (2007e). Eradication of Norway rats on Conception Island (Seychelles) in 2007. Proposed Eradication Method and Timetable. Version 15.06.07. Report Project FFEM Rehabilitation of Islands Ecosystems. Island Conservation Society, Seychelles

Rocamora, G. and Labiche, A. (2007). Preliminary report on the eradication of Rattus norvegicus conducted on Conception in August-September 2007. Annex 1. pp 2-8. *In* G. Rocamora and A. Jean-Louis. Annexes. Final report to the FFEM secretariat: 4th year of operation (1st May 08 to 30th June 09), synthesis for the four years and perspectives. FFEM Project Rehabilitation of Island Ecosystems, Island Conservation Society, Seychelles.

Rocamora, G. (2008). Observations and recommendations for Anonyme Island management, 15.03.08. Internal report for Anonyme Island. Island Conservation Society, Seychelles

Rocamora, G. and Henriette Payet, E. (2008). Conservation introductions of the Seychelles white-eye on predator-free rehabilitated islands of the Seychelles archipelago, Indian Ocean. pp viii+284. *In* P. S. Soorae. Global re-introduction perspectives: re-introduction case studies from around the globe. IUCN/SSC Re-introduction Specialist Group, Abu Dhabi, UAE.

Rocamora, G. and Jean-Louiis, A. (2008). Réhabilitation des Ecosystèmes Insulaires Rapport annuel au secrétariat du FFEM. Troisieme année d'opérations 1/05/07 au 30/04/08. Island Conservation Society. Seychelles.

Rocamora, G. and Galman, G. (2009). Conservation introduction to Frégate Island of the Leaf-Insect *Phyllium bioculatum*, a rare invertebrate endemic to Seychelles. FFEM Project 'Réhabilitation des Ecosystèmes Insulaires'. Island Conservation Society (Seychelles).

Rocamora, G. and Gerlach, J. (2009). Reintroduction proposal for the Seychelles Black Terrapin Pelusios subniger parietalis to Aride Island. Projet FFEM Réhabilitation des Ecosystèmes Insulaires. Island Conservation Society, Seychelles.

Rocamora, G. and Henriette, E. (2009). Seychelles white-eye *Zosterops modestus* Conservation Assessment & Action Plan. 2009-2013. FFEM Project 'Réhabilitation des Ecosystèmes Insulaires'. Island Conservation Society & Ministry of Environment, Natural Resources & Transport (Seychelles). 33 pp.

Rocamora, G. and Jean-Louiis, A. (2009). Final report to the FFEM secretariat : 4th year operation (1st May 08 to 30th June 09), synthesis for the four years and perspectives. FFEM Project Rehabilitation of Island Ecosystems, Island Conservation Society, Seychelles. 85 pp.

Rocamora, G. and Labiche, A. (2009a). Variation of abundance of land birds and reptiles on Grande Ile and Grand Polyte (Cosmolédo atoll) between the pre and post rat eradication periods (2005 to 2008). pp 75-82. *In* G. Rocamora and A. Jean-Louis. Annexes. Final report to the FFEM secretariat: 4th year of operation (1st May 08 to 30th June 09), synthesis for the four years and perspectives. FFEM Project Rehabilitation of Island Ecosystems, Island Conservation Society, Seychelles.

Rocamora, G. and Labiche, A. (2009b). Variation of abundance of land birds and reptiles on Conception between the pre and post rat eradication periods (2005 to 2008). pp 66-74. *In* G. Rocamora and A. Jean-Louis. Annexes. Final report to the FFEM secretariat: 4th year of operation (1st May 08 to 30th June 09), synthesis for the four years and perspectives. FFEM Project Rehabilitation of Island Ecosystems, Island Conservation Society, Seychelles.

Rocamora, G. and Labiche, A. (2009c). Results of landbird and reptile monitoring on North Island during the pre and post rat eradication (2005-2008). pp 57-65. *In* G. Rocamora and A. Jean-Louis. Annexes. Final report to the FFEM secretariat: 4th year of operation (1st May 08 to 30th June 09), synthesis for the four years and perspectives. FFEM Project Rehabilitation of Island Ecosystems, Island Conservation Society, Seychelles.

Rocamora, G. and Labiche, A. (2009d). Preliminary report on the eradication of Rattus norvegicus conducted on Conception in August-September 2007. Annex 1.

Pp. 2-8. In Rocamora G. and Jean-Louis A. Annexes. Final report to the FFEM secretariat: 4th year of operation (1st May 08 to 30th June 09), synthesis for the four years and perspectives. FFEM Project Rehabilitation of Island Ecosystems, Island Conservation Society, Seychelles.

Rocamora, G. and Laboudallon, V. (2009). Seychelles Black Parrot Coracopsis (nigra) barklyi Conservation Assessment & Action Plan. 2009-2013. FFEM Project 'Réhabilitation des Ecosystèmes Insulaires'. Island Conservation Society & MENRT (Seychelles).

Rocamora, G. and Yeatman-Berthelot, D. (2009). Dicruridae (Drongos). pp 172-223. *In* J. del Hoyo, A. Eliott and C. D. Handbook of the Birds of the World. Volume 14. Bush-shrikes to Old World Sparrows. Lynx Edicions, Barcelona.

Rocamora, G., Vanherck, L., Labiche, A. and Dufrenne, A. (2009). Conservation introduction to North Island of the Seychelles Black mud terrapin Pelusios subniger parietalis. Report n°1: Transfers & monitoring activities undertaken between July 2008 & June 2009. Projet FFEM Réhabilitation des Ecosystèmes Insulaires, Island Conservation Society, Seychelles.

Rocamora, G. (2010a). Island Restoration. pp 178-183. *In* A. Skerrett, P. Thelma and J. Skerrett. Outer Islands of Seychelles. Zil Elwannyen Sesel. Camerapix, Nairobi, Kenya.

Rocamora, G. (2010b). Oiseau-lunettes et autres espèces menacées: bilan des actions menées aux Seychelles. CEPA magazine. 21: 19-23.

Rocamora, G. and Henriette, E. (2011). The Recovery of the Seychelles White-eye: a partnership success story combining island restoration, species management and capacity development. pp 22-23. *In* Government of Seychelles. Fourth National Report to the United Nations Convention on Biological Diversity. Department of Environment, Victoria, Seychelles.

Rocamora, G., Henriette, E. and Labiche, A. (2013). Rat Control Schemes in Breeding Areas of the Seychelles White-Eye in Mahé : The Concept of Biological (or "Mainland") Islands. pp 19-20. *In* M. Salamolard, F. X. Couzi and A. Duncan. Seminar Proceedings. Large Scale Rat Control: Lessons Learned, Results and Evaluation. LIFE+ CAP DOM EU Programme. Parc National de la Réunion, SEOR. LPO France Editions (Rochefort).

Rocamora, G., Henriette, E., Labiche, A. and Galman, G. (2013). Effects and Benefits of Rat Eradication on Biodiversity in the Seychelles pp 43-44. *In* M. Salamolard, F. X. Couzi and A. Duncan. Seminar Proceedings. Large Scale Rat Control: Lessons Learned, Results

and Evaluation. LIFE+ CAP DOM EU Programme. Parc National de la Réunion, SEOR. LPO France Editions (Rochefort).

Rocamora, G. and Laboudallon, V. (2013). Coracopsis barklyi Seychelles Black Parrot. pp *In* R. J. Safford and A. F. A. Hawkins. The Birds of Africa. Volume VIII: The Malagasy Region. Christopher Helm, London.

Rocamora, G. (2015). Biosecurity protocols for protected areas and islands of high biodiversity value in Seychelles. GEF project Mainstreaming Prevention and Control Measures for Invasive Alien Species (IAS) into Trade, Transport and Travel across the Production Landscape Report UNDP/GEF/Government of Seychelles.

Rocamora, G., Labiche, A., Henriette, E., Galman, G. and Russell, J. (in prep). Rat eradication and ecosystem recovery in Seychelles: a review.

Rocamora, G. and Nolin, R. (unpublished). Results from the 2011-2012 Myna eradication attempt on Grande Soeur.

Rocamora Magali (2015). The silent invaders threatening natural World Heritage. IUCN blog. 06 Oct. 2015. https://portals.iucn.org/blog/2015/10/06/the-silent-invaders-threatening-natural-world-heritage/

Rocha, S., Harris, D.J., Perera, A., Silva, A., Vasconcelos, R. and Carretero, M.A. (2009). Recent data on the distribution of Lizards and Snakes of the Seychelles. Herpetological Bulletin. 110: 20-32.

Rodrigues, A.S.L., Brooks, T.M., Butchart, S.H.M., Chanson, J. and Cox, N. (2014). Spatially Explicit Trends in the Global Conservation Status of Vertebrates. PLoS ONE. 9 (11). DOI: e113934. doi: 10.1371/journal.pone.0113934.

Ruffino, L., Zarzoso-Lacoste, D. and Vidal, E. (2015). Assessment of invasive rodent impacts on island avifauna: methods, limitations and the way forward. WildLife Research. 42 (185–195). Published online June 2015. http://dx.doi.org/10.1071/WR15047.

Rural Development Service. (2006). Rats: options for controlling infestations. Technical Advise Note 34. Retrieved 18th May, 2012. www.nationalrural.uk/upload/rusource/883.pdf. United Kingdom.

Russell, J.C. (2011). Indirect effects of introduced predators on seabird islands. pp 261-279. *In* C. P. H. Mulder, W. B. Anderson, D. R. Towns and P. J. Bellingham. Seabird islands: ecology, invasion, and restoration. Oxford University Press, New York.

Russell, J.C. (2012). Spatio-temporal patterns of introduced mice and invertebrates on Antipodes Island. Polar Biology. 35 (8): 1187-1195.

Russell, J.C. and Clout, M.N. (2004). Modelling the distribution and interaction of introduced rodents on New Zealand offshore islands. Global Ecol. And Biogeography. 13: 497-507.

Russell, J.C. and Holmes, N.D. (2015). Tropical island conservation: rat eradication for species recovery. Biol. Conserv. 185: 1-7.

Russell, J.C., Beaven, B.M., MacKay, J.W.B., Towns, D.R. and Clout, M.N. (2008). Testing island biosecurity systems for invasive rats. Wildlife Res. 35: 215-221.

Russell, J.C., Lecomte, V., Dumont, Y. and Le Corre, M. (2009). Intraguild predation and mesopredator release effect on long-lived prey. Ecological modelling. 220 (8): 1098-1104.

Russell, J.C., Miller, S.D., Harper, G.A., MacInnes, H.E., Wylie, M.J. and Fewster, R.M. (2010). Survivors or reinvaders? Using genetic assignment to identify invasive pests following eradication Biological Invasions 12: 1747-1757. DOI 10.1007/s10530-009-9586-1.

Russell, J.C., Ringler, D., Trombini, A. and Le Corre, M. (2011). The Island Syndrome and population dynamics of introduced rats. Oecologia. 167: 667-676.

Russell, J.C., Sataruddin, N.S. and Heard, A.D. (2014). Over-invasion by functionally equivalent invasive species. Ecology. 95 (8): 2268-2276.

Russell, J.C., Towns, D.R., Anderson, S.H. and Clout, M.N. (2005). Intercepting the first rat ashore. Nature. 437: 1107.

Ryall, C. (1986). Killer crows stalk the Seychelles New Scientist. 86: 48-49.

Saavedra, S. (2010). Eradication of invasive Mynas from islands. Is it possible? . Aliens: The Invasive Species Bulletin. Newsletter of the IUCN/SSC Invasive Species Specialist Group 29: 40-47.

Safford, R.J. and Hawkins, A.F.A., Eds. (2013). The Birds of Africa. Volume VIII: The Malagasy Region. Christopher Helm, London.

Samaniego-Herrera, A., Anderson, D. P., Parkes, J. P. and Aguirre-Munoz, A. (2013). Rapid assessment of rat eradication after aerial baiting. J. Appl. Ecol. 50: 1415-1421.

Samaniego-Herrera, A., A., A.-M., Méndez-Sánchez, F., Rojas-Mayoral, E. and Cárdenas-Tapia, A. (2015). Pushing the boundaries of rodent eradications on tropical islands: Ship rat eradication on Cayo Centro, Banco Chinchorro, Mexico. 27th International Congress for Conservation Biology. Montpellier, France, 2-6 August 2015.

Samways, M., Hitchins, P., Bourquin, O. and Henwood, J. (2010). Tropical island recovery. Cousine Island, Seychelles. Wiley-Blackwell, Chichester, UK.

Samways, M.J., Hitchins, P.M., Bourquin, O. and Henwood, J. (2010). Restoration of a tropical island: Cousine Island, Seychelles. Bio-

diversity and Conservation. 19 (2): 425-434.

Sands, T. (2012). Wildlife in Trust: A Hundred Yearsof Nature Conservation. Elliott and Thompson, London

Savy, J. (2006). News from Bird. Birdwatch. April to June 2006. 59: 5-6.

SCBD. (2002). Guiding Principles for the Prevention, Introduction and Mitigation of Impacts of Alien Species that Threaten Ecosystems, Habitats or Species. In COP Decision VI/23: Alien species that threaten ecosystems, habitats or species. Retrieved 20th March 2012. http://www.cbd.int/decision/cop/?id=7197.

Scoop Media. (2014). Experimental Antidote For 1080 Successful. http://www.scoop.co.nz/stories/GE0204/S00020.htm.

Seabrook, W. (1987). Examination of the impact of the feral cat Felis catus (L.) on the fauna of Aldabra Atoll, Seychelles, with recommendations on management. Unpublished report to Seychelles Islands Foundation.

Seabrook, W. (1989). Feral cats (Felis catus) as predators of hatchling green turtles (Chelonia mydas). J. Zool. Lond. 219: 83-88.

Seabrook, W. (1990). The impact of the feral cat (Felis catus) on the native fauna of Aldabra Atoll, Seychelles. Revue Ecologie (Terre Vie). 45: 135-145.

Seddon, P.J., Armstrong, D.P. and Maloney, R.F. (2007). Developing the Science of Reintroduction Biology. Conservation Biology. 21 (2): 303-312. doi:10.1111/j.1523-1739.2006.00627.x http://www.blackwell-synergy.com/doi/abs/10.1111/j.1523-1739.2006.00627.x

Sellier, R., Mahieu, N. and Angebault, J.Y. (1975). Les chenilles urticantes: biologie et importance économique et médicale. Bulletin de la Société des Sciences Naturelles de l'Ouest de la France. 73: 29-41.

Senterre, B. (2009). Invasion risk from climbing and creeping plant species in Seychelles. Consultancy report, Ministry of Environment-UNDP-GEF project, 86 pp.

Senterre, B., Gerlach, J. and Mougal, J. (2009). Old growth mature forest types and their fl oristic composition along the altitudinal gradient on Silhouette Island (Seychelles) – the telescoping effect on a continental mid-oceanic island. Phytocoenologia. 39 (2): 157-174.

Senterre, B., Rocamora, G., Bijoux, J., Mortimer, J.A. and Gerlach, J. (2010a). Seychelles biodiversity metadatabase. Output 5: Priority Gap Analysis on Seychelles' Biodiversity knowledge and information. Consultancy Report, Ministry of Environment-UNDP-GEF project, Victoria, Seychelles, 135 pp + 134 pp of appendices.

Senterre, B., Rocamora, G., Bijoux, J., Mortimer, J.A. and Gerlach, J. (2010b). Seychelles biodiversity

metadatabase. Output 3: Report on agreed Objectives and Criteria for national biodiversity inventories. Consultancy Report. Ministry of Environment-UNDP-GEF project, Victoria, Seychelles. 102 pp.

Senterre, B., Henriette, E., Vel, T. and Gerlach, J. (2012). Seychelles Key Biodiversity Areas - Output 4: Site selection and methodology for inventories. Consultancy Report, Ministry of Environment-UNDP-GEF project, Victoria, Seychelles.

Senterre, B., Henriette, E., Chong Seng, L., Gerlach, J., Mougal, J. and Rocamora, G. (2013). Seychelles Key Biodiversity Areas - Output 6: Patterns of conservation value in the inner islands. Consultancy Report, Ministry of Environment-UNDP-GEF project, Victoria, Seychelles.

Senterre, B. and Kaiser-Bunbury, C. (2014). Conception of an integrated database on the flora, fauna and vegetation of the Seychelles. Consultancy Report, Government of Seychelles, United Nations Development Programme, Victoria, Seychelles, 111 pp.

SERI (2004). Society for Ecological Restoration International Science & Policy Working Group. 2004. The SER International Primer on Ecological Restoration. www.ser.org & Tucson: Society for Ecological Restoration International.

Seychelles Natural History Museum & PCA. (2015). The Seychelles Plant Gallery. Retrieved 1st May, 2015. www.seychellesplantgallery.com. Project GEF/UNDP/PCA/SNHM, Victoria.

Shah, N.J. (2001). Eradication of alien predators in Seychelles: an example of conservation action on tropical islands. Biodiversity and Conservation. 10 (7): 1219-1220.

Shah, N.J. (2006). Ecological restoration of islands in the Seychelles. Conservation Evidence. 3: 1-2

Shah, N.J. (2008). Managing expectations and outcomes for successful Re-introductions and multiple stakeholders in Seychelles. Avian Biology Research. 1 (1): 31.

Sharp, T. and Saunders, G. (2005). Humane pest animal control: codes of practice and standard operating procedures. Humane pest animal control: Methods of euthanasia. NSW Department of Primary Industries, Australia

Sharp, T. and Saunders, G. (2008). GEN001 Methods of Euthanasia.' (NSW Department of Primary Industries: Orange, NSW) www.dpi.nsw.gov.au/__data/assets/pdf_file/0004/57253/gen-001.pdf.

Shaw, R., Witt, A., Cock, M., Pollard, K., Thomas, S. and Romney, D. (2014). Safeguarding the environment, food security and livelihoods from invasive species using biological control. CABI impact case study series No.4. http://dx.doi.org/10.1079/CABICOMMA-64-58.

Shiels, A.B., Pitt, W.C., Sugihara, R.T. and Witmer, G.W. (2014). Biology and impacts of Pacific Island

invasive species 11. The black rat, Rattus rattus (Rodentia: Muridae). Pac. Sci. 68 (2): 145-184.

Shine, C. (2008). A Toolkit for Developing Legal and Institutional Frameworks for Invasive Alien Species. Global Invasive Species Programme, Nairobi

Sidewinder Pty Ltd. (2011). Backpack Herbicide Injector Cactus - Succulents. Retrieved 12 December, 2011. http://www.treeinjectors.com/html/herbackspear.html.

SIF (2015). SIF Newsletter. January 2015. Seychelles Islands Foundation. www.sif.sc.

Silva, A., Rocha, S., Gerlach, J., Rocamora, G., Dufrenne, A. and Harris, D.J. (2010). Assessment of mtDNA genetic diversity within the terrapins Pelusios subniger and Pelusios castanoides across the Seychelles islands. Amphibia-Reptilia. 31 (4): 583-588.

Simara, H., Valentin, T., Schumacher, E. and Kueffer, C. (2008). Habitat Restoration in Mare aux Cochons (Morne Seychellois National Park) – Part Two. Kapisen (9).

Simberloff, D. (2011). Non-natives: 141 scientists object. Nature. 475: 36.

Simberloff, D.S. and Rejmánek, M. (2011). Encyclopedia of Biological Invasions. University of California Press, Berkeley and Los Angeles

Simberloff, D. (2013a). Invasive Species. What everyone needs to know. Oxford University Press.

Simberloff, D. (2013b). Biological invasions: Much progress plus several controversies. Contributions to Science 9:7-16. Institut d'Estudis Catalans, Barcelona, Catalonia. DOI: 10.2436/20.7010.01.158 ISSN: 1575-6343 www.cat-science.cat.

Simberloff, D. and Vitule, J.R.S. (2014). A call for an end to calls for the end of invasion biology. Oikos. 123 (4): 408-413.

Singleton, G.R., Hinds, L.A., Krebs, C.J. and Spratt, D.M. (2003). Rats, mice and people: rodent biology and management. ACIAR Monograph. No. 96: 564.

Skerrett, A. (2010). Outer Island Conservation. pp 174-175. In A. Skerrett, P. Thelma and J. Skerrett. Outer Islands of Seychelles. Zil Elwannyen Sesel. Camerapix, Nairobi, Kenya.

Skerrett, A., Bullock, I. and Disley, T. (2001). Birds of Seychelles. Christopher Helm (Publishers) Ltd, London. 320 pp.

Skerrett, A., Betts, M., Bullock, I., Fisher, D., Gerlach, R., Lucking, R., Phillips, J. and Scott, B. (2007). A Checklist of the Birds of Seychelles. Seychelles: Lulu.com.

Slip, D.J., Veitch, C.R. and Clout, M.N. (2003). Control of the invasive exotic yellow crazy ant (Anoplolepis gracilipes) on Christmas Island, Indian Ocean. Turning the tide: the eradication of invasive species: Proceedings of the International Conference on eradication of island invasives., IUCN-The World Conservation Union.

Smith, A.T. and Boyer, A.F. (2008). Oryctolagus cuniculus. The IUCN Red List of Threatened Species. Version 2014.3. Retrieved 17th April 2015. www.iucnredlist.org.

Smith, V.R., Avenant, N.L. and Chown, S.L. (2002). The diet and impact of house mice on a sub-Antarctic island. Polar Biology. 25: 703-715.

Somers-Yeates, R., Hodgson, D., McGregor, P.K., Spalding, A. and French-Constant, R.H. (2013). Shedding light on moths: shorter wavelengths attract noctuids more than geometrids. Biological Letters. 9: 20130376. http://dx.doi.org/10.1098/rsbl.2013.0376.

Soubeyran, Y. (2008). Espèces exotiques envahissantes dans les collectivités françaises d'outre-mer. Etat des lieux et recommandations. Collection Planète Nature, Comité français de l'UICN, Paris, France

Soubeyran, Y., (coord.) (2010). Gestion des espèces exotiques envahissantes. Guide pratique et stratégique pour les collectivités françaises d'outre-mer. Comité français de l'UICN, Paris

Soubeyran, Y., Caceres, S. and Chevassus, N. (2011). Les vertébrés terrestres introduits en outre-mer et leurs impacts. Guide illustré des principales espèces envahissantes. Comité français de l'UICN et ONCFS, Paris

SRBC. (2011). Seychelles Bird Records Committee. Retrieved 15th November, 2011. http://www.seychellesbirdrecordscommittee.com/.

SRBC. (2014). Seychelles Bird Records Committee. Retrieved 15th August 2014. http://www.seychellesbirdrecordscommittee.com/.

St. Clair, J. (2011). The impacts of invasive rodents on island invertebrates. Biological Conservation. 144 (1): 68-81.

Stanley, M.C. (2004). Review of the efficacy of baits used for ant control and eradication. Landcare Research Contract Report: LC0405/044. Ministry of Agriculture and Forestry, Wellington, New Zealand.

Stoddart, D.R. (1971). Settlement, development and conservation of Aldabra. Phil. Trans. Roy. Soc. Lond. B. 260: 611-628.

Stoddart, D.R. (1981). History of goats in the Aldabra Archipelago. Atoll Research Bulletin. 255: 23-26.

Stoddart, D.R. (1984). Biogeography and Ecology of the Seychelles Islands. Junk Publishers, The Hague

Stohlgren, T.J. (2007). Measuring Plant Diversity : Lessons from the Field. Oxford University Press, DOI: 10.1093/acprof:oso/9780195172331.001.0001.

Strahm, W. (1983). Rodrigues: can its flora be saved? Oryx. 17 (03): 122-125.

Streito, J., Ollivier, J. and Beaudoin-Ollivier, L. (2004). Two new pests of Coconut (Cocos nucifera L.) for the fauna of the Comoros: Aleurotrachelus atratus Hempel, 1922 and Paraleyrodes bondari Peracchi, 1971 (Hemiptera, Aleyrodidae). (Deux ravageurs nouveaux du Cocotier (Cocos nucifera L.) pour la faune des Comores: Aleurotrachelus atratus Hempel, 1922, et Paraleyrodes bondari Peracchi, 1971 (Hemiptera, Aleyrodidae).) Bulletin of the Entomological Society of France. 109 (1): 67-72.

Šúr, M., van de Crommenacker, J. and Bunbury, N. (2013). Assessing effectiveness of reintroduction of the flightless Aldabra rail on Picard Island, Aldabra Atoll, Seychelles. Conservation Evidence 10: 80-84.

Sutcliffe, R., Calabrese, L. and Underwood, A. (2012). Aride annual report 2010. Aride Island Nature Reserve. Island Conservation Society, Seychelles

Tan, R. (2001). Changeable lizard, Calotes versicolor http://www.naturia.per.sg/buloh/verts/changeable_lizard.htm.

Tan, Z., Wu, Y., Lin, G., Wu, B., Liu, H., Xu, X., Zhou, W., Pu, G. and Zhang, M. (1984). Study on identification and synthesis of insect pheromone. XVII. The sex pheromone of Euproctis similis xanthocampa. Acta Chimica Sinica. 42: 1178-1181.

Tassin, J. (2010). Le réchauffement climatique va-t-il conduire les petites îles à être englouties sous les invasions biologiques? . VertigO - la revue électronique en sciences de l'environnement. 10 (3). Accessed on 18th August 2014. http://vertigo.revues.org/10546; DOI : 10.4000/vertigo.10546.

Tassin, J. (2014). La grande invasion. Qui a peur des espèces invasives? Editions Odile Jacob, Paris. 216 pp.

Tassin, J. and Sarrailh, J.-M. (2007). Essences forestières et invasions : des systèmes de prédiction de plus en plus fiables. Bois et Forêts des Tropiques. 292 (2): 71-79.

Tatayah, R.V.V., Haverson, P., Wills, D. and Robin, S. (2007). Trial of a new bait station design to improve the efficiency of rat Rattus control in forest at Black River Gorges National Park, Mauritius. Conservation Evidence. 4: 20-24.

Tatayah, R.V.V., Birch, D., Haverson, P., Khadun, A. and Zuel, N. (2007): Successful transport and quarantine of materials using sealable plastic barrels, Round Island, Mauritius. - Conservation Evidence 4: 13-15.

Tatayah, R.V.V., Malham, J. and Haverson, P. (2007). The use of copper strips to exclude invasive African giant land-snails Achatina spp. from echo parakeet Psittacula eques nest cavities, Black River

Gorges National Park, Mauritius. Conservation Evidence 4:6-8.

Taylor, R.H. (1978). Distribution and interactions of rodent species in New Zealand. pp In P. R. Dingwall, I. A. E. Atkinson and C. Hay. The ecology and control of rodents in New Zealand nature reserves. Wellington, New Zealand.

Taylor, D. and Katahira, L. (1988). Radio telemetry as an aid in eradicating remnant feral goats. Wildlife Society Bulletin. 16: 297-299.

Thomas, B.W. and Taylor, R.H. (2002). A history of ground-based rodent eradication techniques developed in New Zealand, 1959-1993. pp 301-310. In C. R. Veitch and M. N. Clout. Turning the tide: the eradication of invasive species. IUCN, Gland, Switzerland and Cambridge, UK.

Thomaz, S.M., Agostinho, A.A., Gomes, L.C., Silveira, M.J., Rejmanek, M., Aslan, C.E. and Chow , E. (2012). Using space-for-time substitution and time sequence approaches in invasion ecology. Freshwater Biology 57: 2401-2410.

Thorsen, M. and Shorten, R. (1997). Attempted eradication of Norway rats during initial stages of an invasion of Frégate Island, Seychelles. Independent report to Birdlife International, Frégate Island Resorts Ltd, Department of Conversation and National Parks, New Zealand Department of Conservation, Mauritian Wildlife Foundation. 30pp.

Thorsen, M., Shorten, R., Lucking, R. and Lucking, V. (2000). Norway rats (Rattus norvegicus) on Fregate Island, Seychelles: the invasion; subsequent eradication attempts and implications for the island's fauna. Biological Conservation. 96 (2): 133-138.

Tideman, C.R. and King, D.H. (2009). Practicality and humaneness of euthanasia of pest birds with compressed carbon dioxide (CO2) and carbon monoxide (CO) from petrol engine exhaust. Wildlife Research. 36: 522-527.

Tollenaere, C., Brouat, C., Duplantier, J.-M., Rahalison, L., Rahelinirina, S., Michel Pascal, M., Moné, H., Mouahid, G., Leirs, H. and Cosson, J.-F. (2010). Phylogeography of the introduced species Rattus rattus in the western Indian Ocean, with special emphasis on the colonization history of Madagascar. J. Biogeogr. 37: 398-410.

Torchin, M.E. and Mitchell, C.E. (2004). Parasites, pathogens, and invasions by plants and animals. Frontiers in Ecology and the Environment. 2: 183-190.

Towns, D.R., Daugherty, C.H. and Atkinson, I.A.E., Eds. (1990). Ecological restoration of New Zealand islands. Conservation Sciences Publication No. 2. Department of Conservation, Wellington, NZ.

Towns, D.R. and Broome, K.G. (2003). From small Maria to massive Campbell: forty years of rat

eradications from New Zealand islands. New Zealand Journal of Zoology. 30: 377-398.

Towns, D., Atkinson, I.A.E. and Daugherty, C.H. (2006). Have the harmful effects of introduced rats on islands been exaggerated? Biological Invasions. 8: 863-891.

Towns, D.R., Byrd, G.V., Jones, H.P., Rauzon, M.J., Russell, J.C. and Wilcox, C. (2011). Impacts of introduced predators on seabirds. pp 56-90. In C. P. H. Mulder, W. B. Anderson, D. R. Towns and P. J. Bellingham. Seabird islands: ecology, invasion, and restoration. Oxford University Press, New York.

Triolo, J. (2005). Guide pour la restauration écologique de la végétation indigène. Office National des Forêts, Direction Régionale de La Réunion. ONF - Région Réunion - Europe, La Réunion, France

Tropilab Inc. (2011). Adenanthera pavonina - Red sandalwood. Retrieved 11 December, 2011. http://www.tropilab.com/adenan-pav.html.

Tu, M., Hurd, C. and Randall, J.M. (2001). Weed Control Methods Handbook. The Nature Conservancy.

Tuttle, N.C., Beard, K.H. and Pitt, W.C. (2009). Invasive litter, not an invasive insectivore, determines invertebrate communities in Hawaiian forests. Biological invasions. 11 (4): 845-855.

Tye, A. (2009). Guidelines for invasive species management in the Pacific: a Pacific strategy for managing pests, weeds and other invasive species. SPREP, Samoa

UFAW. (undated). Guiding Principles in the Humane Control of Rats and Mice. Retrieved 18th May, 2012. www.ufaw.org.uk/rodents.php.

UNDP-GEF-GOS (2010). Biosecurity Programme IAS Prioritisation Workshop Report, 12 November 2010. 8pp. Care House, Mahe, Seychelles.

University of California Integrated Pest Management. (2011). Exotic and Invasive pests. Retrieved 23th November, 2011. http://www.ipm.ucdavis.edu/.

Uranie, S. (2015). Stinging hairy caterpillars: Seychelles seeks overseas analysis to confirm invasive species. Seychelles News Agency article of 20th February 2015.

Valéry, L., Fritz, H. and Lefeuvre, J.-C. (2013). Another call for the end of invasion biology. Oikos. 122 (8): 1143-1146.

Van aarde, R., Ferreira, S., Vassenaar, T. and Erasmus, D.G. (1996). With the cats away the mice may play. South African Journal of Science. 92: 357-358.

van der Woude, J. and Wolfs, P. (2009). Monitoring and studying the Seychelles warbler. Denis Island field report 2009. Centre for Ecological and Evolutionary Studies, University of Groningen, The Netherlands. School of Biological Sciences, University of East Anglia, Norwich, England

van der Woude, J. and S., P. (2010). Monitoring and studying the Seychelles warbler. Denis Island field report June-August 2010. Centre for Ecological and Evolutionary Studies, University of Groningen, The Netherlands; School of Biological Sciences, University of East Anglia, Norwich, England. Nature Seychelles, Roche Caiman, Mahé, Seychelles

van Dinther, M., Bunbury, N. and Kaiser-Bunbury, C.N. (2015). Trial of herbicide control methods for sisal (Agave sisalana) in the arid island environment of Aldabra Atoll, Seychelles. Conservation Evidence. 12: 14-18.

Van Leeuwen, J., Froyd, C.A., Van der Knapp, P., Coffey, E., Tye, A. and Willis, K.J. (2008). Fossil pollen guides conservation in the Galapagos. Science. 322: 1206.

Vanherck, L. and Rocamora, G. (2009). Summary progress report on the Myna eradication on North Island. pp 5-7 Annexes. In G. Rocamora and A. Jean-louis. Réhabilitation des Ecosystèmes Insulaires. Rapport annuel au secrétariat du FFEM. Quatrieme année d'opérations 1/05/08 au 30/06/09. Island Conservation Society, Seychelles.

Varnham, K. (2010). Invasive rats on tropical islands: Their history, ecology, impacts and eradication. RSPB Research Report No. 41. Royal Society for the Protection of Birds, Sandy, Bedfordshire, UK. ISBN 978-1-905601-28-8

Vega, L. (2013). Seychelles Fody Foudia sechellarum. pp 876-879. In R. J. Safford and A. F. A. Hawkins. The Birds of Africa. Volume VIII: The Malagasy Region. Christopher Helm, London.

Veitch, C.R., Clout, M.N. and Towns, D.R. (2011). Island invasives: eradication and management. Gland, Switzerland, IUCN.

Vié, J., Hilton-Taylor, C. and Stuart, S.N. (2008). Wildlife in a changing world: an analysis of the 2008 IUCN Red List of Threatened Species. IUCN, Gland, Switzerland

Vielle, M. (1999). Vascular wilt in Takamaka (Calophyllum inophyllum) and the bark beetle Cryphalus trypanus. Phelsuma. 7.

Vilà, M., Basnou, C., Pyšek, P., Josefsson, M., Genovesi, P., Gollasch, S., Nentwig, W., Olenin, S., Roques, A., Roy, D., Hulme, P.E. and DAISIE partners. (2010). How well do we understand the impacts of alien species on ecosystem services? A pan-European cross-taxa assessment. Frontiers in Ecology and the Environment. 8: 135-144.

Vitlin, L.M. and Artem'ev, M.M. (1987). Field trials of spawning tropical fishes in the control of mosquito larvae in the south of the USSR Medicinskaâ parazitologiâ i parazitarnye bolezni. 4: 64-68.

Vololomboahangy, R. and Goodman, S. (2008). Tenrec ecaudatus. The IUCN Red List of Threatened Species. (IUCN SSC Afrotheria Specialist Group - Tenrec Section) Version

2014.3 Retrieved 17th April 2015. www.iucnredlist.org.

von Brandis, R.G. (2007). The Aldabra goat eradication program (2007) Final Report. Unpublished report to Seychelles Islands Foundation. Seychelles Islands Foundation.

von Brandis, R.G. (2012). Rehabilitation of abandoned coconut plantations at D'Arros Island, Republic of Seychelles. Ocean & Coastal Management. 69: 1-7.

Wainhouse, D., Murphy, S., Greig, B., Webber, J. and Vielle, M. (1998). The role of the bark beetle Cryphalus trypanus in the transmission of the vascular wilt pathogen of takamaka (Calophyllum inophyllum) in the Seychelles. Forest Ecology and Management. 108 (3): 193-199.

Wakamura, S., Yasuda, T., Hirai, Y., Tanaka, H., Doki, T., Nasu, Y., Shibao, M., A., Y. and Kadono, K. (2007). Sex pheromone of the oriental tussock moth Artaxa subflava (Bremer) (Lepidoptera: Lymantriidae): Identification and field attraction. Applied Entomology and Zoology. 42 (3): 375-382.

Wakamura, S., Yasuda, T., Ichikawa, A., Fukumoto, T. and Mochizuki, F. (1994). Sex attractant pheromone of the tea tussock moth, Euproctis pseudoconspersa (Strand) (Lepidoptera: Lymantriidae): identification and field attraction. Applied Entomology and Zoology. 29 (3): 403-411.

Walker, K. and Elliott, G. (1997). Effect of the poison brodifacoum on non-target birds on the Chetwode Islands. Ecological Management. . 5: 21-28. Department of Conservation, Wellington, NZ.

Wallace, A.R. (1880). Island life. Macmillan, London

Wanless, R., Angel, A., Cuthbert, R.J., Hilton, G.M. and Ryan, P.G. (2007). Can predation by invasive mice drive seabird extinctions? Biology Letters. 3: 241-242.

Wanless, R.M., Cunningham, J., Hockey, P.A.R., Wanless, J., White, R.W. and Wiseman, R. (2002). The success of a soft-release reintroduction of the flightless Aldabra rail (Dryolimnas [cuvieri] aldabranus) on Aldabra Atoll, Seychelles. Biological Conservation. 107 (2): 203-210.

Warman, S. and Todd, D. (1984). A biological survey of Aride Island Nature Reserve, Seychelles. Biological Conservation. 28: 51-71.

Watari, Y., Caut, S., Bonnaud, E., Bourgeois, K. and Courcham, F. (2011). Recovery of both a mesopredator and prey in an insular ecosystem after the eradication of rodents: a preliminary study. pp 377-383. In C. R. Veitch, M. N. Clout and D. R. Towns. Island Invasives: Eradicationand Management. Proceedings of the International Conference on Island Invasives. IUCN, Gland, Switzerland and Auckland, New Zealand: CBB.

Watson, J., Warman, C., Todd, D. and Laboudallon, V. (1992). The Seychelles magpie robin Copsychus

sechellarum: ecology and conservation of an endangered species. Biological Conservation. 61 (2): 93-106.

Weaving, A.J.S. (1980). Observations on Hilda patruelis Stal.(Homoptera: Tettigometridae) and its infestation of the groundnut crop in Rhodesia. Journal of the Entomological Society of Southern Africa. 43 (1): 151-167.

Wegmann, A. (2008). Land Crab Interference with Eradication Projects: Phase I - Compendium of Available Information. Pacific Invasives Initiative. The University of Auckland, New Zealand, 30 pp.

Wegmann, A., Braun, J. and Neugarten, R. (2008a). Pacific rat Rattus exulans eradication on Dekehtik Island, Federated States of Micronesia, Pacific Ocean Conservation Evidence. 5: 23-27.

Wegmann, A., Braun, J. and Neugarten, R. (2008b). Ship rat Rattus rattus eradication on Pein Mal Island, Federated States of Micronesia, Pacific Ocean. Conservation Evidence. 5: 28-32.

Wegmann, A., Buckelew, S., Howald, G., Helm, J. and Swinnerton, K. (2011). Rodent eradication campaigns on tropical islands: novel challenges and possible solutions. pp 239-243. In C. R. Veitch, M. N. Clout and D. R. Towns. Island invasives: eradication and management. IUCN, Gland, Switzerland.

Wendling, B., Engelhardt, U., Adam, P.A., Alcindor, R., Louange, A., Rosine, G. and Zialor, V. (2004). Pilot study of management of black-spined sea urchin populations around the granitic islands of the Seychelles with an objective of restoration of the coral reef ecosystem Unpublished report for GEF-SEYMEMP (= Annex 6 of the Final Report for the GEF-Seychelles Marine Ecosystem Management Project) 30 pp.

Wendling, B., Rosine, G., Adam, P.-A., Zialor, V. and Louange, A. (2003). Status of carnivorous fishes and recent hard coral recruitment rates around outer islands (Seychelles). An additional survey to the Seychelles Marine Ecosystem Management Project (SEYMEMP). Unpublished technical report in French.

Wetterer, J.K. (2007). Biology and impacts of Pacific Island invasive species. 3. The African big-headed ant, Pheidole megacephala (Hymenoptera: Formicidae). Pacific Science. 61 (4): 437-456.

Wheater, C. and Penny, A. (2003). Studying invertebrates. Naturalists' Handbooks 28. Richmond Publishing. 120 pp.

Whittaker, R.J. and Fernández-Palacios, J.M. (2007). Island biogeography (2nd ed.). Oxford University Press, Oxford

Wildlife Ethics Committee (2013). Euthanasia of research animals in the field policy. Department of Environment, Water and Natural

Resources. Zoos South Australia, South Australia Museum, Government of South Australia.

Williams, P.A. (2003). Proposed guidelines for weed-risk assessment in developing countries. pp 71-112. In R. Labrada. FAO Expert Consultation on Weed Risk Assessment, Madrid, Spain, 11-13 June, 2002. FAO, Rome.

Witt, A. (2014). Fighting invasive alien species to safeguard food security. The Economist. 10.11.14. New-York & London.

Wittenberg, R. and Cock, M.J.W., Eds. (2001). Invasive Alien Species: A Toolkit of Best Prevention and Management Practices. CAB International, Wallingford, Oxon, UK xii - 228 pp.

Working for wetlands. (2011). Best management practices. Retrieved 13 December, 2011. http://www. sanbi.org/sites/default/files/documents/documents/wfwbestmanageprac.pdf.

Wright, D.J., Spurgin, L.G., Collar, N.J., Komdeur, J., Burke, T. and Richardson, D.S. (2014). The impact of translocations on neutral and functional genetic diversity within and among populations of the Seychelles warbler. Molecular Ecology. 23: 2165-2177.

Yasuda, T., Wakamura, S. and Arakaki, N. (1995). Identification of sex attractant pheromone components of the tussock moth, Euproctis taiwana (Shiraki) (Lepidoptera: Lymantriidae). Journal of Chemical Ecology. 21 (11): 1813-1822.

Yeandle, M. (2009). Barn Owl Eradication on Aride Island, Seychelles. Final report. Projet FFEM Réhabilitation des Ecosystèmes Insulaires. Island Conservation Society, Seychelles.

Yongmo, W., Feng, G., Xianghui, L., Feng, F. and Lijun, W. (2005). Evaluation of mass-trapping for control of tea tussock moth Euproctis pseudoconspersa (Strand) Lepidoptera: Lymantriidae) with synthetic sex pheromone in south China. International journal of pest management. 51 (4): 289-295.

Young, L.C., VanderWerf, E.A., Lohr, M.T., Miller, C.J., Titmus, A.J., Peters, D. and Wilson, L. (2013). Multi-species predator eradication within a pest-proof fence at Ka`ena Point, Hawai`i. Biological Invasions. 15 (6). DOI 10.1007/s10530-013-0479-y.

Zalucki, M.P., Clarke, A.R. and Malcolm, S.B. (2002). Ecology and behaviour of first instar larval Lepidoptera. Annual Review of Entomology. 47: 361-393.

Zavaleta, E.S., Hobbs, R.J. and Mooney, H.A. (2001). Viewing invasive species removal in a whole-ecosystem context. Trends in Ecology and Evolution. 16: 454-459.

ANNEX: Example of rat abatement and biosecurity protocols (North Island)

(from Climo & Rocamora, 2006/ ICS FFEM project)

MAHÉ BOAT LOADING PROCEDURES

The long term eradication of rats on North Island requires two things, the successful eradication of existing rats and the prevention of rats reinvading the island. If rats have a chance to reinvade they will. One of the biggest risks to North Island is rat reinfestation through transport to the island of goods and materials. A quarantine procedure for goods coming to the island is essential to prevent rodents arriving accidentally with transported goods.

Measures to stop reinvasion on the island itself is a final precaution. Most of the actual responsibility to ensure that a rat does not stow away on route to North Island will fall on the Mahé Office.

The following procedure details the boat loading process for shopping, staff luggage and containerised goods. It aims to make the risk of rat reintroduction from these sources non-existent.

BOAT LOADING PROCEDURE SHOPPING AND STAFF
- All island shopping, staff shopping and staff baggage must be packed into the plastic boxes provided before transfer to the boat. Until this system is in place, all independent salesmen and key Mahé personel must adhere to the following when transporting goods to the quay and subsequent loading:
    - Do not at any time leave boxes standing unattended on jetty during loading;
    - Pipes to be checked for rodents (flow water inside for those that cannot be checked by view), and ends subsequently closed off before putting them on the boat:
    - Extra care should be taken with cardboard (carton) boxes and bags: these should be carefully checked for chew marks and droppings, and subsequently for hidden invader animals.
    - After checking, cardboard boxes and bags should be immediately properly taped close.
    - Do not load unauthorized plants or animals.
    - Planks to be treated before loading.
- If possible, boats not to stay moored at the jetty overnight. Any loaded boats left in Mahé overnight must have baited bait stations aboard. Cones/bottles to be attached on the mooring ropes. Baiting stations to be set up around the moored boat, on the jetty. Bait blocks must be replaced weekly (barges) or 2-weekly (boats).
- All containers must be fumigated (with Aluminium phosphide) prior to transport to the island.
- Island Contractors must be informed of the importance to inspect their gear for seeds, rodents or insects before transport to the Island.

MAHÉ STORAGE OF GOODS
- Angel Fish complex care taker to place and subsequently monitor and re-bait bait stations around the complex.

RESPONSIBILITIES
- Ensuring shopping is packed in plastic boxes – [WHO].
- Ensuring all boats leaving from Mahé have bait stations aboard, and bait blocks are regularly changed (use of log book) – [WHO].
- Ensuring contractors are informed of the need to inspect goods for seeds, rodents and insects before transport – GM.
- Supervision of the placing of baiting stations, and their subsequent rebaiting – Environment Assistant & Mahé Staff. Angel Fish complex to be involved.
- Ensuring Mahé office and boats/barges have a sufficient supply of bait and bait stations – Environment Officer.
- Process running smoothly and alteration of procedure if necessary – Environment Officer.

BOAT (SHOPPING & STAFF LUGGAGE) & BARGE (CARGO) UNLOADING PROCEDURES

(from Climo & Rocamora, 2006/ ICS FFEM project)

A proper unloading procedure is essential for the safe unpacking and checking of goods for any rodents that may accidentally come in with transported goods.

The following procedures detail the process for unloading barges (containing cargo) and boats (containing shopping & staff luggage) and aim to make the risk of rodent reintroduction – or the invasion by alien invader animals – from these sources non-existent.

BARGE/ BOAT UNLOADING PROCEDURE
- All boats/ barges coming to North Island must have permanent bait stations aboard. Any loaded boats/ barges overnighting at Mahé prior to travel to North Island must have bait stations placed on board, and poison blocks are to be replaced regularly in boats, and weekly or more often if required in barges (wet blocks to be replaced). Small boats need one bait station, larger boats need 2-3 bait stations.
- Unbaited Elliott traps in wooden boxes (imitations of sheltered places to attract escaping rodents on exposed beaches during offloading) to be set up on the beach during offloading of barges containing food items, cardboard boxes and building material with cavities.
- The only cargo items exempted (as per agreement of the General Manager and Environmental Officer), are listed below (*). All other goods, including staff shopping and baggage, must go to the rodent-proof room.
- The rodent-proof trailer must be used to transport goods (cargo, shopping and staff luggage) to the rodent-proof room. For smaller loads, a gator with rodent-proof canopy can be used instead.
- Goods should be packed in rodent-proof sealed containers on Mahé before loading where possible.
- Future cargo should be transported as whole containers rather than being destuffed on Mahé – this will enable fumigation to be undertaken prior to offloading on North Island.

(*) EXEMPTIONS OF CARGO ITEMS THAT DO NOT NEED TO GO TO THE RODENT-PROOF ROOM

The list of exemptions given below are due to the negligible risk of harbouring rodents or the difficulty of transporting items to the rodent-proof room. Items that are not taken to the rodent-proof room have to be checked on the beach. This is the responsibility of the Maintenance Manager and Security Officer.

- Drums containing petrol, oil, chlorine and hazardous chemicals (if not packed in cardboard boxes)
- Large containers of cleaning chemicals (if not packed in cardboard boxes)
- Crates of drinks
- Cling wrapped packets of water
- Cement
- Aggregate
- Timber planks
- Diesel
- Gas cylinders
- Large unpacked machinery (motor compartments need to be checked on the beach for mice and poison blocks put inside at Mahé prior to transport)
- Cling wrapped tins of chemical products (e.g. paint)
- Tyres
- Copper
- Metal and PVC piping – large rolls of piping need to be checked on Mahé and the ends plugged immediately after inspection. These can be checked using compressed air or flushed water.
- Rolls of building plastic
- Shade cloth

RESPONSIBILITIES

- Ensuring that bait stations are placed on incoming boats/barges and re-baited timely (weekly for barges) – Skipper and [WHO] (Mahé Office). Incoming boats' baiting to be verified by the Dive Center Manager (for shopping boats and charter boats), Security Officer and Environment Department's staff.
- Offloaded goods not to be placed on the beach but directly loaded into rodent-proof trailer and transported to rodent-proof room where goods and staff bags are inspected behind closed door – Maintenance Manager, Security Officer and Central Store Manager.
- Process running smoothly and alteration of procedure if necessary – Environmental Officer.

RODENT (& OTHER PESTS) PROOF ROOM PROCEDURES (NORTH ISLAND, MARCH 2012)

OFFLOAD ZONE

1. Trailer enters and both steel doors are carefully closed, making sure that the bottom rubber is properly sealed and doors are locked (by securing the chain into the metal holders). Only then can the trailer doors be opened and can the offload begin.
2. All goods are to be offloaded directly into the pest control room whilst the 2nd door to the purchase control room remains closed.

3. Only when all goods are offloaded and both pest control doors are closed, can the roller doors be opened again and the trailer removed.

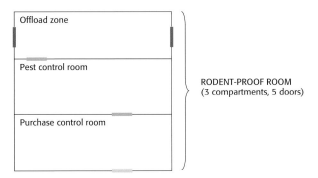

Offload zone: drive-through with metal roller doors.
Pest control room: quarantine room where goods are checked for pests.
Purchase control room: room where cargo, already checked for pests, is stored, before further dispatching to the different department's stores.

4. This zone is for transport only; strictly nothing to be stored here (maintain easy visibility).
5. Stick ("rat bat") to be kept close by; metal walk-in traps to be set along walls. No objects to be attached to the walls (avoid building "ladders" for animals to climb up which will hinder killing them).

PEST CONTROL ROOM

6. Both doors to be closed and checking of all goods to be done immediately.
7. Once all goods are inspected, they are to be immediately moved into the purchase control room.
8. During barge offloads with large cargo volumes to be processed, an exception can be made with checked cargo moved into the purchase control room whilst checking is still ongoing; in such case, the purchase control room's door to the stores should be kept closed.
9. This room is a transit room where only goods awaiting pest inspection can remain for as short as possible. Strictly nothing to be stored here. To maintain easy visibility, all discarded packaging is to be removed.
10. Stick ("rat bat") to be kept close by; metal walk-in traps and bait stations to be permanently placed along the walls.
11. Broken bait stations and traps are to be reported immediately to the Environment Office for replacement.
12. The poison blocks in the bait stations are to be replaced every 2 weeks by Environment Office. Blocks to be checked for chew marks. Metal walk-in traps to be kept operational at all times (if door closed, trap is to be shaken to detect if an animal was caught, in which case Environment Office needs to be called immediately).

PURCHASE CONTROL ROOM

13. No goods are to be moved from the pest control room to the purchase control room before pest inspection in the pest control room. Doors are to be kept closed until

the inspection has been finalized.

14. Damage to the rodent-proof room needs to be immediately reported for repair by the Central Store Manager/ Security Officer.

15. In charge of supervising the correct implementation of the alien invader avoidance procedures at all times, are:
    - for staff luggage only: Security Officer.
    - for all other cargo: Central Stores Manager.

16. Any pest animals or suspicious signs of pest animals are to be reported immediately to the Environment Officer.

NO DEVIATION FROM THIS PROCEDURE WITHOUT THE APPROVAL OF THE GENERAL MANAGER AND THE ENVIRONMENT OFFICER.

CENTRAL STORE MANAGER:
FURTHER SPECIFICATION OF TASKS RE. PEST AVOIDANCE

Rodents being a common pest on Mahé, daily cargo sent to North Island therefore requires stringent implementation of procedures to avoid accidental reintroduction. Avoidance procedures are to be followed at all stages of the transport: on Mahé, on the beach during offloads, as well as during rodent-proof room cargo deliveries and subsequent inspections. Together with the Mahé Logistics Officers, skippers offload teams and security officers, the Central Store Manager therefore plays a key role in keeping North island free of rodents or other potential pests (plants or animals), and will receive training from the Environment Officer during the first week of his/her assignment.

His/her responsibility will include all possible actions to ensure detection, immediate reporting and subsequent immediate destruction of any dangerous pest animal/ plant escaping from cargo arriving in the rodent-proof room, by:

- *Thorough cargo inspections: all goods arriving in the rodent-proof room require inspection before release to the relevant departments. Only the Environment Officer can exempt goods from a search. For container offloads, assistance will be given by trained Environment staff. During cargo searches, packaging and goods need to be searched for hidden animals or plants, or any suspicious signs (droppings, chew marks, chewed nesting material) of a rodent or another animal.*

- *Correct through-fare of goods: once goods have been declared safe, they need to be removed from the pest control room before arrival of next cargo, unless a qualified person remains in the room to ensure checked goods are not contaminated by pest animals escaping from newly arrived cargo into already opened checked boxes still present in the room. Otherwise, all cargo needs to be re-checked before release.*

- *Reporting: any pest animal/ plant or suspicious sign of an unwanted pest that came on the island via cargo, needs to be immediately destroyed and subsequently reported to the Environment Officer. If assistance is required with the destruction, the Environment Officer/ Landscape Manager need to called upon immediately, whilst doors need to remain closed at the place of sighting (rodent-proof room or stores) for other staff, and cargo movement needs to be ceased until the Environment Officer has given go-ahead to resume normal work.*

- *Maintaining rodent-proof room efficiency: any damage or malfunction in the rodent-proof room, jeopardizing its effectiveness to contain an unwanted pest animal escaping from cargo inside the room, needs to be reported in writing to the Maintenance Manager, with the Environment Manager copied in. Maintaining the work place includes regular checks of walls and door mechanisms. Assist the Environment staff with keeping traps and bait stations in rodent-proof room and stores operational.*

Boat cargo brought in the pest-proof trailer is unloaded in the offload zone of this rodent (and other pests) proof room, and brought into the neighbouring pest-control room to be checked.
G. Rocamora (upper left), L. Vanherck (upper right), C.G. Havemann (bottom)

'OFFLOAD ZONE' CHECK LIST (MARCH 2012)

OFFLOAD ZONE	Responsible*	INSPECTION DATE: DONE BY:	
		TICK BOX	COMMENTS
Seal both steel doors carefully before opening the trailer door and offloading.	All		
Offload all goods directly into the pest control room with door to purchase room closed.	All		
When all goods are offloaded: close Pest room door, then open steel doors before trailer exits.	All		
Nothing (including removed packaging) stored in Offload zone.	All		
"Rat bat " within easy reach	CSM		
Metal walk-in traps operational	CSM, E		

'PEST-PROOF ROOM' CHECK LIST (MARCH 2012)

PEST CONTROL ROOM			
Close both steel doors and the door to the purchase control room when bringing new cargo in from Offload zone.	All		
Check all goods immediately for pests/suspicious signs, in closed room.	CSM: shopping boats E: barges. S-staff bags.		
After checking for pests/ suspicious signs, move checked goods immediately into Purchase control room.	CSM: shopping boats E: barges		
No clutter: also remove discarded packaging after each inspection.	CSM, E		
Close door to Purchase room again before receiving new unchecked goods.	CSM, E		
Transit room only. (No goods to be stored)	CSM		
"Rat bat" within easy reach	CSM		
Doors are to be kept closed until the inspection has been finalized.	S, A		
Permanent rat bait stations rebaited every 14 days	E		
Metal walk-in traps operational	CSM, E		

PURCHASE CONTROL ROOM			
Only pest-free goods can move to the Purchase control room.	CSM		
Any pest animals or suspicious signs (see photos) are to be reported immediately to the Environment Officer.	All		

General rules for entire "Pest-proof room"			
Report damage immediately	CSM, S, E		
Obey at all times to person in charge: -Central Stores Manager. - Staff luggage only: Security Officer.	- All		
No changes to rules without permission of Environment officer	All		

*CSM = Central Store Manager, S = Security Officer, E = Environment Office, All = all entering "Rodent-proof room"

Useful contacts regarding IAS management in Seychelles

Government services, parastatals & public trusts

Name	Tel. contact	Email contact(s)	Website
Ministry of Environment, Energy and Climate Change - Environment Department			http://www.env.gov.sc/
Ministry of Health - Public Health Department	(248) 4 38 80 79	Jude.Gedeon@health.gov.sc geralda.didon@health.gov.sc	http://www.health.gov.sc/
Islands Development Company	(248) 4 38 46 40	ceo@idc.sc	http://www.idc.sc/
National Botanical Gardens Foundation	(248) 4 29 53 00	admin@nbgf.sc	http://www.bgci.org/garden.php?id=1383
Seychelles Agricultural Agency	(248) 4 67 64 50	ceo.saa@gov.sc ceosecsaa@gov.sc	https://www.facebook.com/pages/Seychelles-Agricultural-Agency/139599822820889
Seychelles Islands Foundation	(248) 4 32 17 35	info@sif.sc	http://www.sif.sc/
Seychelles Marine Park Authority	(248) 4 22 51 14	info@snpa.sc	http://www.snpa.sc/
Seychelles Veterinary Department	(248) 4 28 59 50	seyvet@seychelles.net	
Natural History Museum – National Herbarium	(+248) 4 321 333	nathismus@seychelles.sc charles6422@hotmail.com	http://www.seychellesplantgallery.com/

Non Governmental Organizations

Name and website	Tel. contacts	Email contacts	Website
Green Islands Foundation	(248) 4 28 88 29	gm@gif.sc	http://greenislandsfoundation.blogspot.com/
Island Conservation Society (Seychelles)	(248) 4 37 53 54	ceo@ics.sc	http://www.islandconservationseychelles.com/
Marine Conservation Society of Seychelles	(248) 4 26 15 11	info@mcss.sc	http://www.mcss.sc/
Nature Protection Trust of Seychelles		gerlachs@btinternet.com	http://islandbiodiversity.com/nptsindex.htm#NPTS
Nature Seychelles	(248) 4 60 11 00	nature@seychelles.net	http://www.natureseychelles.org/
Plant Conservation Action Group (Seychelles)	(248) 4 24 11 04	pca.seychelles@gmail.com	http://www.pcaseychelles.org/
Terrestrial Restoration Action Society of Seychelles	(248) 2 51 33 70	trass.seychelles@gmail.com	http://www.trass.org.sc/

CONTACTS
Biotope éditions
22 boulevard Maréchal Foch – BP 58
34140 Mèze – France
Phone: 00 33 (0)4 67 18 65 39
Fax: 00 33 (0)4 67 18 46 29
edition@biotope.fr
www.biotope-editions.com

Informations on our books, extracts and online shop:
www.leclub-biotope.com

Biotope Océan Indien
910 chemin Lagourgue
97440 Saint-André – La Réunion
Tél. : + 262 (0)2 62 46 67 75
oceanindien@biotope.fr
www.biotope.fr

Publications scientifiques du Muséum
CP 41 – 57 rue Cuvier
75231 Paris cedex 05 – France
Phone: 00 33 (0)1 40 79 48 05
Fax: 00 33 (0)1 40 79 38 40
diff.pub@mnhn.fr
sciencepress.mnhn.fr

Island Biodiversity and Conservation center
University of Seychelles,
P.O. Box 1348, Anse Royale, Seychelles
IBC@unisey.ac.sc

Cover photo: Grande Soeur and Petite Soeur, two privately owned islands free of rats and cats since 2010. Raymond Sahuquet / STB

Design and production: Biotope
Production editor: **Julien Marmayou**
Page layout: **Julien Marmayou, Béatrice Garnier**
Image processing: **Loïc Simon**

Printed by **PBtisk a.s.** (Czech Republic)
Dépôt légal : décembre 2015

FSC
www.fsc.org
MIX
Paper from
responsible sources
FSC® C004378